Aquatic Oligochaete Biology VIII

Developments in Hydrobiology 158

Series editor
H. J. Dumont

Aquatic Oligochaete Biology VIII

Proceedings of the 8th International Symposium on Aquatic Oligochaeta, held in Bilbao, Spain, 18–22 July 2000

Edited by

Pilar Rodriguez[1] & Piet F. M. Verdonschot[2]

[1] *Department of Animal Biology and Genetics, University of Basque Country, Bilbao, Spain*
[2] *Alterra, Green World Research, Department of Freshwater Ecology, Wageningen, The Netherlands*

Reprinted from Hydrobiologia, volume 463 (2001)

Springer Science+Business Media, B.V.

Library of Congress Cataloging-in-Publication Data

International Symposium on Aquatic Oligochaeta (8th : 2000 : Bilbao, Spain)
Aquatic oligochaete biology VIII : proceedings of the 8th International Symposium on Aquatic Oligochaeta, held in Bilbao, Spain, 18-22 July 2000 / edited by Pilar Rodriquez and Piet F.M. Verdonschot.
p. cm. -- (Developments in hydrobiology ; 158)
"Reprinted from Hydrobiologia."
ISBN 978-94-010-3887-4 ISBN 978-94-010-0597-5 (eBook)
DOI 10.1007/978-94-010-0597-5
1. Oligochaeta--Congresses. 2. Aquatic invertebrates--Congresses. I. Rodriquez, Pilar. II. Verdonschot, Piet F. M. III. Hydrobiologia. IV. Title. V. Series.

QL391.A6 I57 2000
592'.64176--dc21

2001050246

ISBN 978-94-010-3887-4

Printed on acid-free paper

This volume is dedicated to
Ralph O. Brinkhurst
in recognition of his
life-long contribution
to the knowledge of
aquatic oligochaetes.

TABLE OF CONTENTS

PART TWO: DIVERSITY AND DISTRIBUTION

PART THREE: MORPHOLOGY AND EMBRYOLOGY

PART FOUR: FEEDING BIOLOGY AND REPRODUCTION

PART FIVE: ECOTOXICOLOGY AND BIOACCUMULATION

PART SIX: ECOLOGY

Hydrobiologia **463:** xi–xii, 2001.
P. Rodriguez & P.F.M. Verdonschot (eds), Aquatic Oligochaete Biology VIII.

Preface

The Eighth Symposium on Aquatic Oligochaeta held in Bilbao (18th to 22nd July, 2000) was organized by the Department of Animal Biology and Genetics of the University of the Basque Country (UPV/EHU). The attendance of 65 scientists from 30 different countries, through a participation of 140 contributors in poster and oral communications, gives an idea of the width and scope of this meeting, as well as of the efforts of the organization to bring together most of the specialists in the field of aquatic oligochaete research. The involvement of experts in various branches of biology promoted the discussion of different perspectives and concepts in a range of specialized fields in oligochaete biology. The symposium was organized by a team of enthusiastic researchers: Dr Pilar Rodriguez (Lecturer in Animal Biology), Dr Maite Martinez-Madrid (Associate Lecturer in Animal Physiology) and Mr Jesus Arrate (PhD student). Additional support in the conference rooms and during social activities was provided by Mr Jurgi Salaverria and Mr Dietmar Fernandez. The joint effort the organizing team, as well as the helpful suggestions and collaboration of many other colleagues around the world, especially the Scientific Committee, including Dr Stuart R. Gelder (organizer of the previous symposium) and Dr Mark Wetzel (responsible for the web page of the aquatic oligochaetologists, http://www.inhs.uiuc.edu:80/ mjwetzel/ISAOB dir.html), all contributed to the success of the Symposium.

Participants were formally welcomed by Dr J. Zuñiga, Vice-Rector of Research, Dr E. Navarro (Director of the Department of Animal Biology and Genetics) and Dr Aurora Alonso (Dean of the Faculty of Economics, where the conference was held). The program consisted of a total of 10 scientific sessions which covered a wide range of branches of knowledge dealing with the biology of aquatic oligochaetes: toxicity, physiology and reproduction studies, ecology and community studies, oligochaetes as vectors in whirling-disease, systematics and phylogeny, zoogeography, morphology and ultrastructure. One of the invited lecturers, Dr Peter Chapman (EVS Consultants, Canada), opened the symposium and presented an important overview of the present state of, and future recommendations for the use of oligochaetes in risk assessment in the aquatic environment. Dr Mark Siddall (American Museum of Natural History, USA), another invited lecturer, opened the session on Systematics and Phylogeny with a very interesting proposal regarding the systematic position of the hirudineans within the Oligochaeta, using data obtained independently from molecular and morphological characters.

As a novelty, during the symposium there were simultaneous short courses which dealt with the use of programs for personal computers on phylogeny as well as community studies. Dr Siddal was responsible for the course on “Phylogeny”, and Dr Trefor B. Reynoldson together with Dr Piet F. M. Verdonschot and Drs Rebi Nijboer were responsible for the course on “Multivariate Analysis Techniques in Community Studies”. About 30 scientists attended both courses. The symposium was an important forum for all the participants who benefited from personal discussions with other specialists. The fruits of these interchanges will undoubtedly consist of more interdisciplinary research in the future.

As part of the social activities of the participants in the symposium, the visits to the Bilbao Guggenheim Museum and to the Atapuerca Paleontological site, where important remains of *Homo heidelbergensis* and *Homo antecessor* have been found, widened everyone's perspectives. The visit to Atapuerca was completed by an informal dinner at Laguardia (La Rioja) during a pleasant summer evening.

The symposium contributions, as well as the discussions during the sessions and the informal gatherings, provided participating oligochaetologists with the opportunity to learn from each other, to exchange ideas and knowledge, and to inspire and be inspired. The scientific fields of special interest concerned the role of oligochaetes in ecological and ecotoxicological risk assessment. Whirling-disease studies were shown to be a useful tool for taxonomy, on the basis of fish–worm species interactions. The decreasing attention which has been paid to taxonomy in recent years is a commonly felt shortcoming. Meanwhile, the use of oligochaetes in applied ecology and water management continues to increase. However, the role of ecological knowledge at the species level requires continuing attention to the field of taxonomy. One of the challenging new fields which will undoubtedly develop over the coming years is the role of oligochaetes in improving purification processes in waste-water treatment plants.

The foregoing symposium reports published in the series "Developments in Hydrobiology" could only have become available due to a great contribution by some of our outstanding colleagues. Therefore, we wish to express special thanks to the previous editors G. Bonomi, C. Erseus, R. O. Brinkhurst, R. J. Diaz, J. L. Kaster, T. B. Reynoldson, K. A. Coates and B. M. Healy for the substantial work they performed.

For more than 20 years, the oligochaete community of dedicated researchers gathered together to talk about a subject which most of the rest of the world laughs about. Ralph Brinkhurst was one of the early engines of this movement. Many followed him during all these years, and the recent advances in both the systematics and biology of aquatic oligochaetes cannot be separated from Ralph's work. The large number of young researchers who were present in Bilbao is clearly indicative of the continuous growth of our knowledge of oligochaetes. A fresh generation is developing new ideas on this important group of organisms. If the world understood the importance of oligochaete function in aquatic and other ecosystems, the laugh would surely become a sigh of admiration.

We would like to express our sincere thanks to all authors and numerous referees who have worked hard and made it possible to compose this volume. It was a great deal of work, but in the words of our predecessor, John Stephenson, "*apart from any question of our success or failure, we do believe that the task we have undertaken was worth doing*". Finally, we would like to thank to the agency ACTS (Academic Consulting and Translating Systems) (acts@euskalnet.net) for revising the English of many of the papers in this volume. Financial support for the organization of the symposium and the publication of these proceedings has been provided by the University of Basque Country (UPV/EHU), the Basque Government (Department of Education, University and Research 2/39; Department of Environment), the Spanish Government (Ministry of Education and Culture PGC2000-2234-E), the European Community (INCO Project: ICA1 1999 60015 *Tubifex*) and the Spanish Association of Limnology (AEL).

PILAR RODRIGUEZ & PIET F. M. VERDONSCHOT
Editors
May 2001

Hydrobiologia 463: xiii–xvi, 2001.

List of participants

Mónica João de Barros **AMORIM,** Instituto do Ambiente e Vida, Dpt. de Zoologia, Universidade de Coimbra, P-3004-517 Coimbra, Portugal.
mjamorim@zoo.uc.pt
mjamorim@iav.uc.pt

Laura **ARMENDARIZ,** Departamento Científico Zoología de Invertebrados, Facultad de Ciencias Naturales y Museo de Universidad Nacional de La Plata, Paseo del Bosque s/n, 1900 La Plata, Argentina.
paulaar@netverk.com.ar

Jesús Angel **ARRATE,** Departamento Biología Animal y Genética, Facultad de Ciencias, Universidad del País Vasco, Apdo. 644, E-48080 Bilbao, Spain.
ggbarjoj@lg.ehu.es

Naime P. **ARSLAN,** Osmangazi Üniversitesi Fen-Edebiyat Fak., Biyoloji Bölümü Meselik, 26480 Eskisehir, Turkey.
narslan@ogu.edu.tr

Bastian Roelof **BUYS,** Wageningen University Sub-Department of Environmental Technology, P.O. Box 8129, 6700 EV Wageningen, The Netherlands.
bas.buys@algemeen.mt.wau.nl

Katherine **BEAUCHAMP,** University of California, Department of Medicine and Epidemiology, Davis, CA 95616, U.S.A.
kbeaucha@ucdavis.edu

Regine **BOENSCH,** Institut für Angewandte Ökologie, Lindenweg 3, D-18184 Neu Broderstorf, Germany.
sordyl@t-online.de

Silvia **BOI,** Universitá degli Studi di Milano, Sezione di Zoologia e Citologia, 26 Via Celoria, I-20133 Milano, Italy.
silvia.boi@unimi.it

Eric **BRUNS,** ECT Oekotoxikologie GmbH, Böttgerstraße 2–14, D-65439 Flörsheim Main, Germany.
e-bruns@ect.de

Kristin **BRUST,** Institute of Hydrobiology, Dresden University of Technology, Mommsenstrasse 13, D-01062 Dresden, Germany.
kristin.brust@mailbox.tu-dresden.de

Sandra **CASELLATO,** Dipartimento di Biologia, Univ. Padova, Via Colombo 3, I-35121 Padova, Italy.
sandra@civ.bio.unipd.it

Inés Irma **CÉSAR,** Departamento Científico Zoología de Invertebrados, Facultad de Ciencias Naturales y Museo de Universidad Nacional de La Plata, Paseo del Bosque s/n, 1900 La Plata, Argentina.
icesar@museo.fcnym.unlp.edu.ar

Kathryn A. **COATES,** Bermuda Biological Station for Research, Inc., 17 Biological Lane, Ferry Reach, St. Georges, GE-01, Bermuda.
kcoates@bbsr.edu

Rut **COLLADO,** Universidade da Coruña, Departamento de Bioloxía Animal e Bioloxía Vexetal, Campus da Zapateira s/n, E-15071 A Coruña, Spain.
rutco@udc.es

Chiara **CRISTONI,** Centro di Scienze Ambientali, Università Degli Studi de Bologna, Via Tombesï Dall'Ova 55, I-48100 Ravena, Italy.
c.cristoni@delfino.ambra.unibo.it

Sergio **CUADRADO,** Departmento de Bioloxía Animal e Bioloxía Vexetal, Facultad de Ciencias, Universidade da Coruña, Campus da Zapateira s/n, E-15071 A Coruña, Spain.

Peter **CHAPMAN,** E.V.S. Consultants, 195 Pemberton Avenue, North Vancouver, B.C. V7P 2R4, Canada.
pchapman@attglobal.net

Sheila S. **CHERIAN,** Chippy, All Saint's College, Trivandrum – 695011, Kerala, India.
chippy@md4.vsnl.net.in

Robert J. **DIAZ,** Virginia Institute of Marine Science, College of William and Mary, Gloucester Point, VA 23062. U.S.A.
diaz@vims.edu

Elzbieta **DUMNICKA,** Institute of Freshwater Biology, Polish Academy of Sciences, Slawkowska 17, 31-016 Krakow, Poland.
dumnicka@zbw.pan.krakow.pl

Philipp **EGELER,** ECT Oekotoxikologie GmbH, Böttgerstraße 2–14, D-65439 Flörsheim Main, Germany.
ph-egeler@ect.de

Christer **ERSÉUS,** Swedish Museum of Natural History, Department of Invertebrate Zoology, Box 50007, SE-104 05 Stockholm, Sweden.
christer.erseus@nrm.se

Steve **FEND,** US Geological Survey, M/S 465, 345 Middlefield Road, Menlo Park, CA 94025, U.S.A.
svfend@usgs.gov

Marco **FERRAGUTI,** Universitá degli Studi di Milano, Sezione di Zoologia e Citologia, Via Celoria 26, I-20133 Milano, Italy.
ferraguti@unimi.it

Stuart **GELDER,** Department of Biology, University of Maine at Presque Isle, 181 Main Street, Presque Isle, ME 04769-2888, U.S.A.
gelder@polaris.umpi.maine.edu

Olav **GIERE,** Zoological Institute and Zoological Museum, University of Hamburg, Martin-Luther-King-Platz 3, D-20146 Hamburg, Germany.
olav.giere@zoologie.uni-hamburg.de

Patricia **GILLIS,** Aquatic Ecosystem Restoration Branch, National Water Research Institute, 867 Lakeshore Road, Burlington, Ontario, L7R 4A6 Canada.
patty.gillis@cciw.ca

Lena **GUSTAVSSON,** Swedish Museum of Natural History, Department of Invertebrate Zoology, Box 50007, SE-104 05 Stockholm, Sweden.
lena.gustavsson@nrm.se

Brenda **HEALY,** Department of Zoology, University College, Belfield, Dublin 4, Ireland.
bhealy@indigo.ie

René **HESSLING,** Universität Osnabrück, FB5, Biologie/Chemie, Abt. Spezielle Zoologie, Barbarastraße 11, D-49069 Osnabrück, Germany.
hessling@biologie.uni-osnabrueck.de

Jacques **JUGET,** Université Claude Bernard (Lyon I), Laboratoire de Biologie Animale et Ecologie, 43 Boulevard du 11 Novembre 1918, F-69622 Villeurbanne Cedex, France.
deschate@univ-Lyon1.fr

Billie **KERANS,** Montana State University, Department of Biology, Bozeman, MT 59717, U.S.A.
bkerans@montana.edu

Deedee **KATHMAN,** Aquatic Resources Center, P.O. Box 345, College Grove, TN 37046, U.S.A.
aquatres@ix.netcom.com

Michael **LAFONT,** CEMAGREF. B.E.A. Division, 3 bis Quai Chauveau, CP 220, F-69336 Lyon Cedex 09, France.
michel.lafont@cemagref.fr

Claus **LINDEGAARD,** Freshwater Biological Laboratory, University of Copenhagen, 51 Helsingørsgade, DK-3400 Hillerød, Denmark.
clindegaard@zi.ku.dk.

Roberto **MAROTTA,** Universitá degli Studi di Milano, Sezione di Zoologia e Citologia, Via Celoria 26, I-20133 Milano, Italy.

Patrick **MARTIN,** Royal Belgian Institute of Natural Sciences, Freshwater Biology, 29 Rue Vautier, B-1000 Brussels, Belgium.
patrick.martin@naturalsciences.be

Enrique **MARTINEZ ANSEMIL,** Departmento de Bioloxía Animal e Bioloxía Vexetal, Facultad de Ciencias, Universidade da Coruña, Campus da Zapateira s/n, E-15071 A Coruña, Spain.
eansemil@udc.es

Maite **MARTINEZ-MADRID,** Departamento de Biología Animal y Genética, Facultad de Ciencias, Universidad del País Vasco, Apdo. 644, E-48080 Bilbao, Spain.
ggpmamam@lg.ehu.es

Göran **MILBRINK,** Department of Animal Ecology, Evolutionary Biology Centre, Uppsala University, Norbyv. 18D, S-752 36 Uppsala, Sweden.
goran.milbrink@ebc.uu.se

Tetsuya **NARITA,** Centre for Ecological Research, Kyoto University, 509 Ohtsuka, Kamitanakami-Hirano, Otsu 520-2113, Japan.
narita@ecology.kyoto-u.ac.jp

Rebi C. **NIJBOER,** Alterra, Green World Research, Team Freshwater Ecology, P.O. Box 47, 6700 AA Wageningen, The Netherlands.
r.c.nijboer@alterra.wag-ur.nl

Akifumi **OHTAKA,** Department of Natural Science, Faculty of Education, Hirosaki University, Hirosaki 036-8560, Japan.
ohtaka@cc.hirosaki-u.ac.jp

Pietro **OMODEO,** Dipartimento di Biologia Evolutiva, Via Mattioli 4, I-53100 Siena, Italy.
rota@unisi.it

Andrea **PASTERIS,** Università degli Studi di Bologna, Centro de Scienze Ambientali, Via Tombesi Dall'Ova 55, I-48100 Ravenna, Italy.
pasteris@ambra.unibo.it

Momir **PAUNOVIC,** Institute for Biological Research, Department of Hydroecology and Water Protection, 29 novembra 142, 1100 Beograd, Yugoslavia.
mpaunovi@ibiss.bg.ac.yu

Adrian **PINDER,** Wildlife Research Centre, Department of Conservation and Land Management, P.O. Box 51, Wanneroo 6946, Western Australia
adrianp@calm.wa.gov.au

Charlotte **RASMUSSEN,** Western Fisheries Research Center, USGS-Biological Resources Division, 6505 NE 65th Street, Seattle, WA 98115, U.S.A.
charlotte_rasmussen@usgs.gov

Trefor B. **REYNOLDSON,** Acadia Centre for Estuarine Research, 23 Westwood Avenue, Box 115, Acadia University, Wolfville, Nova Scotia B0P 1X0, Canada.
trefor.reynoldson@acadiav.ca

Geta **RISNOVEANU,** Faculty of Biology, Department of Systems Ecology, Spl. Independentei 91–95, 76201 Bucharest, Romania.
rgeta@bio.bio.unibuc.ro

Pilar **RODRIGUEZ,** Departamento de Biología Animal y Genética, Facultad de Ciencias, Universidad del País Vasco, Apt. 644, E-48080 Bilbao, Spain.
ggprorop@lg.ehu.es

Emilia **ROTA,** Università di Siena, Dipartimento di Biologia Evolutiva, Via Mattioli 4, I-53100 Siena, Italy.
rota@unisi.it

Beatrice **SAMBUGAR,** Museo Civico di Storia Naturale, Lungadige Porta Vitoria 9, I-37100 Verona, Italy.
beatrice@easynet.it

Jana **SCHENKOVA,** Department of Zoology and Ecology, Masaryk University, Kotlárská 2, 61137 Brno, Czech Republic.
schenk@sci.muni.cz

Rüdiger M. **SCHMELZ,** Universität Osnabrück, FB5 Biologie/Chemie, Systematic Zoology, D-49069 Osnabrück, Germany.
schmelz@biologie.uni-osnabrueck.de

Takashi **SHIMIZU,** Molecular and Cellular Interactions, Division of Biological Sciences, Hokkaido University, Sapporo 060-0810, Japan.
stak@sci.hokudai.ac.jp

Mark **SIDDALL,** Division of Invertebrate Zoology, American Museum of Natural History, Central Park West at 79th Street, New York, NY 10024-5192, U.S.A.
siddall@amnh.org

Alice M. **TAKEDA,** Universidade Estadual de Maringá-Nupelia, Av. Colombo 5790, CEP 87020-900, Maringa-PR, Brazil.
alice@nupelia.uem.br

Tarmo **TIMM,** Institute of Zoology and Botany, Võrtsjärv Limnological Station, EE-61101 Rannu, Tartumaa, Estonia.
ttimm@zbi.ee

Piet F. M. **VERDONSCHOT,** Alterra, Green World Research, Team of Freshwater Ecology, P.O. Box 47, 6700 AA Wageningen, The Netherlands.
p.f.m.verdonschot@alterra.wag-ur.nl

Doris **VIDAL,** Department of Integrative Biology, University of California at Berkeley, Berkeley, CA 94270, U.S.A.
pingui@socrates.berkeley.edu

Hongzhu **WANG,** State Key Laboratory of Freshwater Ecology and Biotechnology, Institute of Hydrobiology, Chinese Academy of Sciences, Wuhan, China.
wanghz@ihb.ac.cn

Mark J. **WETZEL,** Illinois National History Survey, Center for Biodiversity, 172 Natural Resources Building, 607 East Peabody Drive, Champaign, IL 61820, U.S.A.
mjwetzel@uiuc.edu

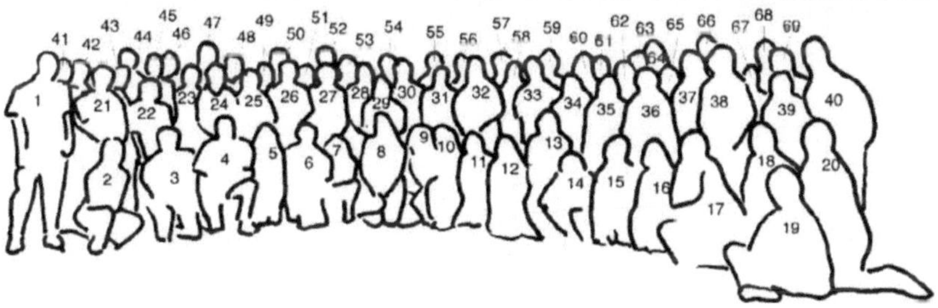

Group photograph taken at the VIII International Symposium on Aquatic Oligochaeta held in Bilbao, Spain, 18–22 July, 2000. Numbers in the line drawing indicates the position of the attendees:

(1) Dietmar Fernandez
(2) Maite Martinez-Madrid
(3) Akifumi Ohtaka
(4) Peter Chapman
(5) Patricia Gillis
(6) Tetsuya Narita
(7) Monica Amorim
(8) Naime Arslan
(9) Rut Collado
(10) Pilar Rodriguez
(11) Reet Laugaste
(12) Andrea Gelder
(13) Katherine Beauchamp
(14) Katye Dilke
(15) Helen Valdez
(16) Doris Vidal
(17) Dr Diaz's wife
(18) Robert Diaz
(19) Steve Fend
(20) Billie Kerans
(21) Sheila Cherian
(22) Alice Takeda
(23) Chiara Cristoni
(24) Kristin Brust
(25) Lena Gustavsson
(26) Takashi Shimizu
(27) Deedee Kathman
(28) Jana Schenková
(29) Gillian Gelder
(30) René Hessling
(31) Regine Boensch
(32) Trefor Reynoldson
(33) Tarmo Timm
(34) Stuart Gelder
(35) Elzbieta Dumnicka
(36) Beatrice Sambugar
(37) Sergio Cuadrado
(38) Kathryn Coates
(39) Miriam Donadoni-Omodeo
(40) Rüdiger Schmeltz
(41) Andrea Pasteris
(42) Enrique Martínez-Ansemil
(43) Sandra Casellato
(44) Piet Verdonschot
(45) Claus Lindegaard
(46) Rebi Nijboer
(47) Mark Wetzel
(48) Göran Milbrink
(49) Laura Armendariz
(50) Luis Andrés Ciappesoni
(51) Inés César
(52) David Evans
(53) Mark Siddall
(54) Olav Giere
(55) Jurgi Salaverria
(56) Hongzhu Wang
(57) Phillipp Egeler
(58) Adrian Pinder
(59) Christer Erséus
(60) Emilia Rota
(61) Patrick Martin
(62) unknown
(63) Marco Ferraguti
(64) Bastian Buys
(65) Charlotte Rasmussen
(66) Eric Bruns
(67) Saelon Renkes
(68) Jesús Arrate
(69) Pietro Omodeo

(11), (12), (14), (15), (17), (29), (39), (50), (52), (62) and (67) are accompanying persons.

Hydrobiologia **463:** xix–xx, 2001.
P. Rodriguez & P.F.M. Verdonschot (eds), Aquatic Oligochaete Biology VIII.

Some personal memories of Dr Gilberto Righi (1937–1999)

Gilberto Righi was a tough, methodical and jolly man. He loved to tell jokes and stories during coffee breaks in the morning and in the afternoon, but otherwise he could always be found in his worm-room, constantly from eight in the morning to eight at night. There he would only tolerate interruptions to minister his appointed classes and to smoke his pipe. If eventual visitors were fortunate enough to seek him during his smoking hours, they could benefit from some extra minutes of attention, while Gilberto talked among puffs of his beloved pipe. These were good occasions to hear stories about giant worms living in the interior of the State of São Paulo, or of the exciting discovery of green, square worms in the region of the Pantanal in the State of Mato Grosso. For disbelieving minds, he would joyfully plunge into his secret, private, personal collections of worms. Undergraduates under his supervision, like myself, quickly learned that these were the precious moments to discuss any advances or problems encountered in our novice attempts to do research.

Gilberto was a self-made person, with some very strong ideas. He declined to write most of his papers in English, even though he enjoyed studying languages, including English, German, Russian and aboriginal Tupi, simply because his native language was Portuguese. He felt that if anybody really wanted to read his papers, they should translate them, in the same way that he, as a self-taught person, had learned to read papers in German and Russian that interested him. Gilberto was also a big and athletic type, who practised sports regularly, including pedalling his way to and from his worm-lab by bike. His passing away from a heart attack at night, after a normal day of routine work, took everybody by surprise. I consider myself very fortunate to have started my scientific career under the guidance of Dr Gilberto Righi. Above all, I have learnt from him that prosperity in research comes from being methodical, persistent, independent and above all hardworking.

MARTIN L. CHRISTOFFERSEN
Dpt. de Sistematica e Ecologia,
Universidade Federal da Paraiba,
Campus I,
58051-900 João Pessoa,
P.B., Brazil

Scientific profile of Dr Gilberto Righi

From 1963 to 1998, Dr Gilberto Righi produced more than one hundred papers and 4 books. Although the first contributions dealt mainly with marine molluscs and crustaceans, the study of both terricolous and aquatic oligochaetes soon attracted the attention of the author. Dr Righi has been one of the most important contributors to the knowledge of the family Glossoscolecidae, with more than 30 papers describing new taxa or with notes on morphology or distribution of the species within this family. He also contributed to the systematics of the families Haplotaxidae, Enchytraeidae, Naididae and Tubificidae of Brazil and other countries in South and Central America and described with M.E. Varela the new family Narapidae from the river Paraná (Argentina). Below is a selected bibliography of the author.

Selected publications

1973. Sobre três espécies brasileiras de Enchytraeidae (Oligochaeta). Bol. Zool. Biol. Mar. (n.s.) 30: 469–482.

1973. On *Pristina minuta* (Oligochaeta, Naididae) from Brazilian soil and its epizoic *Rhabdostyla pristinis*, sp.n. (Ciliata, Epistyfidae). Zool. Anz. 191: 295–300.

1974. Notas sobre os Oligochaeta Enchytraeidae do Brasil. Papéis Avulsos Zool., S. Paulo 28: 127– 145.

1974. *Marionina pituca*, sp.n. (Oligochaeta, Enchytraeidae) from Brazilian soil and its intestinal parasite, *Buetschliella marioninae*, sp.n. (Ciliata, Astomata). Zool. Anz. 192: 414–419.

1974. Alguns Sporozoa (Gregarinida e Haplosporida) de Oligochaeta (Naididae e Enchytraeidae) brasileiras. Papéis Avulsos Zool., S. Paulo 28: 185–195.

1975. Algumas Enchytraeidae (Oligochaeta) brasileiras. Ciênc. & Cult. 27: 143–150.

1979. Marine Oligochaeta (Tubificidae and Enchytraeidae) from the Caribbean Sea. Stud. Fauna Curaçao 58: 44–68 (with E. Kanner).

1981. Alguns Oligochaeta cavernicolas do Equador. Papéis Avulsos Zool., S. Paulo 34 (22): 235–249.

1981. Notas sobre Enchytraeidae (Oligochaeta) brasileiras. Rev. Bras. Biol. 41: 427–430.

1983. *Narapa bonettoi*, gen.n., sp.n. (Oligochaeta, Narapidae, fam. n.) de água doce da Argentina. Rev. Asoc. Cienc. nat. Lit. 14: 7–15 (with M. E. Varela).

1984. *Pontoscolex* (Oligochaeta, Glossoscolecidae), a new evaluation. Stud. neotrop. Fauna 19: 159–177.

1984. Manual de identificação de invertebrados límnicos do Brasil. Oligochaeta. CNPq, Brasilia, 48 pp.

1984. On some earthworms (Oligochaeta, Glossoscolecidae) from the Sierra Nevada de Santa Marta. In van der Hammen, T. & P. M. Ruiz (eds), Studies on Tropical Andean Ecosystems, 2J. Cramer, Berlin: 455–478.

1985. Dois novos Oligochaeta da Amazônia venezuelana. Papéis Avulsos Zool., S. Paulo 36: 23–30.

1987. Relations among the male genital fields in *Rhinodrilus* and its associated, Glossoscolecidae, Oligochaeta. Zool. Anz. 219: 154–158.

1988. Dois novos Microdrili, Oligochaeta, terrestres da Amazônia. Papéis Avulsos Zool., S. Paulo 36: 315–321.

1988. *Pontoscolex (P.) pydanieli*, sp.n. (Oligochaeta, Glossoscolecidae) and its parasite *Pessoaelia pontoscolecis*, gen.n., sp.n. (Eugregarinida, Aikinetocystidae). Stud. neotrop. Fauna 23 (2): 71– 76.

1989a. *Pristina longidentata* e a taxonomia das Naididae, Oligochaeta. Rev. Bras. Biol. 49: 409– 414 (with V. Hamonui).

1989b. Mass migration and life cycle adaptation – a survival adaptation. Amazoniana 11: 23–30 (with J. Adis).

1995. Distribution of earthworms species on the two flanks of the Central Cordillera (Transecto Parque Los Nevados, Colombia) (with T. van der Hammen). In van der Hammen, T. & A. G. Santos (eds), Studies on Tropical Andean Ecosystems, 4, Cramer. Berlin: 475–483.

1995. Colombian earthworms. In van der Hammen, T. & A. G. Santos (eds), Studies on Tropical Andean Ecosystems, 4, Cramer. Berlin: 484–607.

1996. Some Venezuelan Oligochaeta: Glossoscolecidae and Octochaetidae. Rev. Suisse Zool. 103: 677– 684.

1998. Oligochaeta Naididae of the West Indies and adjacent regions, I, II and III. Stud. neotrop. Fauna (in press) (with V. Hamonui).

1998. Earthworms of the Ilha de Maracá. In Milliken, W. & J. Ratter (eds), The Biodiversity and Environment of an Amazonian Rainforest. John Willey & Sons, Chichester: 391–397.

The complete scientific contribution of Dr Righi can be found in Megadrilogica 7 (8): 55–60, by Catalina C. de Mischis & John W. Reynolds (1999).

PILAR RODRIGUEZ
Dpt. Animal Biology and Genetics,
University of Basque Country UPV/EHU,
Bilbao, Spain

Hydrobiologia **463:** xxi–xxiii, 2001.
P. Rodriguez & P.F.M. Verdonschot (eds), Aquatic Oligochaete Biology VIII.

In memoriam

Professor Perry C. Holt, PhD (1910–1999)

Professor Perry C. Holt, PhD, passed away in the Spring of 1999 at the age of 87. He devoted his research career of 45 years to the study of branchiobdellidans; the freshwater clitellate annelids that are frequently found on crayfishes in the Holarctic region. He attended many academic meetings to raise zoologists' awareness of branchiobdellidans, and to stress that they are not parasites! In 1988, Perry presented a review of branchiobdellidan characters at the 4th International Symposium on Aquatic Oligochaetes held at the Louisiana State University in Baton Rouge, Louisiana, U.S.A. He used this work to refute strenuous claims that branchiobdellidans were leeches (Holt, 1989a).

Perry's interest in branchiobdellidans started when his mentor and life-long friend, Horton H. Hobbs, Jr, suggested branchiobdellidans as a suitable doctoral thesis subject. The status of branchiobdellidan knowledge at that time consisted of three major reviews, detailed microanatomical descriptions of four species, and many vague and incomplete descriptions of approximately 50 species in eight genera. From the works of Prof. H. Yamaguchi in Japan, Perry established the standard of morphological character descriptions needed to advance the systematics of the taxon. Between publishing his thesis work (Holt, 1949) and his last new species description (Holt, 1989b), Perry, together with his colleague, Richard L. Hoffman, added 8 new genera and 75 new species of branchiobdellidans to the taxon. Although very much aware of branchiobdellidans in Eurasia, Perry recognized the enormous amount of work that needed to be done in North America and so focused the majority of his efforts on that region.

The increase in the number of new species was matched in importance by his redescriptions of existing species. From these works he was able to propose that the branchiobdellidans should be raised from family rank to ordinal level, as Branchiobdellida (Holt, 1965). The move emphasized the difference of the branchiobdellidans from, and made them equal to, the oligochaetes and leeches. This paper was followed by a general review of branchiobdellidan biology in which Perry proposed the first evolutionary tree for the group (Holt, 1968). The review and evolutionary tree were updated (Holt, 1986) to include new information, and he grouped the genera into five families within the Branchiobdellida. He also used his combined knowledge to address the question of where was the locus of the ancestral symbiotic branchiobdellidan. His last publication (Holt & Opell, 1993) was co-authored with Brent D. Opell on a checklist and illustrated key of the Cambarincolidae.

Perry recognized that accurate alpha taxonomy is a prerequisite for any meaningful studies involving a taxon. As a result he became involved with surveys involving branchiobdellidans, especially those occurring in the southern Appalachians (Hobbs et al., 1967; Holt, 1969), and in caves in the eastern United States (Holt, 1973). He also wrote a long-forgotten article (Holt, 1963c) entitled, "*The New Taxonomy?*", a note on the first International Conference on Taxonomic Biochemistry, Serology, and Physiology held at Lawrence, Kansas, September 4–6, 1962. This commentary on the meeting emphasized that these "new" tools for comparative investigations are of no value unless the species being studied have been described and named. Forty years later – with the addition of comparative ultrastructure, gene sequences, and other character suites being used in cladistic analyses – the need for the application of accurate alpha taxonomy has never been greater. Recent developments using 18S rDNA and mitochondrial cytochrome *c* oxidase subunit I to study the relationships of leeches, branchiobdellidans, acanthobdellidans and lumbriculids have shown very strong evidence that these taxa form a monophyletic group, and that each of these taxa should be considered an order. These molecular data appear to have resolved the impasse of over 100 years based on morphological characters as to whether branchiobdellidans are oligochaetes or leeches, or form an independent taxon of equal rank. Undoubtedly, Perry would find this a highly amusing development as it supports his elevation of branchiobdellidans to ordinal rank over 35 years ago!

Perry C. Holt left a significant legacy to those studying branchiobdellidans; he set the necessary high standard for species descriptions, he deposited many specimens for future reference in the National Museum of Natural History, Smithsonian Institution, Washington D.C., and proposed a clear, justified taxonomy of the branchiobdellidans. Whatever advances are made in branchiobdellidan research in the next 100 years, they will owe much to the critical groundwork established through the dedicated research of Perry C. Holt.

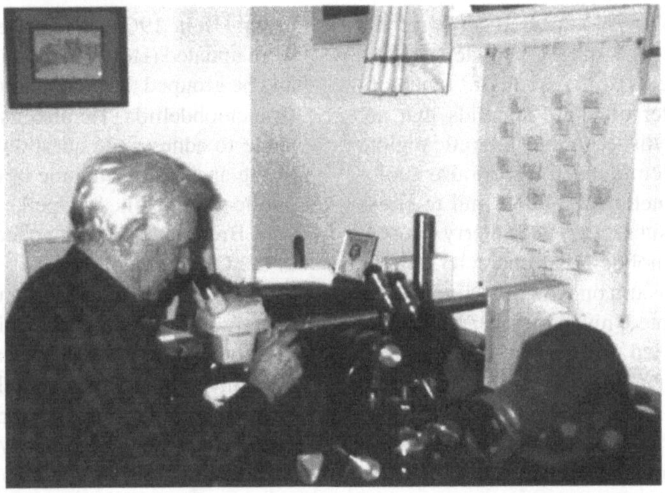

Perry C. Holt, Ph.D. working at home with his 1986 "evolutionary tree" of the branchiobdellidans on the right.

Publications

Berry, J. W. & P. C. Holt, 1959. Reactions of two species of Branchiobdellidae (Oligochaeta) to high temperature and low oxygen tensions. Virginia Agri. Exp. Stat., Tech. Bull. 141: 1–11.

Hobbs, H. H. Jr, P. C. Holt & M. Walton, 1967. The crayfishes and their epizootic ostracod and branchiobdellid associates of the Mountain Lake, Virginia, region. Proc. U.S. nat. Mus. 123: 1–84.

Holt, P. C., 1949. A comparative study of the reproductive systems of *Xironogiton instabilius* (Moore) and *Cambarincola philadelphica* (Leidy) (Annelida, Oligochaeta, Branchiobdellidae.) J. Morph. 84: 535–572.

Holt, P. C., 1951. The genera *Xironodrilus* and *Pterodrilus* in North America, with notes on other North American genera of the family Branchiobdellidae. Unpublished doctoral dissertation, University of Virginia.

Holt, P. C., 1953. Characters of systematic importance in the family Branchiobdellidae (Oligochaeta). Virginia J. Sci. 4: 57–61.

Holt, P. C., 1954. A new branchiobdellid of the genus *Cambarincola* (Oligochaeta, Branchiobdellidae) from Virginia. Virginia J. Sci. 5: 168–172.

Holt, P. C., 1955. A new branchiobdellidae of the genus *Cambarincola* Ellis, 1912 (Oligochaeta, Branchiobdellidae) from Kentucky. J. Tenn. Acad. Sci. 30: 27–31.

Holt, P. C., 1960a. The genus *Ceratodrilus* Hall (Branchiobdellidae, Oligochaeta) with the description of a new species. Virginia J. Sci. 11: 53-77.

Holt, P. C., 1960b. On a new genus of the family Branchiobdellidae (Oligochaeta). Am. Midl. Nat. 64: 169–176.

Holt, P. C., 1963a. A new branchiobdellid (Branchiobdellidae: *Cambarincola*). J. Tenn. Acad. Sci. 38: 97–100.

Holt, P. C., 1963b. The systematic position of the Branchiobdellidae. Am. Zool. 3: 204.

Holt, P. C., 1963c. The new taxonomy? A note on the First International Conference on Taxonomic Biochemistry, Serology, and Physiology held at Lawrence, Kansas, September 4–6, 1962. Assoc. southest. Biologists 10: 15–22.

Holt, P. C., 1964a. A new branchiobdellid (Annelida) from Costa Rica. Tulane Stud. Zool. 12: 1–4.

Holt, P. C., 1964b. The distribution of the branchiobdellids: some zoogeographical problems. Am. Zool. 4: 438.

Holt, P. C., 1964c. The anatomy of the Phoenix. Assoc. southest. Biologists 11: 27–31.

Holt, P. C., 1965a. On *Ankyrodrilus*, a new genus of branchiobdellid worms (Annelida). Virginia J. Sci. 16: 9–21.

Holt, P. C., 1965b. The systematic position of the Branchiobdellidae (Annelida: Clitellata). Syst. Zool. 14: 25–32.

Holt, P. C., 1967a. *Oedipodrilus oedipus*, n.g., n. sp. (Annelida, Clitellata: Branchiobdellida). Trans. Am. microsc. Soc. 86: 58–60.

Holt, P. C., 1967b. Status of the genera *Branchiobdella* and *Stephanodrilus* in North America with description of a new genus (Clitellata: Branchiobdellida). Proc. U.S. nat. Mus. 124: 1–10.

Holt, P. C., 1967c. Notes on the natural history of science. Assoc. southest. Biologists 14: 71–74.

Holt, P. C., 1968a. The Branchiobdellida: epizootic annelids. Biologist 50: 79–94.

Holt, P. C., 1968b. New genera and species of branchiobdellid worms (Annelida: Clitellata). Proc. biol. Soc. Wash. 81: 291–318.

Holt, P. C., 1968c. The genus *Pterodrilus* (Annelida: Branchiobdellida). Proc. U.S. nat. Mus. 125: 1–44.

Holt, P. C., 1969. The relationship of the branchiobdellid fauna of the southern Appalachians. In Holt, P. C. (ed.), The Distributional History of the Biota of the Southern Appalachians. Part I: Invertebrates, Vol. 1. Research Division Monograph, Virginia Polytechnic Institute: 191–219.

Holt, P. C., 1973a. Epigean branchiobdellids (Annelida: Clitellata) from Florida. Proc. biol. Soc. Wash. 86: 79–104.

Holt, P. C., 1973b. A summary of the branchiobdellid (Annelida: Clitellata) fauna of Mesoamerica. Smithson. Contr. Zool. 142: 1–40.

Holt, P. C., 1973c. Branchiobdellids (Annelida: Clitellata) from some eastern North American caves, with descriptions of new species of the genus *Cambarincola*. Int. J. Speleol. 5: 219–256.

Holt, P. C., 1973d. A free-living branchiobdellid (Annelida: Clitellata)? Trans. Am. microsc. Soc. 92: 152-153.

Holt, P. C., 1973e. Am emeneded description of *Cambarincola meyeri* Goodnight (Clitellata: Branchiobdellida). Trans. Am. microsc. Soc. 92: 677–682.

Holt, P. C., 1974a. An emendation of the genus *Triannulata* Goodnight, 1940, with the assignment of *Triannulata montana* to *Cambarincola* Ellis 1912 (Clitellata: Branchiobdellida). Proc. biol. Soc. Wash. 87: 57–72.

Holt, P. C., 1974b. The genus *Xironogiton* Ellis, 1919 (Clitellata: Branchiobdellida). Virginia J. Sci. 25: 5–19.

Holt, P. C., 1975. The genera *Oedipodrilus* and *Sathodrilus*: some Pacific drainage species. Am. Zool. 15: 831 (Abstr.).

Holt, P. C., 1976. The branchiobdellid (Annelida: Clitellata) associates of astacoidean crawfishes. In Avault, J. W., Jr. (ed.), Freshwater Crayfish. Proc. Internat. Symp. on Crayfish, Louisiana. State Univ. Press, Baton Rouge: 337–396.

Holt, P. C., 1977a. An emendation of genus *Sathodrilus* Holt 1968 (Annelida: Branchiobdellida), with the description of four new species from the Pacific drainage of North America. Proc. biol. Soc. Wash. 90: 116–131.

Holt, P. C., 1977b. A gill-inhabiting new genus and species of the Branchiobdellida (Annelida: Clitellata). Proc. biol. Soc. Wash. 90: 726–734.

Holt, P. C., 1978. The reassignment of *Cambarincola elevatus* Goodnight, 1940 (Clitellata: Branchiobdellida) to the genus *Sathodrilus* Holt, 1968. Proc. biol. Soc. Wash. 91: 472–483.

Holt, P. C., 1981a. A resume of the members of the genus *Cambarincola* (Annelida: Branchiobdellida) from the Pacific drainage of the United States. Proc. biol. Soc. Wash. 94: 675–695.

Holt, P. C., 1981b. New species of *Sathodrilus* Holt, 1968 (Clitellata: Branchiobdellida) from the Pacific drainage of the United States, with the synonymy of *Sathodrilus virgiliae* Holt, 1977. Proc. biol. Soc. Wash. 94: 848– 862.

Holt, P. C., 1982. A new species of the genus *Cambarincola* (Clitellata: Branchiobdellida) from Illinois with remarks on the bursa of *Cambarincola vitreus* Ellis, 1919, and the status of *Sathodrilus* Holt, 1968. Proc. biol. Soc. Wash. 95: 251–255.

Holt, P. C., 1984a. On some branchiobdellids (Annelida: Clitellata) from Mexico with the descriptions of new species of the genera *Cambarincola* and *Oedipodrilus*. Proc. biol. Soc. Wash. 97: 34–42.

Holt, P. C., 1984b. A new species of the genus *Cambarincola* (Clitellata: Branchiobdellida) from California. Proc. biol. Soc. Wash. 97: 544–549.

Holt, P. C., 1986. Newly established families of the order Branchiobdellida (Annelida: Clitellata) with a synopsis of the genera. Proc. biol. Soc. Wash. 99: 676–702.

Holt, P. C., 1988a. The correct name of *Ceratodrilus orphiorhysis* Holt, 1960 (Annelida: Branchiobdellida). Proc. biol. Soc. Wash. 101: 308.

Holt, P. C., 1988b. Four new species of cambarincolids (Clitellata: Branchiobdellida) from the southeastern United States with a redescription of *Oedipodrilus macbaini* (Holt, 1955). Proc. biol. Soc. Wash. 101: 794–808.

Holt, P. C., 1989a. Comments on the classification of the Clitellata. Hydrobiologia 180: 7–15.

Holt, P. C., 1989b. A new species of the Cambarincolid genus *Sathodrilus* from Missouri, with the proposal of a replacement name for *Adenodrilus* Holt, 1977 (Clitellata: Branchiobdellida). Proc. biol. Soc. Wash. 102: 738–741.

Holt, P. C. & R. L. Hoffman, 1959. An emended description of *Cambarincola macrodonta* Ellis, with remarks on the diagnostic characters of the genus (Oligochaeta: Branchiobdellae). J. Tenn. Acad. Sci. 34: 97–104.

Holt, P. C. & B. D. Opell, 1993. A checklist of and illustrated key to the genera and species of the Central and North American Cambarincolidae (Clitellata: Branchiobdellida). Proc. Biol. Soc. Wash. 106: 251–295.

Holt, P. C. & A. M. Weigl, 1979. A new species of *Xironodrilus* Ellis, 1918 from North Carolina (Clitellata: Branchiobdellida). Brimleyana 1: 23–29

Holt, P. C. & W. A. Whitaker, 1955. Notes on Branchiobdellidae of South Indian Creek, Unicoi County, Tennessee. J. Tenn. Acad. Sci. 30: 132.

My thanks to Mark J. Wetzel for his help and suggestions in preparing this work.

STUART R. GELDER
University of Maine at Presque Isle,
Maine 04769, U.S.A.

Hydrobiologia **463:** 1–11, 2001.
P. Rodriguez & P.F.M. Verdonschot (eds), Aquatic Oligochaete Biology VIII.

Descriptions of three *Pristina* species (Naididae, Clitellata) from Amazonian forest soils, including *P. marcusi* sp. nov.

Rut Collado[1] & Rüdiger M. Schmelz[2]
[1] *Universidade da Coruña, Departamento de Bioloxía Animal e Bioloxía Vexetal, A Coruña, Spain*
Fax: 34 981 167065. E-mail: rutco@udc.es
[2] *Universität Osnabrück, FB 5, Biologie/Chemie, Spezielle Zoologie, Osnabrück, Germany*
E-mail: schmelz@biologie.uni-osnabrueck.de

Key words: taxonomy, new species, Oligochaeta, Naididae, *Pristina*, Brazil, live observation

Abstract

Three species of the oligochaete genus *Pristina* (Naididae, Clitellata), one of them new to science, are described from soil and litter samples originating from terra firme sites of the primary rain forest at the Brazilian Agroforestry Research Facility in Central Amazonia: *P. marcusi* sp. nov., *P. jenkinae* (Stephenson, 1931) and *P. notopora* Černosvitov, 1937. Investigations were carried out on live and preserved material, with emphasis on non-chaetal, non-sexual characters of the live anatomy, usually neglected in recent Naididae taxonomy. *P. marcusi* sp. nov. is characterized by simple-pointed and bayonet-shaped needle chaetae and by a regular pattern of ellipsoid intra-stomachal cavities. *P. notopora* is redescribed in full detail for the first time since its original description, including the exceptional dorsal opening of the nephridia.

Introduction

As in most naidid taxa, species of the genus *Pristina* are currently determined and separated by differences in the shape, distribution and number of the chaetae (Brinkhurst, 1971; Brinkhurst & Marchese, 1989; Kathman & Brinkhurst, 1998), although other informative traits of the soft-bodied anatomy are known for many species. The confinement to the chaetae has practical reasons, because they are the only characters that remain unaltered by fixation and because live observation is difficult, time-consuming and often impossible, for example under tropical conditions (Grimm, 1987). However, species-separating differences of chaetal patterns are minute and often difficult to ascertain. Intra-specific (Pop, 1973; Kasprzak & Szczesny, 1976) and even intra-individual (Martínez-Ansemil & Giani, 1982) variations in taxonomically important details – partly inducable experimentally by changing the environmental conditions (Loden & Harman, 1980) – enhance the uncertainty of species identifications that rely on the chaetae alone (Grimm, 1986).

We experienced the same uncertainty with the chaetotaxy in Naididae working on a collection of ethanol-preserved specimens originating from soil and litter samples of an experimental study area, a primary rain forest plot at the Brazilian Agroforestry Research Facility (EMBRAPA-CPAA: Empresa Brazileira de Pesquisa Agropecuária, Centro de Pesquisa Agroflorestal na Amazônia Ocidental) close to Manaus, Brazil. Specimens in this collection, especially those belonging to *Pristina* spp., displayed a large range of minute differences in the chaetal patterns that were difficult to interpret in taxonomical terms and neither the number of species found nor their identity could be determined with certainty. Fortunately, soil and litter samples from the same area could be obtained with flourishing populations of *Pristina* sp., which gave us the opportunity to examine the animals alive, and reproducing cultures could be established with some of these animals for intraspecific comparisons. The information that we obtained on internal organs such as nephridia, coelomocytes and structures of the alimentary truct together with the possibility to compare observations on preserved and live material facilitated the separation and identification of species. Two new species of the genus have already been described (Collado & Schmelz, 2000). In this paper, we describe *P. marcusi* sp. nov., and we present redescriptions of *P. notopora* and *P. jenkinae*. There are only few previous records of *P. notopora*, but *P. jenkinae* has been recorded many times, and after the revision of Kathman (1985) and the synonymization of a number of species therein, its distribution is cosmopolitan. However, more than one species may be involved. Many anatomical details of *P. notopora* including its exceptional dorsal location of the nephridiopores are re-investigated here for the first time after its original description. A comparative overview of the species found at the EMBRAPA site, together with a critical discussion of the non-chaetal, non-sexual characters and their usefulness in the taxonomy of *Pristina*, will be published elsewhere.

Materials and methods

A detailed description of the study area and the methods of investigation are given in Collado & Schmelz (2000). Our taxonomic investigations are part of the project SHIFT ENV 52 (Studies on Human Impact on Floodplains and Forests in the Tropics) addressing the development of methods for sustainable land use in Amazonian rain forests (Römbke et al., 1999; Römbke & Meller, 1999).

The oligochaetes were extracted from soil and litter samples using the hot/wet funnel technique established for enchytraeids by O'Connor (1955) and investigated with stereolupe ($\times 10 - \times 40$) and microscope ($\times 100 - \times 1000$), using interference contrast (Nomarski) optics. Specimens were examined alive and as preparations (stained whole mounts, squash preparations). Chaetal details of some specimens were examined using SEM microscopy. All microscopical observations including the live observations were documented photographically. Figures were drawn from photographs of live specimens and from fixed and whole-mounted animals with the help of a drawing tube. Some of the determined specimens were cultured at room temperature on a mixture of 0.5% Agar–Agar and mineral soil from the habitat in polystrol Petri dishes. All worms survived for more than 3 weeks, a first culture of *P. jenkinae* reproduced for 3 months, a second one for more than 12 months now, and *P. notopora* for more than 2 years now.

Type and other reference material is deposited at the Instituto Nacional de Pesquisas da Amazônia (INPA), Manaus, Brazil, and at the Hamburgisches Zoologisches Museum und Institut (ZIM), Germany. Reference specimens are stained and whole mounted or ethanol-preserved. They have been examined alive prior to fixation.

Descriptions

***Pristina marcusi* sp. nov.** (Figs 1a–f and 4h,i)

Type locality: Primary forest plot of the SHIFT project ENV 52 study area, 02° 53′ 47″ S, 59° 59′ 45″ W; at 29 km on the road Manaus-Itacoatiara close to the agroforestry research station EMBRAPA-CPAA in the state of Amazonas, Brazil.

Type material: Holotype: whole mounted specimen, INPA 113. Paratypes: 4 whole mounted specimens, INPA 114, ZIM Ol 14158.

Other material examined: Twenty fixed specimens, 5 of them observed alive prior to fixation.

Description: All specimens sexually immature. Living specimens 2–3 mm long and 100–150 μm wide; distinctly wider in mid-body-region than in head and anal region; measured lengths in fixed specimens 780–1150 μm, widths 107–275 μm at VI, 93–310 μm at XII. Usually 18–28 segments, budding zone after XV or XVI (but one specimen observed with 21 segments in the first and 17 in the second zooid, and another with 19+9, respectively). Seven segments regenerated anteriorly. Living specimens colourless but not transparent, no reflections of coelomocytes. Epidermis with elongate, transversely oriented hyaline bodies from IX on (Fig. 1e), 2 per segment, in the longitudinal line of the ventral chaetae bundles, equidistant from the respective anterior and posterior bundle. Prostomium conical, 41–85 μm long, 61–100 μm wide at base, as long as wide or longer in living specimens; no proboscis. No eyes.

Dorsal chaetae (Figs 1a and 4i) from II, one hair and one needle per bundle. Hairs smooth light-microscopically but very finely hispid at $\times 3000$ magnification (SEM); measured lengths 80–192 μm, but intraindividual variation smaller, e.g. 118–133 μm in a small specimen and 130–186 μm in a large specimen; hairs shorter in II than in the rest, lengths from III on varying irregularly; thickness 1.8–2.2 μm. Needles thick, bayonet-shaped, without nodulus but with a kink at 2/5 from the distal tip; proximally straight, distally curved, with simple pointed tip, measured lengths 31–68 μm, intraindividual variation smaller,

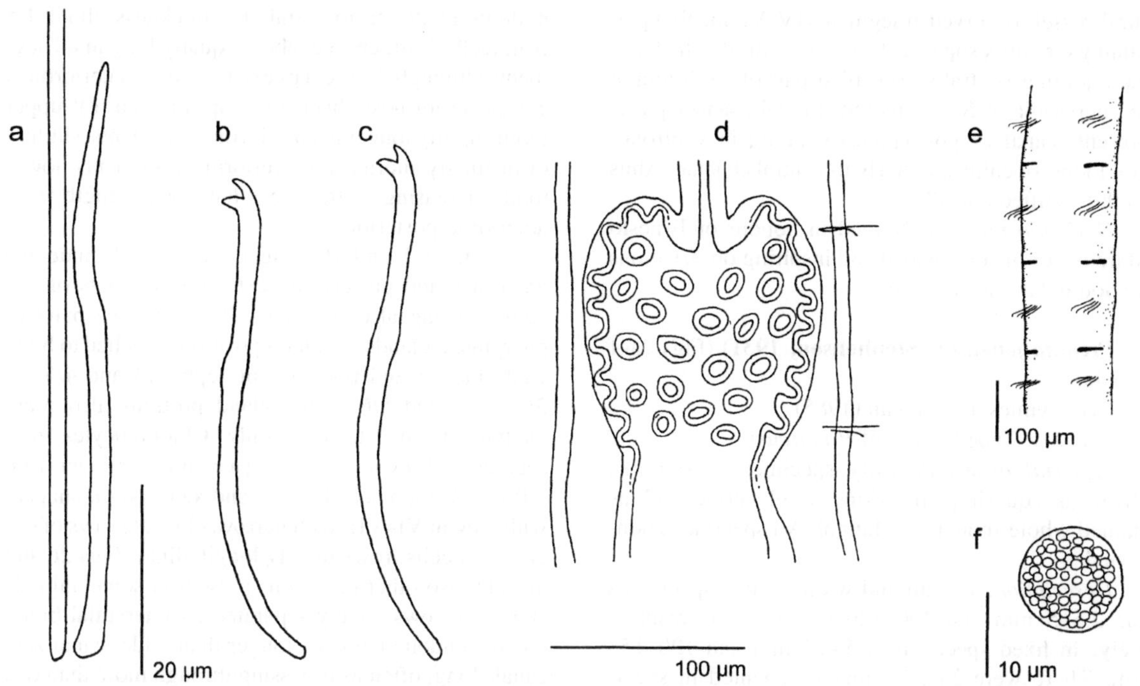

Figure 1. Pristina marcusi sp. nov. (a) dorsal needle chaeta (right) and proximal shaft of hair chaetae (left); (b) ventral chaeta in II; (c) ventral chaeta in X; (d) stomach *in vivo*, dorsal view, head region on top, hair chaetae shortened; (e) hyaline bodies on epidermis *in vivo*; (f) coelomocyte.

e.g. 53–65 μm (large specimen) or 34–43 μm (small specimen); proximally thicker than hairs (2.5–3 μm).

Ventral chaetae (Figs 1b,c and 4i) from II, bifid crotchets only, from II-V 4–8 chaetae per bundle, in II usually 6, teeth short and equally long (Fig. 1b); from VI on upper teeth increasingly shorter and thinner, lower increasingly thicker (Fig. 1c). The change in form is inconspicuous from V to VI but better marked from VII on; here 4–6 chaetae per bundle, usually 5; in posteriormost segments often 3–2 chaetae per bundle. Measured lengths 35–56 μm, intraindividual variation smaller; e.g. 47–56 μm (large specimen) and 38-41 μm (small specimen). Ventral chaetae in II often shorter than in the rest.

Brain in I and II, concave but not incised posteriorly, anterior margin not seen. Pharynx in II and III. Pharyngeal glands in IV and V, partly attached to septa 4/5 and 5/6 dorsally, partly elongate and free ventrolaterally. Chloragocytes beginning in IV with single cells; continuous layer from IX. Stomach (Figs 1d and 4h) in 1/2VI and VII, occasionally in VII only, conspicuous, widening abruptly, usually telescoped anteriorly by the terminal part of the oesophagus that projects into stomachal lumen, anterior stomachal walls bulged anteriad; epithelium thickened in the transition zone. Stomachal endothelium with regularly distributed, presumably intracellular cavernous invaginations of the stomachal lumen, not forming canals; cytoplasma surrounding invaginations granulated. No blood plexus visible between endothelial cells, hence cell boundaries not distinguished either. Posterior transition zone between stomach and intestine contractile, so stomach and intestine sometimes separated by a narrow constriction, sometimes not separated at all, with intestine as wide a stomach (comp. Fig. 4h), depending of the state of contraction of the transition zone; the latter state rarely seen in fixed specimens. Coelomocytes (Figs 1f and 4h) numerous, spherical, diameter ca. 12 μm, filled with distinctly visible but colourless and non-refractile spherical bodies, cells colourless grey under transmittent light, invisible at top light, i.e. without whitish reflections. Blood colourless, dorsal vessel projecting anteriorly to brain, branching into circumoesophageal commissural vessels that pass ventro-posteriad to reunite into ventral blood vessel in V; three simple transverse commissural vessels observed in segments IV–VI, the first pair joining circumoesophageal vessels ventrally, the latter pairs joining ventral vessel. First pair of nephridia in IX, nephridia in X also paired; funnel (=anteseptale) present, canal (= postseptale) winding in ventrolateral loops, opening anteriorly to ventral chaetae. Anus slightly shifted dorsally.

Derivatio nominis: Named in honour of Ernesto Marcus, for his unequalled and inspiring observations on South American naidids.

***Pristina jenkinae* (Stephenson, 1931)** (Figs 2a–f and 4a,b)

Synonymies in Kathman (1985)

Locality: Type locality of *P. marcusi*.

Material observed: Fifty specimens, examined alive, as squash preparations in glycerine and as stained whole-mounts in Malinol. All specimens non-sexual.

Description: Length and width of live specimens ca. 2.5–5 mm and 100–120 (−150) μm, respectively; in fixed specimens 1.5–2.5 mm and 120–158 μm. There were 25–33 segments counted in specimens without budding zone. Animal chains with 1 or 2 zooids; budding zone observed in XIII and XVIII. Specimens (top light, stereolupe) mottled whitish, with a roughly segmental pattern, caused by aggregations of coelomocytes at septa. Prostomium conical, 1.5 times longer than wide in living specimens moving freely in water, rounded in fixed specimens, 54–63 μm long, 70–84 μm wide at base. No eyes. No glandular modifications of epidermis observed.

Dorsal chaetae (Fig. 2a) from II, one hair and one needle per bundle; hairs smooth, 1.5 μm thick proximally, measured lengths 70–247 μm, usually 150–220 μm. All hairs posterior to VII more or less of same length, from II to VII increasing in length. Needles (Fig. 2a) stout, proximally straight, curved in the distal third, nodulus absent, distal third set off from the rest by a slight kink; needle length 39–60 μm, usually 46–52 μm, shortest in II and III, those in IV and V not larger than in the rest, bifid, teeth parallel, not curved, upper tooth (i.e. here the tooth of the convex side of the curve of the distal third of the needle) distinctly shorter and thinner than lower, usually 1/2 to 2/3 as long and thick; in II and III teeth occasionally subequal and almost equally thick.

Ventral chaetae (Fig. 2b–d) from segment II on, bifid crotchets, sigmoid, length 37–50 μm; in II usually 5 per bundle, with teeth equally long; from III 4–5, rarely 6, posteriorly 3–2 per bundle, with distal nodulus at 2/5 from distal tip; thickness about 1.5 μm; teeth short, curved, about equally long in all segments although differences exist between anteriormost and posteriormost chaetae (comp. Fig. 2b, c): upper tooth slightly longer in anteriormost segments, slightly or distinctly shorter in posteriormost segments, lower tooth increasing in thickness but not in length from anterior to posterior.

Brain in I and II, longer than broad, indented anteriorly and posteriorly, extending to roots of circumoesophageal connectives. Two or three pairs of pharyngeal glands, in lateral position, attached to body wall anterior and posterior of septa 3/4 and 4/5, or 3/4, 4/5, and 5/6, some gland portions also more medial, attached to septa only. Chloragocytes from segment V, brownish, forming a continuous layer from VII or XI. Stomach (Figs 2e and 4a,b) onion-shaped, widening in VI–VII; with narrow, elongate rows of endothelial cells; rows mostly longitudinal (Figs 2e and 4b) but also other directions present in some animals (Fig. 4a); rows widely separated by interstitial blood plexus; endothelial cells longer than wide, canalized; canals long, often as if passing through more than one cell, or one canal per cell, comma-shaped, opening into the stomachal lumen by a small porus. Posterior end of stomach often without marked constriction, usually in wide connection with following parts of intestine. Intestine widened in IX and X, gradually narrowing towards the posterior end. Coelomocytes (Figs 1f and 4a,b) numerous, spherical, diameter 10–15 μm, partly or completely filled with cristalline, refractile granules of uneven surface and much differing size, from granules up to 1.5 μm diameter down to splitters of less than 0.5 μm length. Colour of granules yellowish brown or even blue; overall colour of coelomocyte aggregations dark brownish grey, but white or whitish under top light. In some animals additionally smaller cells present devoid of granules and filled with few spherical, non-refractile vesicles. Blood faintly red, dorsal blood vessel embedded in chloragocyte tissue posterior to VI, pulsating in VI, projecting until first segment anterior of brain, branching into circumoesophageal commissural vessels that pass ventro-posteriad to reunite into ventral bood vessel in V; six simple transverse commissural vessels observed in segments II–VII; the first 4 joining circumoesophageal vessels ventrally. All commissurals in the posterior part of the respective segment, between chaetae and septum. First nephridia in IX, paired, following nephridia unpaired, not in all segments; funnel conspicuous, stalked; stalk granulated, duct in ventro-

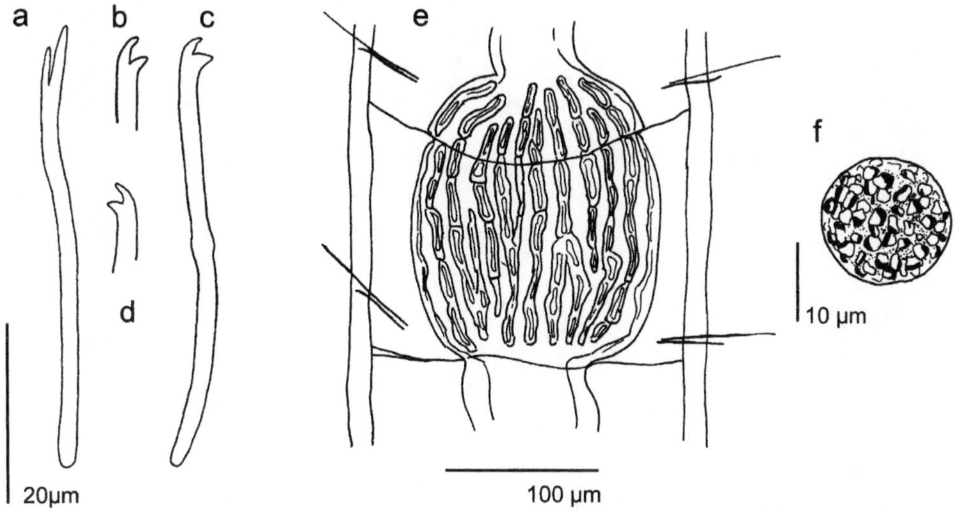

Figure 2. Pristina jenkinae (Stephenson, 1931). (a) dorsal needle; (b) tip of ventral chaeta in II; (c) ventral chaeta in XXX, (d) tip of ventral chaeta in mid-segment; (e) stomach with canals *in vivo*, diagrammatic, dorsal view, head region on top; (f) coelomocyte *in vivo*.

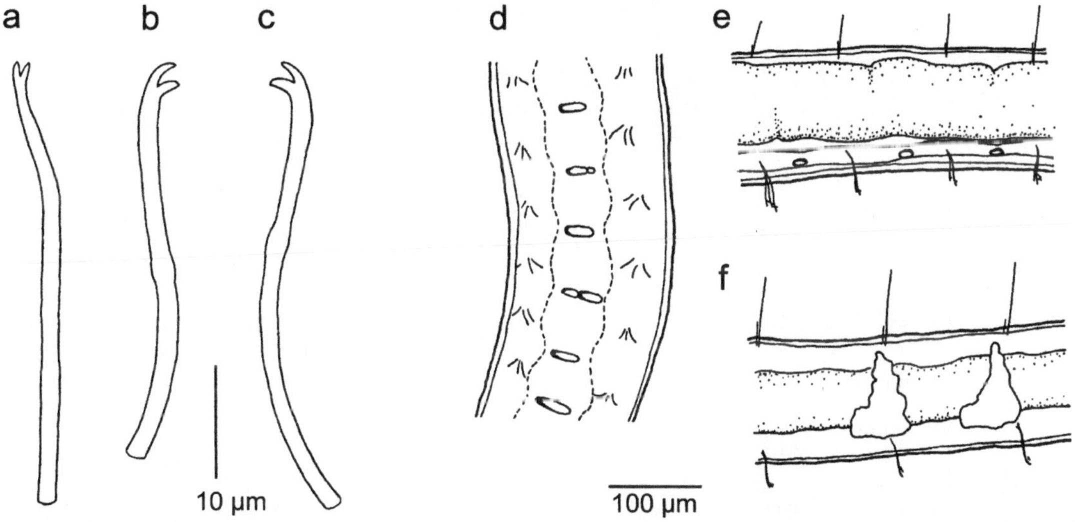

Figure 3. Pristina notopora Černosvitov, 1937. (a) dorsal needle; (b) ventral chaeta in II; (c) ventral chaeta in XI; (d) hyaline segmental bodies of nerve chord, ventral view; (e) same as (d), lateral view; (f) nephridia with dorsal projections as seen in ethanol-preserved specimens, lateral view.

lateral loops, opening anterior to ventral chaetae with an elongate terminal vesicle with hardly thickened walls.

***Pristina notopora* Černosvitov, 1937** (Figs 3a–f, 4C–H)

Pristina notopora Černosvitov, 1937: 140–141, Figs 13–16.

Locality: Type locality of *P. marcusi* sp. nov.

Material examined: Fifty specimens alive and 15 fixed, 10 as squash preparations in glycerine and 5 as stained whole-mounts in Malinol. All specimens nonsexual. Paratypes BMNH (British Museum of Natural History, London) Cat. No. 1949.3.1.871–876: 4 specimens on slides as unstained whole mounts, 3 specimens ethanol-preserved, 1 of them submature, stained and whole-mounted for this study, later destained and retransferred to ethanol.

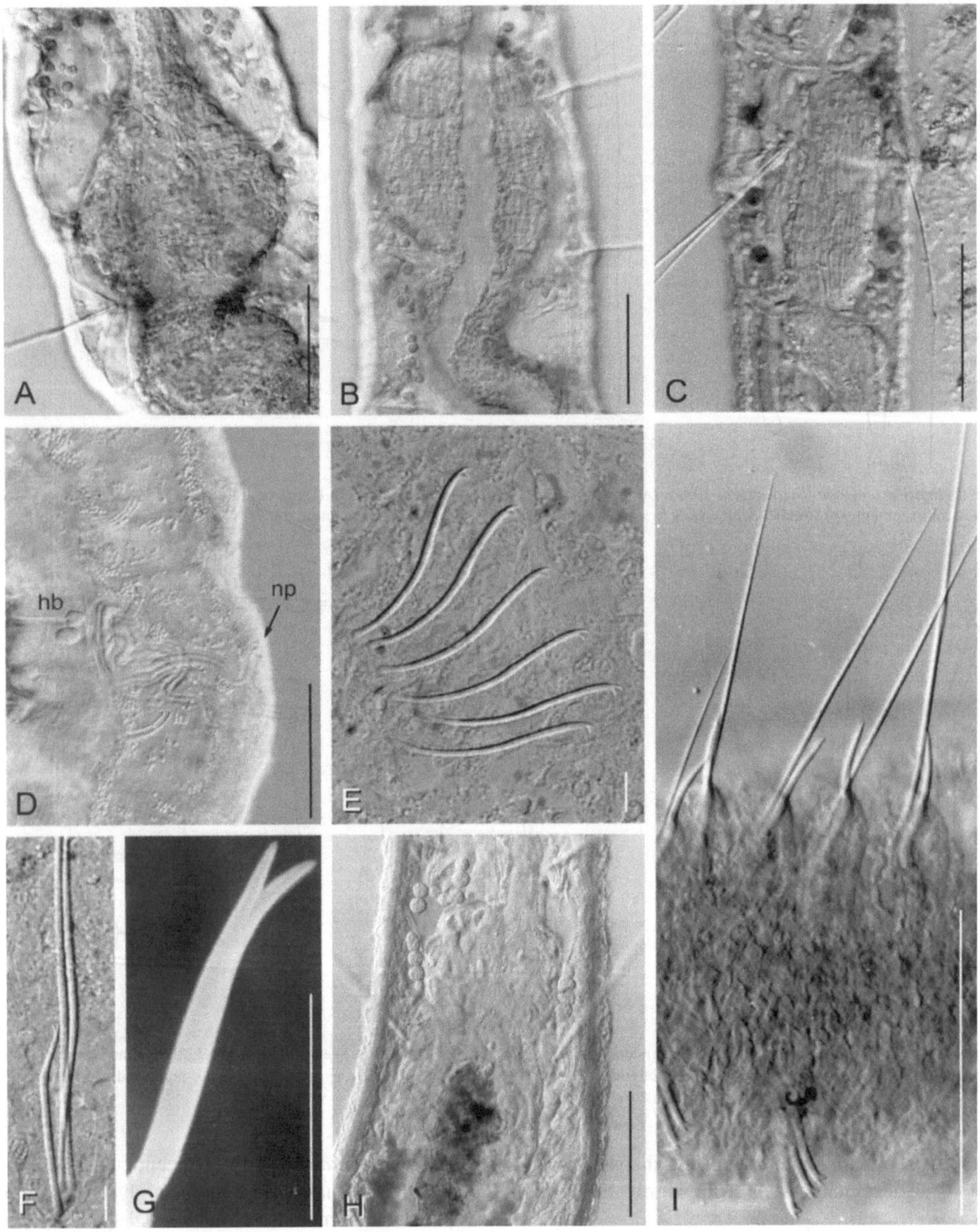

Figure 4. ***Pristina jenkinae*** (A,B), ***Pristina notopora*** (C–G) and ***Pristina marcusi*** (H,I). (A) Stomach and coelomocytes of *P. jenkinae*, dorsal view; (B) same as (A), different specimen, here dorsal vessel visible; (C) fusiform 'stomachal' dilatation and coelomocytes of *P. notopora*; (D) *P. notopora*, lateral view, nephridial canal with dorsal nephridiopore (np) and hyaline bodies (hb) inside nerve chord; (E–G) chaetae of *P. notopora*, (E) ventrals in VI, (F) dorsals in VI, (G) needle tip; (H) stomach and coelomocytes of *P. marcusi*, dorsal view; (I) *P. marcusi*, dorsal and ventral chaetae close to posterior end, growth zone. (A–D,H) from living specimens, (E,F) from squash preparation, (G) SEM photograph, I from whole-mounted ethanol-preserved specimen. Head region on top in (A–D,H), to the right in (I). Scale: (A–D,H,I) 100 μm, (D) 50 μm, (E,F,G) 5 μm.

Description: Length of living specimens 2–3 mm, width 50–100 μm, 50 μm in posteriormost segments, length of fixed specimens 1–1.6 mm, width 100–160 μm. There were 22–30 segments counted in specimens without budding zone. First budding zone behind XVII–XXI, second budding zone, when present, after XXVI. Animals with an overall reddish-brown tinge caused by blood and chloragocytes and with a roughly segmental pattern of opaque white fields caused by aggregations of coelomocytes. Prostomium conical or rounded, without proboscis, slightly longer than broad at base (up to 1.5 times longer when animals stretch forward), as long as broad in fixed specimens, with ciliated sense organs. No eyes.

Dorsal chaetae (Figs 3a and 4f,g) from II, usually 2 hairs and needles per bundle (occasionally 1 each); hairs hispid, some hairs apparently smooth, 78–165 μm long, shorter in anterior segments, increasing in length posteriorly; ca. 1.5 μm thick at base, strongly tapering distally; needles 29–38 μm long, sharply bent at distal fourth or fifth, without or with faint nodulus, bifid, teeth fine, of equal length, dividing at an acute angle, or almost parallel.

Ventral chaetae (Figs 3b,c and 4e) from II, 28–39 μm long, shorter in II and III than in the rest, 4–6 bifid crotchets anteriorly, 3 (4) posteriorly; upper tooth slightly longer than lower in anteriormost segments (Fig. 3b), as long as lower in most segments (Fig. 3c), slightly shorter with lower tooth thicker than upper in posteriormost segments; nodulus medial in II, shifted distally in the rest.

Brain in I and II, incised anteriorly and posteriorly, extending to roots of circumoesophageal connectives, length 90 100 μm, slightly longer than broad, but variable in shape and size relations. Pharyngeal pad large, in III. Pharyngeal glands in III – VI, but those in III not well distinguishable; paired cell aggregations attached to body wall laterally in IV and V and to septa 4/5 and 5/6, often left and right portions united here; elongate lumps moving freely ventrolaterally in V, attached only anteriorly; also, a pair of latero-ventral spherical aggregations in VI, projecting from septum 5/6 posteriorly. No stomach with intestinal epithelial cavities or canals detected, but intestine slightly widened in VII (Fig. 4c), spindle-shaped, and with longitudinal stripes caused by the periintestinal blood system that is narrowed into a plexus here, and the arrangement of chloragocytes that follow its pattern. Intestine narrow in VIII, enlarging in IX, full width from X. Single chloragocytes from IV, continuous layer from VI, cells flat, granula spherical, greenish, of about the same size as coelomocyte granula, but lighter in colour. Blood fairly reddish, blood vessels thick. Dorsal vessel projecting anteriorly to brain, branching into two circum-oesophageal commissural vessels that pass ventro-posteriad to reunite into ventral blood vessel in IV; five more pairs of simple commissural vessels present: in III behind the brain, in IV and in V, anteriorly to septal lumps of pharyngo-oesophageal gland cells, and in VI and VII, directly anteriorly and posteriorly to stomachal dilatation; all five pairs located in the hind part of the segment, anteriorly to but not connected with the septa. Commissural vessel of segment III uniting with circum-oesophageal vessels, commissural vessles from IV to VII uniting into ventral blood vessel, which is free, i.e. not connected with the intestine. Dorsal blood vessel in close contact with intestine and embedded in chloragocytes throughout its course up to segment IV, rising in hindmost segments, pulsating and larger from IX. Periintestinal blood sinus voluminous, in VII and X constricted to a longitudinal network (plexus) of branching and reuniting vessels. Coelomocytes (Fig. 4c) numerous, dark through transmittent light, diameter about 9–12 μm, with granula of roundish but irregular outline, like sand grains, neither spherical nor elongated, colour greenish brown. First nephridium in IX; nephrostome apparently absent, no anteseptal part seen, no ciliar movement observed, postseptale attached to septa ventrolaterally by a small solid mass; canals aggregated in large ventrolateral loops that surround another solid mass, terminally projecting dorsad and opening with a small bladder-like thicker-walled dilatation at the dorsal surface, anteriorly or laterally of dorsal chaetae (Fig. 4d). Only one nephridium per segment observed, most nephridia on the left body side only. In preserved specimens, nephridial mass with a broadly tapering dorsal projection when viewed laterally (Fig. 3f). Peculiar agglomerations of conspicuous hyaline subspherical masses observed in a number of animals, situated ventrally inside the nerve chord, close to the body wall (Figs 3d,e and 4d), from IX to XVIII–XXII, also visible in ethanol-preserved specimens; absent in some specimens.

Sexual organs (from 1 submature paratye specimen only): developing sperm seen in VIII and IX, developing clitellum dorsally in transverse rows. Male reproductive system not seen. Spermathecae paired and separate, in VII, ca. 3x as long as wide, projecting freely posteriad into the coelomic cavity, location of spermathecal openings in the line of ventral chaetal bundles, in the anteriormost region of VII, immediately behind septum of VI/VII, openings marked by a conspicuous hemispherical cuticular invagination.

Discussion

Pristina marcusi

Genus level: Presence of hairs and needles dorsally from II with needles very much unlike the ventral chaetae are found in *Stephensoniana* Černosvitov, 1938, *Bratislavia* Košel, 1976 and *Pristina* Ehrenberg, 1828. *Pristinella* Brinkhurst, 1985 as a fourth naidid genus that fits these characters has been invalidated and re-combined into *Pristina* by Collado & Schmelz (2000). The three genera are mainly separated by the number of anterior segments formed during budding and the position of the sexual organs: 4 and IV/V, respectively, in *Stephensoniana*, 5 and V/VI in *Bratislavia*, 7 and VII/VIII in *Pristina*. The generic identity of the new species as *Pristina* is assured by the number of anteriorly regenerated segments. Furthermore, the presence of coelomocytes and the absence of a thin sheath incrusted with foreign matter covering the body wall disagrees with the generic diagnosis of *Stephensoniana*; species of *Bratislavia* have no specialized stomach and their mode of asexual reproduction is by fragmentation, not by budding.

Species level: *P. marcusi* is most easily distinguished from all other *Pristina* species without proboscis by the shape of the needle chaetae: stout, thicker than hairs, bayonet-shaped, simple-pointed. Three more species of *Pristina* without proboscis possess simple-pointed needles, *P. menoni* (Aiyer, 1930), *P. longisoma* Harman, 1977 and *P. orghidani* Botea, 1983. The needles of *P. menoni* resemble those of *P. marcusi* in general shape (although, according to Sperber (1948) and Grimm (1986) they are bifid with much reduced distal tooth), but they are much thinner (1.7 μm and not 2.5–3 μm). Other characters that distinguish *P. menoni* from *P. marcusi* are: brain deeply incised anteriorly and posteriorly (not simply concave posteriorly), coelomocytes black (not pale), stomachal dilatation slight and inconspicuous (not large and abrupt), stomach with branching canals (not with circular invaginations), a maximum of 5 chaetae ventrally (not 8), often 2 hairs and 2 needles per bundle dorsally (not 1 only). The needles of *P. longisoma* are not bent and they extend distally into a very thin and sharply pointed tip. Other distinguishing characters are: hairs much longer (1400 μm in II, 400–1000 μm in the rest), segment number much higher (more than 100), five segments regenerated anteriorly, no stomach in VI/VII, only a gradual gut dilatation in IX–X, reproduction presumably by fragmentation. The generic position of this species is uncertain (Harman, 1977; Brinkhurst & Marchese, 1989). The needles of *P. orghidani* are without a kink, and the simple-pointed needles are replaced by bifid needles in posterior segments. Another difference, the location of the stomachal dilatation in IV–V, is doubtful. The description of *P. orghidani* is rather poor in detail.

Among the *Pristina* species with a proboscis, three of them have simple-pointed needle chaetae: *P. longiseta* Ehrenberg 1828, *P. proboscidea* Beddard 1896 and *P. macrochaeta* Stephenson, 1931. Apart from the proboscis the separating characters are as follows: needle chaetae straight and very fine (all three); hairs of III extremely elongated (*P. longiseta*); dorsal chaetae from III or IV (*P. macrochaeta*); ventral chaetae of II longer and thicker than the rest (*P. proboscidea* and *P. macrochaeta*); all ventral chaetae with upper tooth longer than the lower (all three).

Apart from the chaetae, living specimens of *P. marcusi* are easily recognized by the pale and non-refractile coelomocytes, common in *Nais* but rare in *Pristina* (Schmelz, pers. obs.), and by the peculiar stomach with regularly distributed ellipsoid invaginations. The latter trait is also conserved in specimens that have been fixed in Bouin's fluid.

P. marcusi bears some superficial resemblance to the specimens identified as *P. jenkinae* also found in the soil samples (see below), with respect to size and to the general appearance of the chaetae. When the needle chaetae of *P. jenkinae* are not in lateral position and the large proximal tooth overlays and hides the thin distal tooth, they are easily taken for those of *P. marcusi*. However, the differences in the coelomocytes, the stomachal canals, and finally the needles themselves when observed properly, clearly separate the species.

Pristina jenkinae

The Brazilian specimens agree with the original description of the species (Stephenson, 1931) and the redescription of Kathman (1985), except for the differences in the sharpness of the needle teeth tips (pointed in Stephenson (1931: Fig. 1), rather blunt in Kathman (*op. cit.*: Fig. 1), intermediate in our study), for which certainly more the artists' interpretation than the nature of the object is responsible. Among the

Pristina species without proboscis, *P. jenkinae* is most easily recognized by the needle chaetae with parallel bifid teeth and the upper tooth much shorter and thinner than the lower, but according to Kathman (*op. cit.*), who compared a number of specimens identified by her and others as *P. jenkinae*, there is much variation in the needle teeth length, both absolutely (from 1.5 to 10 μm, see *op. cit.*: Table 1) and relatively (from distal tooth 1/2 as long as the proximal tooth to equally long, see *op. cit.*: Fig.1). These variations were the main reason for synonymizing *P. idrensis* with *P. jenkinae* (Kathman, 1985). Furthermore, *P. taita* and *P. nothofagi* were declared synonymous with *P. jenkinae*. In our material, the distal tooth is constantly much thinner and 1/2–2/3 as long as the proximal tooth, with the only exception of the anteriormost needles in segment II and III, where the needles of some specimens have short and almost equally long teeth. Whereas the latter finding conforms with Kathman's observations, we are not so sure about the synonymies established. Overlap in one character must not necessarily imply that the species in question are synonymous. It can also mean that the characters used are not appropriate to separate the species. Kathman relies on chaetal characters, because this is mainly what the original description of *P. jenkinae* contains. The original descriptions of *P. idrensis*, *P. nothofagi* and *P. taita*, however, include more characters than discussed by Kathman. When they are compared with the morphological account of *P. jenkinae* presented here, many details are congruent (e.g. shape of the brain, location of pharyngeal glands, colour of coelomocytes, pattern of commissural blood vessels, location and shape of nephridia) but two characters are different: (1) In *P. idrensis*, the needles in IV and V are longer and stouter than in other segments (Sperber 1948: 220f.), but not so in *P. taita* and not in our specimens, either. This difference, should it prove to be species-separating, would revalidate *P. idrensis* and *P. taita* and transform *P. jenkinae* into a *species inquirenda*, as its respective character state is unknown, and types are apparently lost (Reynolds & Cook, 1976). (2) The stomachal cavities of *P. nothofagi*, as inferrable from Figure 1 in Stout (1958), are much reminiscent of the circular invaginations in *P. marcusi*, whereas the stomachal canals of *P. taita* as drawn in Stout (1956, Fig. 2) are similar to what is found in our Brazilian *P. jenkinae*. These differences in the stomachal canals of *P. taita* and *P. nothofagi* may be judged as accidentally created by the artist as they are not referred to in the text, but their correspondence to the differences between *P. marcusi* and our *P. jenkinae* remains striking and may indicate, as in the latter two species, that *P. taita* and *P. nothofagi* are different species. In our view, the question whether these differences further contribute to the large intraspecific variability of the species or whether they indicate the existence of different species under the current name of *P. jenkinae sensu* Kathman, is still open and requires comparative investigations of living or well-preserved specimens from geographically distant regions including the type localities.

Pristina notopora

The identity of our specimens is beyond doubt as we were able to confirm most of the anatomical details of the original description, all of which, except the chaetae, have been re-examined here for the first time. This applies to the brain, the coelomocytes, the blood colour, the absence of a canalized stomach and especially to the nephridia with their exceptional dorsal location of the nephridiopores. Our observations agree in all details with Černosvitov's account: "*les canaux néphridiens longs et etroits avec une petite dilatation terminale près du grand orifice externe situé un peu en avant des faisceaux dorsaux*" (1937: 141). There is also agreement in all details on the chaetae except that they are slightly smaller in our specimens.

The absence of a nephridial funnel, not mentioned in the original description, is a further peculiarity of the species. Černosvitov describes a "small and fusiform postseptale" (*loc. cit.*) attached to the septum, apparently the 'solid mass' of our description. A functioning nephridial funnel can be distinguished in live specimens by the ciliar movements and also by the movements of coelomic particles in its vicinity. We never observed any of that neither anterior to nor inside the body with which the nephridium is attached to the septum. Ciliar movements in the loops of the canals, however, were distinct. Funnels are also absent in the nephridia of *Chaetogaster* (Vejdovský, 1884).

Two more differences are notable: First, the budding zone is located more anteriorly than described by Černosvitov (XVII–XIX instead of XXVI). The intraspecific variability of this character, however, has been shown in other species of *Pristina*. In *P. jenkinae*, for example, it is behind XIX (Stephenson, 1931) or XI–XIII (Sperber, 1948); in *P. americana* it is behind XXX–VL (Černosvitov, 1937) or XXV–XXXIII (Marcus, 1943). Second, the transition from esophagus into intestine is described differently. Our specimens have a fusiform dilatation in segment VII, and a gradual widening of the intestine in IX. In Černosvitov's description, the fusiform dilatation extends over VIII and IX, and the intestine widens abruptly in XI. Not mentioned by Černosvitov are the anterior commissural blood vessels and the enigmatic midventral bodies below the ventral nerve chord. The latter do not show in all specimens. Judging on their location and on their non-cellular and hyaline appearance they could be neurosecretory products. Their outline is transversely elongate, and they are occasionally subdivided into two or more bodies like separating lipid droplets. These bodies, when present, are very conspicuous in live specimens and do not disappear in ethanol-preserved material.

The types of *P. notopora*, though rather ill-preserved, exhibit the following characters with sufficient clarity: (1) The partly dried-out squash preparations show chaetal details, especially the distal tips of the ventrals and the dorsal needles; they are exactly as redescribed here, the variability in the needles included (see below). (2) In the ethanol-preserved specimens, the sexual one included, the dorsal extension of the nephridia could be traced by focusing the amorphous mass (to which the nephridia are usually coagulated by fixation and long storage) ventrally and then tracing it dorsad. (3) The single sexual specimen is not fully mature and most of the organs of interest are invisible in the obligatorily non-destructive whole-mount preparation – sections would perhaps do better. However, the presence of morulae in 2 segments suggests that the sperm vesicle is large when fully developed, and the presence of spermathecae gives a third example (with *P. silvicola* Collado & Schmelz, 2000 and *P. osborni* Walton, 1905 (Erséus & Grimm, 1998)) of a *Pristina* species that lacks a proboscis but possesses spermathecae, a character combination that contributed to the invalidation of *Pristinella* as a genus separate from *Pristina* (Collado & Schmelz, 2000).

The needles of *P. notopora* have a peculiar shape: the proximal 3/4 or 4/5 are straight and of equal length throughout; the distal fourth or fifth is then distinctly set off by a marked kink and a subsequent reduction in diameter. The teeth in turn are bent back at about the same angle so as to restore the direction or orientation of the proximal shaft. The kink may (Fig. 4f) or may not (Fig. 3a) show a nodulus-like enlargement, and the needle teeth can be almost parallel (Fig. 3a) or distinctly diverging at an acute angle (Fig. 4f,g). These variations are also present in the type specimens. The hispid nature of the hair chaetae is rather fine and was not always seen. It shows best when the hairs are strongly bent and the fringes stick out at the convexity. Smooth hairs were found only occasionally in specimens that otherwise had hispid hairs.

P. notopora has been recorded from Brazil (Montanholi-Martins & Takeda, 1999), Guyana (Stacey & Coates 1996), Peru (Harman et al., 1988), Africa (Grimm, 1974), Spain (Martínez-Ansemil 1982), and Italy (Di Chiara Paoletti & Sambugar, 1996). All material was preserved and all clues to the identification, if present, confine to chaetal characters. Only Martínez-Ansemil (1982) reports on an attempt to find the dorsal nephridiopores in his specimen from Spain, without success. The needle teeth of his specimen are also smaller (*op. cit.*: Fig. 5a) and the upper teeth of the ventrals in II (*op. cit.*: Fig. 5c) are longer than in our and the type specimens of *P. notopora*; so it was probably a species other than *P. notopora*. The identification of Harman et al. (1988) of specimens from Peru is probably false, as the needle teeth are stated to diverge widely. The needle teeth of Grimm's single specimen of *P. notopora* from Africa (1974: Fig. 6 a,b) are within the variation range discovered here. All records outside South America are based on one specimen each and require confirmation.

Acknowledgements

We thank several unnamed coworkers of the project SHIFT-ENV 52 for collecting naidid specimens. Jörg Römbke, Frankfurt, Germany, repeatedly provided us with soil samples from the study site and gave informations on the study area. The British Museum of Natural History (BMNH) London, provided type material of *P. notopora*. W. Westheide kindly supported our studies. This paper contributes to the Project ENV 52 "Soil Fauna and Litter Decomposition" in the framework of the German-Brazilian research programme SHIFT: "Studies on Human Impact on Floodplains and Forests in the Tropics".

References

Brinkhurst, R. O., 1971. Naididae. In Brinkhurst, R. O. & R. O. Jamieson (eds), Aquatic Oligochaeta of the World. Oliver & Boyd, Edinburg: 304–443.

Brinkhurst, R. O. & M. Marchese, 1989. Guide to the freshwater aquatic Oligochaeta of South and Central America. Asociación de Ciencias Naturales del Litoral, Santo Tomé, Argentina: 179 pp.

Černosvitov, L., 1937. Notes sur les Oligochaeta (Naididées et Enchytraeidées) de l'Argentine. Ann. Mus. Argent. 39: 135–157.

Collado, R. & R. M. Schmelz, 2000. *Pristina silvicola* and *Pristina terrena* spp. nov., two new soil-dwelling species of Naididae (Oligochaeta, Annelida) from the tropical rain forest near Manaus, Brazil, with comments on the genus *Pristinella*. J. Zool. Lond. 251: 509–516.

Di Chiara Paoletti, A. & B. Sambugar, 1996. Aquatic Oligochaeta in Italy, with special reference to Naididae. Hydrobiologia 334: 37–49.

Erséus, C. & R. Grimm, 1998. *Pristina proboscidea* and *Pristinella osborni* (Oligochaeta, Naididae) from a freshwater creek near Darwin, Northern Territory, Australia, with descriptions of the genital organs of both species. The Beagle, rec. mus. art gall. north. terr. 14: 149–158.

Grimm, R., 1974. Einige Oligochaeten aus Nigeria, dem Tschad und der Zentralafrikanischen Republik. Mitt. hamb. zool. Mus. Inst. 71: 95–114.

Grimm, R., 1986. Beiträge zur Systematik der afrikanischen Naididae (Oligochaeta). III. Untersuchungen zur qualitativen und quantitativen Chaetotaxonomie der Naididae. Mitt. hamb. zool. Mus. Inst. 83: 101–105.

Grimm, R., 1987. Contributions towards the taxonomy of the African Naididae (Oligochaeta). IV. Zoogeographical and taxonomical considerations of African Naididae. Hydrobiologia 155: 27–37.

Harman, W. J., 1977. Three new species of Oligochaeta (Naididae) from the southeastern U.S.. Proc. Biol. Soc. Wash. 90: 483–490.

Harman, W. J., R. O. Brinkhurst & M. Marchese, 1988. A contribution to the taxonomy of the aquatic Oligochaeta (Naididae) of South America. Can. J. Zool. 66: 2233–2242.

Kasprzak, K. & B. Szczesny, 1976. Oligochaetes (Oligochaeta) of the river Raba. Acta Hydrobiol. Krakow 18: 75–87.

Kathman, R. D., 1985. Synonymy of *Pristinella jenkinae* (Oligochaeta. Naididae). Proc. Biol. Soc. Wash. 98: 1022–1027.

Kathman, R. D. & R. O. Brinkhurst, 1998. Guide to the freshwater oligochaetes of North America. Aquatic Ressources Center, College Grove, Tennessee, U.S.A.: 264 pp.

Loden, M. S. & W. J. Harman, 1980. Ecophenotypic variation in setae of Naidiae (Oligochaeta). In Brinkhurst, R. O. & D. G. Cook (eds), Aquatic Oligochaete Biology. Plenum Press, New York: 33–40.

Marcus, E., 1943. Sôbre Naididae do Brasil. Bol. Fac. Fil. Cienc. Letr. Univ. Sao Paulo 7: 3–247.

Martínez-Ansemil, E. & N. Giani, 1982. Contribución al conocimiento del género *Pristina* (Oligochaeta, Naididae) en la Península Ibérica. Bol. R. Soc. Española Hist. Nat. (Biol.) 80(3–4): 249–260.

Montanholi–Martins, M. C. & A. M. Takeda, 1999. Communities of benthic oligochaetes in relation to sediment structure in the upper Paraná river, Brazil. Stud. Neotrop. Fauna Environ. 34: 52–58.

O'Connor, F. B., 1955. Extraction of enchytraeid worms from a coniferous forest soil. Nature 175: 815–816.

Pop, V., 1973. Deux espèces nouvelles de *Pristina* (Naididae, Oligochaeta) et leurs affinités. Stud. Univ. Babes-Bolyai (Biol.) 1: 79–89.

Reynolds, J. W. & D. G. Cook, 1976. Nomenclatura Oligochaetologica. The University of New Brunswick, Fredericton, New Brunswick: 217 pp.

Römbke, J. & M. Meller, 1999. Applied research on Enchytraeidae in Central Amazonia: project approach, methodology and first results. In Schmelz, R. M. & K. Sühlo (eds), Newsletter on Enchytraeidae No. 6. Proceedings of the 3rd International Symposium on Enchytraeidae, Osnabrück, Germany. Universitätsverlag Rasch, Osnabrück, Germany: 69–75.

Römbke, J., H. Höfer, C. Martius, B. Förster, E. Franklin, M. Garcia & L. Beck, 1999. Die Rolle der Bodenfauna beim Streuabbau in Primär- und Sekundärwäldern und einer Polykulturplantage in Amazonien (SHIFT Projekt ENV 52): Methodische Überlegungen. In Oehlmann, J. & B. Markert (eds), Ökotoxikologie: Ökosystemare Ansätze und Methoden. ecomed, Landsberg, Germany: 268–275.

Sperber, C., 1948. A taxonomic study of the Naididae. Zool. Bid. Upps. 28: 1-296.

Stacey, D. F. & K. A. Coates, 1996. Oligochaetes (Naididae, Tubificidae, Opistocystidae, Enchytraeidae, Sparganophilidae and Alluroididae) of Guyana. Hydrobiologia 334: 17–29.

Stephenson, J., 1931. Oligochaeta from Burma, Kenya and other parts of the world. Proc. Zool. Soc. London 1931: 33–92.

Stout, J. D., 1956. Aquatic Oligochaetes occurring in forest litter. I. Trans. Roy. Soc. N. Z. 84(1): 97–102.

Stout, J. D., 1958. Aquatic Oligochaetes occurring in forest litter. II. Trans. Roy. Soc. N. Z. 85(2): 289–299.

Vejdovský, F., 1884. System und Morphologie der Oligochaeten. Prag: 172 pp.

Hydrobiologia **463:** 13–22, 2001.
P. Rodriguez & P.F.M. Verdonschot (eds), Aquatic Oligochaete Biology VIII.

A new lumbriculid genus and species from North America (Clitellata, Lumbriculidae)

Steven V. Fend[1] & Daniel L. Gustafson[2]
[1]*U.S. Geological Survey, Menlo Park, California, U.S.A.*
Fax: 650 329 4463. E-mail: svfend@usgs.gov
[2]*Montana State University, Bozeman, Montana, U.S.A.*

Key words: Clitellata, Oligochaeta, Lumbriculidae, taxonomy, description

Abstract

Secubelmis limpida n. sp., n. gen., is described from Montana, U.S.A. The monotypic genus resembles the lumbriculid genera *Rhynchelmis* and *Tatriella* in having atria in X, spermathecae in VIII, and semi-prosoporous male ducts. It differs from *Tatriella* in having paired atria and spermathecae, and from *Rhynchelmis* in having petiolate atria. The single prostate gland is unique within the Lumbriculidae. The new species appears to have a restricted range in the northern Rocky Mountains.

Introduction

The oligochaete family Lumbriculidae currently includes 10 genera known to occur in North America, although there is some disagreement regarding generic definitions (Kathman & Brinkhurst, 1998). The Nearctic fauna, with about 35 previously described species, appears depauperate compared with the more than 120 Palearctic species. Regarding the circum-Pacific part of this range, this disparity has been attributed in part to the greater geographical diversity of the Asian side (Timm, 1999). However, the recent discovery of several Nearctic species of *Rhynchelmis*, a genus of rather conspicuous, epigean species, suggests that the fauna has been underreported (Fend & Brinkhurst, 2000). Many of the Palearctic species are small, cryptic worms, described from groundwater or hyporheic habitats. Until recently, metazoans of those habitats have received comparatively little study in North America (Ward & Voelz, 1994); nevertheless, new lumbriculids occasionally turn up in subsurface waters (e.g. Cook, 1975; Rodriguez & Coates, 1996). Additionally, much of the lumbriculid study in the western part of North America has been by collectors interested in megadriles (Eisen, 1888; Altman, 1936; McKey-Fender, 1988), and has consequently been biased toward larger organisms.

Although studies of freshwater metazoans typically focus on surface waters, and oligochaetes are often not identified to species, recent macroinvertebrate collections from studies of water quality and fish disease have produced many new lumbriculid records from western North America. Collections of oligochaetes from several sites in Montana included a new species of Lumbriculidae that resembles the widespread genus *Rhynchelmis* in having spermathecal pores in VIII and male pores in X. However, the structure of the reproductive organs differs from all Nearctic *Rhynchelmis* species in several respects, and suggests closer affinity with European lumbriculids.

Materials and methods

Worms were collected in multiple-habitat samples from small streams in the Rocky Mountains (Montana, U.S.A.). Collections were made by disturbing the substratum with a shovel in front of a dip net having 0.5 mm mesh. Samples were elutriated to remove inorganic debris, using a bucket and a 0.25 mm mesh net. The samples were first treated with dilute alcohol to anaesthetize the organisms, and then fixed in Kahle's fluid. Samples were sorted in the lab with a dissecting microscope, and worms were then preserved in 70%

ethanol. Sagittal sections were approximately 7 μm and transverse sections were 10 μm; all were stained in hematoxylin and eosin Y. Dissections and whole worms were dehydrated through an alcohol series, cleared in methyl salicylate, and slide mounted in Kleermount. Some of the mounted specimens were first stained with hematoxylin or borax carmine, and partially de-stained with acidified alcohol.

Length and diameter at segment X were measured with a dissecting microscope to the nearest 0.1 and 0.01 mm, respectively; chaetae and internal characters were measured with a compound microscope to about 2 μm for small structures, or 10 μm for large structures. Reproductive structures were measured on worms having well-developed genital pores and mature eggs; measurements of other structures were made on worms that were at least partially mature, with gonads and genital pores. Blood vessels were observed in cleared whole worms and transverse sections, as live specimens were unavailable. Drawings were made with a drawing tube.

Type specimens were deposited in the United States National Museum (USNM). Specimens of several potentially related Palearctic species were examined for comparison: *Tatriella slovenica* Hrabě, 1936 2 sectioned and 2 whole mounts, from Lake Uljuste, Estonia, and Lake Terek"javr, Kola Peninsula, Russia, T. Timm collection; *Rhynchelmis olchonensis* Burov & Kozhov, 1932 from Lake Baikal, 2 sectioned, Czech National Museum; *Rhynchelmis alyonae* Martin et al., 1998 and *R. shamanensis* Martin et al., 1998 from Lake Baikal, several dissected, P. Martin collection; *Rhynchelmis paraolchonensis* Giani & Martínez–Ansemil, 1984 2 sectioned, from France, N. Giani collection.

Segments are designated with Roman numerals; intersegmental septa are designated with Arabic numerals corresponding to the adjacent segments, as '3/4'. Abbreviations used in the illustrations are: at – atrium; br – brain; f1 – anterior male funnel; f2 – posterior male funnel; gl – gland; gu – gut; ov – ovary; mu – muscle; pe – penis; ph – pharynx; pr – prostate; sp – spermatheca; spa – spermathecal ampulla; spd – spermathecal duct; t1 – anterior testis; t2 – posterior testis; v1 –anterior vas deferens; v2 – posterior vas deferens; vn – ventral nerve cord. Map coordinates are given in decimal degrees.

Results

Taxonomy

Secubelmis n. gen.

Diagnosis: Small worms without proboscis. Posterior lateral blood vessels absent. Chaetae two per bundle, sigmoid, with simple points. Testes paired in IX and X, ovaries paired in XI. Atria paired, with male pores behind ventral chaetae in X; spermathecae paired, with pores behind ventral chaetae in VIII. Atrium petiolate, with narrow, ectal duct; ental ampulla with circular muscle layer. Single, pyriform prostate gland inserts at ental end of atrial ampulla. Male pores with penes in shallow sacs, surrounded by oblique muscles and glands. Vasa deferentia 2 per atrium, posterior vasa do not penetrate septum 10/11 into XI.

Type species: Secubelmis limpida n. sp. (monotypic)

Etymology: Secubo (Latin), to live or sleep alone. Refers to the isolated distribution and empty spermathecae of the type species.

Secubelmis limpida n. sp. (Figs 1–4)

Holotype: A whole mounted specimen in Kleermount, unstained. USNM 1000124.

Type Locality: Montana, U.S.A.: Granite Co.: Ranch Creek at Rock Creek Road, 46.5808° N, 113.6708° W, elevation 1207 m, 9.IV.1997. Collected by D. L. Gustafson.

Paratypes: All collected at the type locality. Four whole mounts, unstained, 9.IV.1997. One whole mount, stained with hematoxylin, 5.VI.1998. Two sagittally sectioned specimens; 2 dissected on slides (1 stained with carmine); 2 whole mounts (1 stained with carmine), 8.V.1998. One transversely sectioned worm; 1 whole mount, stained with hematoxylin, 30.X.1998. USNM 1000125–1000130.

Additional Material: Montana, U.S.A.: Beaverhead Co.: Brays Canyon Creek, 45.2013° N, 113.1745° W, elevation 1032 m. Two whole mounts, 30.X.1999. Buffalo Creek, 45.1561° N, 113.1896° W, elevation 2194 m. One whole mount, 3.X.1999. Dingley Creek, 45.4406° N, 113.0741° W, elevation 2368 m. Two whole mounts, 4.IX.1998. Reservoir Creek, 45.1432° N, 113.1890° W, elevation 2194 m. One whole mount, 2 dissected, 1 transversely sectioned, 11.V.1997. 45.1410° N, 113.1924° W, elevation 2138 m. One transversely sectioned, 3.X.1999. Granite Co.: Ranch Creek at Grizzly Camp, 46.5756° N, 113.6596° W, elevation 1275 m. Ten whole mounts

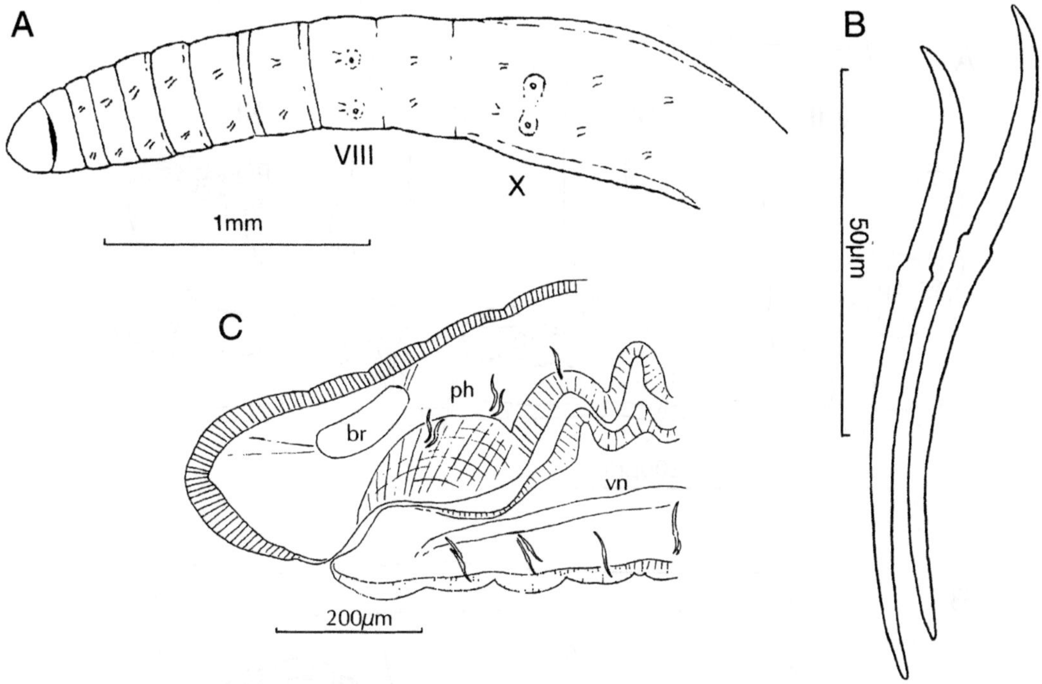

Figure 1. Secubelmis limpida. (A) External features, ventral view. (B) A pair of chaetae from mid-body region. (C) Anterior end, from a whole mount. Abbreviations used in the figures are listed in the 'Materials and methods' section.

on 4 slides, 22.VI.1997. Ranch Creek at Rock Creek Road (type locality). One whole mount, 7.XI.1999. Lincoln Co.: North Fork Keeler Creek at mouth, 48.3444° N, 115.9024° W, elevation 795 m, 1 whole mount, 16.VIII.1998. All specimens collected by D. L. Gustafson.

Etymology: limpidus (Latin), clear or transparent. Refers to the clean, unpolluted habitat.

Description: Length of preserved worms 6–14 mm; diameter 0.31–0.45 mm in X, maximum diameter to 0.50 mm. Prostomium rounded to nearly conical (Fig. 1 A,C). Chaetae paired in 4 bundles per segment, simple-pointed, sigmoid, with nodulus slightly distal, about 1/3 (32–39%) of chaetal length from the tip (Fig. 1B). Chaetal length as little as 60 μm in II, 80–100 μm in clitellar region, 70–95 μm posteriorly.

Epidermis in anterior segments 12–20 μm thick; in clitellum 20–40 μm. Clitellum distinct in some specimens, from (IX) X to XII (1/2 XIII). Epidermis in prostomium thickened to 25–40 μm. Circular muscle of body wall 3–5 μm thick; longitudinal muscles 25–30 μm thick, in transverse sections forming a nearly continuous ring, weakly divided by chaetal and lateral lines. Septa indistinct at 1/2 and 2/3; prominent elsewhere. Pharynx with well-defined, dorsal pad in II–III (Fig. 1C). Pharyngeal glands in IV–VI.

First pair of nephridia in VII. Nephridia absent in VIII–XII, usually paired in XIII, and occur irregularly in posterior segments. Each nephridium with small anteseptal funnel, a postseptal expansion (diameter near 50 μm) containing small ducts, a short duct to a nephropore anterior to the ventral chaetae, and a long, convoluted duct that may pass through several anterior and posterior segments ventral to the gut, possibly joining ducts of nephridia in other segments. There may also be a branch or loop entering the dorsal part of the segment. Both intersegmental (ventral) and dorsal branches may form dense masses of convoluted tubules.

Vascular system without dorsolateral vessels in segments posterior to genital pores. Main dorsal and ventral vessels prominent, without secondary longitudinal vessels. One pair of commissures in at least some anterior segments; those in XI typically entering egg sacs. Gut plexus begins in about VII.

Male pores paired in X, spermathecal pores paired in VIII, all on chaetal line between ventral chaetal bundles and posterior intersegmental groove (Figs 1A

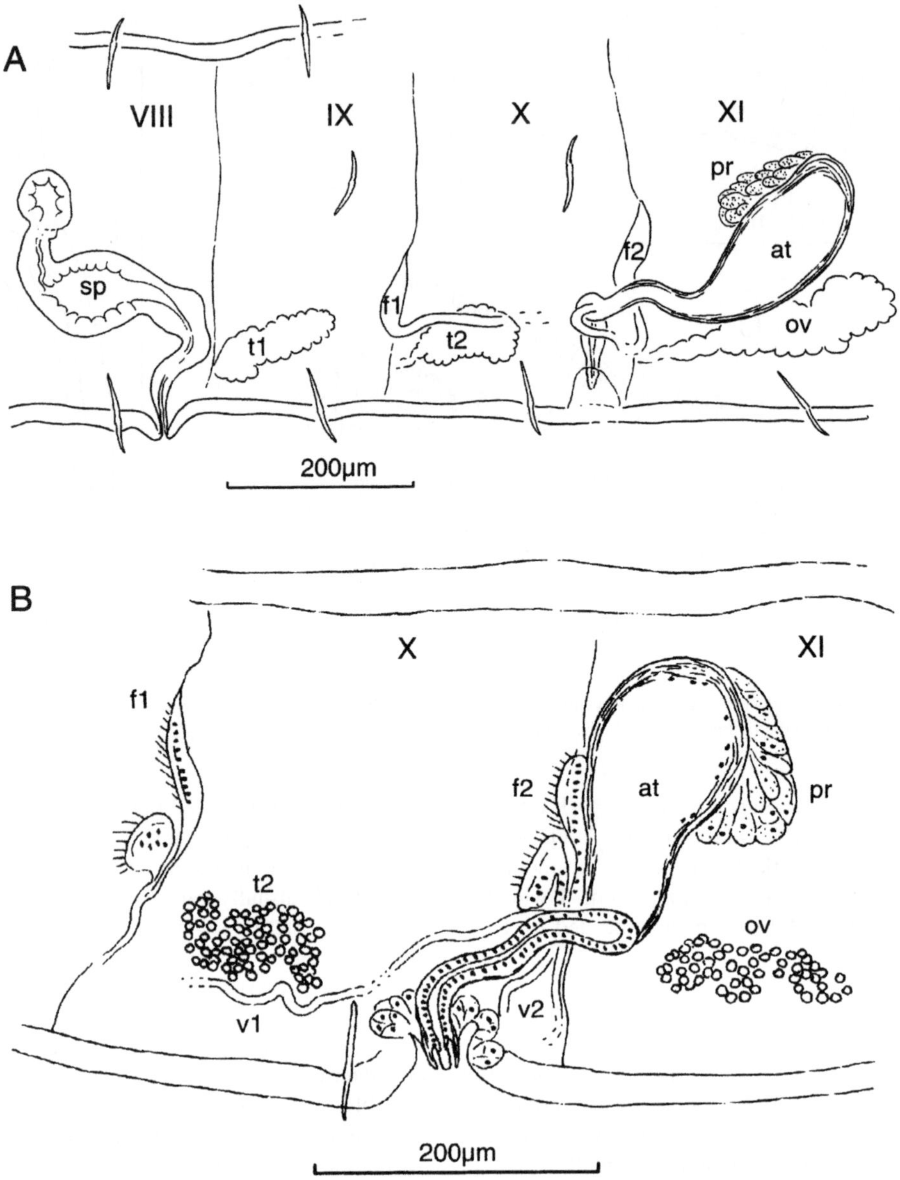

Figure 2. Secubelmis limpida. (A) Reproductive organs, sagittal view, from a whole mount. (B) Atrium and male ducts, reconstructed from sections.

and 2A). Female pores paired, on chaetal line at intersegmental groove 11/12. Paired testes in IX and X, ovaries in XI. Sperm sacs usually paired, extending back from X to about XIV. Egg sacs extend to about XVI.

Spermathecal pores on low, broadly conical papillae (Fig. 2A). Ectal duct of spermatheca a narrow tube, 80–160 μm long and 23–33 μm wide; wall of duct with tightly packed, columnar lining cells (Figs 3B and 4D). Spermathecal ampulla an irregular, elongate sac having thick, granular or vacuolar lining cells and a very thin muscle layer. Length of ampulla to 300 μm, diameter to 95 μm. All observed specimens with ampullae entirely in same segment as pores. No sperm

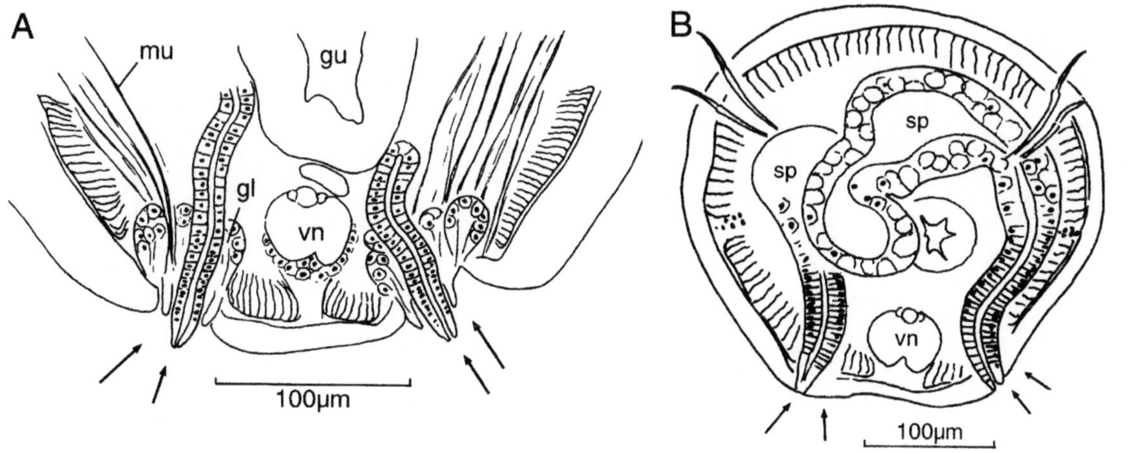

Figure 3. Secubelmis limpida, transverse views reconstructed from sections. Arrows indicate positions of ventral chaetae. (A) Male pores in X. (B) Spermathecae in VIII.

was present in spermathecae of any examined worms, including specimens with clitellum and mature eggs.

Male funnels not convoluted, about 100 μm wide, paired on 9/10 and 10/11; posterior funnels may fold back a short distance into sperm sac. Vasa deferentia thin and often indistinct, 7–12 μm wide; not obviously ciliated, and epithelial lining cells indistinct in the middle portion. Posterior vasa deferentia enter X directly, without forming a loop in XI (Figs 2B and 4B). No sperm present on funnels or in vasa deferentia in the available material. Most specimens with sperm sacs containing sperm morulae, but no mature sperm; some mature worms without sperm sacs. Irregular, paired masses of cells associated with the anterior septum, above testes in IX and X, are possibly vestigial gonadal sacs.

Atria may be entirely in X or pass through septum 10/11 into XI, or rarely, through 9/10. Ectal portion of atrium a narrow duct, 130–200 μm long and 15–32 μm wide, with an outer layer of loose epithelium and nearly cuboidal lining cells. Both vasa deferentia join near the middle of this duct, and travel toward the ental ampulla under the muscle layer. Point of entry of vasa into lumen uncertain; they are not visible in the wall of the atrial ampulla. Ental portion of atrium forms an ovoid ampulla 140–240 μm long and up to 130 μm wide. Atrial ampulla with distinct muscle layer, about 5–6 μm thick, with most fibers arranged transversely to the longitudinal axis; inner epithelium of ampulla thin and cells often indistinct, although nuclei are usually visible (Fig. 4B). Prostate cells granular, in a single, pyriform bundle, about 100 μm long, the narrow end attaching apically or subapically to the atrial ampulla (Figs 2B and 4A). Each male pore in a shallow depression or penis sac about 40 μm wide. Penis sac surrounded by a ring of pyriform, multicellular glands about 100 μm wide; individual glands about 40 μm tall (Figs 3A and 4C). The glands are interspersed with prominent, oblique muscles extending from ventral to lateral sides of the body wall, spreading over the posterior half of segment X. A thin collar of epidermal tissue surrounds a penis, about 20 μm long, apparently formed by a short extension of the muscular atrial wall and extensions of the epithelial cells (Fig. 4C).

Remarks: The absence of sperm in the male funnels, atria, or spermathecae of all apparently mature specimens suggests that *S. limpida* is parthenogenetic. The male reproductive system does not appear to develop fully in the observed populations of the new species, and the reduction appears variable among individuals and among populations. Twenty-four of the listed specimens, representing six collection sites and dates from April to October, appeared fully mature, with a well-developed clitellum and mature eggs. Large testes were present in those worms, and the sperm sacs, when present, contained what appeared to be sperm morulae, but no mature sperm. These features suggest that these individuals were not at a stage of resorption of genitalia. All three specimens from Reservoir Creek lacked the apparent sperm morulae and sperm sacs. The thin vasa deferentia were difficult to observe; they appeared to be degenerate or even absent in some specimens.

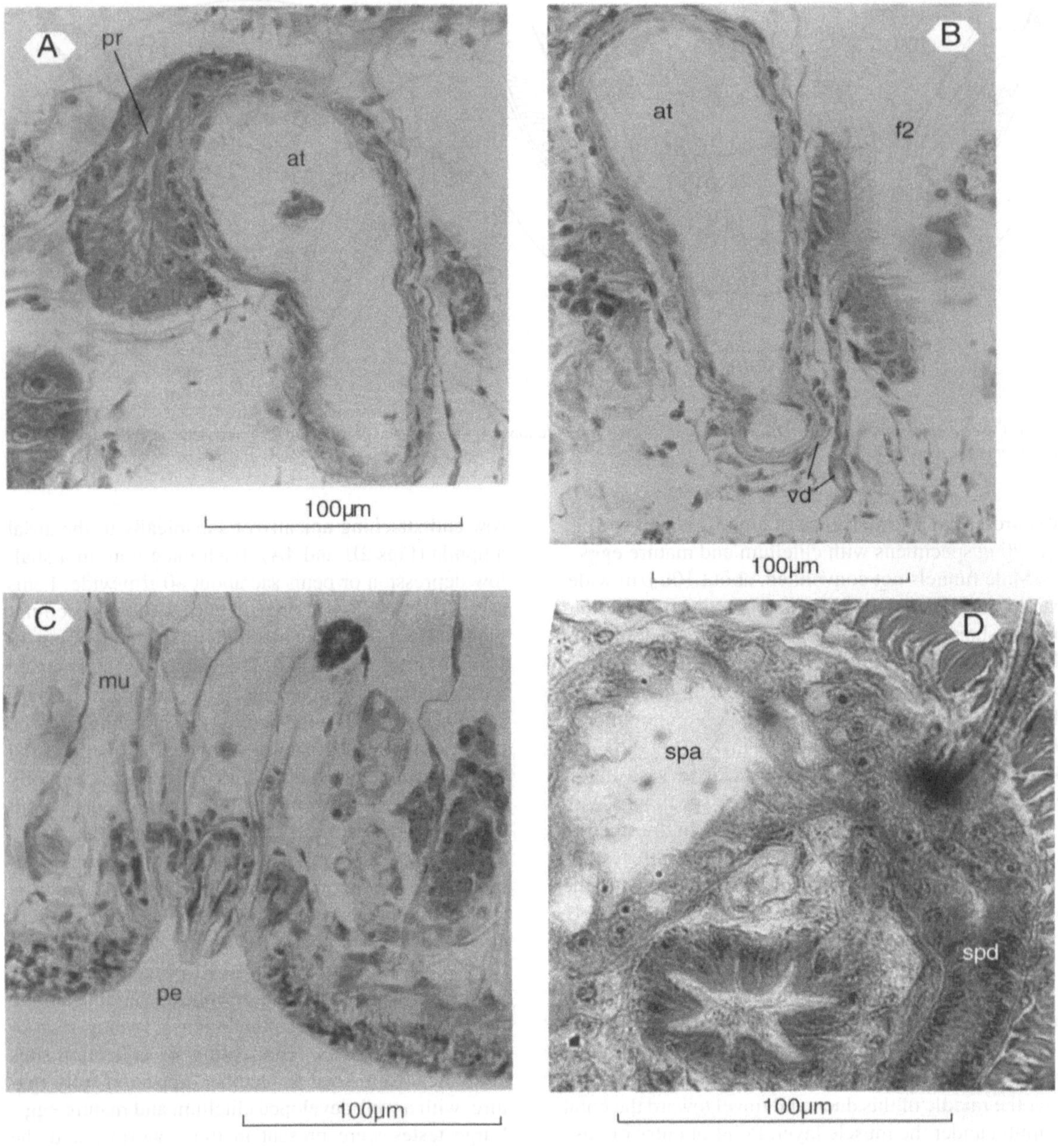

Figure 4. Secubelmis limpida, photographs. (A) Atrial ampulla and prostate, sagittal section in XI. (B) Atrial ampulla, posterior male funnel, and vas deferens in X and XI, sagittal section. (C) Male pore in X, sagittal section. (D) Spermatheca in VIII, transverse section.

Additional paired spermathecae in VII occurred in one of the more than 40 mature or partially mature specimens. Another specimen had a single spermatheca in VIII and another in IX. Within-population variation in spermathecal number is common in lumbriculids. A single extra spermatheca may occur in VII in some *Rhynchelmis* species (Fend & Brinkhurst, 2000), in VIII in *Tenagodrilus musculus* Eckroth & Brinkhurst, 1996, and spermathecae may be present or absent in

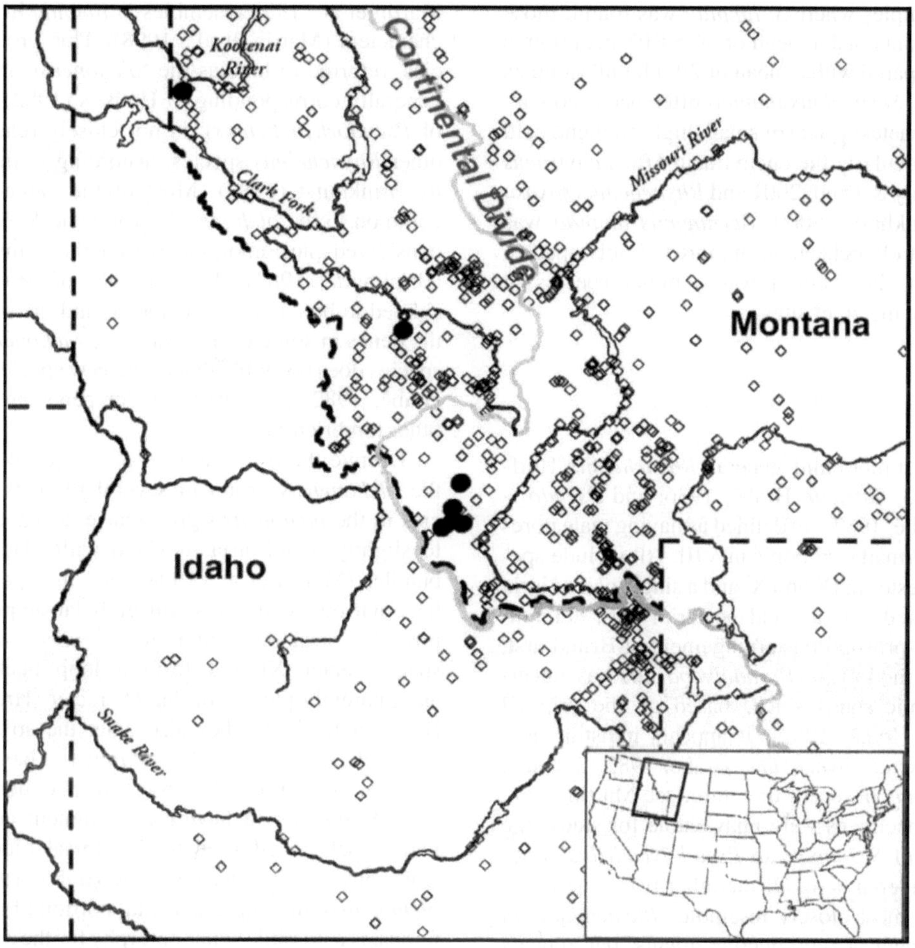

Figure 5. Map showing location of sites. Sites where *S. limpida* was found are indicated by filled circles, sites where the species was not found are indicated by gray diamonds.

Tatriella slovenica Hrabě, 1936 (Timm & Popchenko, 1978).

Distribution and ecology

All records of *S. limpida* are from western Montana, in 13 of 1252 samples taken in a 1977–2000 survey of salmonid-inhabiting streams in Montana, Idaho, and Wyoming (Fig. 5). All of the six streams in which the species occurred fall within Omernik's (1987) Northern Rockies ecoregion. Collection sites span the Continental Divide and three major drainages, the Missouri, Clark Fork, and Kootenai rivers. The sites are small (width 1–4 m), pristine, cold streams in watersheds dominated by coniferous forest, at elevations ranging from about 2150 m in the south to 800 m in the north. These streams have high gradients with coarse substratum, mostly boulders and cobble. Gravel and sand are limited to protected areas and silt occurs only in rare backwater areas. Specimens of *S. limpida* were collected in patches of sand and gravel entrained by larger particles.

Collection sites of the new species are characterized by highly diverse invertebrate communities dominated by EPT (Ephemeroptera, Plecoptera and Trichoptera) taxa, indicating high quality habitat. The entire benthic community sample was identified in 11 of the 13 samples where *S. limpida* was found; those collections contained a mean of 36.5 EPT taxa (range 29–42), compared with a mean of 29.0 for all samples in the region. *Secubelmis limpida* often occurred with other oligochaetes typical of small, high-gradient, cold streams, particularly the enchytraeid *Mesenchytraeus rhithalis* Healy & Fend, 2001 and *Rhynchelmis saxosa* Fend & Brinkhurst, 2000. *Secubelmis limpida* was the dominant oligochaete at the three closely spaced, southernmost sites, but it was a minor species in samples from the other sites.

Discussion

Three other lumbriculid genera, *Rhynchelmis* Hoffmeister, 1843, *Tatriella* Hrabě, 1936, and *Pseudolycodrilus* Hrabě, 1982, are defined as having male pores in X and spermathecal pores in VIII. All include species having testes in IX and X, and a single pair of atria with associated anterior and posterior vasa deferentia (the semi-prosoporous arrangement of Brinkhurst, 1999). The monotypic *Pseudolycodrilus* has uncertain taxonomic status since, based on the original description (Hrabě, 1982), it appears indistinguishable from *Pseudorhynchelmis* Hrabě, 1982 (included in *Rhynchelmis* in recent literature, see Martin et al., 1998). As specimens were unavailable for study, the relationship of *S. limpida* to *Pseudolycodrilus* is not further considered here. Of the other two genera, the new species most closely resembles *Rhynchelmis* in the arrangement of reproductive organs, but appears more closely related to *Tatriella* on the basis of their structure.

The paired spermathecal pores in VIII and paired male pores in X would assign the new species to the genus *Rhynchelmis* as defined by Cook (1971). In most recent literature, *Rhynchelmis* is broadly defined to include the genera *Sutroa* Eisen, 1888, *Rhynchelmoides* Hrabě, 1936, and *Pseudorhynchelmis*. *Secubelmis limpida* is distinguished from all Nearctic *Rhynchelmis* s. lat. (and most Palearctic species) by the lack of a proboscis, the presence of two pairs of testes, and the distinct pharyngeal pad (see Fend & Brinkhurst, 2000). It is also distinguished from most *Rhynchelmis* by its small size, lack of a connection between the spermathecae and gut, and relatively short atria. A group of atypical Palearctic species, including *Rhynchelmis olchonensis* Burov & Kozhov, 1932 *R. paraolchonensis* Giani & Martínez–Ansemil, 1984 *R. alyonae* Martin et al., 1998 and *R. shamanensis* Martin et al., 1998 resembles *S. limpida* in the above characters (Martin et al., 1998). This group of species, referred to here as the '*olchonensis* group' (but generally corresponding to Hrabě's (1982) definition of *Pseudorhynchelmis*), is not closely related to the other *Rhynchelmis* species, according to the analysis by Brinkhurst (1989). Most of the character states common to the *olchonensis* group and *S. limpida* are considered plesiomorphic within the Lumbriculidae (Brinkhurst, 1989). The presence of penes is considered to be a derived character, and the structure of the penes in some of the examined *olchonensis* group species does resemble that of the new species (see also Hrabě, 1982). However, similar penes are found in other lumbriculids.

Despite these similarities, *S. limpida* differs from the *olchonensis* group in several respects. All species in the *olchonensis* group have nearly cylindrical to slightly ovoid atria, covered with many prostate bundles (Martin et al., 1998), and the spermathecae have narrow ectal ducts and globular ampullae. The posterior vasa deferentia of all *olchonensis* group species enter XI and form a loop back into X, an apparently plesiomorphic character (Hrabě, 1984; Brinkhurst, 1989); they also enter the atria apically. The *olchonensis* group, like *S. limpida*, has a reduced vascular system, but the observed specimens differed from *S. limpida* in having a prominent dorsointestinal vessel, in addition to the dorsal blood vessel. One unusual character noted by Hrabě (1982) for *R. olchonensis*, the thickened ental portion of the vas deferens, is a possible synapomorphy for the *olchonensis* group, but is not present in *S. limpida*. However, the absence of the latter character may simply relate to the incomplete development of the male reproductive system in the new species. *Secubelmis* is separated from all *Rhynchelmis* s. lat. by the petiolate atrium and the single prostate gland.

The genus *Tatriella* is distinguished from *S. limpida*, as well as all *Rhynchelmis* species, by the combination of unpaired, median atria and spermathecae. However, the two *Tatriella* species resemble *S. limpida* in having petiolate atria with a long, narrow ectal duct and a distinct ampulla. *Tatriella slovenica* Hrabě, 1936, a widespread Palearctic species (T. Timm, pers. com.), shares with *S. limpida* several additional characters that are potentially derived: the thin vasa deferentia enter the muscle layer in the ental portion of the atrial duct, the male pores are on a small pad of thickened epidermal cells with a fan-like array of muscle fibers extending to the lateral body wall (Fig.

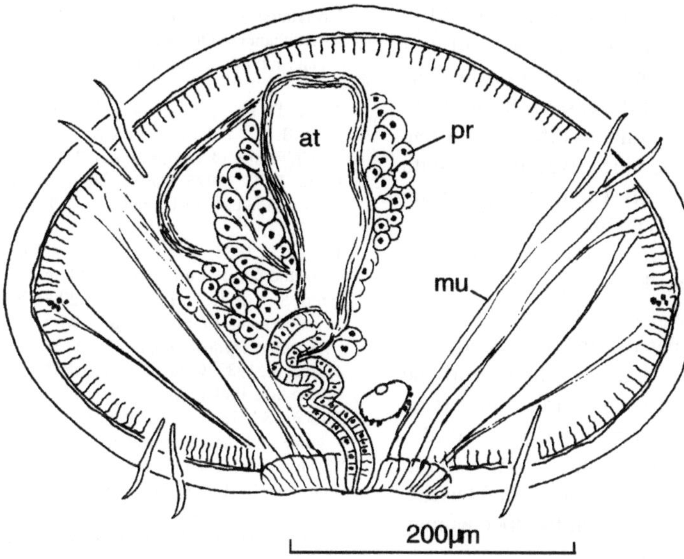

Figure 6. Tatriella slovenica, transverse view of atrium and male pore in X, reconstructed from sections.

6), and the posterior vasa deferentia enter X without a posterior loop into XI. The coverage of prostates is unusual in both species, but *T. slovenica* has a continuous layer of prostate bundles in the ental portion of the atrial duct (Fig. 6), instead of a single large bundle at the ental end of the ampulla. *Tatriella slovenica* lacks a distinct penis and glands at the male pore, but *Tatriella longiatriata* Popchenko, 1975, from the Onega Lake drainage, is illustrated as having a penis resembling that of the new species. However, the prostates of *T. longiatriata* form a continuous layer, covering the ental part of the atrium, and the vasa deferentia join the atria entally.

Clearly derived characters relating the new species to either *Tatriella* or the *olchonensis* group of *Rhynchelmis* are limited. The semi-prosoporous male ducts of these three taxa appear to be plesiomorphic (Brinkhurst, 1989, 1999), and this is also the most common arrangement within the Lumbriculidae. Although the placement of the spermathecae two segments anterior to the atria is apparently a derived character, the phylogenetic analysis by Brinkhurst (1989) suggests that spermathecal position has undergone many reversals, and may be of limited significance in lumbriculid phylogeny. Of the potential synapomorphies based on the structure of organs, only the petiolate atria (shared with *Tatriella*) appear to be distinctive within the family. However, the atria of *Eclipidrilus*, *Kincaidiana freidris* Cook, 1966, and some *Trichodrilus* are also petiolate. The single prostate gland per atrium is the only obvious autapomorphy distinguishing *Secubelmis* from all similar genera, and it appears to be unique within the family. Autapomorphies that distinguish the two taxa most likely to be sister groups include the expanded vasa deferentia of most *olchonensis* group species and the unpaired genitalia of *Tatriella*.

The phylogenetic affinities of *S. limpida* remain uncertain, with the weight of morphological evidence weakly suggesting an association with *Tatriella*. Although generic distinctions based only on paired versus unpaired genitalia have been rejected within both *Rhynchelmis* (Brinkhurst & Cook, 1966) and *Eclipidrilus* (Wassell, 1984), assigning the new species to *Tatriella* would redefine that genus to include species with paired genitalia, weakening the distinction between *Tatriella* and the already paraphyletic *Rhynchelmis*. Given the high level of endemism in lumbriculids relative to other aquatic oligochaetes (Timm, 1999), a single Nearctic species without clear affinities based on plausible synapomorphies is a weak argument for modifying groupings of Palearctic taxa. Therefore, we take the conservative approach of assigning the new species to a separate genus.

The known populations of *S. limpida* inhabit an area that has apparently been relatively unaffected by events of the last 15 Ma. All records are from the area south of the continental ice sheets, east of the

Columbia River basalt flows and Glacial Lake Missoula floods, north of the volcanic eruptions and lava flows of the Snake River Plains, and west of the Yellowstone Plateau glaciers and the eastern front of the Rocky Mountains. This area contains other monotypic genera that appear to be endemic. Examples include the ephemeropteran *Caurinella* Allen, 1984, and the trichopterans *Sericostriata* Wiggins et al., 1985, *Goereilla* (Denning, 1971) and *Rossiana* Denning, 1953. River connections with the known area of distribution have a complex history, and the collection sites have previously been associated with both eastern (Atlantic) and western (Pacific) drainages. In general, the Continental Divide has moved eastward, transferring headwater tributaries to western drainages (e.g. Janecke et al., 2000). Therefore, although *S. limpida* currently inhabits both sides of the Continental Divide, former river connections suggest that the species may be derived from a prior eastern fauna.

Acknowledgements

W. Granath assisted with the field expedition that first discovered this new genus, and the Montana Department of Fish, Wildlife and Parks (contract no. 960207) supported that sampling effort. We thank T. Timm for the loan of *T. slovenica* specimens and an English translation of the description. P. Martin loaned *R. alyonae* and *R. shamanensis*, located other material, and provided an English translation of *R. paraolchonensis*. N. Giani loaned *R. paraolchonensis*, and A. Kurka (Czech National Museum, Prague) loaned *R. olchonensis*. R. Brinkhurst provided advice at all stages of this project, including a review of the manuscript. We thank R.D. Kathman, P. Rodriguez, T. Timm and C. Erséus for suggesting many improvements to the manuscript. S. Renkes assisted with the graphics.

References

Altman, L. C., 1936. Oligochaeta of Washington. Univ. Wash. Publ. Biol. 4: 1–137.

Brinkhurst, R. O., 1989. A phylogenetic analysis of the Lumbriculidae (Annelida, Oligochaeta). Can. J. Zool. 67: 2731–2739.

Brinkhurst, R. O., 1999. Lumbriculids, branchiobdellidans and leeches: an overview of recent progress in phylogenetic research on clitellates. Hydrobiologia 406: 281–290.

Brinkhurst, R. O. & D. G. Cook, 1966. Studies on the North American aquatic Oligochaeta III: Lumbriculidae and additional notes and records of other families. Proc. Acad. nat. Sci. Phila. 118: 1–33.

Cook, D. G., 1971. Family Lumbriculidae. In Brinkhurst, R. O. & B. G. M. Jamieson (eds), Aquatic Oligochaeta of the World. University of Toronto Press, Toronto: 200–285.

Cook, D. G., 1975. Cave-dwelling aquatic oligochaeta (Annelida) from the eastern United States. Trans. am. micros. Soc. 94: 24–37.

Eisen, G., 1888. On the anatomy of *Sutroa rostrata*, a new annelid of the family Lumbriculina. Mem. Calif. Acad. Sci. 2: 3–8.

Fend, S. V. & R. O. Brinkhurst, 2000. New species of *Rhynchelmis* (Clitellata, Lumbriculidae), with observations on the Nearctic species. Hydrobiologia 428: 1–59.

Hrabě, S., 1982. Contribution to the knowledge of Oligochaeta from the lake Baikal. Vest. Cs. spol. zool. 46: 174–193.

Hrabě, S., 1984. Two atavistic characters of some Lumbriculidae and their importance for the classification of Oligochaeta. Hydrobiologia 115: 15–17.

Janecke, S. U., C. J. VanDenburn, J. J. Blankenau & J. W. M'Gonigle, 2000. Long-distance longitudinal transport of gravel across the Cordilleran thrust belt of Montana and Idaho. Geology 28: 439–442.

Kathman R. D. & R. O. Brinkhurst, 1998. Guide to the freshwater oligochaetes of North America. Aquatic Resources Center, College Grove, Tennessee: 264 pp.

Martin, P., M. Ferraguti & I. Kaygorodova, 1998. Description of two new *Rhynchelmis* species (Oligochaeta, Lumbriculidae) from Lake Baikal (Russia), using classical morphological and ultrastructural spermatozoal characters. Ann. Limnol. 34: 283–293.

McKey-Fender, D. & W. M. Fender, 1988. *Phagodrilus* gen. nov. (Lumbriculidae): systematics and biology of a predaceous oligochaete from western North America. Can. J. Zool. 66: 2304–2311.

Omernik, J. M., 1987. Ecoregions of the conterminous United States. Ann. am. Ass. Geograph. 77: 118–125.

Popchenko, V. I., 1975. A new species of the genus *Tatriella* (Oligochaeta, Lumbriculidae) from water bodies of Karelia. Zool. Zh. 54: 295–297.

Rodriguez, P. & K. A. Coates, 1996. A new American *Stylodrilus* species (Lumbriculidae, Oligochaeta). Can. J. Zool. 74: 92–96.

Timm, T., 1999. Distribution of freshwater oligochaetes in the west and east coastal regions of the North Pacific Ocean. Hydrobiologia 406: 67–81.

Timm, T. & V. Popchenko, 1978. Maloscetinkovye cervi (Oligochaeta) vodemov Murmanskoj oblasti. In Seasonal phenomena in fresh-water biology. Hydrobiological Researches VII. Academy of Sciences of the Estonian SSR, Institute of Zoology and Botany, Tartu: 71–132.

Ward, J. V. & N. J. Voelz, 1994. Groundwater fauna of the South Platte River System, Colorado. In Gibert, J., D. L. Danielopol & J. A. Stanford (eds), Groundwater Ecology. Academic Press, San Diego: 391–423.

Wassell, J. T., 1984. Revision of the lumbriculid oligochaete *Eclipidrilus* Eisen, 1881, with descriptions of three subgenera and *Eclipidrilus* (*Leptodrilus*) *fontanus* n. subg., n. sp. from Pennsylvania. Proc. biol Soc. Wash. 97: 78–85.

Hydrobiologia **463:** 23–28, 2001.
P. Rodriguez & P.F.M. Verdonschot (eds), Aquatic Oligochaete Biology VIII.

Taxonomical survey of some stygobiont oligochaetes from the eastern part of France, including description of a new species

Jacques Juget[1] & Michel des Châtelliers[1,2]
[1] *UMR CNRS 5023 Ecologie des Hydrosystèmes Fluviaux, Université Claude Bernard Lyon 1, F-69622 Villeurbanne Cedex, France*
Fax: 33 4 72 43 1141. E-mail: deschate@univ-Lyon1.fr
[1,2] *Hydrobiologie et Ecologie Souterraines, Université Claude Bernard Lyon 1, F-69622 Villeurbanne Cedex, France*

Key words: aquatic Oligochaeta, *Gianius*, *Trichodrilus*, taxonomy, subterranean waters, Lyon area

Abstract

Aquatic Oligochaeta from various habitats of subterranean waters in the Lyon area were examined. This preliminary taxonomical survey takes the families Tubificidae and Lumbriculidae into account. A new species of Tubificidae (subfamily Phallodrilinae), *Gianius cavealis*, is described. It was recorded from a subterranean gallery 'Le Trou des Voleurs' (Poncin, Southern Jura), and with regard to the genital setae, it is close to *G. riparius* (Giani & Martínez, 1981) which is also know from southwestern Europe. The study of specimens of *Trichodrilus campoyi* Rodriguez, 1988, a lumbriculid cohabiting with *G. cavealis*, leads to a discussion of taxonomical criteria in this species, based on a comparison with literature data. Moreover, new material of *T. tenuis* Hrabě, 1960 collected from several sampling sites allows us to supplement previous descriptions with information on the structure of the spermathecae.

Introduction

Many specimens of stygobiont Oligochaeta were collected during recent investigations (spanning the period 1997–2000) in various habitats of subterranean waters (caves, subterranean galleries, closed down mines) within the Lyon area. Tubificids and lumbriculids are the most common cavernicolous oligochaetes in the sampling area. Among the family Tubificidae, two members of the Phallodrilinae, *Gianius cavealis*, n. sp. and *Abyssidrilus cuspis* (Erséus & Dumnicka, 1988) were found, the latter originally described from wells in Central Italy and recently reported from various subterranean waters of Italy, Slovenia and France (Dumnicka, 1990; Sambugar et al., 1999). Two taxa of Rhyacodrilinae, *Rhyacodrilus falciformis* Bretscher, 1901, associated with groundwater seeps (Timm et al., 1996) and *Rhyacodrilus* cf. *balmensis* Juget, 1959, originally collected from a cave in the Jura mountains were also identified in the sampling area. A fifth species, *Tubificoides galarzai* Giani & Rodriguez, 1988 (Tubificinae) and its geographical intraspecific variation will be the subject of a further publication with the collaboration of Rodriguez.

Among the Lumbriculidae, three species belonging to the genus *Trichodrilus*, *T. campoyi* Rodriguez, 1988, *T. tenuis* Hrabě, 1960 and *T. cernosvitovi* Hrabě, 1937 were found. The study of *T. campoyi*, cohabiting with *Gianius cavealis* in the locality called 'Le Trou des Voleurs' in the Southern Jura (Département de l'Ain) permits a discussion of the taxonomic characters used to distinguish this species from *T. strandi* Hrabě, 1936 in this paper. Moreover, new material of *T. tenuis* Hrabě, 1960 collected from several sampling sites, especially 'Braizieu' and 'Montgelas' where several hundreds of specimens were collected, allows us to supplement previous descriptions that lacked information on the structure of the spermathecae. This study presents taxonomical data about this collection.

Materials and methods

The taxa mentioned above were collected in two sec-

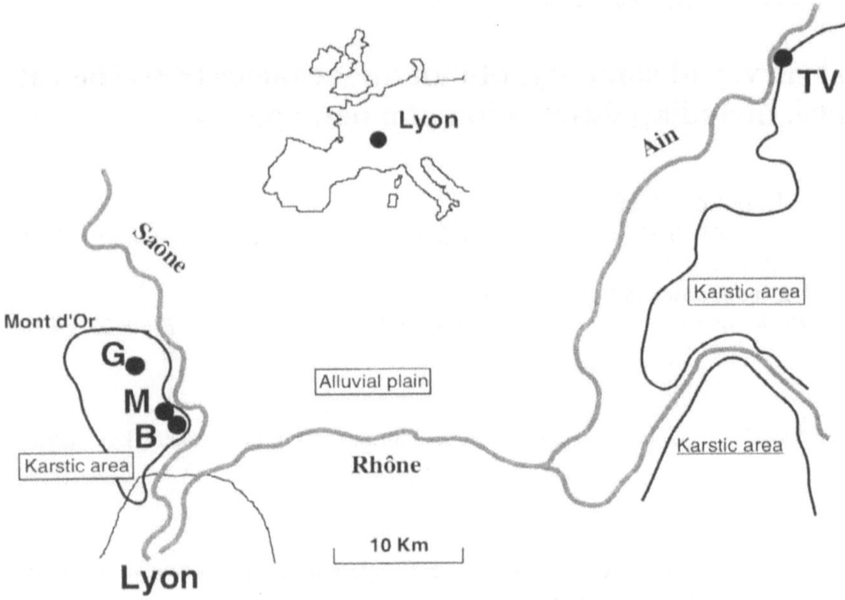

Figure 1. Location of the study sites. The position of the four stations are indicated by full circles (B: Braizieu, G: Gambins, M: Montgelas, TV: Trou des Voleurs).

tors (Fig. 1), the 'Mont d'Or' near Lyon (Département du Rhône) and Southern Jura (Département de l'Ain). The subterranean galleries called 'Montgelas' (4° 49′ 08″ E; 45 °49′ 33″ N), 'Braizieu' (4° 49′ 14″ E; 45° 49′ 24″ N) and 'Les Gambins' (4° 47′ 52″ E; 45° 51′ 07″ N), are in the first sector and the site called 'Le Trou des Voleurs' (5° 23′ 26″ E; 46° 05′ 01″ N) in Southern Jura. These sampling sites belong to karstic networks in limestone regions.

Physical and chemical variables such as water temperature (comprised between 10.9 and 12.2 °C), specific conductance (comprised between 555 and 768 μS cm^{-1}), pH (close to pH 8) and dissolved oxygen (saturation close to 100%) do not show significant space and time variability, neither among localities nor among samples within the same locality.

Granulometric analysis of the sediments also shows a relative homogeneity, insofar as sands constitute the exclusive or prevailing component of the texture. Nevertheless, the localities can be classified in two categories: the galleries of 'Braizieu' and 'Montgelas' with silty or fine sands (particles <200 μm prevailing), and the galleries 'Les Gambins' and 'Le Trou des Voleurs' with more than 70% coarse sands (>200 μm). Differences in sediment organic content estimated by loss on ignition of dry weight at 550 °C are related to texture, the loss on ignition being greater as the sediment is finer with calculated extreme values of 2.15% ('Les Gambins') and 11% ('Braizieu').

Oligochaetes were sampled with a Surber net (200 μm mesh size), fixed in Bouin's fluid and most of them stained in paracarmine and whole mounted in Canada balsam. The worms were fixed in Bouin's fluid, stained in paracarmine and whole mounted in Canada balsam. Some worms were studied as whole mounts in glycerol; some transverse and longitudinal sections were also made and stained in trichrome de Masson.

The type material of *Gianius cavealis* is deposited in the Museum d'Histoire Naturelle de Lyon (MHNL), France.

Systematic account

Gianius cavealis n. sp. (Fig. 2)

Holotype: MHNL 44002191, a mature specimen stained in paracarmine and whole mounted in Canada balsam.

Type locality: Puddle within 'Le Trou des Voleurs', a cave near Poncin (5° 23′ 26″ E; 46° 05′ 01″ N) (Southern Jura). Collected by M. des Châtelliers, M. J. Turquin & M. Meyssonnier (Spéléo-club de Villeurbanne), June 17, 1999.

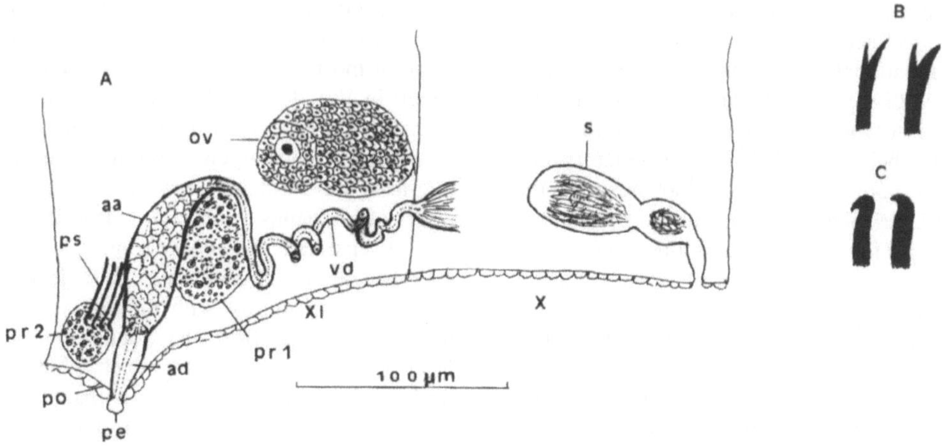

Figure 2. Gianius cavealis n. sp. (A) Lateral view of spermatheca, ovary and male genitalia in segments X-XI. (B) somatic setae. (C) penial setae. Abbreviations: aa – atrial ampulla; ad – atrial duct; ov – ovary; pe – protrusible penis; po – porophore; pr1 – anterior prostate gland; pr2 – posterior prostate gland; ps – penial setae; s – spermatheca; vd – vas deferens.

Paratypes: MHNL 44002192–44002199, whole-mounted specimens from type locality.

Etymology: The specific name *cavealis* is Latin for 'cooped up in a hollow' and refers to the habitat of this species.

Description: Length 4.3–6.6 mm, 28–42 segments, diameter at V 150–210 μm, at XI 160–260 μm (in fixed, whole-mounted, slightly compressed, sexually mature specimens). Inconspicuous clitellum. Prostomium rounded, slightly shorter than or as long as broad (57–78 μm long, 57–91 μm wide). Secondary annulation conspicuous on segments II–IV, (V).

All somatic setae bifid, 29–42 μm long, 1–2 μm thick, 2-4 (5) per bundle anteriorly, 2–3 (4) per bundle posteriorly, with upper tooth thinner and shorter than lower tooth (upper tooth more reduced in preclitellar segments), and with inconspicuous nodulus (ratio tip-nodulus/total length: 0.47–0.50). Ventral setae of XI modified as penial setae; each bundle contains (3) 4 (5) single-pointed, straight but ectally hooked setae, the length (34–36 μm) and maximum thickness (1–2 μm) of which are close to those of somatic setae.

Male and spermathecal pores opening in ventral setal lines, male pores in posterior third of XI on apex of a conical non-retractile porophore; spermathecal pores near intersegmental furrow IX–X.

Pharyngeal glands in IV–VI, dorso-lateral to alimentary tract. Seminal vesicle extending anteriorly into IX, ovisac in XII–XIII. Vasa deferentia very coiled, longer than atrium, about 200–230 μm long and 5–13 μm wide, entering atria apically. Atria elongate with narrow lumen, well-developed inner epithelium and slender outer layer of muscles (1–2 μm thick). Each atrium is composed of a curved ental part about 60 μm long and up to 28 μm wide, and an ectal, somewhat bulbous part, about 40 μm long and up to 20 μm wide ending in a protrusible penis. Two apparently non-pedunculate prostates, ental one up to 75 μm wide located near junction between vas deferens and atrium, second one smaller (diameter up to 47 μm) located at ectal end of atrium.

Each spermatheca consists of a short duct (17–28 μm long, up to 13–19 μm wide) and bipartite ampulla; ental part of ampulla 56–112 μm long, 32–82 μm wide, ectal one smaller and more rounded (28–39 μm long, 21–37 μm wide). Sperm in indistinct masses or bundles, present in both ampullar parts or absent from the smaller one.

Remarks: The described species confirms the diagnosis of the genus *Gianius* by Erséus (1992) in his generic revision of the subfamily Phallodrilinae. This refers, for instance, to the elongated and curved, ectally tapering, shape of the atrium with the vas deferens entering its apical end, to the discrete prostate glands, the anterior one located near the entrance of the vas deferens, the posterior one near the ectal end of the atrium, and to the position of the spermathecal pores in line with the ventral setae, the plesiomorph state of this latter character being suggested by its existence in all limnic species of *Gianius*.

G. cavealis is closely related to *G. riparius* (Giani & Martínez-Ansemil, 1981) as indicated by the number and the shape of penial and somatic setae, number of segments, secondary annulation and location of the spermathecal and male pores, but it is clearly distinct particularly with regard to the length of the vasa deferentia (at least twice as long as the atria), the ectal part of the atria being modified into an ejaculatory duct with the male pores on conical porophores, and the shape of the spermathecae (characterized by bipartite ampullae, sperm filling both or only the larger part of each ampulla).

The habitat of the new species (medium-sized sands from a cave, the 'Trou des Voleurs' where this species can reach 70% of the oligochaete fauna) also contrasts with the type locality of *G. riparius* (pebbles and gravel of superficial sediments from the River Tambre, Galicia, Spain), (Giani & Martínez-Ansemil, 1981) and the River Dordogne and Rhône (Lafont, 1989, Lafont & Durbec, 1990). However, significant differences regarding the length of penial setae (twice as long as the somatic setae) in the French material of *G. riparius* are noteworthy, in comparison with the Spanish specimens. New data on anatomical structures of French material would permit a specification of the affinities of *G. cavealis* and *G. riparius*.

The three other species confined to limnic habitats and included in the genus *Gianius* can be separated from *G. cavealis*, according to their penial setae: ventral setae on XI not modified into genital setae in *G. labouichensis* (Rodriguez & Giani, 1989), 4–8 penial setae some of which are normally forked and some others having the ectal end single-pointed and slightly bent in *G. aquaedulcis* (Hrabě, 1960), 7–8 fan-shaped bifid penial setae and thick muscular layer of the atrium in *G. crypticus* (Rodriguez & Giani, 1989).

The marine *Gianius* species known to date differ from the freshwater group either by spermathecal pores in lateral lines, for example *G. cristolatus* (Erséus, 1983), *G. densespectinis* (Erséus, 1987), *G. monnioti* (Erséus, 1987), or by the unusual location of spermathecae and spermathecal pores in segment XI instead of X, for example *G. eximius* Erséus,1997. These four marine species differ particularly from *G. cavealis* by their heavily muscular atria.

Trichodrilus campoyi Rodriguez, 1988 (Fig. 3)

Several mature specimens of this species, collected at the same station as *Gianius cavealis,* allowed a comparison of these specimens from Southern Jura with those collected by P. Rodriguez from the Northern Iberian Peninsula. Morphological characters from the new material and those published by Rodriguez & Giani (1994) are compared in Table 1. This table emphasises the great similarity in the range and the mean of most characters. The presence of secondary annuli in V and the bulbous appearance of the ectal spermathecal duct confirm this point of view.

An essential divergence appears in the atrial structure and size. In their review of the taxonomic characters of the genus *Trichodrilus*, Rodriguez & Giani (1994) separated *T. campoyi* from *T. strandi* Hrabě, 1936 because of the absence of both ejaculatory duct and penis. In contrast with the Spanish population, our new material from the Southern Jura (Fig. 3) lacks penis but has ejaculatory ducts which are clearly distinguishable from the atrial ampullae. The atrial ampulla (83–109 μm long) and ejaculatory duct (42–52 μm long) of the new material differ in relative length from those of the type material of *T. strandi* (ampulla 60 μm long, ejaculatory duct 70 μm long). The atrial musculature and the size of the ampulla are similar to the original material (Table 1), although the size is closer to the lower range of measurements.

These new data demonstrate the importance of the shape of the atrium as a diagnostic character for this species and at the same time seem to illustrate the impact of geographic isolation on the differentiation of the populations.

Trichodrilus tenuis Hrabě, 1960 (Fig. 4)

According to remarks by Rodriguez & Giani (1994), many descriptions of lumbriculids lack information regarding details of the spermathecae, even though the shape of the spermathecae can be an important diagnostic character.

In his description of the type collection of *T. tenuis*, Hrabě (1960) gives this information: '*the spermathecae are composed of very long ovoid ampullae which reach the next segment and of short distinctly separated narrow ducts*'. In accordance with the description by Hrabě, the two pairs of spermathecal orifices in our new specimens are situated closely behind the ventral setae on segments XI and XII. Figure 4 shows a spermatheca of one of our worms from 'Braizieu'. It consists of a short duct (47–50 μm long and up to 28 μm wide at middle) expanding into a long ampulla at least five times as long and at least twice as wide (47–56 μm) as the spermathecal duct. The ampulla, which is filled with a large bundle of sperm, shows three peculiarities: (1) One to several strongly marked folds affect the course of the ampullae as well as the bundle of sperm; (2) the thick wall of the spermathecal ampulla shows a more or less regular

Table 1. Comparison of some taxonomic characters of *Trichodrilus campoyi* in a French population from Southern Jura (present study) and a Spanish population (Rodriguez & Giani, 1994)

Characters	Southern Jura			Basque Country		
	Range	Mean	*sd (n)*[a]	Range	Mean	*sd*
Body Ø (μm)	340–690	493	92.7 (10)	235–614	388.9	175.2
Prostomium length (μm)	90–182	131	30.3(9)	73–207	135.5	33.4
Length of setae (μm)	75–122	101	12.5(12)	73–112	85.3	11.2
Pharyngeal glands (up to)	V or VI			V, VI or VII		
Tegument:						
Longitudinal muscl. (μm)	23–34	27	4.9 (5)	6.8–42.5	22.1	10.6
Epidermis (μm)	16–28	19	4.8 (5)	(2.5)12–25	17.1	4.4
Genitalia:						
Atrium length (μm)[b]	125–161	145	16.3 (4)	80–188	141.7	31.6
Atrium Ø (μm)	47–73	61	8.1 (7)	39–98	72.5	20.7
Atrial muscles (μm)	≈1.5–≈4	2.5	0.8 (6)	≈2–4.2	2.8	0.9
Atrial epithelium (μm)	18–39	28	8.8 (4)	15–41	25.7	9.5
Prostate max. length (μm)	18–49	32	8.7 (4)	14–93	39.3	27.6
Spermatheca						
Ampulla length (μm)	182–300	234	51.2 (4)	136–406	216.7	96.1
Ampulla max. Ø (μm)	73–130	99	24.4 (4)	96–196	140	37.8
Duct length (μm)	117–180	146	25.9 (4)	119–≈308	179.9	60.3
Ampulla position	X or XI			X, XI or XII		

[a] Number of studied specimens.
[b] Atrial ampulla + ejaculatory duct in French population.

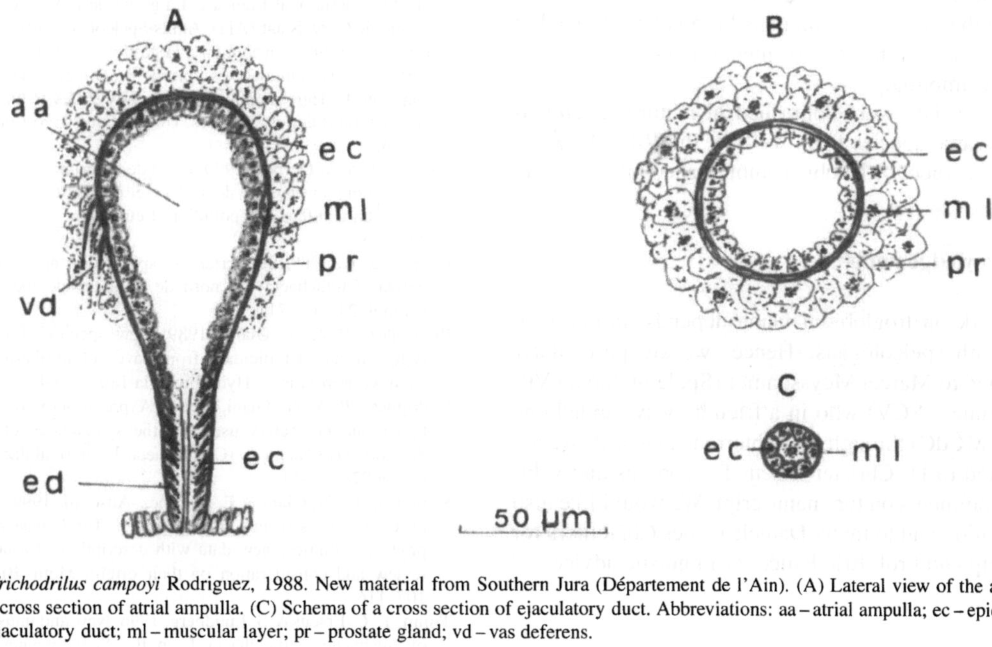

Figure 3. Trichodrilus campoyi Rodriguez, 1988. New material from Southern Jura (Département de l'Ain). (A) Lateral view of the atrium. (B) Optical cross section of atrial ampulla. (C) Schema of a cross section of ejaculatory duct. Abbreviations: aa – atrial ampulla; ec – epidermal cells; ed – ejaculatory duct; ml – muscular layer; pr – prostate gland; vd – vas deferens.

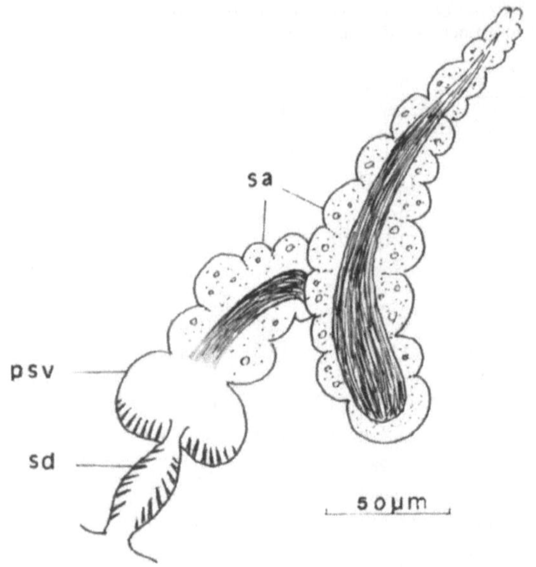

Figure 4. Trichodrilus tenuis Hrabě, 1960. New material from 'Mont d'Or' (Département du Rhône). Lateral view of spermatheca. Abbreviations: psv – pseudovestibule; sa – spermathecal ampulla; sd – spermathecal duct.

sequence of constrictions and expansions suggesting a vacuolate structure, and (3) the distal part of the ampulla is differentiated as a pseudovestibule with inner epithelium of conspicuous high cells arranged in a circle around the zone of junction between the duct and the ampulla.

These data were confirmed in mature specimens of the same species from wells of the Rif (Morocco) (Juget & Yacoubi-Khebiza, unpublished information).

Acknowledgements

All work on troglobiont fauna depends on cooperation with speleologists. Hence, we are particularly indebted to Marcel Meyssonnier (Spéléo-Club de Villeurbanne – SCV) who in a friendly way guided one of us (MCdC) through the subterranean world. We are indebted to Dr Christer Erséus for copious and valuable comments on the manuscript. We would like also to mention and to thank Danièle C. des Châtelliers for her help and Prof. Eric Pattée for linguistic advice.

References

Dumnicka, E., 1990. Oligochaetes from subterranean waters of Italy and Greece. Mém. Biospéol. 17: 163–168.

Erséus, C., 1983. Deep-sea *Phallodrilus* and *Bathydrilus* (Oligochaeta, Tubificidae) from the Atlantic Ocean, with description of ten new species. Cah. Biol. mar. 24: 125–146.

Erséus, C., 1987. Seven new marine species of *Phallodrilus* (Oligochaeta, Tubificidae) from various parts of Europe, and a reexamination of the type species *Phallodrilus parthenopaeus* Pierantoni. J. nat. Hist. 21: 915–931.

Erséus, C., 1992. A generic revision of the Phallodrilinae (Oligochaeta, Tubificidae). Zool. Scr. 21: 5–48.

Erséus, C., 1997. Additional notes on the taxonomy of the marine Oligochaeta of Hong Kong, with a description of a new species of Tubificidae. In Morton, B. (ed.), Marine Flora and Fauna of Hong Kong and Southern China, Hong Kong. Proc. 8th. Int. Mar. Biol. Workshop. Hong Kong Univ. Press, 1995: 37–52.

Erséus, C. & E. Dumnicka, 1988. A new *Phallodrilus* (Oligochaeta: Tubificidae) from subterranean waters in central Italy. Stygologia 4: 116–120.

Giani, N. & E. Martínez-Ansemil, 1981. Observaciones acerca de algunos Tubificidae (Oligochaeta) de la península ibérica, con la descripción de *Phallodrilus riparius* n. sp. Annls Limnol. 17: 201–209.

Giani, N. & P. Rodriguez, 1988. Description de quelques espèces nouvelles de Tubificidae (Oligochaeta) de grottes et de sources karstiques de la péninsule ibérique. Stygologia 4: 121–137.

Hrabě, S., 1936. *Trichodrilus strandi* n. sp. ein neuer vertreter der Höhlen Lumbriculiden. Festschr. Prof. Dr Embrik Strand, Riga, 1: 404–407.

Hrabě, S., 1937. Contribution à l'étude du genre *Trichodrilus* (Oligochaeta, Lumbriculidae) et description de deux espèces nouvelles. Bull. Mus. r. Hist. nat. Belg. 13: 1–23.

Hrabě, S., 1960. Oligochaeta limicola from the collection of Dr S. Husmann. Publ. Fac. Sci. Univ. Brno, Tchécoslovaquie 415: 245–277.

Juget, J., 1959. Recherches sur la faune aquatique de deux grottes du Jura méridional Français: La grotte de la Balme (Isère) et la grotte de Corveissiat (Ain). Annls Spéléol. 14: 391–401.

Lafont, M., 1989. Contribution à la gestion des eaux continentales: utilisation des Oligochètes comme descripteurs de l'état biologique et du degré de pollution des eaux et des sédiments. Thèse de doctorat d'état es-sciences, UCBL, Lyon 1: 311 pp + annexes 92 pp.

Lafont, M. & A. Durbec, 1990. Essai de description biologique des interactions entre eaux de surface et eau souterraine: vulnérabilité d'un aquifère à la pollution d'un fleuve. Annls Limnol. 26: 119–129.

Rodriguez, P., 1988. Sur certaines espèces de Lumbriculidae (Annelida: Oligochaeta) du nord de la péninsule ibérique. Annls Limnol. 24: 203–211.

Rodriguez, P. & N. Giani, 1989. New species of *Phallodrilus* (Oligochaeta, Tubificidae) from caves of northern Spain and southwestern France. Hydrobiologia 180: 57–64.

Rodriguez, P. & N. Giani, 1994. A preliminary review of the taxonomic characters used for the systematics of the genus *Trichodrilus* Claparède (Oligochaeta, Lumbriculidae). Hydrobiologia 278: 35–51.

Sambugar, B., N. Giani & E. Martínez–Ansemil, 1999. Groundwater oligochaetes from southern-Europe. Tubificidae with marine phyletic affinities: new data with description of a new species, review and consideration on their origin. Mém. Biospéol. 26: 107–116.

Timm, T., C. Erséus & S. Lundberg, 1996. New and unusual records of freshwater Oligochaeta from the Scandinavian Peninsula. Nordic J. Freshwat. Res. 72: 15–29.

Hydrobiologia **463:** 29–38, 2001.
P. Rodriguez & P.F.M. Verdonschot (eds), Aquatic Oligochaete Biology VIII.

A preliminary study of oligochaetes in Poyang Lake, the largest freshwater lake of China, and its vicinity, with description of a new species of *Limnodrilus*

Hongzhu Wang & Yanling Liang
State Key Laboratory of Freshwater Ecology and Biotechnology, Institute of Hydrobiology, Chinese Academy of Sciences, Wuhan 430072, China
Tel: 86 27 87647719. Fax: 86 27 87647664. E-mail: wanghz@ihb.ac.cn

Key words: Tubificidae, Naididae, freshwater oligochaetes, taxonomy, new species, Poyang Lake

Abstract

The oligochaete fauna of the largest freshwater lake of China, Poyang Lake, has never been investigated before. On the basis of a preliminary survey in the lake and its vicinity in 1997–1999, 25 species belonging to 20 genera and 5 families are recorded. Among them, one genus, *Cernosvitoviella* (Enchytraeidae), and two species, *Bratislavia unidentata* (Naididae), *Potamothrix bedoti* (Tubificidae), are recorded from China for the first time. Another tubificid, *Limnodrilus paramblysetus* sp. nov., is new to science. It is similar to *L. amblysetus* Brinkhurst et al., 1990 in chaetal shape but differs in having short hooded penis sheaths.

Introduction

Located in the northern Jiangxi Province at the junction of the middle and lower reaches of Changjiang (Yangtze) River, Poyang Lake (28° 25′–29° 45′ N, 115° 48′–116° 44′ E) covers an area of 3283 km^2 at a water-level of 21.7 m ASL, with a maximum length of 173 km, a maximum width of 73 km and a shore development (D_L) of 5.9. It is a fluvial lake formed about 6000 years ago, and receives flows from five rivers, undergoing a great change of water level within a year (Zhu & Zhang, 1997). Being the largest freshwater, shallow lake in the country and one of the two large lakes still connecting the main river course of Changjiang, Poyang Lake plays an important role in maintaining aquatic biodiversity. However, no faunistic work on oligochaetes has been carried out in this water body. During the limnological investigation of Poyang Lake and its vicinity in 1997–1999, oligochaete specimens were collected mainly from the quantitative samples and reported below. This includes a new species of *Limnodrilus* and two new records for China. One additional new record, *Potamothrix bedoti* from Houhu Lake of Wuhan, is also given herein.

Materials and methods

Quantitative bottom samples were collected with a weighted Petersen grab (1/16 m^2) and sieved with a 420 μm sieve, and then worms were sorted manually in a white porcelain dish and preserved in 10% formalin. For some qualitative samples, live oligochaetes were sorted under dissecting microscope, or extracted by the wet-funnel method and fixed in 10% formalin. Specimens were prepared by two methods, i.e. glycerine mounts and Canada balsam mounts. For the former, oligochaetes were saturated with glycerine under cover slips sealed with asphalt around their edges. For the latter, worms were stained in borax carmine and mounted in Canada balsam. Measurements of chaetae and internal organs refer to glycerine and balsam mounts, respectively, and drawings were made using a camera lucida. The types and other specimens are deposited in the Specimen Room of Invertebrates, Institute of Hydrobiology (IHB), Chinese Academy of Sciences (CAS), in Wuhan.

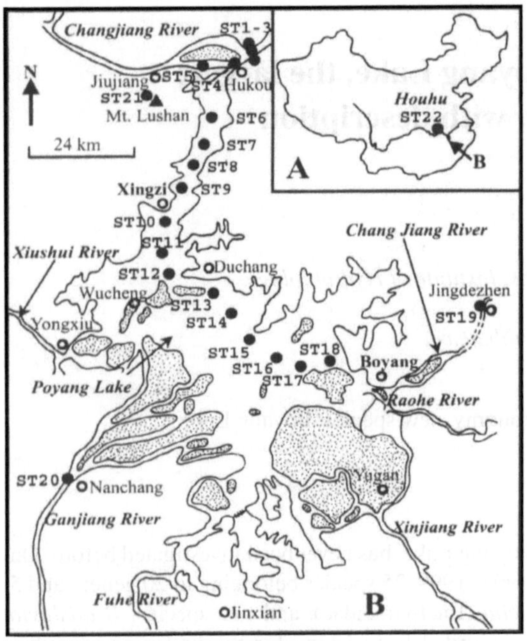

Figure 1. Sampling stations in Poyang Lake and its vicinity. (A) China, (B) Poyang Lake.

Description of study sites

The stations (Fig. 1, ST1–19) in Poyang Lake and adjacent Changjiang were sampled by the authors at five times, in October, 1997, April and November, 1998, and May and July, 1999. Station 2 is omitted, as no sample was collected due to the rapid current. Environmental factors associated with these stations are given for the different times when oligochaetes were collected. The specimens from Chang Jiang River, Ganjiang River (tributaries of Changjiang) and Mt. Lushan (Fig. 1, ST19–21) were collected by the authors in June, 1999. In Houhu Lake (Fig. 1, ST22), Wuhan, the sample was taken by Dr Z. Gong in April, 1999. Macrophytes were not found at any stations.

ST1. Changjiang River, N of Balijiang transect, 29° 47.90′ N, 116° 16.49′ E, mainly silt. 17 Apr 1998: 6.0 m depth, 0.3 m s^{-1} flow; 7 May 1999: 6.0 m depth, 0.9 m s^{-1} flow.

ST3. S of the above transect, 29° 47.71′ N, 116° 17.77′ E. 28 Oct 1997: clay, 5.0 m depth, 0.2 m s^{-1}flow; 14 Nov 1998: mainly sand, 2.0 m depth, 0.8 m s^{-1}flow.

ST4. N of S branch of Changjiang near Shizhongshan in Hukou, 29° 45.12′ N, 116° 13.11′ E, mainly silt. 14 Nov 1998: 9.0 m depth, 0.3 m s^{-1} flow; 7 May 1999: 7.2 m depth; 12 Jul 1990: 17.3 m depth, 0.3 m s^{-1} flow.

ST5. S of S branch of Changjiang, W of Hukou, 29° 46.32′ N, 116° 09.96′ E, sand, 16.0 m depth, 0.5 m s^{-1}flow, 12 Jul 1999.

ST6. Outlet of Poyang lake, W of Xieshan, 29° 39.87′ N, 116° 09.97′ E, clay. 16 Apr 1998: 2.5 m depth, 0.4 m s^{-1}flow; 13 Nov 1998: 11.0 m depth, 0.3 m s^{-1} flow.

ST7. Lake outlet, between Hamashi and Xieshan, 29° 37.03′ N, 116° 08.26′ E, mainly sand, 10.0 m depth, 0.2 m s^{-1} flow, 11 Jul 1999.

ST8. Lake outlet, SE of Hamashi, 29° 32.26′ N, 116° 06.73′ E, mainly clay, 1.5 m depth, 29 Oct 1997.

ST9. Lake outlet, NE of Xingzi, 29° 28.25′ N, 116° 05.57′ E, mainly sand. 13 Nov 1998: 2.5 m depth, 0.4 m s^{-1} flow; 9 Jul 1997: 13.4 m depth, 0.2 m s^{-1} flow.

ST10. Lake outlet, S of Xingzi, 29° 24.27′ N, 116° 03.78′ E. 29 Oct 1997: 3 m depth, 0.3 m s^{-1} flow; 16 Apr 1998: mainly sand, 4.3 m depth, 0.2 m s^{-1} flow; 6 May 1999: clay, 5.8 m depth, 0.5 m s^{-1} flow.

ST11. S end of lake outlet, 29° 19.92′ N, 116° 04.33′ E, mainly clay. 16 Apr 1998: 4.2 m depth, 0.2 m s^{-1} flow; 13 Nov 1998: 5.0 m depth, 0.6 m s^{-1}flow; 6 May 1999: 4.5 m depth, 0.4 m s^{-1} flow.

ST12. Lake proper, W of Duchang, 29° 16.32′ N, 116° 06.23′ E, mainly silt. 13 Nov 1998: 11.0 m depth, 0.6 m s^{-1}flow; 6 May 1999: 11.0 m depth, 0.6 m s^{-1} flow; 11 Jul 1999: 8.5 m depth, 0.2 m s^{-1} flow.

ST13. Lake proper, E of Wugongshan, 29° 14.45′ N, 116° 11.73′ E. 30 Oct 1997: 4.8 m depth, 0.3 m s^{-1} flow; 15 Apr 1998: sand with clay, 2.8 m depth, 0.2 m s^{-1} flow; 5 May 1999: silt, 8.9 m depth, 0.2 m s^{-1} flow.

ST14. Lake proper, NW of Huangjinzui, 29° 14.27′ N, 116° 14.32′ E. 30 Oct 1997: 3.5 m depth, 0.2 m s^{-1} flow; 12 Nov 1998: mainly silt, 2.8 m depth, 0.5 m s^{-1} flow; 5 May 1999: mainly silt, 3.5 m depth, 0.2 m s^{-1} flow; 9 Jul 1999: clay, 8.1 m depth, 0.1 m s^{-1} flow.

ST15. Lake proper, SW of Huangjinzui, 29° 11.80′ N, 116° 16.08′ E. 15 Apr 1998: clay, 1.9 m depth, 0.1 m s^{-1} flow; 15 Nov 1998: mainly sand, 5.0 m depth, 0.6 m s^{-1} flow; 9 Jul 1999: clay, 7.9 m depth, 0.1 m s^{-1} flow.

ST16. Lake proper, SW of Zhupaoshan, 29° 09.22′ N, 116° 18.34′ E. 15 Apr 1998: mainly sand, 5.3 m depth, 0.3 m s^{-1} flow; 9 Jul 1999: clay, 7.6 m depth, 0.1 m s^{-1} flow.

ST17. Lake proper, NW of Nanshancaozhou, 29° 06.34′ N, 116° 20.15′ E, mainly clay. 15 Apr 1998: 4.8 m depth, 0.5 m s^{-1} flow; 12 Nov 1998: 4.0 m depth, 0.6 m s^{-1} flow; 5 May 1999: 7.3 m depth, 0.5 m s^{-1} flow.

ST18. Lake proper, N of Nanshancaozhou, 29° 04.59′ N, 116° 22.29′ E. 12 Nov 1998: silt, 3.5 m depth, 0.7 m s^{-1} flow; 5 May 1999: clay, 5.6 m depth.

ST19. Chang Jiang River inside Jingdezhen City, 29.3° N, 117.2° E, shallow area close to bank, cleftstone with attached algae, 11 Nov 1998.

ST20. Ganjiang River near Nanchang City, 28.6° N, 115.9° E, shallow area close to bank, clay, 14 Jun 1999.

ST21. Various localities in Mt. Lushan, 29.5° N, 115.9° E, mainly brooks, 15–17 Jun 1999.

ST22. Houhu Lake, Wuhan, 30° 33′ N, 114° 23′ E, lentic waters, silt, 2.2 m depth, Apr 1999.

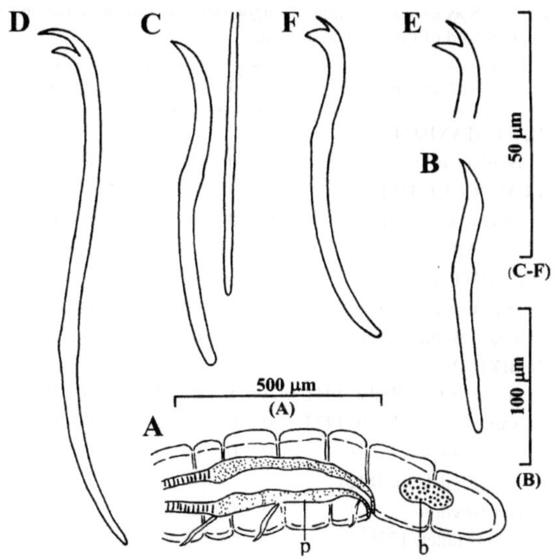

Figure 2. Haplotaxis sp.: (A) lateral view of anterior segments, (B) chaeta (VII). *Bratislavia unidentata*: (C) hair and needle (VII), (D–F) ventral chaetae (III, VI, VII, respectively). Abbreviations: b – brain; p – pharynx.

Results

Species composition

A total of 25 species, representing 20 genera of 5 families, have been identified from Poyang Lake and its vicinity (Table 1). Among them, *Limnodrilus paramblysetus* is new to science. One genus, *Cernosvitoviella*, and two species, *Bratislavia unidentata*, *Potamothrix bedoti* are recorded from China for the first time. *Nais elinguis* has never been recorded from eastern China before, although it is a cosmopolitan species (Sperber, 1948) and was known from northwestern China, i.e. Xinjiang (Sinkiang) Uygure Autonomous Region (Liang, 1964).

Systematic account

In this part, we describe new species, new records, and rare and other taxa with some taxonomic significance. Enchytraeid species are excluded, considering that these specimens are either immature, or not sufficient for the determination of details.

HAPLOTAXIDAE

***Haplotaxis* sp.** (Fig. 2 A,B)

Material examined: 1 balsam mount from ST15.

Brief description: Specimen incomplete and immature, length > 60 mm, width 0.3 mm, segments > 190. Prostomium cylindrical, separated from peristomium by a transverse groove (Fig. 2 A). Segments V–XXIX conspicuously 2-ringed (Fig. 2 A); XXX–XLII inconspicuously 3–4(5)-ringed. No dorsal chaetae. Ventral chaetae 1 per bundle, absent in the right of III, occasionally with 1 short replacement chaeta anteriorly; sigmoid, simple-pointed, with nodulus (Fig. 2 B); chaetae in II and III smaller, 97–104 μm long, 6–8 μm thick; from IV onwards 194–206 μm long, 10–12 μm thick. Chromophilous masses in X–XI.

Remarks: Previously, the family Haplotaxidae has only been described from subterranean waters or rivers in China (Brinkhurst et al., 1990; Wang, 1995). The occurrence of *Haplotaxis* in Poyang Lake is the first record of this family in a Chinese shallow lake.

NAIDIDAE

Bratislavia unidentata (Harman, 1973) (Fig. 2 C–F)

Pristina unidentata Harman, 1973: 161–163, Figure 2; Harman, 1974: 17.

Bratislavia unidentata (Harman). Harman & Loden, 1978: 541–544, Figure 1; Kathman & Brinkhurst, 1998: 92–93; Brinkhurst & Marchese, 1989: 114–115.

Material examined: 3 glycerine and 5 balsam mounts from ST19.

Brief description of new material: Specimens immature, length 4–9 mm, segments 40–80 plus undifferentiated tails. No stomach. Pharyngeal glands in III-VI. Hairs (0) 1 (2) per bundle, 80–120 μm long,

Table 1. Species composition of oligochaetes in Poyang Lake and its vicinity. 1, presence; 0, absence; #, new record of the genus for China; *, new record of the species for China

Stations	1	3	4	5	6	7	8	9	10	11	12	13	14	15	16	17	18	19	20	21	22
HAPLOTAXIDAE																					
Haplotaxis sp.	0	0	0	0	0	0	0	0	0	0	0	0	0	1	0	0	0	0	0	0	0
LUMBRICULIDAE																					
Lumbriculus variegatus (Müller, 1774)	0	0	0	0	0	0	0	0	0	0	0	0	0	0	0	0	0	0	0	1	0
ENCHYTRAEIDAE																					
#*Cernosvitoviella* sp.	0	0	0	0	0	0	0	0	0	0	0	0	0	0	0	0	0	0	0	1	0
Mesenchytraeus sp.	0	1	0	0	0	0	0	0	0	0	0	0	0	0	0	0	0	0	0	0	0
NAIDIDAE																					
Paranais frici Hrabě, 1941	0	1	0	0	0	0	0	0	0	0	0	0	0	0	0	0	0	0	0	0	0
Nais elinguis Müller, 1773	0	0	0	0	0	0	0	0	0	0	0	0	0	0	0	0	0	0	0	1	0
Slavina appendiculata d'Udekem, 1855)	0	0	0	0	0	0	0	0	0	0	0	1	0	0	0	1	0	1	0	1	0
Dero digitata (Müller, 1773)	0	0	0	0	0	0	0	0	0	0	0	0	0	0	0	0	1	1	0	0	0
Aulophorus furcatus (Müller, 1773)	0	0	1	0	0	0	0	0	0	0	0	0	0	0	0	0	0	0	0	0	0
**Bratislavia unidentata* (Harman, 1973)	0	0	0	0	0	0	0	0	0	0	0	0	0	0	0	0	0	1	0	0	0
Pristinella jenkinae (Stephenson, 1931)	0	0	0	0	0	0	0	0	0	0	0	0	0	0	0	0	0	0	1	1	0
Pristina longiseta Ehrenberg, 1828	0	0	0	0	0	0	0	0	0	0	0	0	0	1	0	0	0	0	0	0	0
TUBIFICIDAE																					
Tubifex sp.	0	0	0	0	0	0	0	0	0	1	0	0	1	0	0	0	1	0	0	1	0
Limnodrilus hoffmeisteri Claparède, 1862	0	0	0	0	0	0	0	0	0	0	0	0	0	0	0	0	0	0	0	1	1
Limnodrilus cf. *grandisetosus* Nomura, 1932	1	0	1	0	0	0	0	0	0	0	0	0	1	0	0	0	1	0	0	0	1
Limnodrilus paramblysetus sp. nov.	1	0	0	0	0	0	0	0	0	0	0	0	0	0	0	0	0	0	0	0	0
#**Potamothrix bedoti* (Piguet, 1913)	0	0	0	0	0	0	0	0	0	0	0	0	0	0	0	0	0	0	0	0	1
Teneridrilus mastix (Brinkhurst, 1978)	0	1	0	1	0	0	0	1	1	1	1	0	0	1	1	0	1	0	1	0	0
Spirosperma nikolskyi (Lastockin, 1953)	0	0	0	0	0	0	0	0	0	0	0	1	1	1	0	1	0	0	0	0	0
Aulodrilus limnobius Bretscher, 1899	0	0	0	0	0	0	0	0	0	1	0	0	0	0	0	0	0	0	0	0	0
Aulodrilus pluriseta (Piguet, 1906)	0	0	0	0	0	0	0	0	0	0	0	0	1	0	0	0	1	0	0	0	0
Aulodrilus pigueti Kowalevski, 1914	0	0	0	0	0	0	0	1	1	1	0	0	1	0	0	0	0	0	0	0	0
Aulodrilus pectinatus Aiyer, 1928	0	0	0	0	0	0	0	0	1	0	0	0	0	0	0	0	0	0	0	0	0
Bothrioneurum sp.	0	0	0	0	0	0	0	0	0	0	0	0	0	1	0	0	0	0	1	0	0
Branchiura sowerbyi Beddard, 1901	1	1	1	0	1	1	1	0	1	1	1	1	1	0	0	1	0	0	0	1	1
Species number	3	4	3	1	1	1	1	2	3	5	2	3	6	5	1	3	5	3	3	8	4

maximally 1.8 μm thick, without serration. Needles 1 (2) per bundle, simple-pointed and sigmoid, with nodulus distal (Fig. 2 C). Ventral chaetae in II–V 4–5 per bundle, longer and slightly thinner than the rest, with nodulus proximal and upper teeth about twice as long as lower (Fig. 2 D); those following 3–4 per bundle, with nodulus distal and upper teeth thinner, as long as (Fig. 2 E) or shorter than lower (Fig. 2 F).

Remarks: This species was originally described as *Pristina unidentata* from Oklahoma and Texas (Harman, 1973), and later reassigned to *Bratislavia* on the basis of genital organs (Harman & Loden, 1978). The new specimens, which were numerous at the locality, coincide well with the descriptions of original and Surinamese material (Harman, 1973, 1974), but there are some minor differences. The Chinese worms have thicker hair chaetae (1.8 μm *versus* 1 μm), and more chaetae in the anteriormost segments (4–5 *versus* 3–4 per bundle).

Distribution: Central China (new record), Eastern USA, Surinam, Argentina.

TUBIFICIDAE

Limnodrilus cf. ***grandisetosus*** Nomura, 1932

Limnodrilus grandisetosus Nomura, 1932: 511, Figures 1–5, Pls. XIII–XVII; Yamaguchi, 1940: Figure 6; Liang, 1979: 277; Ohtaka, 1985: Figure 8 A–E, I; Brinkhurst et al., 1990: Figure C.

Material examined: 2 glycerine mounts from ST1 and ST4, respectively.

Brief description: Specimens immature, slender and coiled, length 9–24 mm, segments 44 up to more than 100. Prostomium conical. Dorsal chaetae 2–4 per bundle anteriorly, with upper teeth reduced or absent; 1–2 per bundle posteriorly, with upper teeth about twice as long as lower. Ventral chaetae 1–2 per bundle. Anterior ventral chaetae bifid, or sometimes simple-pointed, usually longer and thicker than dorsal; giant chaetae in V–VIII (IX), with blunt teeth. Posterior ventral chaetae similar to dorsal.

Remarks: Although simple-pointed chaetae have never been observed in *L. grandisetosus*, the presence of giant chaetae suggests that the present specimens could be tentatively identified as this species.

In terms of the ectal ends of the penis sheaths, there are two different descriptions of *L. grandisetosus* in Chinese material. Yamaguchi (1940) recorded this species from northeastern China and stated that the penis sheaths are hooded, and Liang (1979) later noted the same for the specimens from Huama Lake along middle Changjiang. However, Brinkhurst et al. (1990)

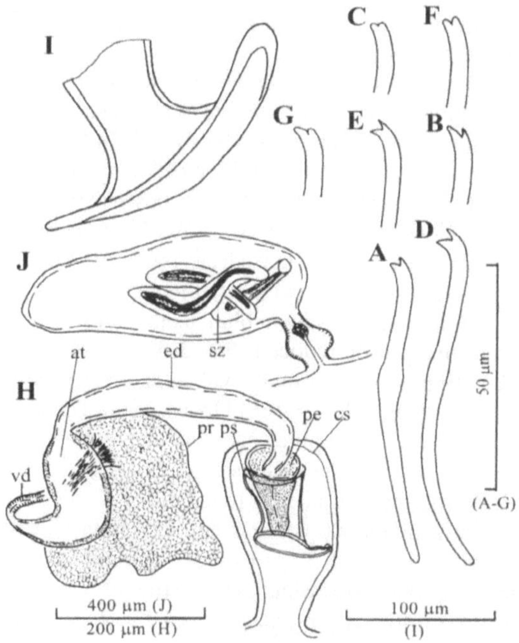

Figure 3. Limnodrilus paramblysetus sp. nov. (A–E) dorsal chaetae (III, IV, V, XIII, XL, respectively), (F, G) ventral chaetae (VI), (H) male duct, (I) penis sheath, (J) spermatheca. Abbreviations: at – atrium; cs – copulatory sac; ed – ejaculatory duct; pe – penis; pr – prostate; ps – penis sheath; sz – spermatozeugmata; vd – vas deferens.

reported that the penis sheaths in this species have rounded triangular plate-like heads, an observation based upon three Chinese specimens and one Japanese worm. This fits the descriptions of Japanese material including type specimens (Nomura, 1932; Ohtaka, 1985). For the time being, we do not know whether the penis sheaths were similar or not in the Chinese and Japanese material of Brinkhurst's study. Obviously, a further study is needed to determine whether this variation is interspecific or intraspecific, or just due to an observational problem.

Distribution: Japan, China, Malaysia.

Limnodrilus paramblysetus sp. nov. (Fig. 3)

Holotype: IHB AHA1998001a, mature specimen in balsam mount.

Type locality: China, Changjiang River, N of Balijiang, 17 Apr 1998 (ST1).

Paratypes: 7 mature specimens: IHB AHA 1998001b-d, 3 balsam mounts (2 dissected), and IHB AHA1999001a, 4 specimens in 10% formalin, all from type locality on 17 Apr 1998 and 7 May 1999, respectively.

Etymology: Named *paramblysetus* for the resemblance with *amblysetus* with regard to chaetae.

Description: Dimensions of the formalin fixed individuals: length 15–23 mm (holotype: 18 mm), width 0.6–0.8 mm anteriorly, 0.4–0.6 mm posteriorly; segments 54–97 (holotype: 54). Prostomium conical. Clitellum over 1/2X–XII. Chaetae bifid, 66–80 μm long, 2.4–2.9 μm thick, with distal nodulus. Bifids 5–7 per bundle anteriorly, with blunt teeth, and subequal teeth or upper teeth reduced (Fig. 3 A–C, F–G); 3–5 per bundle posteriorly, usually with upper teeth thinner than and as long as lower (Fig. 3 D–E). Ventral chaetae absent from XI. Male pores paired in line with ventral chaetae in mid-XI. Spermathecal pores paired in line with ventral chaetae, anterior to middle of X.

Male genitalia paired. Vas deferens long and irregularly coiled, 27–39 μm wide, entering atrium apically. Atrium somewhat oblong, straight or curved (Fig. 3 H), 140–200 μm long, maximally 66–125 μm wide. Prostate gland large, entering atrium medially. Ejaculatory duct 320–820 μm long, 45–70 μm wide. Penis conical, enclosed in copulatory sac; cuticularized penis sheath cylindrical and hooded, 97–109 μm long, minimally 47–58 μm wide, maximally 66–79 μm wide and 183–214 μm wide at hood (Fig. 3 I); spiral muscles appear to be absent. Spermathecal ducts spindle-shaped, 145–195 μm long, 80–160 μm wide, with vestibules; ampullae oblong, 440–730 μm long, 200–300 μm wide, with spermatozeugmata in lumina (Fig. 3 J).

Remarks: This species was initially identified as *Limnodrilus amblysetus* Brinkhurst et al., 1990, because of the bluntness of the teeth in some chaetae (Wang et al., 1999), but a careful examination later revealed distinct differences between the two taxa. In *L. paramblysetus* sp. nov., all chaetae are bifid, but in *L. amblysetus*, some simple-pointed chaetae are also present (Brinkhurst et al., 1990; Ohtaka & Nishino, 1999). Moreover, the penis sheaths of the former are hooded with a cylindrical shaft, whereas those of the latter are spade-shaped with a flared shaft. Referring to measurements of the penis sheaths, the new species has the shortest sheaths and the lowest length–width ratio in comparison with *L. amblysetus*, as well as with *L. grandisetosus* Nomura, 1932 and *L. silvani* Eisen, 1879 (Table 2). In the latter two species, the penis sheaths are also spade-shaped as in *L. amblysetus*.

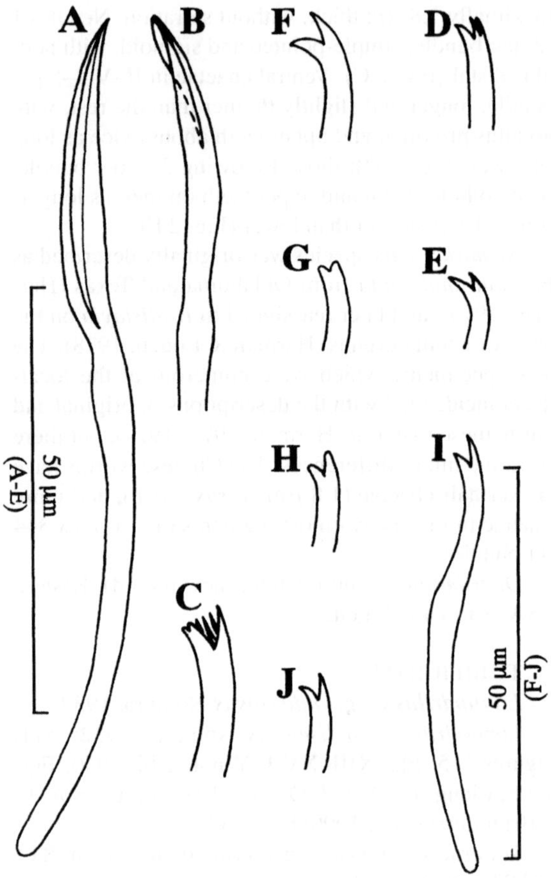

Figure 4. Potamothrix bedoti: (A, B) spermathecal chaetae (VIII), (C) dorsal bifid (XIII). *Aulodrilus pluriseta*: (D) dorsal bifid (VI), (E) ventral chaeta (XII). *Teneridrilus mastix*: (F–I) dorsal bifids (II, II, VII, XV, respectively), (J) ventral chaeta (XXIV).

Distribution: Known only from type locality in mid-lower Changjiang, China.

Potamothrix bedoti **(Piguet, 1913) (Fig. 4 A–C)**

Tubifex (Ilyodrilus) bedoti Piguet, 1913: 124–126, Figure 4 c–e.

Euilyodrilus bedoti (Piguet). Timm, 1970: 65, Figure 12.

Potamothrix bedoti (Piguet). Timm, 1999: 98–99.

Material examined: 1 balsam mount from ST22.

Brief description of new material: Specimen partially mature, length 8 mm, segments 60. Prostomium slightly conical. Hairs 0–4 per bundle, without serration. Dorsal bifids 2–3 per bundle, 80–108 μm long, nodulus distal, usually with 2–3 intermediate teeth;

Table 2. Comparison of penis sheaths of *Limnodrilus paramblysetus* sp. nov. and some allied species. L – length; W_{max} – maximum width in shaft; W_{min} – minimum width. For the measurements, means and ranges (in parentheses) are given

	L. paramblysetus sp. nov.	*L. amblysetus* Brinkhurst et al.	*L. grandisetosus* Nomura	*L. silvani* Eisen
L (μm)	103(97–109)	328(320–335)	227(185–277)	306(303–307)
W_{max}(μm)	72(66–79)	180(170–190)	114(88–158)	69(66–72)
W_{min}(μm)	52(47–58)	60(55–65)	43(36–62)	46(44–48)
L/W_{max}	1.4	1.8	2.1	4.5
L/W_{min}	2.0	5.5	5.5	6.7
W_{max}/W_{min}	1.4	3.0	2.7	1.5
Reference	The present paper	Measured from Brinkhurst et al., 1990	Ohtaka, 1985	Ohtaka, 1985

upper teeth slightly longer than lower, and lower teeth thicker and truncated posteriorly, sometimes secondarily branched (Fig. 4 C). Ventral chaetae 2–5 per bundle, with nodulus distal and upper teeth as long as but thinner than lower. Two spermathecal chaetae modified in the right of VIII, one developed, the other developing (Fig. 4 A–B); those on the left side unmodified, 3 per bundle. Chromophilous masses in VIII–IX.

Remarks: The identification was mainly based upon spermathecal chaetae and its unusual position (Timm, 1970, 1999). Many specimens from Houhu Lake have similar chaetal morphology, and may be juveniles of *P. bedoti*. The abnormal position of the spermathecal chaetae and the presence of numerous immature worms probably indicate that the fragmentation is a prevalent means of reproduction of this species in Houhu. *P. bedoti* may be widespread in China, but easily overlooked owing to the rarity of maturation.

Distribution: Europe, North America and Asia (Changjiang Basin and northwards).

Teneridrilus mastix (Brinkhurst, 1978) (Fig. 4 F–J)

Ilyodrilus mastix Brinkhurst, 1978: 2171–2173, Figure 3; Erséus & Qi, 1985: 193–194, Figure 1.

Teneridrilus mastix (Brinkhurst). Holmquist, 1985: 332–334, 336–341, 357–360, Figures 18, 21 D–E, 22E, 23 K–L, 24 P–R and 31A; Erséus et al., 1990: 840–841; Ohtaka & Nishino, 1999: 38–40, Figures 5–6.

Material examined: 5 balsam mounts and 3 glycerine mounts from ST16, ST8-12, ST14 and ST20.

Brief description of new material: Specimens partially mature, length 7.5–9.5 mm, segments 41–61. Prostomium reduced. Hairs short and bayonet-shaped, usually absent in the first two or three anterior segments and 1–3 per bundle in the following. Dorsal and ventral bifids usually with upper teeth longer than (Fig. 4 I), or as long as (Fig. 4 F, J), lower, but sometimes teeth short and blunt anteriorly (Fig. 4 G–H). Bifids in II, (1) 2 per bundle, somewhat stout or normal, with sharp or blunt teeth (Fig. 4 F–G); those in III (0) 2 (3) per bundle. From IV onwards, dorsal bifids 1–4 per bundle, ventral chaetae 2–5 per bundle. Eggs present. Ovaries one pair in XI. Male ducts undeveloped.

Remarks: This species is widespread in Zhujiang (Pearl) River and the lakes along Changjiang (Erséus & Qi, 1985; Liang, et al., 1995). The new material fits the recent revision of *T. mastix* by Ohtaka & Nishino (1999), but the blunt teeth of some anterior chaetae are reported for the first time in this species. In the material from Zhujiang, the bifid chaetae of segment II were stated to be stout with upper teeth shorter than lower (Erséus & Qi, 1985; Ohtaka & Nishino, 1999), but the corresponding chaetae of the present specimens are hardly thicker than those of other segments, and their teeth are either modified or unmodified.

Distribution: North America, Japan, China, Indonesia.

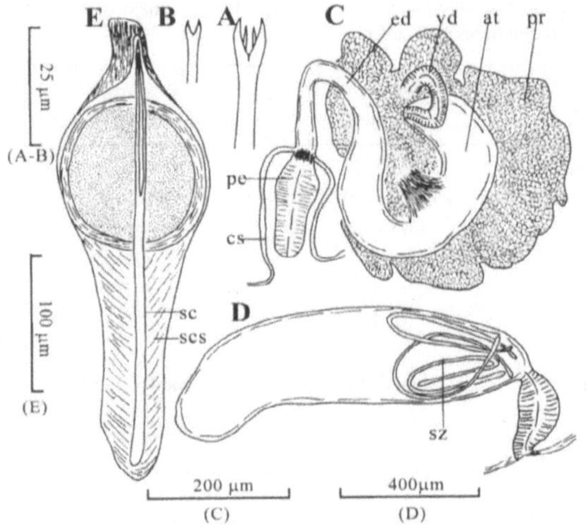

Figure 5. Spirosperma nikolskyi. (A, B) chaetae (anterior, posterior, respectively), (C) male duct, (D) spermatheca, (E) spermathecal chaeta. Abbreviations: sc – spermathecal chaeta; scs – sac of spermathecal chaeta; others as for Figure 3.

Spirosperma nikolskyi (Lastockin, 1953) (Fig. 5)

Peloscolex nikolskyi Lastockin. Chekanovskaya, 1962: 353–354, Figure 180; Brinkhurst & Jamieson, 1971: 507–508, Figure 8.20 K–M; Liang, 1979: 277, Figure 2.

Orientodrilus nikolskyi (Lastockin). Holmquist, 1978: 189, 195, 197, 200–206, Figure 3, Figure 9 I; Holmquist, 1979: 49, Figure 15.

Spirosperma nikolskyi (Lastockin). Brinkhurst, 1981: 1058, Figure 3.

Material examined: 2 mature specimens in balsam mounts (dissected) and 5 immature specimens in glycerine mounts, all from ST15.

Brief description of new material: Length 30–45 mm, segments 85–115. Cuticular papillae numerous. Sensory papillae 2 rows per segment anteriorly, with one row along with chaetal bundle and the other between bundles; 3–4 rows per segment in middle part and 2 rows per segment posteriorly. Hairs not serrated, (3) 4 (6) per bundle anteriorly, 1–4 per bundle posteriorly. Dorsal bifids up to 5 per bundle anteriorly, prongs equal, with few intermediate teeth and without nodulus; 2–3 per bundle posteriorly, thinner and shorter than anterior dorsals, and usually without intermediate teeth and also without nodulus. Ventral chaetae with 1–2 simple-pointed chaetae and 1 bifid chata per bundle in II–IV, and with 1–2 bifids per bundle in following segments. Ventral bifids with upper teeth longer than and as thick as lower in II to about VII, and with upper teeth much shorter and thinner than lower in following segments. Spermathecal chaetae 1 per bundle, straight and simple-pointed, entirely enclosed in chaetal sacs. Vasa deferentia long, entering horse-shore shaped atria apically; prostate glands large, attached to median parts of atria; ejaculatory ducts distinct; penes cylindrical, with copulatory sacs. Spermathecae one pair in X, with long duct and long spermatozeugmata.

Remarks: Since Liang (1979) recorded this species from China for the first time, it has been found to be common in the Changjiang and Zhujiang Basins, but to be rarely mature. The present specimens coincide well with Liang's description except for some intraspecific variation. First, the ejaculatory ducts are longer than those described by Liang. Second, the spermathecal ducts were very short according to Liang, which could be due to the loss of the ectal parts in his dissection. Third, Liang counted two spermathecal chaetae per bundle, as opposed to one per bundle in the new material. Fourth, the glandular sacs of the spermathecal chaetae, which were previously reported only from the Japanese form (Brinkhurst, 1981), were not noted by Liang.

Distribution: Europe, North America and Asia (Zhujiang Basin and northwards).

Aulodrilus pluriseta (Piguet, 1906) (Fig. 4 D–E)

Aulodrilus pluriseta (Piguet). Brinkhurst & Jamieson, 1971 (*partim*): 525–526, Figure 8.23 E; Kathman & Brinkhurst, 1998: 184–185; Timm, 1999: 90–91.

Material examined: 1 glycerine mount from ST14.

Brief description of new material: Specimen immature, length 10 mm, segments 60. Hairs 2–5 per bundle, bayonet-shaped. Dorsal bifids 2–7 per bundle, usually with upper teeth duplicated and shorter than, or as long as, lower (Fig. 4 D). Ventral chaetae 3–9 per bundle, with upper teeth about half as long as lower (Fig. 4 E).

Remarks: This species is common in China, but was formerly overlooked (Liang, 1979). In Chinese material, the upper teeth of dorsal bifids have been found to be variable, i.e. they are either bifurcate as mentioned above, or normal with similar shape and size as the lower teeth (Wang, 1995).

Distribution: Cosmopolitan (not recorded in South America).

Discussion

Historically, all the lakes along Changjiang were connected to the mainstream of the river, and a complex potamo-lacustrine ecosystem was thus formed. However, during 1950–1970s, most of them were separated from the river trunk by damming. This has probably resulted in the loss of habitat and species diversity. As one of the two large lakes maintaining a connection with the mainstream, Poyang Lake has more natural conditions, and could be a shelter for species that have disappeared from other waters. Although the material of this study was largely from quantitative samples in the lake center, it is still shown by the occurrences of rare taxa, such as *Haplotaxis, Bratislavia* and *Limnodrilus paramblysetus* sp. nov., that Poyang Lake region is an invaluable pool of aquatic oligochaetes. Therefore, to increase our faunistic knowledge about Chinese oligochaetes, a continued scrutiny in Poyang Lake seems well justified.

Acknowledgements

The project was supported by the Chinese Academy of Sciences (KZ951-A1-102; KZ951-B1-104) and the Education Bureau, and by the Ministry of Science and Technology of China (96-008-02-03). We are indebted to Drs Tarmo Timm and Akifumi Ohtaka for valuable comments, to Mr Zhicai Xie for collecting assistance, to Ms Qiaoqing Xu and Dr Zhijun Gong for kind help. Special thanks are due to Prof. Christer Erséus for financing the first author to participate in the Aquatic Oligochaete Symposium in Bilbao, and for careful review of the manuscript.

References

Brinkhurst, R. O., 1978. Freshwater Oligochaeta in Canada. Can. J. Zool. 56: 2166–2175.

Brinkhurst, R. O., 1981. A contribution to the taxonomy of the Tubificinae (Oligochaeta: Tubificidae). Proc. biol. Soc. Wash. 94: 1048–1067.

Brinkhurst R. O. & B. G. M. Jamieson, 1971. The Aquatic Oligochaeta of the World. Oliver & Boyd, Edinburgh: 860 pp.

Brinkhurst, R. O. & M. R. Marchese, 1989. Guide to the freshwater aquatic Oligochaeta of South and Central America. Climax 6: 1–179.

Brinkhurst, R. O., S. Qi & Y. L. Liang, 1990. The aquatic Oligochaeta from the People's Republic of China. Can. J. Zool. 68: 901–916.

Chekanovskaya, O. V., 1962. Aquatic Oligochaeta of the USSR, Opred Faune SSSR 78(3): 1–513 (English translation 1981 by Amerind Publishing Co. Pvt. Ltd., New Delhi).

Erséus, C. & S. Qi, 1985. Two aberrant Tubificidae (Oligochaeta) from Pearl River in the People's Republic of China. Hydrobiologia 127: 193–196.

Erséus, C., J. K. Hiltunen, R. O. Brinkhurst & D. W. Schloesser, 1990. Redefinition of *Teneridrilus* Holmquist (Oligochaeta: Tubificidae), with description of two new species from North America. Proc. biol. Soc. Wash. 103: 839–846.

Harman, W. J., 1973. New species of Oligochaeta (Naididae) with additional distributional records from Oklahoma and Texas. Southwest. Nat. 18: 151–164.

Harman, W. J., 1974. The Naididae (Oligochaeta) of Surinam. Zool. Verhandl. 133: 1–36.

Harman, W. J. & M. S. Loden, 1978. *Bratislavia unidentata* (Oligochaeta: Naididae), a re-description. Southwest. Nat. 23: 541–544.

Holmquist, C., 1978. Revision of the Genus *Peloscolex* (Oligochaeta, Tubificidae). 1. Morphological and anatomical scrutiny, with discussion on the generic level. Zool. Scr. 7: 187–208.

Holmquist, C., 1979. Revision of the genus *Peloscolex* (Oligochaeta, Tubificidae) 2. Scrutiny of the species. Zool. Scr. 8: 37–60.

Holmquist, C., 1985. A revision of the genera *Tubifex* Lamarck, *Ilyodrilus* Eisen, and *Potamothrix* Vejdovsky & Mràzek (Oligochaeta, Tubificidae), with extensions to some connected genera. Zool. Jb. Syst. 112: 311–366.

Liang, Y. L., 1964. Studies on the aquatic Oligochaeta of China. II. On some species of Naididae from Sinkiang with description of a new species *Allodero prosetosa*. Acta Zool. Sin. 16: 643–652 (in Chinese with English abstract).

Liang, Y. L., 1979. Studies on the aquatic Oligochaeta of China. III. Aquatic Oligochaeta of the Huama Lake. Oceanol. Limnol. Sin. 10: 273–281 (in Chinese with English abstract).

Liang, Y. L., T. H. Wu & Z. C. Xie, 1995. On the current conditions of zoobenthos in Baoan Lake with an assessment of its potential fishery production capacity. In Liang, Y. L. & H. Q. Liu (eds), Resources, Environment and Fishery Ecological Management of Macrophytic Lake (No. 1). Science Press, Beijing: 178–192 (in Chinese with English abstract).

Kathman, R. D. & R. O. Brinkhurst, 1998. Guide to the freshwater oligochaetes of North America. Aquatic Resources Center, College Grove, Tennessee: 264 pp.

Nomura, E., 1932. *Limnodrilus grandisetosus* nov. sp., a freshwater oligochaete. Scient. Rep. Tohoku imp. Univ. (Ser. IV) 7: 511–527.

Ohtaka, A., 1985. Taxonomical revision of three Japanese *Limnodrilus* species (Oligochaeta). Proc. jap. Soc. Syst. Zool. 30: 18–35.

Ohtaka, A. & M. Nishino, 1999. Studies on the aquatic oligochaete fauna in Lake Biwa, central Japan. II. Records and taxonomic remarks of nine species. Hydrobiologia 406: 33–47.

Piguet, E., 1913. Notes sur les Oligochètes. Rev. suisse Zool. 21: 111–146.

Sperber, C., 1948. A taxonomical study of the Naididae. Zool. Bidr. Upps. 28: 1–296.

Timm, T., 1970. On the fauna of the Estonian Oligochaeta. Pedobiologia 10: 52–78.

Timm, T., 1999. A guide to the Estonian Annelida. Estonian Academy Publishers, Tallinn: 208 pp.

Wang, H. Z., 1995. Studies on taxonomy, distribution and ecology of microdrile oligochaetes of China, with descriptions of

two new species from the vicinities of the Great Wall Station of China, Antarctica. D.Sc. Thesis, Institute of Hydrobiology, Chinese Academy of Sciences, Wuhan, China: 184 pp (in Chinese with English summary).
Wang, H. Z., Z. C. Xie, X. P. Wu & Y. L. Liang, 1999. A preliminary study of zoobenthos in the Poyang Lake, the largest freshwater lake of China, and its adjoining reaches of Changjiang River. Acta Hydrobiol. Sin. 23 (supplement): 132–138.
Yamaguchi, H., 1940. Oligochaeta of Manchoukuo. In Kawamura, T. (ed.), Report of the Limnobiological Survey of Kwantung and Manchoukuo. Kwantung Kyoku: 382–394 (in Japanese).
Zhu, H. H. & B. Zhang, 1997. The Poyang Lake. Press of University of Science & Technology of China. Hefei: 349 pp (in Chinese).

Hydrobiologia **463**: 39–47, 2001.
P. Rodriguez & P.F.M. Verdonschot (eds), Aquatic Oligochaete Biology VIII.

New alluroidids (Annelida, Clitellata) from Guyana

Pietro Omodeo[1] & Kathryn A. Coates[2]
[1]*Dipartimento di Biologia Evolutiva, Via P. A. Mattioli 4, I-53100 Siena, Italy*
Tel: 39 0577 232956. Fax: 39 0577 232825. E-mail: rota@unisi.it
[2]*Bermuda Biological Station for Research Inc., 17 Biological Lane, St. Georges, GE 01, Bermuda*

Key words: Alluroididae, *Righiella jamiesoni* n. gen. et sp., *Brinkhurstia donaldi* n. sp., Guyana

Abstract

A new genus and two new species of Alluroididae are described from Guyana and additional data are provided for two previously known species. *Brinkhurstia donaldi* n. sp. is characterized by very long ribbon-like penial chaetae and an unpaired spermatheca with a sperm-filled diverticulum and opening mid-dorsally at 7/8. *Righiella* n. gen. is metandric and without penial chaetae; *R. jamiesoni* n. sp. has a large, unpaired spermatheca (devoid of diverticula) which also opens mid-dorsally at 7/8. *Brinkhurstia donaldi* and *Righiella jamiesoni* were found in running waters in creeks, whereas *Brinkhurstia americana* (Brinkhurst, 1964) and *Kathrynella guyanae* Omodeo, 1996 were collected in anoxic mud in a pool.

Introduction

The aquatic Oligochaeta of South America are poorly known. Thus, being reasonably certain that the species diversity is high in this subcontinent, collecting campaigns in Guyana, Peru and Ecuador were undertaken by K. A. Coates, D. F. Stacey and R. da Silva, with the support of the Ontario Museum and the Javier Prado Museo de Lima (see Coates & Stacey, 1994; Stacey & Coates, 1996). A first taxonomic paper on the Guyana material was published by Omodeo (1996); other results are herewith recorded.

Materials and methods

Sampling area, field methods and preparation of the collections were described by Stacey & Coates (1996). In the present study, the worms were first examined as whole mounts in Canada balsam. Three specimens of *Brinkhurstia donaldi* n. sp. and the single specimens of *B. americana* (Brinkhurst, 1964) and *Righiella jamiesoni* n. gen. et sp. were then dismounted. Their anterior body regions were removed, embedded in paraffin, serially cut, and re-stained for further microscopic examination. Types have been deposited at Invertebrate Zoology, Royal Ontario Museum, Toronto, Canada (ROMIZ).

Systematic account

Family ALLUROIDIDAE Michaelsen, 1900, *emendavit* Jamieson, 1968

Brinkhurstia americana (Brinkhurst, 1964) (Fig. 1)

Alluroides americanus Brinkhurst, 1964: 533–534, Figures 5–7.

Brinkhurstia americanus (Brinkhurst), Jamieson, 1968: 80–81, Figures 3 and 12; Brinkhurst & Jamieson, 1971: 716–717, Figure 14.1E,F; Righi et al., 1978: 11–12, Figures 11–21; Stacey & Coates, 1996: Tables 1 and 2.

Material: ROMIZ I2363, one specimen collected in a swampy forest at Kwabanna, Guyana, Station S62, 7° 34.00′ N, 59° 09.00′ W, coll. Donald F. Stacey, October 11, 1992. Twenty anterior segments removed, longitudinally sectioned and stained with ferric haematoxylin for histological examination.

External characters: Size 26 by 0.6 mm; segment number 109. Clitellum annular, 1/3XII–1/3XVI. Male pores in XIII aligned with chaetae *ab*; spermathecal pore mid-dorsal in anterior of IX. Chaetae rather long

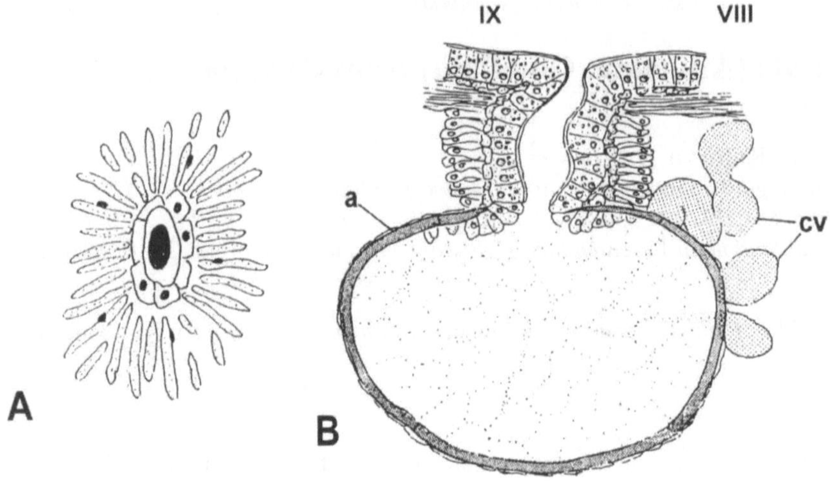

Figure 1. Brinkhurstia americana (Brinkhurst, 1964). (A) Cross section of the muscular sheath enclosing penial chaeta. (B) Spermatheca, opening in the anterior of IX. Abbreviations: a – ampulla, cv – commissural vessel.

and thin, their length grows from II to XI, then becomes smaller, but the caudal chaetae again enlarge. Penial chaetae straight, > 560 by 12.2 μm.

Internal organization: Perienteric blood sinus begins in IX, the intestine begins in XIV (in XVII according to Righi et al., 1978). Long convoluted commissures between dorsal and ventral vessel, the first four pairs partly embedded among the basophilic cells of septal glands, their diameter is 15–25 μm. Capillaries seen only on the prostates and on the nephridia, 3.8 μm in diameter. First pair of nephridia in X.

Testes and ovaries as described by previous authors. Male ducts observed only in the fore part of segments XI and XIII. There they anastomose with the prostates in the ental part at about one third of their length.

The prostates (*'atria'* according to Brinkhurst, 1964 and Righi et al., 1978) are long, convoluted cylindrical glands formed by turgid basophilic cells wrapped by a very thin sheet of fibres, covered by hyaline club-shaped cells containing large granules; in the ental part they are devoid of lumen, whereas in the central and ectal parts they become tubular. A thin cylindrical gland, 'paratrium' according to Righi et al. (1978), is attached to the ectal portion of the prostate; it lacks an internal canal and is covered by thin flat peritoneal cells: it may be the anlage of a substitution prostate. Muscle fibres of the body wall ribbon-like, 2.2–2.5 μm thick.

The penial chaetae have a maximum diameter of 12.5 μm, are almost straight and enclosed by a muscular sheath whose diameter is about 60 μm (Fig. 1A); this sheath is attached, as observed by previous authors, on the anterior face of septum 12/13; such an anterior attachment of the penial chaetae is an unknown condition in terrestrial megadriles. The prostate and the penial chaetae have their outlet within a small terminal chamber covered by muscular fibres.

The unpaired spermatheca (Fig. 1B) has a thick (125 μm) short duct and a globular free ampulla (diameter 225 μm) not described previously; duct lined by cuticle and covered by a sheath of high, hyaline peritoneal cells (Fig. 1B); no sperm are contained in the spermatheca.

Observations: The intestinal contents consist of organic matter mingled with minute sand grains: the sponge spicules and the astomate ciliates, so conspicuous in the gut of *B. donaldi* n. sp. and *Righiella jamiesoni* n. gen. et sp., are absent in this species and in *Kathrynella guyanae* Omodeo, 1996.

The enormous development of the vascular apparatus and the intestinal contents suggest that *B. americana* as well as *K. guyanae* dwell in a milieu very poor in oxygen and devoid of sponge debris (see below).

Brinkhurstia donaldi n. sp. (Figs 2–5)

Brinkhurstia sp. Coates & Stacey, 1994: 81, Table 1, Figure 2.

Brinkhurstia sp. 2. Stacey & Coates, 1996: Table 1 and Figure 4D–F.

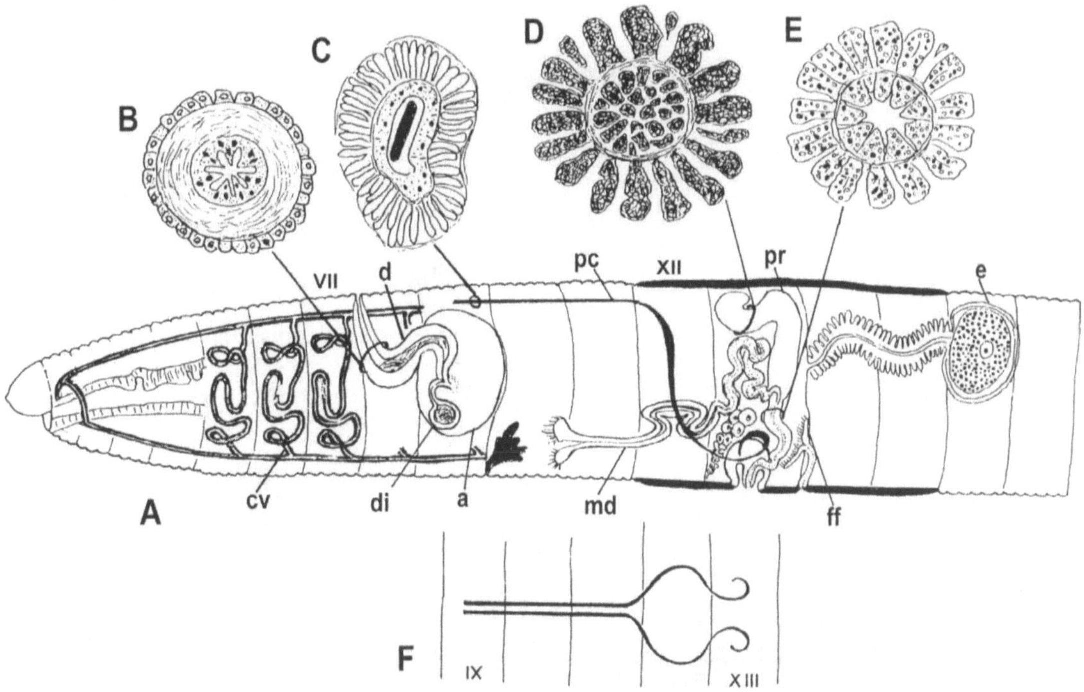

Figure 2. Brinkhurstia donaldi n. sp. (A) Scheme showing arrangement of genital organs and vascular commissures. (B) Cross section through spermathecal duct. (C) Cross section of the muscular sac ensheating penial chaeta. (D) Section through ental part of prostate. (E) Cross section of prostatic duct. (F) Dorsal view of the arrangement of the penial chaetae within IX–XIII. Abbreviations: a – spermathecal ampulla, cv – commissural vessels, d – spermathecal duct, di – spermathecal diverticulum, e – egg, ff – female funnel, md – male duct, pc – penial chaeta, pr – prostate.

Material: Holotype, ROMIZ I4739 (longitudinally sectioned specimen), from Kurupukari creek, flowing into Essequibo River, W of base camp, Station S61, 04° 43′ N 58°, 37′ W, October 11, 1990, coll. D. F. Stacey. Paratypes, ROMIZ I2067, seven specimens (one longitudinally sectioned) from same locality and date as holotype; ROMIZ I2060, four specimens (one longitudinally sectioned) from Amatuk, Station S33, 05° 19.90′ N 59°, 18.90′ W, October 2, 1990, coll. D. F. Stacey.

External characters: Body form slender, cylindrical, tapering posteriorly; length 30–32 mm, width 0.82–0.90 mm. Lateral lines recognizable only on microscope slides: they appear as a series of buttons formed by 6–8 hyaline cells situated amid the longitudinal muscle fibres in a line included between chaetae *b* and *c* (Fig. 4A). Segments multiannulated, 105, 109, 131, 135, 150 in number. Anus terminal.

Clitellum annular over 1/3XII-1/2XV, best developed in XIII-XIV, male pores in the middle of XIII aligned with chaetae *a*; in a few cases the terminal chambers and the penial chaetae protrude through them. Female pores in furrow 13/14, slightly ventral to chaetal line *a*. Spermathecal pore a mid-dorsal cleft in furrow 7/8.

Chaetae sigmoid, slender, unsculptured; the longest ones are in XI and measure respectively: *a* 328 μm, *b* 284 μm, *c* 164 μm, *d* 280 μm; their lengths decrease anteriorly and posteriorly; chaetal formula at XX, *aa*: *ab*: *bc*: *cd*: *dd* = 6.1: 1.0: 9.0: 1.0: 8.2. Cuticle about 2 μm thick.

Internal organization: Septum 3/4 incomplete, septa 4/5–12/13 fairly thick. Septal glands in V through VIII. Perienteric blood sinus commencing in IX. The nephridia are devoid of bladders and begin in XI. Dorsal and ventral vessels have large diameters; commissural vessels in V–XI are tortuous with thin walls, their diameter is about 17 μm; their dorsal loops in V–VIII are embedded in the septal glands; in adult specimens the commissures in X and XI are compressed and hidden by large gregarine gamonts; numerous capillaries may be seen beneath and among

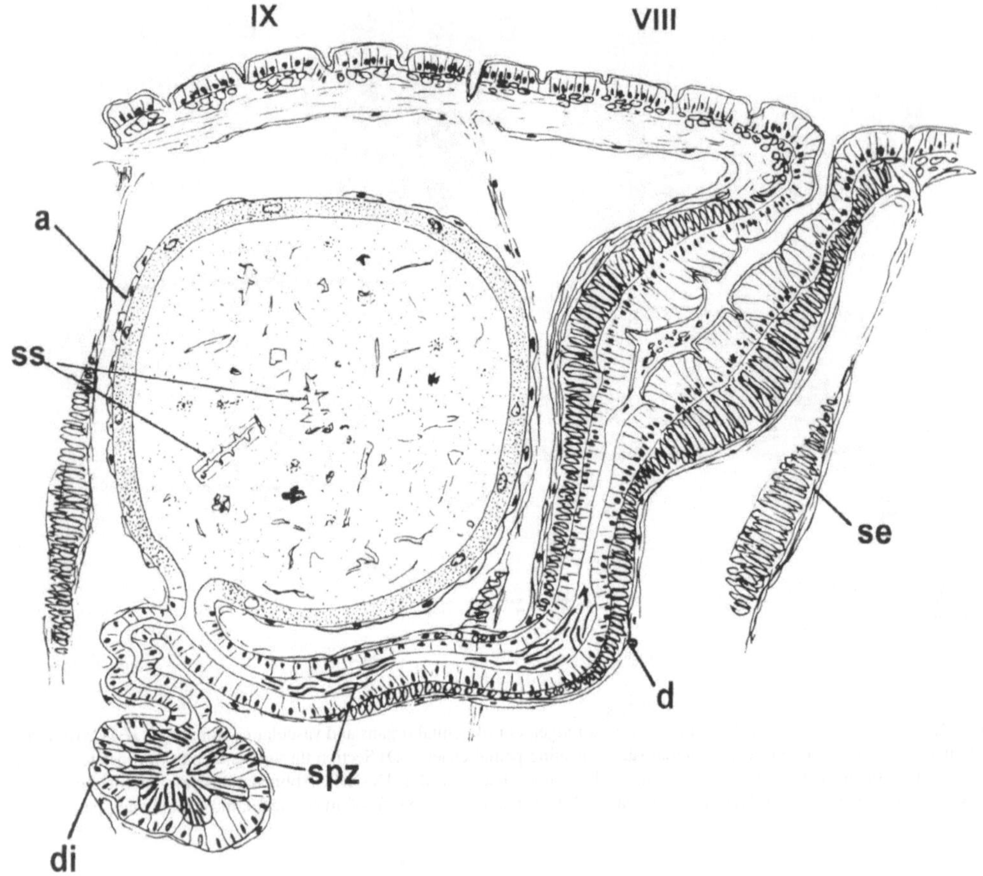

Figure 3. Brinkhurstia donaldi n. sp. Spermatheca, sagittal section. Abbreviations: a – ampulla, d – duct, di – diverticulum, se – septum with muscle fibres, spz – spermatozoa, ss – sponge spicules.

the epidermal cells (Fig. 4A), especially in the clitellar region, and among the glandular cells covering the prostates; their diameter is about 3.5 μm. Muscular fibres of the body wall appear ribbon-shaped, 3 μm thick, 21 μm high.

Testes and seminal funnels in X; male ducts thin in XI and XII; in XIII they become sinuous and filled with sperm (Fig. 2A), forming a highly convoluted epididymis, and finally connect with the prostate at an intermediate point. The prostates are long, convoluted, confined to XIII; in section, they appear formed by an inner layer of epithelial (glandular?) cells surrounded by a thick layer of muscle cells and covered by large, club-shaped gland cells (Fig. 2D); this covering diminishes and then disappears near the outlet into the globular terminal chamber, which opens through as a small penial papilla (Fig. 2E). In each terminal chamber emerges a ribbon-like penial chaeta (Figs 2C and 4B), whose proximal end in adult worms lies in the anterior dorsal part of IX: the two penial chaetae, each ensheathed in a muscular sac (Fig. 2C), run parallel near the mid-dorsal line in IX, X and XI (Fig. 2A,F); then in XII they diverge sidewards and ventralwards, arcing around the terminal chamber in XIII and exiting through it (Fig. 2A; see also Coates & Stacey, 1994: Fig. 2C).

Ovaries in XIII; two egg sacs arise near the female funnels and extend backwards to XVI (Fig. 2A); the histology of the egg sac resembles that of *Kathrynella guyanae* (see Omodeo, 1996: Fig. 1P). Ovaries and egg sacs disappear in fully mature specimens leaving two or three free large ellipsoid eggs (420 by 420 by 300 μm) in the coelom of XV and XVI.

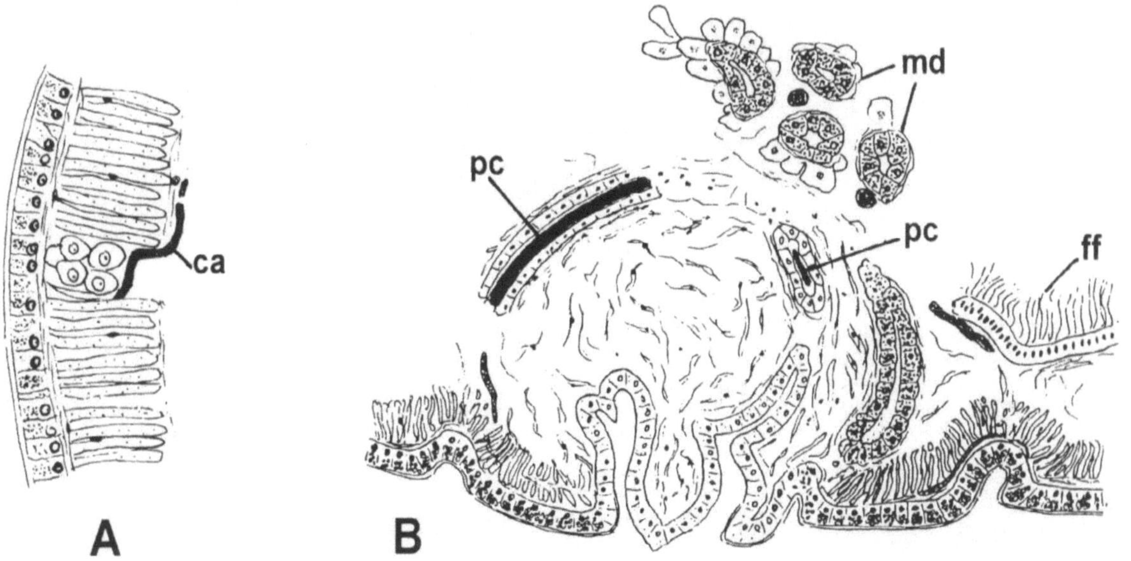

Figure 4. Brinkhurstia donaldi n. sp. (A) Cross section of the body wall at the level of the lateral line, which is accompanied by a capillary vessel. (B) Male atrium receiving penial chaeta and male duct. Abbreviations: ca – capillary vessel, ff – female funnel, md – male duct, pc – penial chaeta.

The spermatheca has a thick, spindle-shaped muscular duct (Figs 2B and 3) that runs to the middle of IX abutting on a very large, free, globular ampulla that almost fills the cavity of this segment. Near the insertion of the duct is attached a pedunculated diverticulum (Figs 2A and 3). Sperm are contained in the diverticulum and in the ental part of the duct.

Observations: The gut contains a mixture of organic matter, a few sand granules, abundant spicules of fresh water sponges (Fig. 5B) and other unidentified mineral matter, it does not contain any diatom shells. Some spicules surely belong to the South American endemic sponge genera *Uruguaya* and *Drulia*. Other spicules can be tentatively ascribed to the genera *Acalle* or *Trochospongilla*. The scarcity of sand granules relative to the sponge spicules is strange and indicates that these worms live among the debris of dead freshwater sponges (spongillites). The gut contains also, in the anterior segments, some astomate ciliates of the genus *Anoplophrya* (Fig. 5A).

The contents of the spermathecal ampulla (Fig. 3) in all the adult specimens are consistent and amazing: dispersed among some organic matter, resembling mucous secretion, is debris of material resembling the intestinal contents: fragmented sponge spicules and a few sand granules.

In fully mature specimens, gregarine gamonts fill almost completely X–XII; in some cases gregarine oocysts fill a part of vasa deferentia.

Taxonomic position: *B. donaldi* n. sp. differs from *B. americana*, the only described congener, in the form, size and peculiar insertion of the penial chaetae. Its spermatheca has a diverticulum and a much longer duct (correlated with the length of the penial chaetae) and opens one segment before.

Righiella **n. gen.**

The new species described below shows such a unique set of characters (see Table 1) that it cannot be ascribed to any known alluroidine genus. It is possible, of course, to modify the diagnosis of some established genus in order to include in it the new taxon, but in this case the boundaries between the established genera would become ill-defined. We are aware that to create a new genus on the basis of a sole specimen is inadvisable, yet, at present, no other procedure seems appropriate.

Diagnosis: Alluroidinae with large coiled prostates having a thin muscular sheath surrounded by a thick glandular sleeve. The vas deferens enters the prostate at the ental tip. Metandric, male pores lateral in XIII, in the place of the absent ventral chaetae. A large unpaired spermatheca formed by duct and am-

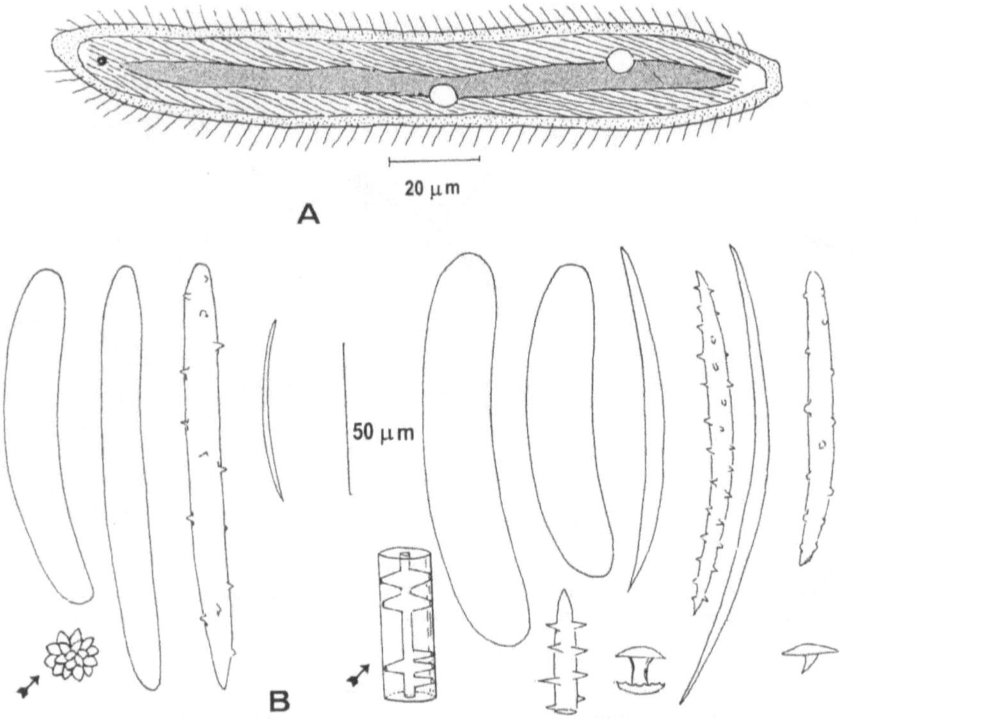

Figure 5. Intestinal and spermathecal contents of the two new Guyana alluroidids, *B. donaldi* n. sp. and *Righiella jamiesoni* n. sp. et gen. (A) *Anoplophrya* sp., an astomatous ciliate commensal in the anterior gut. (B) Skeletal and gemmular siliceous spicules of freshwater sponges from the gut and spermathecal ampullae of *B. donaldi* (left) and *R. jamiesoni* (right). Arrows indicate unidentified objects accompanying the spicules.

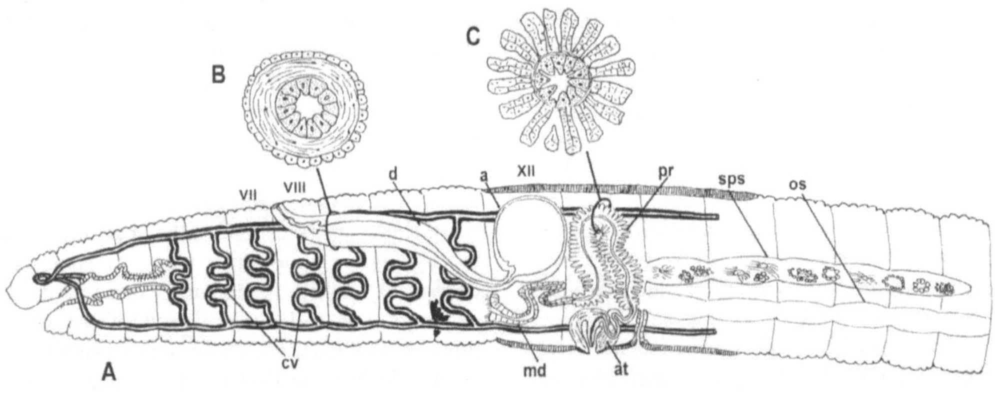

Figure 6. Righiella jamiesoni n. gen et sp. (A) Scheme showing arrangement of genital organs and vascular commissures. (B) Cross section through spermathecal duct; C, Cross section through prostate. Abbreviations: a – spermathecal ampulla, at – male atrium, cv – commissural vessels, d – spermathecal duct, md – male duct, os – egg sac, pr – prostate, sps – sperm sac.

pulla situated in pregonadic segments, opening in the mid-dorsal line. Lateromuscular groove conspicuous.

The new genus differs from the African genera and *Brinkhurstia* in being metandric, from *Kathrynella* by the position of the male and spermathecal pores, and from both American genera for the absence of penial chaetae (Table 1).

Table 1. The most significant characters distinguishing the genera of Alluroidinae

	Alluroides Beddard, 1894	*Standeria* Jamieson, 1968	*Brinkhurstia* Jamieson, 1968	*Righiella* n. gen.	*Kathrynella* Omodeo, 1996
Male pores	XIII	XIII	XIII	XIII	XIV
Spermathecal pore(s)	Mid-dorsal 7/8, 8/9	Lateral 8/9	Mid-dorsal 7/8, IX	Mid-dorsal 7/8	Lateral 6/7
Testes in	X	X	X	XI	XI
Insertion of male ducts	In terminal chamber	In terminal chamber	Half way in the prostates	On the tip of prostates	On the tip of prostates
Penial chaetae	no	no	yes	no	yes
Secondary annulation	no	yes	no	no	yes
Genital papillae	no	no	no	no	yes
Geographic distribution	Central Africa	Central Africa	South America	Guyana	Guyana

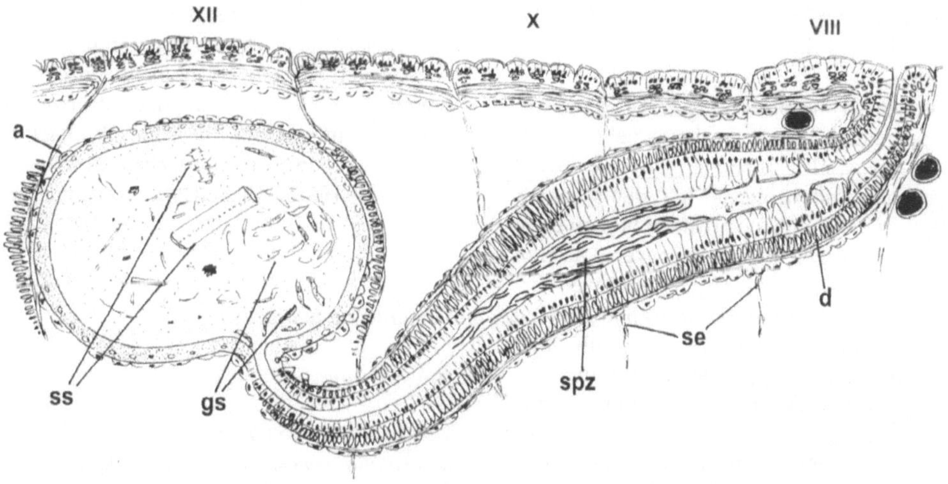

Figure 7. Righiella jamiesoni n. gen et sp. Spermatheca, sagittal section. Abbreviations: a – ampulla, d – duct, gs – gregarine sporozoites, se – septa, spz – spermatozoa, ss – sponge spicules.

Righiella jamiesoni n. sp. (Figs 5–7)

Material: Holotype, ROMIZ I4737, adult specimen collected in a pool, between buried roots, twigs, mud and silt, next to Kurupukari creek flowing into Essequibo River, Guyana, Station S61, 04° 43′ N, 58° 37′ W, coll. D. F. Stacey, October 11, 1990. This specimen was previously stained and mounted whole in Canada balsam on a slide, later the first 24 segments were removed, serially cut and re-stained with ferric haematoxylin. The worm was damaged by manipulation on collection: the compression has smashed some internal organs in X and XI. Further, when serial sections were mounted on slides, about four of them were lost. These problems notwithstanding, the anatomy could be reliably reconstructed.

External characters: Body slender, cylindrical, tapering posteriorly; length 35 mm, width 0.8 mm. Lateral lines evident in IV–XIV. Segments multiannulated, 105 in number. Anus terminal.

Clitellum annular over XII–2/3XV. Male pores in XIII in the place of chaetae *ab* which are absent. Female pores in furrow 13/14 in chaetal line *a*. Spermathecal pore mid-dorsal in furrow 7/8.

Chaetae sigmoid, slender, unsculptured, the longest are in X and measure about 180 μm, their length decreases frontward and caudalward so that they measure 134 μm in V and 156 μm in XVI. Approximate chaetal formula at XX, *aa*: *ab*: *bc*: *cd*: *dd* = 3.0:1.0:4.5:1.0:3.8. No penial chaetae. Cuticle 1.8 μm thick.

Internal organization: Septa 4/5–11/12 fairly thick, septal glands in V–VIII moderately developed. Nephridia without bladders, beginning in XVII. The moniliform intestine begins in XIV and attains its maximum diameter in XVII. Perienteric sinus begins in XIV; commissural vessels sinuous, moderately large (diameter 18 μm) in V–XI (Fig. 6A). Muscular fibres of the body wall ribbon-shaped, ca. 3 μm thick and 25 μm high.

Sperm funnels in the posterior part of XI; testes regressed, apparently situated in the same segment; sperm sacs in XIV–XVIII, filled and partly destroyed by large gregarine trophozoites. Male ducts convoluted in XIII, they insert into the prostates at their ental tips. Prostates confined in XIII; in transverse section they appear formed by an inner epithelial layer surrounded by a thin layer of muscle fibres and covered by long digitiform glands formed by cells with small nuclei (Fig. 6C); this covering disappears in the convoluted ectal portion which terminates with a penis enclosed in the terminal globular chamber. Ovaries and female funnels regressed in the posterior part of XIII, egg sac regressed, in XIV–XXIII? Two eggs in XXII and XXIII, measuring 300 by 400 μm.

The spermatheca (Figs 6 A,B and 7) has a large globular ampulla in XII and a long duct opening in 7/8; the ectal portion of the duct is filled with sperm, the ental portion is narrow. The contents of the ampulla are similar to those described for *B. donaldi* n. sp.: amid mucous matter are mineral particles, fragments of sponge spicules and young gregarine trophozoites.

Intestinal contents: Organic matter and a few sand granules together with fragments of spiculae of fresh-water sponges (Fig. 5B), but these fragments are not as numerous as in *B. donaldi* n. sp. In the gut lumen of XVI through XL are harboured many astomatous ciliates (Fig. 5A): they are spindle shaped, devoid of cytostome, 80–100 μm long, 17 μm wide, have a thick pellicle and a macronucleus almost as long as the cell, 48–50 kineties; they are attributable to *Anoplophrya*, a common commensal of oligochaetes (see Lee et al., 1985).

Parasites: Many monocystid gregarine trophozoites in the male ducts, in the sperm sac, in the coelom and even in the first segments under the brain; gregarine sporocysts in the spermatheca.

Systematic position: This new species lacks penial chaetae, a feature that is otherwise peculiar to the African Alluroidinae, but it is metandric whereas all African species are proandric; it also has more complex spermathecae. *Righiella jamiesoni* also resembles the South American *Kathrynella*, but that metandric genus has the male pores in XIV, has penial chaetae and has two simple spermathecae and thick muscular euprostates (see Table 1).

Genus *Kathrynella* Omodeo, 1996

Supplementary data (based on re-examination of the types of *K. guyanae*): Lateromuscular groove not observed. If present, its elements do not form a continuous line. Segments divided into two unequal annuli, the anterior one, bearing the chaetae, is about 4 times as long as the posterior. Muscle fibres of the body wall ribbon-shaped, 4.3 μm thick and 15 μm high.

Concluding remarks

Biogeography

Until recently, only one species of Alluroididae, *B. americana*, was known from South America. This created the view that the presence of these worms was exceptional in this part of the world. Finding three additional species together with *B. americana* in a small area of Guyana (Omodeo, 1996 and this paper) suggests a different hypothesis: the alluroidids are relatively common and diverse in South America, and the sampling of aquatic megadriles, especially in stagnant waters, is very incomplete.

Systematics

The Alluroidinae resemble the Haplotaxidae (*sensu lato*) in the general body structure, the morphology of the vascular apparatus, the simplicity of the intestine, the presence of long egg sacs and sperm sacs and the frequent mid-dorsal opening of the spermathecae. They show also some characteristics peculiar to a few terrestrial megadriles: complex spermathecae, similar to those of Megascolecoidea, and complex male terminalia, resembling those of Megascolecoidea and Eudriloidea. These points of resemblance, being consistent also in details, can be interpreted as evidence of kinship (Omodeo, 1998, 2000).

On the contrary, the presence of a spermathecal diverticulum filled with sperm, so conspicuous in *B. donaldi* n. sp. as well as in most Acanthodrilidae, and the presence of penial chaetae in South American Alluroidinae as well as in the Acanthodrilidae and Octochaetidae, being sporadic and divergent in details, may be interpreted as convergences and of independent origin.

Acknowledgements

We thank Don Stacey for his outstanding field work and specimen preparations, Renata Manconi for identification of sponge spicules, Enzo Burroni and Gabriella Bonifazi for technical assistance, Emilia Rota for useful comments and help with the preparation of the manuscript.

References

Brinkhurst, R. O., 1964. A taxonomic revision of the Alluroididae (Oligochaeta). Proc. zool. Soc. Lond. 142: 527–536.

Brinkhurst, R. O. & B. G. M. Jamieson, 1971. Aquatic Oligochaeta of the World. Oliver & Boyd, Edinburgh: 860 pp.

Coates, K. A. & D. F. Stacey, 1994. Oligochaetes (Naididae, Tubificidae, Enchytraeidae and Alluroididae) of Guyana, Peru and Ecuador. Hydrobiologia 278: 79–84.

Jamieson, B. G. M., 1968. A taxonomic investigation of the Alluroididae (Oligochaeta). J. Zool., Lond. 155: 55–86.

Lee, J. J., H. H. Seymour & E. C. Bovee, 1985. An illustrated guide to the Protozoa. Society of Protozoologists, Kansas City: 629 pp.

Omodeo, P., 1996. *Kathrynella*, a new oligochaete genus from Guyana. Hydrobiologia 334: 11–15.

Omodeo, P., 1998. History of Clitellata. Ital. J. Zool. 65: 51-73.

Omodeo, P., 2000. Evolution and biogeography of megadriles (Annelida, Clitellata). Ital. J. Zool. 67: 179–201.

Righi, G., I. Ayres & E. C. R. Bittencourt, 1978. Oligochaeta (Annelida) do Istituto Nacional de Pesquisas da Amazonia. Acta Amazonica 8: 1–49.

Stacey, D. F. & K. A. Coates, 1996. Oligochaetes (Naididae, Tubificidae, Opistocystidae, Enchytraeidae, Sparganophilidae and Alluroididae) of Guyana. Hydrobiologia 334: 17–29.

Hydrobiologia **463:** 49–64, 2001.
P. Rodriguez & P.F.M. Verdonschot (eds), Aquatic Oligochaete Biology VIII.

Notes on the diversity and distribution of Australian Naididae and Phreodrilidae (Oligochaeta: Annelida)

Adrian Pinder
Department of Conservation and Land Management, P.O. Box 51, Wanneroo, 6946, Western Australia
E-mail: adrianp@calm.wa.gov.au

Key words: Australia, biogeography, Naididae, Phreodrilidae, oligochaetes

Abstract

This paper provides a summary of the diversity and distribution of Australian naidid and phreodrilid oligochaetes. As for other continents, the Australian naidid fauna consists mostly of cosmopolitan species, although there are indications of greater endemicity than currently recognised. While some naidids are widespread in Australia, others have a northern or a southern bias to their distribution, but few have been recorded in Tasmania. Many new records and species of Phreodrilidae have been documented since the review by Pinder & Brinkhurst (1997a) and these have allowed notions of phreodrilid zoogeography to be refined. The family is still considered particularly diverse in some temperate areas (Tasmania and the far south-west of Western Australia), but surprisingly few species are known from the temperate south-east mainland. Increasingly, new phreodrilid species are being collected from seasonal habitats on granite outcrops in the south-west and from refugial habitats (caves, groundwater and permanent river pools) in drier regions. A complete picture of oligochaete distributions will require much more work and patterns suggested by current data are presented here as hypotheses.

Introduction

Prior to Brinkhurst (1971), there had been few studies of Australian aquatic oligochaetes, amounting to records of only 12 species by Benham (1907), Michaelsen (1907), Goddard (1909ab), Davies (1913), Nichols (1921) and Jackson (1931), although these early papers already hinted at a distinctive fauna for the region. The review by Brinkhurst (1971) and subsequent papers by Jamieson (1978), Brinkhurst & Fulton (1979, 1980), Naidu & Naidu (1980ab) and Brinkhurst (1982, 1984) increased the number of species known from Australia to about 60 and confirmed the presence of a significant endemic component. More recently, Erséus (1993, 1997), Erséus & Grimm (1998), Pinder & Brinkhurst (1997ab, 1998, 2000) and Pinder *et al.* (1998) provided descriptions of many more new species (increasing the total to 94) and some new genera, broadened geographic coverage and started to reveal some interesting zoogeographic patterns. While this work is far from complete (the author has at least 40 undescribed species), the 8th International Symposium on Aquatic Oligochaeta provided an opportunity to present an interim summary of the diversity and distribution of 2 families (Naididae and Phreodrilidae) for which substantial new information has accumulated since reviews by Brinkhurst (1971) and Pinder & Brinkhurst (1994, 1997a). For other limnic families (except for the Enchytraeidae, which have not been studied at all) the publications listed above adequately summarise existing knowledge.

The Australian environment

Australia, which constitutes most of the Australian continental plate, is 3500 km across and spans latitudes 10.5–43.5° S (Fig. 1). Not surprisingly this area encompasses a great diversity of climates, geology and landscapes, reflected in the diversity of aquatic environments. The following account of Australian environments is sufficient only as a brief background to the discussion of naidid and phreodrilid distributions. More complete accounts are provided by Gentilli (1972), Keast (1981) and Dyne (1987). Australia was once part of the southern supercontinent of Gond-

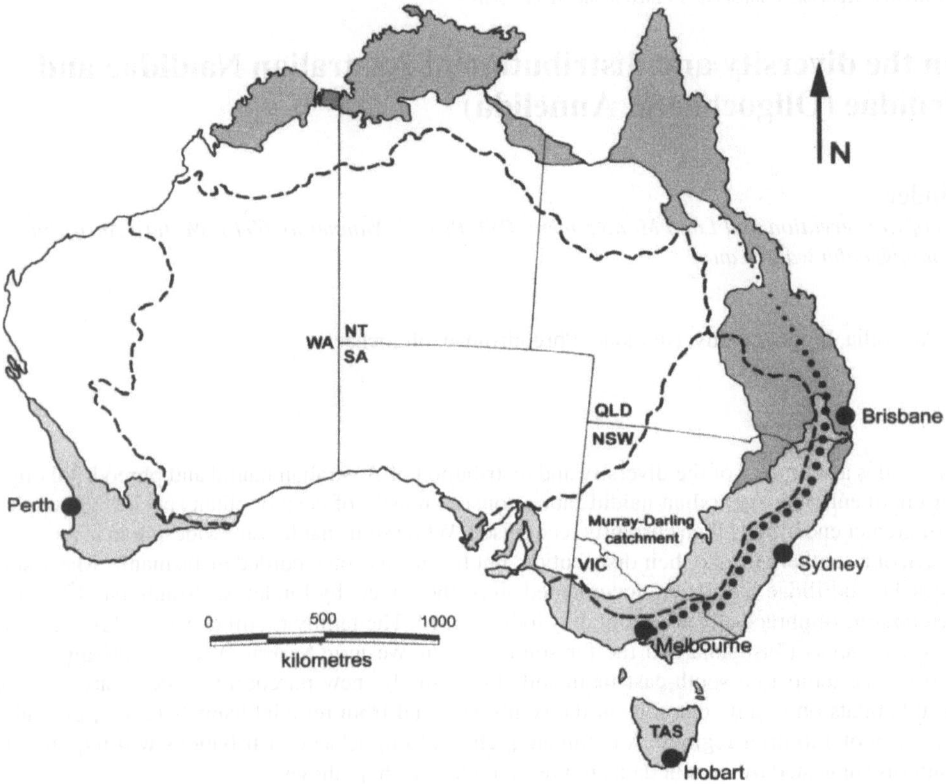

Figure 1. Map of Australia showing state boundaries with solid lines (WA: Western Australia; NT: Northern Territory; Qld: Queensland; NSW: New South Wales; Vic: Victoria; Tas: Tasmania and SA: South Australia), the approximate alignment of the Great Dividing Range (• • •), broad climatic zones (sub-tropical to tropical: dark shading; temperate: light shading and semi-arid to arid: no shading), the division between coastal exorheic drainages and the endorheic/arheic inland and the extent of the Murray-Darling catchment (dashed lines).

wanaland and at the time Australia split from Antarctica, around the beginning of the Tertiary Period, much of the country had a mesic climate and was dominated by temperate forests (Smith, 1982; Frakes et al., 1987; Martin, 1990). From the mid-Tertiary, as Australia moved north, an irregular trend towards more xeric conditions commenced, with the wide climatic fluctuations of the Pleistocene probably having the most discernible influence on the present day distribution of the biota (Bowler, 1982). Temperate climates are now restricted to the south-east mainland (particularly along the Great Dividing Range), Tasmania and coastal areas of the south-west (Fig. 1). Aquatic habitats in these temperate areas have various hydrological regimes but generally have maximum depth/flow in winter and some habitats are permanently wet in the highest rainfall zones. Relief is low across most of the mainland, with the most significant uplands being the Great Dividing Range (Fig. 1), which divides short eastern coastal drainages from the largest exorheic drainage system (the Murray-Darling catchment: Fig. 1). Areas of Tasmania also have significant relief, contributing to the west of the state having the highest rainfall in the country. Most of inland Australia and the mid-south-coast presently has a hot semi-arid to arid climate (Fig. 1). Aquatic habitats in these drier regions are generally seasonal to episodic with little coordinated surface drainage, other than temporary streams in the Murray-Darling catchment (Fig. 1) and endorheic drainage basins mostly feeding into episodic lakes. The lower latitudes of northern Australia are sub-tropical to tropical (Fig. 1), and have aquatic habitats that are mostly non-permanent and fill/flow following summer rainfall, with most rivers drying to isolated permanent pools in winter.

Data sources, limitations and definitions

Most of the distribution data result from identification of specimens (mostly by the author and R. O. Brinkhurst) from projects reported in Timms (1973, 1977, 1978, 1981, 1983), Malipatil & Blyth (1982), Blyth et al. (1984), Marchant et al. (1982), Fulton (1983ab), Doeg (1989), Growns et al. (1992), Horwitz (1997), Kay et al. (1999) and Halse *et al.* (2000). Additional collections are from unpublished surveys and biological monitoring of the inland south-west of Western Australia by the Department of Conservation and Land Management; various regions of Tasmania by the Inland Fisheries Commission and H. B. N. Hynes; the Kakadu region of the Northern Territory by the Alligator Rivers Region Research Institute; mound springs and bores in arid South Australia and Queensland by the South Australian Museum; South Australian wetlands and rivers by the Australian Water Quality Centre, and surveys of south-western Australia, the Kosciusko alpine area of New South Wales and Tasmania by Richard Marchant, Ralph Brinkhurst and the author. The above studies account for about half of the more than 2000 oligochaete records and two-thirds of the more than 750 localities.

A dataset derived from such a diversity of projects inevitably has geographic, habitat and taxonomic biases. For Australian aquatic oligochaete data, the geographic and habitat biases are probably the greater concern as all oligochaetes (except enchytraeids) in a sample have usually been identified rather than just a particular subgroup. There are few or no data for large areas of the country (most notably northern Australia, but even much of New South Wales, Fig. 2) and for particular habitats in some regions (such as lowland wetlands in eastern Tasmania). For interpretation of distribution maps it is important to distinguish absences from lack of data. Thus, in Figures 3–6, sites where some oligochaetes, but not the taxon in question, have been found are shown or can be deduced by comparison with Figure 2 (which shows all localities). Sites where no oligochaetes (or only enchytraeids) were collected have not been included on the author's database. However, the latter would be few in number and would only be located in the regions which already have points on the maps rather than in the large blank areas.

The term 'endemic' is used in its broadest context as a convenient word for 'presently known only from . . .'. Where the 'south-east mainland' is mentioned in the text, this refers to the far south-east of South Australia, Victoria and eastern NSW along and east of the Great Dividing Range. Rainfall given in mm refers to average annual rainfall.

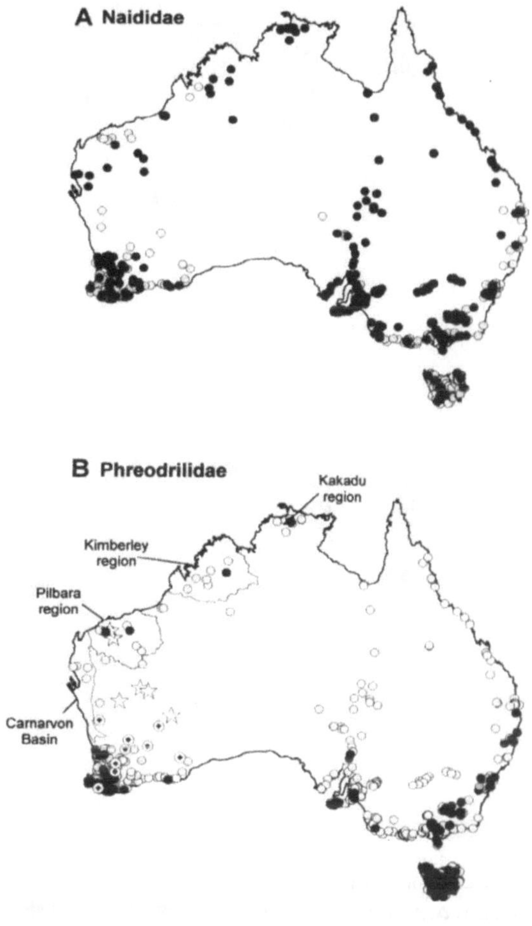

Figure 2. Map of Australia showing (a) records of naidids ●, and localities from which naidids were absent ○; (b) records of phreodrilids from granite outcrops ◉, from groundwater localities ☆, other surface water localities ● and localities from which phreodrilids were absent ○, with some regions mentioned in the text.

Naididae

General

The Naididae appears to be poorly represented in Australia, at both the specific and generic levels. Thirty eight species have been recorded (Table 1), compared with 58 in Africa (Grimm, 1987; Martinez-Ansemil

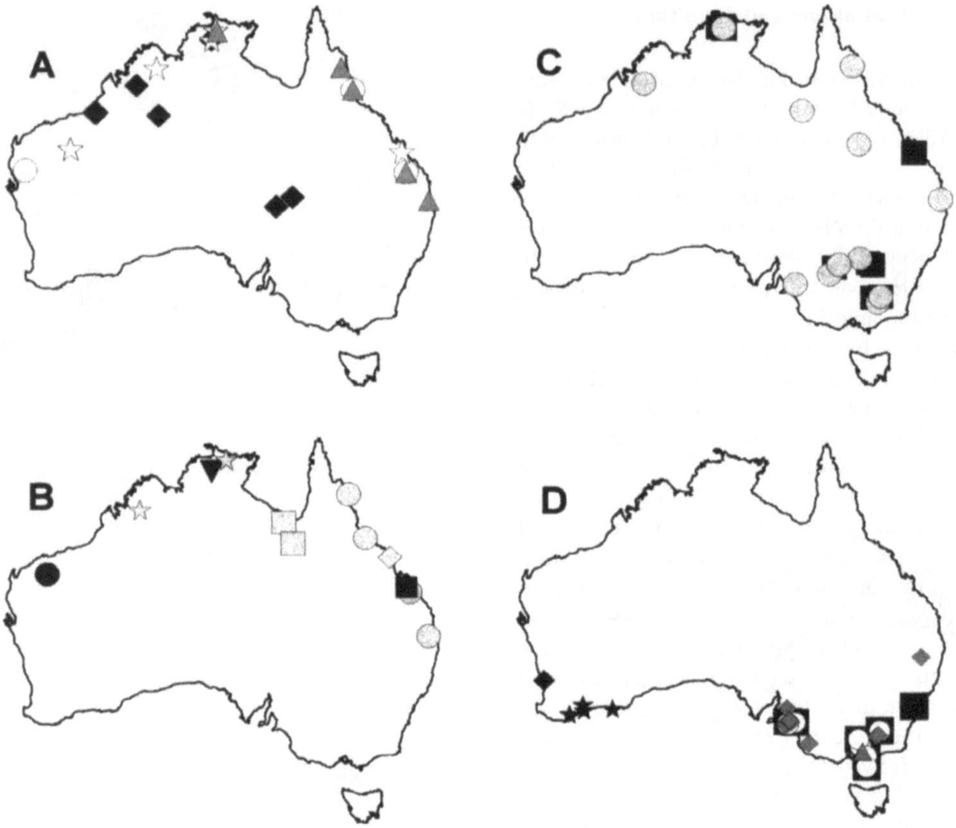

Figure 3. Records of naidids with northern/inland or southern temperate distributions (a) *Allonais ranauana* ♦, *A. paraguayensis* ☆, *A. inaequalis* ▲, *A. pectinata* ○ (b) *Dero vagus* ★, *D.* sp. WA1 ●, *D. pectinata* ■, *D. litoria* ●, *D.* sp. QLD1 ▣, *D. dorsalis* ♦ and both *D.* nr. *cooperi* and ?*Bratislavia* sp. NT1▼; (c) *Branchiodrilus hortensis* ■ and *Dero flabelliger* ●; (d) *Slavina appendiculata* ■, *S. proceriseta* ○, *Chaetogaster limnaei* ♦, *Pristina* sp. VIC1 ▲, *P.* sp. WA1 ♦ and ?*Bratislavia* sp. WA1 ★.

& Giani, 1987), 80 in South and Central America (Brinkhurst & Marchese, 1989), 70 in North America (Kathman & Brinkhurst, 1998), 56 along the Pacific coast of Asia (Timm, 1999) and 60 in Europe (Paoletti & Sambugar, 1996). Only 11 of the world's 22 genera (with *Pristinella* synonymised within *Pristina* by Collado & Schmelz (2000)) are represented in Australia, compared with 21 in both North America and Asia, 19 in Europe and 15 in both South America and Africa. A number of genera (*Uncinais*, *Specaria*, *Ripistes*, *Piguetiella* and *Arcteonais*) are essentially holarctic and their absence from Australia is not surprising. Others (*Vejdovskyella*, *Homochaeta*, *Ophidonais*, *Haemonais*, *Amphichaeta* and *Stephensoniana*) occur in the holarctic and either South America or Africa or both, and it is likely that some of these will be found in Australia. Additional naidids may yet be recorded from Australia, but it remains to be seen whether the Australian fauna will approach the diversity of other continents. Naidids tend to be globally widespread and endemicity is less than 20% for other continents (Timm, 1999). In Australia, only 2 of the 32 named Australian naidids are endemic: *Dero* (*Allodero*) *litoria* Pinder et al. (1999) and *Slavina proceriseta* Pinder & Brinkhurst (1998). However, several species have been collected that are not readily attributable to described species (Table 1) and may represent additional endemic species.

Naidids have been collected from virtually all areas of Australia from which oligochaetes have been identified (Fig. 2A) and many of the cosmopolitan species are widespread. Thus, *Pristina aequiseta*,

Table 1. Distribution of Naididae by state and territory. Australian apital Territory included with NSW and WA split into north (nWA) and south (SWA)

	Species	sWA	NSW	VIC	TAS	SA	nWA	NY	QLD
Allonais	*inaequalis* (Steph.)							+	+
	paraguayensis (Mich.)						+	+	+
	pectinata (Stephenson)						+		+
	ranauana (Mich. et Boldt)					+	+	+	
Branchiodrilus	*hortensis* (Stephenson)		+			+		+	+
?Bratislavia	sp. NT1							+	
	sp. WA1	+							
Chaetogaster	*diaphanus* (Gruithuisen)	+		+		+			
	diastrophus (Gruithuisen)	+	+	+		+			
	limnaei von Baer			+		+			
Dero	*digitata* (Müll)	+	+	+	+	+		+	+
	dorsalis Ferronniere								+
	litoria Pinder, Hill et Green								+
	nivea Aiyer	+		+	+	+		+	+
	nr. *cooperi* Steph.							+	
	pectinata Aiyer								+
	flabelliger (Steph.)		+	+		+	+	+	+
	furcata (Müll)	+	+	+		+	+	+	
	vagus Leidy							+	
	sp. WA1	+							
	sp. QLD1								+
Nais	*bretscheri* Mich.	+		+		+			
	communis / variabilis Piguet	+	+	+	+	+			+
	elinguis (Müll.)			+		+			+
	pseudobtusa Piguet								+
Paranais	*frici* Hrabe			+					
	litoralis (Müll.)	+		+	+	+			
Pristina	*aequiseta* Bourne	+	+	+		+		+	+
	longiseta Ehrenberg	+	+	+		+	+	+	+
	probiscidae Beddard	+		+		+		+	+
	jenkinae (Steph.)	+		+		+			
	osborni (Walton)			+		+		+	
	sp. WA1	+							
	sp. VIC1			+					
Slavina	*appendiculata* d'Udekem		+	+		+			
	proceriseta Pinder et Brink.			+		+			
Stylaria	*lacustris* (Linnaeus)		+	+		+			

Pristina longiseta, *Pristina proboscidea*, *Pristina osborni*, *Dero digitata*, *Dero furcata*, *Dero nivea*, *Nais elinguis* and *Stylaria lacustris* occur in most mainland states, except that the latter 2 species are yet to be collected in Western Australia. Other species are not so widespread, with some having a northern or southern distribution in Australia. To some extent, this distinction between the north and south is a gradation rather than a disjunction, with the interior of Australia having elements of both southern and northern faunas together with some of the widespread species.

Northern species

Species known only from northern Australia are the 4 *Allonais*, 6 of the *Dero* species, including species of the subgenus *Dero* (nr *cooperi*, *pectinata*, *dorsalis*, sp. WA1 and sp. QLD1) and subgenus *Aulophorus* (*vagus*), *Nais pseudobtusa* and an unidentified species tentatively known as ?*Bratislavia* NT1 (Fig. 3 A,B). Most of these are known only from the north and north-east coastal areas but most of the *Allonais* (*ranauana*, *pectinata* and *paraguayensis*) extend south into the more arid interior (Fig. 3A). *Dero* WA1 is known only from one groundwater sample in the Pilbara region (Fig. 3B). *Branchiodrilus hortensis* and *Dero (Aulophorus) flabelliger* also seem to have a northern and inland distribution, with southern records only from the Murray-Darling catchment (indicated in Fig. 1) rather than more temperate coastal areas of the south-east (Fig. 3C). *Dero* (*Allodero*) *litoria* (Fig. 3B) is an endosymbiont in the Green Tree Frog (*Litoria caerula*), but whether it is specific to this north-eastern frog species is not known. The tendency for *Allonais*, *Branchiodrilus* and *Dero* (particularly the subgenera *Aulophorus* and *Allodero*) to have a distributional bias towards lower latitudes has been recognised elsewhere (Sperber, 1948; Timm 1980, 1999; Grimm, 1987; Paoletti & Sambugar, 1996). Grimm (1987) and Paoletti & Sambugar (1996) also considered *Pristina* to be a tropical genus in Africa and a southern (warm climate) species in Italy, respectively, but this was based on how commonly the genus was encountered in collections rather than overall ranges of species or regional species richness. In Australia, the only *Pristina* that is possibly more common in the north than in the south is *P. longiseta*, though even this is far from rare in the south. Most other *Pristina* are widespread or are known only from southern Australia so far (Fig. 3D).

Southern species

Numerous species have been recorded only from the south-east and/or the south-west (Table 1). Most of these have been recorded in both the south-west and south-east. By contrast, *P. frici* (not shown in Fig. 3), *S. proceriseta*, *S. appendiculata*, *C. limnaei* and an unnamed species (*Pristina* VIC1) are known only from the south-east, while 2 un-named species (*Pristina* WA1 and ?*Bratislavia* WA1) have been recorded only from the south-west (Fig. 3D). Whether this apparent east–west restriction persists once naidid distributions are better mapped remains to be seen. If the 3 un-named species are new and both these and *S. proceriseta* are restricted to the south-west or south-east, this would fit a pattern of east–west endemism found within many invertebrate groups. Of course, some of the above species may turn out not to be restricted to the south once more northern sites are studied. This has already been shown for 2 species (*P. osborni* and *S. lacustris*) previously known only from southern Australia (Pinder & Brinkhurst, 1994) but now recorded from the Northern Territory (Erséus & Grimm, 1998; P. McEvoy, pers. comm.).

Tasmania

A striking anomaly in the distribution of naidids in Australia is their apparent low diversity in Tasmania, where there are just 10 records of *N. communis/variabilis*, a few estuarine records of *P. ?litoralis* and 1 record each of *D. nivea*, *D. digitata* and *Chaetogaster* sp. The family has been found at only 9.5% of 147 Tasmanian sites from which oligochaetes have been identified, compared to 56% of 564 mainland sites. Tasmania is much smaller than the mainland, but even in eastern Victoria, which is comparable in size and climate to Tasmania, 21 species have been recorded, with 1 or more species at 67% of the 40 sites. The absence of *B. hortensis* and *D. flabelliger*, which only seem to occur in the inland regions of the south-east, is not surprising, but the apparent absence of species that are very widespread on the mainland or at least common in the temperate south-east, such as most *Chaetogaster* and *Pristina*, *N. elinguis*, *S. lacustris* and *S. appendiculata*, is more perplexing. This low diversity in Tasmania may be real, and attributable to lack of suitable habitat, Tasmania's isolation and/or more severe climates during glacial times, but could equally reflect some sampling deficiency.

Learner et al. (1981) noted that the family is often "particularly prominent as components of the fauna of streams with stony substrata'. On the mainland several species (*N. communis/variabilis*, *N. elinguis*, *C. diaphanus*, *D. digitata*, *D. nivea*, several *Pristina* and *S. proceriseta*) have been collected from stony, mostly upland, streams. However, in equivalent Tasmanian streams only *N. communis/variabilis* has been collected, despite a strong sampling bias towards such habitats in the state. A possible explanation is that several of the south-east mainland rivers (particularly in Victoria) have been sampled more intensively in space and time than most Tasmanian rivers and species lists have expanded with repeated sampling. Also, worms from mainland sites have mostly been picked from preserved samples using microscopes, whereas most of those from Tasmania have been picked alive in the field.

More explicable is the lack of Tasmanian records of species (*P. longiseta*, *P. aequiseta*, *D. nivea*, *D. digitata*, *D. furcata*, *S. lacustris* and *C. diastrophus*) that have been collected predominantly from lentic waters and lowland rivers on the mainland. There are few data on oligochaetes from Tasmanian lowland rivers and most of the lentic sites that have been sampled are either coloured dystrophic swamps or the deeper parts of large lakes and reservoirs. However, few naidids penetrate into the depths of large water bodies (Learner et al. 1978) and the dystrophic swamps, largely dominated by emergent plants such as button grass, are possibly unsuitable for most naidids, many of which feed on diatoms, other algae and microflora and are often most abundant amongst submerged macrophytes (Learner *et al.* 1978). Significantly, all Tasmanian records of *Dero* and *Chaetogaster* were from 2 coastal wetlands with submerged macrophytes in the north-east. Further sampling in shallow, non-dystrophic freshwater wetlands with abundant macrophytes, such as floodplain wetlands, littoral areas of lowland lakes and reservoirs or coastal wetlands of the east and north-east may produce additional naidids in Tasmania.

For the present, additional sampling of likely naidid habitats and examination of existing ecological collections would seem to be a more circumspect approach, before concluding that naidid diversity in Tasmania is low and constructing hypotheses to explain this based on past environments and isolation of Tasmania. Palaeo-environmental hypotheses have been of little value for explaining the absence of other aquatic invertebrates from the state (Campbell, 1981). Besides, other oligochaete groups (phreodrilids and tubificids) appear to be particularly diverse in Tasmania, despite the island's history, and New Zealand has at least 9 cosmopolitan naidids (twice that of Tasmania) despite its greater isolation and history of extensive glaciation, volcanism and inundation.

Summary

From the above, it can be hypothesised that (1) the naidid fauna of Australia is depauperate at the specific and generic levels and mostly consists of cosmopolitan species, (2) there is a strong latitudinal influence on naidid species occurrence on the mainland, (3) there are some differences between the south-western and south-eastern naidid faunas and (4) there are fewer naidid species in Tasmania than on equivalently well surveyed areas of the mainland.

Phreodrilidae

General

Pinder & Brinkhurst (1997a) listed 27 phreodrilid species for Australia and subsequent surveys have revealed at least another 15 in Western Australia and 9 more in Tasmania, including the first estuarine species (Pinder & Erséus, 2000), so the number of species in Australia will certainly exceed 50 (Table 2). This is far greater than the 4 species known from South America, 7 from New Zealand, 6 from Africa and several from sub-antarctic islands, although oligochaetes remain poorly surveyed over significant parts of these landmasses. While most of the Australian species are endemic, *Phreodrilus branchiatus* also occurs in South America, *Insulodrilus lacustris* in South America and (possibly) New Zealand and *Antarctodrilus niger* in Africa and South America.

The global distribution of the family indicates a Gondwanan origin, although there is evidence of more recent dispersal (Jamieson, 1981; Pinder & Brinkhurst, 1997a). Within Australia, aquatic invertebrate groups with a presumed Gondwanan origin are often restricted to, or are at least more diverse in, the temperate southern regions: particularly Tasmania, the south-east mainland (especially along the Great Dividing Range) and, often to a lesser extent, south-western Australia. Increasingly however, Gondwanan elements (particularly Crustacea) are also being collected in refugial habitats in more arid parts of the country (Poore & Humphreys, 1998; Knott & Halse, 1999; Wilson & Keable, 1999). This distribution pattern is generally attributed to increasing aridity since the mid-Tertiary (Bowler, 1982; Martin 1990) and contraction of a mesic adapted biota to residual temperate regions or to refugial habitats. The overall distribution of the Australian Phreodrilidae generally fits this pattern (Fig. 2B, Table 2), with high diversity in Tasmania (29 species, 23 endemic) and the south-west mainland (17 species, 15 endemic), except that surprisingly few species are known from the south-east mainland (9 species, 4 endemic). The few records from the more arid inland and tropical north are mostly from granite outcrop pools or from refugial habitats such as groundwater aquifers, permanent river pools and caves.

Widespread species

Only 5 species (*P. branchiatus*, *Phreodrilus diemenensis*, *Antarctodrilus proboscidea*, *I. lacustris* and

Table 2. Distribution of Phreodilidae by state and territory. Australian Capital Territory included with NSW and WA split into north (nWA) and south (sWA)

	Species	sWA	NSW	VIC	TAS	SA	nWA	NT	QLD
Antarctodrilus	*acanthaseta* Pinder et Brink.				+				
	horwitzi Pinder et Brink.	+							
	micros Pinder et Brink.	+							
	niger (Beddard)				+				
	palustris (Brink. et Fulton)				+				
	probiscidea (Brink et Fulton)		+	+	+	+			
	uniseta Brink.				+				
Phreodrilus	*branchiatus* (Beddard)		+	+	+				
	diemenensis Pinder et Brink.			+	+				
	melaleucensis Pinder et Brink.				+				
	mitodes Pinder et Brink.				+				
Nesodrilus	*southwellensis* Pinder et Brink.				+				
Astacopsidrilus	*fusiformis* Goddard		+						
	jamiesoni Brink.								+
	myothyros Pinder et Brink.				+				
	notabilis Goddard		+						
	novus Jackson	+							
	plumaseta (Brink. et Fulton				+				
	ostiensis Pinder et. Erséus				+				
Insulodrilus	*bifudus* Pinder et Brink.	+							
	breviatria (Brink. et Fulton)				+				
	lacustris (Benham)	+		+	+	+			
	nudus (Brink. et Fulton)	+		+	+				
	magnaseta (Brink. et Fulton)				+				
	parviseta Pinder et Brink.				+				
	unisetoides Pinder et Brink.				+				
Phreodriloides	*notabilis* Benham		+						
Tasmaniaedrilus	*tasmaniaensis* Goddard sp. inq.				+				
Others	At least 24 new species	12			9	+?	3		
	Unidentified							+	

Insulodrilus nudus) can be considered widespread. These all occur in Tasmania and the south-east mainland and the 2 *Insulodrilus* also occur in the south-west of Western Australia. Most are common within their range, except for *P. diemenensis* which is known only from 1 site in Victoria and 1 site in Tasmania. Other Australian phreodrilids appear to have more restricted distributions.

Tasmania

Twenty nine phreodrilids are known from Tasmania. This figure includes the 5 widespread species listed above plus *A niger* and 23 species known only from Tasmania (including 9 undescribed). Most of these have been collected from upland streams, dystrophic wetlands and from the deeper parts of lakes, although there is a sampling bias towards such environments in the state. There are too few records of most species to comment on fine-scale biogeography within the state, but some coarse patterns are becoming apparent. Firstly, 4 of the 5 species that also occur on the mainland are widespread in Tasmania, whereas almost all of the species endemic to the state are each known from only 1 or 2 localities. Secondly, 14 of the 23 endemic species have been collected only in western Tasmania, which is cooler and wetter (>1200 mm pa, Figure 4) than most of eastern Tasmania. Of the remaining endemic species, all but 1 are either from a number of deep lakes on the Central Plateau (4 species, including 2 shared with western Tasmania)

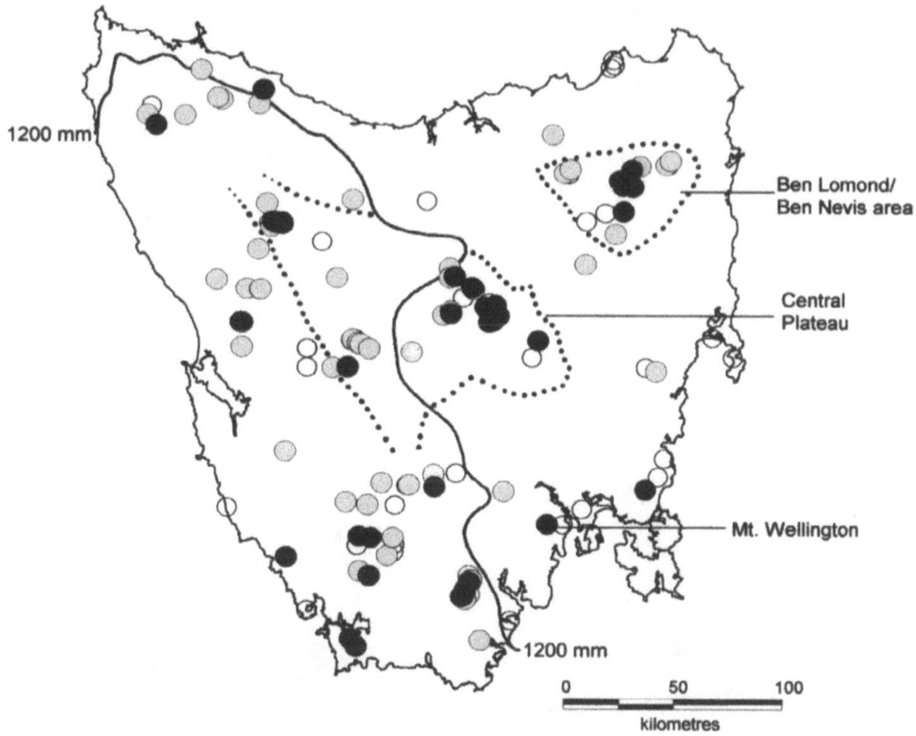

Figure 4. Map of Tasmania showing records of phreodrilid species endemic to Tasmania ●, of any of the five phreodrilids not endemic to Tasmania and localities from which phreodrilids were absent ○. 1200 mm average annual rainfall isohyet shown.

or from Mount Wellington near Hobart (1 species) or from the Ben Lomond/Ben Nevis region in the north-east (3 species, 1 shared with western Tasmania). These areas, marked on Figure 4, are all upland areas (>800 m) with fairly high rainfall (>900 mm). Thus far, the only Tasmanian endemic known from lowland areas of the east (a rain-shadow area) is an undescribed species from a forest stream in the far south-east. This apparently higher diversity of the family in more mesic areas of Tasmania could be interpreted as an artefact of the collecting bias in favour of western and upland areas (Fig. 4). Alternatively, greater diversity in mesic areas could be presented as a hypothesis to be tested by further sampling, particularly as there are corroborative patterns in other invertebrate groups. In particular, Shiel et al. (1989) coined the term Tyler's Line to denote a similarly greater endemism of rotifers in central and western Tasmania compared to the east, citing east–west differences in habitat type (salinity, colour and pH) as an explanation. Mesibov (1994, 1996) shows that this east–west division, which is essentially a "congruence of climatic, geologic, edaphic and vegetational change" (Tyler, 1992), also has significance for other invertebrate groups, though usually as a divide between allopatric species.

The mainland south-east

Compared to Tasmania and south-western Australia, the south-east mainland seems to have few phreodrilid species, just the 5 widespread species (as above) and 4 or 5 endemics. All of the widespread species occur in eastern Victoria. In addition, *A. proboscidea* and *P. branchiatus* extend along the Great Dividing Range to north-eastern New South Wales, *A. proboscidea* and *I. lacustris* also occur on Kangaroo Island (off South Australia), with *A. proboscidea* also found in streams of the Mt. Lofty Ranges on the adjacent mainland (Fig. 5). The only phreodrilids collected between eastern Victoria and the Mt. Lofty Ranges are a couple of immature unidentified specimens from a sinkhole near Mount Gambier in South Australia (Fig. 5), although a large proportion of sites sampled in western

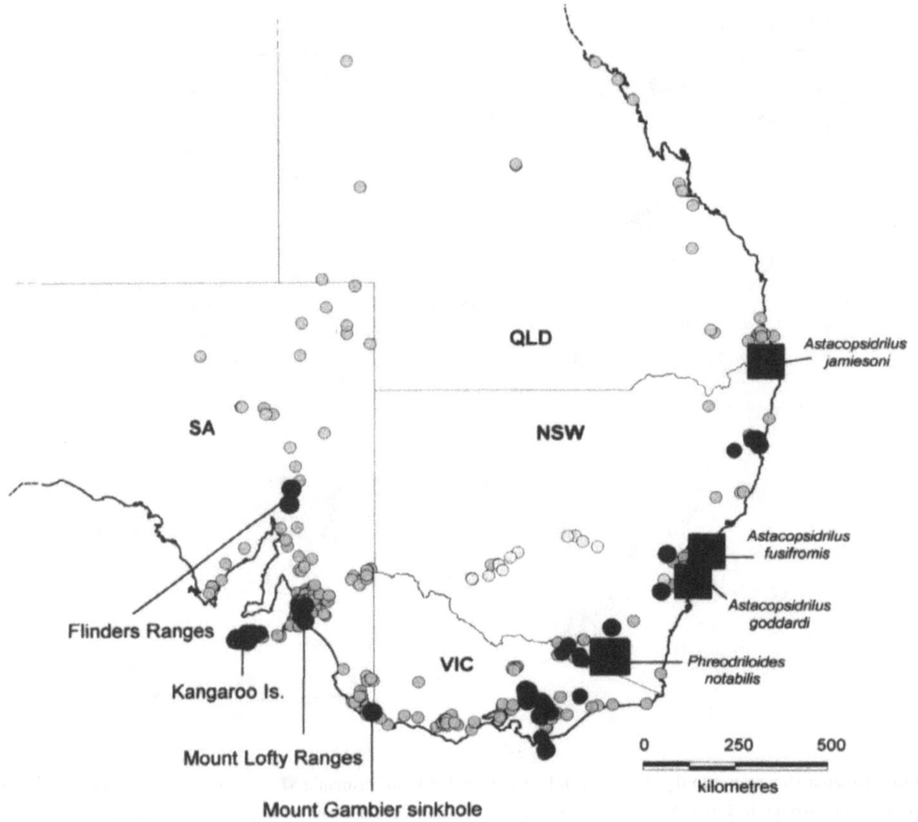

Figure 5. Map of south-eastern mainland Australia showing records of any of the five widespread phreodrilids not endemic to the south-east ●, of phreodrilids endemic to the south-east (species as indicated) ■ and localities from which phreodrilids were absent ◎.

Victoria are saline and unsuitable for the generally salt-intolerant phreodrilids. The only other South Australian records (and the only records from arid areas of the south-east) are of unidentified phreodrilids (1 possibly new) from 2 seasonal streams in the arid Flinders Ranges (Fig. 5). Invertebrates of adjacent Flinders Ranges streams have been shown to take refuge in the permanently wet hyporheic zone during the long summer dry periods (Cooling & Boulton, 1993) and this may also be true for the phreodrilids. There are very few species (*Phreodriloides notabilis* and 3 species of *Astacopsidrilus*) known only from the south-east mainland. The former is from lakes and streams of the Kosciusko alpine area of New South Wales, while the *Astacopsidrilus* are ectocommensal on *Euastacus* crayfish and have been collected only from coastal New South Wales or south-eastern Queensland (Fig. 5). The low diversity in the south-east mainland is unusual. Most aquatic invertebrate groups that have a southern temperate Australian bias to their distribution, such as most stonefly groups and some caddisflies (Hynes, 1989; Jackson, 1998), psephenid beetles (Davis, 1986), blepharicerid dipterans (Zwick, 1981) and some parastacid decapods (Horwitz, 1990), tend to have similar or greater diversity and endemicity in the south-east mainland compared to Tasmania. There are no ready explanations for the lack of phreodrilids in the mainland south-east other than inadequate collecting.

The number of localities from which oligochaetes have been identified is about the same for Tasmania and the south-east mainland and some mainland sites (especially freshwater rivers) have been sampled more intensively than most Tasmanian sites. However, the south-east mainland localities are mostly middle to lower reaches of larger rivers (especially the LaTrobe,

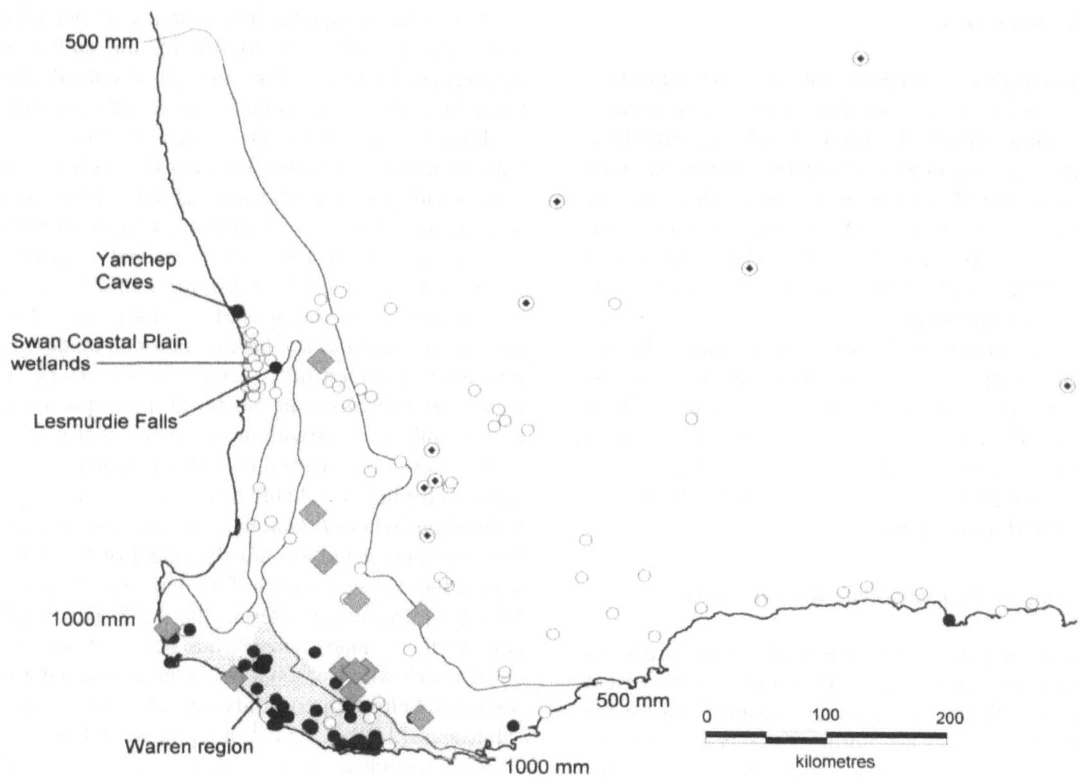

Figure 6. Map of south-western Australia showing records of phreodrilids from granite outcrops ◉, from other habitats •, except for where *I. bifidus* was the only phreodrilid present ◆ and localities from which phreodrilids were absent ○. 1000 and 500 mm average annual rainfall isohyets shown.

Thomson and Mitta Mitta Rivers), alpine lakes in the Kosciusko region of New South Wales (albeit with one of the endemic species) and largely disturbed streams in the greater Sydney area. It is possible that more widespread collecting will reveal further south-eastern phreodrilid diversity. The most likely places to find further phreodrilids should be the most mesic environments, such as streams and swamps in the most upland and/or high rainfall areas of eastern Victoria (including the lowland humic swamps of East Gippsland) and south-east New South Wales but possibly also streams of the Otway Ranges in western Victoria. Examination of specimens from these areas is certainly required before looking of explanations for the region's low phreodrilid diversity in palaeoenvironmental scenarios. Pleistocene glacial periods would have reduced aquatic habitat availability, at least in upland and inland areas (Bowler, 1983; DeDekker, 1986), but such events were more severe in central and western Tasmania and even the south-west of Western Australia would periodically have become more arid during glacial times. In any case, *P. notabilis* is known only from Pleistocene glacial lakes and associated streams of the Kosciusko Plateau and some Tasmanian phreodrilids are known only from glacially affected areas.

Western Australia and the Northern Territory

Until recently, *Astacopsidrilus novus* was the only phreodrilid known from Western Australia. Pinder & Brinkhurst (1997a) revealed the presence of the 2 widespread *Insulodrilus* (*lacustris* and *nudus*) and 6 new species (3 of which remain undescribed) and subsequent surveys have found another 12 undescribed species, bringing the total to 21 (Table 2). Eighteen of these (16 endemics plus *I. lacustris* and *I. nudus*) are known only from the south-west and 3 are known only from the Pilbara region (indicated in Fig. 2B).

The Warren region

Nine of the 18 species known only from the south-west (plus *I. nudus*) have been collected only from streams and shallow, mostly dystrophic heath and shrubland wetlands in a small area of the far south-west coast known as the Warren Region, sensu Thackway & Cresswell (1995) (Fig. 6). Other species also occur in, but are not restricted to, this region. This is the coolest and highest rainfall (1000–1400 mm) region of southern Western Australia, where aquatic habitats tend to be less seasonal than elsewhere in the state. This region also supports numerous other endemic flora and fauna, many of which also have Gondwanan origins (Main & Main, 1991; Hopper, 1992; Wardell-Johnson & Roberts, 1993; Trayler et al. 1996; Morgan et al. 1996; Wardell–Johnson & Horwitz, 1996; Hopper et al., 1996; Horwitz, 1997).

The south-west outside of the Warren region

In less mesic parts of the south-west, the occurrence of phreodrilids is more patchy. Phreodrilids were absent from 23 wetlands on the Swan Coastal Plain (in the Perth environs 200 km north of the Warren region, Fig. 6) sampled by Growns ct al. (1992). While this area has reasonably high rainfall (>800 mm), summers are hotter and drier than in the Warren region and wetlands are thus more seasonal. These wetlands are of holocene origin (Playford et al., 1976), though this itself should not be a reason to expect absence of phreodrilids, as similarly-aged coastal wetlands in the Warren region have numerous phreodrilid species. The only records of the family from the Swan Coastal Plain are of *I. lacustris* and *Antarctodrilus micros* from streams in caves north of Perth. Both of these have otherwise been found only in surface and cave streams in the Warren region. The type locality of *A. novus* (Lesmurdie Falls) is also near Perth but in a range of low hills (the Darling Range) running east of the coastal plain. This site is also the only locality outside of the Warren region from which another species (phreodrilid sp. WA3) has been found. *A. novus* was collected from a 'winter wet pool' at the foot of the cliff face together with blind (and so presumably phreatic) phreatoicid isopods (Jackson, 1931) and the WA3 specimen was collected in very small puddles on mossy ledges maintained by water seeping from the cliff. This suggests that both species are associated with groundwater at this site, although WA3 does occur in surface waters in the Warren region.

A survey of aquatic invertebrates of the inland south-west is being undertaken by the author and colleagues. In this region, rainfall steeply declines away from the coast, so that there is little permanent freshwater further than 100 km inland. Most aquatic habitats are seasonal saline drainage channels and wetlands, with fewer freshwater habitats. Phreodrilids were found at about half of the 14 freshwater sites (a seasonal stream and a few seasonal or permanent wetlands) in the 500–1000 mm rainfall zone (the western and southern edges of the study area). These were all *Insulodrilus bifidus* except for an unidentified immature specimen of a second species at one site. *Insulodrilus bifidus* is the most widespread phreodrilid in the south-west (as with most species, it also occurs in the Warren region) and is evidently able to tolerate seasonal drying of its habitat. As can be seen in Figure 6, localities where *I. bifidus* is the only phreodrilid to have been collected are mostly inland of the Warren region but not east of the 500 mm rainfall isohyet. Of the 41 freshwater sites in the <500 mm rainfall zone further inland, phreodrilids occurred at only 3, all of which were seasonal pools on granite outcrops, discussed below. Phreodrilids were absent from the 91 saline sites (TDS > 3 g l^{-1}, not plotted on Fig. 6).

As mentioned above, several new phreodrilids have recently been recorded from moss beds, associated seepages and temporary pools on granite outcrops (Fig. 6) (Pinder et al., 2000). These have been collected in both the Warren region (1 species from 1 outcrop) and the inland south-west (up to 5 species from 9 outcrops) by the author and D. H. D. Edward (The University of Western Australia). These are the only records of phreodrilids from aquatic habitats from the lower rainfall (<500 mm) inland south-west (Fig. 6). These habitats are completely dry for much of the year, apart from rare deep pools and are thus very different from most phreodrilid habitats, which tend to be permanent or at least likely to have moist refuges during drier periods.

Northern Australia

Further north, in the arid Carnarvon Basin (rainfall 200–300 mm), along the mid-west coast (Fig. 2B), only naidids, enchytraeids and tubificids were recorded from the mostly temporary wetlands sampled by Halse et al. (2000). Further north still, rivers in the hot semi-arid Pilbara region (rainfall 200–300 mm) and subtropical monsoonal Kimberley region (rainfall 400–1800 mm) (Fig. 2B) are generally seasonal to episodic, with water remaining in isolated pools when flow ceases. A recent biomonitoring project (Kay et al., 1999) revealed 3 new phreodrilids in water trickling through cobble beds associated with permanent groundwater-fed pools of 2 Pilbara rivers and an immature phreodrilid from a permanent pool in the Kimberley region. Samples from another 13 Pilbara and Kimberley sites (including other permanent pools) contained only naidids, but with hundreds of samples still unprocessed the picture is far from complete in the north-west. The only other phreodrilids from surface waters of non-temperate Australia are from the granite outcrops of the south-west and streams of the Flinders Ranges (both discussed above), plus a fragment of what appears to be a phreodrilid from a stream on an extensive coastal floodplain of the Northern Territory's Kakadu region (Fig. 2B). Forty one other collecting events on the floodplain, including 25 from the same stream (Magela Creek), resulted in records of numerous naidids and tubificids but no other phreodrilids. The floodplain streams originate on an eroded sandstone escarpment with many springs and seepages. These phreatic habitats support a high diversity of phreatoicid isopods (G. Wilson, Australian Museum, pers. comm.) and are possibly the origin of the phreodrilid.

Another new finding is that phreodrilids have been collected from calcrete groundwater aquifers in buried Tertiary palaeorivers in central Western Australia and the Pilbara region by W. F. Humphreys (Western Australian Museum) and B. Knott (The University of Western Australia) (Fig. 2B). These co-occur with other phreatic fauna, some of which also have Gondwanan affinities (Poore & Humphreys, 1998). Unfortunately, only immature specimens have been recovered so far. Phreodrilids were absent from a group of mound springs in the Kimberley region (sampled by A. Storey, The University of Western Australia) and from numerous bores and springs within the Lake Eyre catchment (sampled by W. Zeidler, South Australian Museum). However, the aquifer from which the latter arise is hotter, deeper and much older than the Pilbara aquifers (Wopfner & Twidale, 1971; Habermehl, 1980). Phreodrilids have also been collected from groundwater in New Zealand (Beddard, 1891) and north Africa (Giani et al., 1995).

The above records from warmer arid and subtropical regions of the country have only recently been documented and suggest that a lesser emphasis on the family's southern distribution is now required. However, phreodrilids in these non-temperate regions appear to be patchily distributed as they still seem to require permanent aquatic environments, with the notable exception of granite outcrop species. These records therefore refine, rather than invalidate, the model of phreodrilids as a mesic-adapted family of Gondwanan origin.

Summary

Considering the above, it can be hypothesised that (1) phreodrilids are still more common and diverse in southern temperate Australia, with Tasmania (especially the western and upland areas) and south-western Australia (especially the Warren region) having the highest diversity, (2) the south-east mainland has a depauperate phreodrilid fauna compared to other temperate areas, (3) drought resistant life-stages are largely restricted to species occurring on granite outcrops in southern Australia and (4) other phreodrilids in inland and northern Australia occupy sites and habitats that provide refuges from aridity.

Conclusions

The above summaries stand as interim conclusions about the broader distributions of the respective families in Australia. The main conclusion, however, is that better geographic coverage and more taxonomic work is required before anything like a robust picture of Australian oligochaete distributions can be presented.

Acknowledgements

I would like to acknowledge Dr S. A. Halse (Dept. Conservation and Land Management, Western Australia) and Dr R. Marchant (Museum of Victoria) for financial assistance to attend the 8th International Symposium on Aquatic Oligochaeta and support for this research in innumerable ways. Much of the field and laboratory work has been funded by grants from the Australian Biological Resources Study to Dr R. O. Brinkhurst and Dr R. Marchant. Data from the wheatbelt region of Western Australia is derived from a biological survey which is part of the State Salinity Strategy. I would also like to thank my mentor and compadre Dr Ralph Brinkhurst for a wonderful decade of collaboration. Sources of specimens are acknowledged in previous publications, with the notable exception of Paul McEvoy of the Australian Water

Quality Centre, South Australia who kindly provided a large dataset from that state.

References

Beddard, F. E., 1891. Anatomical descriptions of two new genera of aquatic Oligochaeta. Trans. r. Soc. Edinb. 36: 273–303.

Benham, W. B., 1907. On the Oligochaeta from the Blue Lake, Mount Kosciusko. Rec. aust Mus. 6: 251–264.

Blyth, J.D., T. J. Doeg & R. M. St Clair, 1984. Response of the macroinvertebrate fauna of the Mitta Mitta River, Victoria, to the construction and operation of Dartmouth Dam. 1. Construction and initial filling period. Occ.Pap. Mus. Vic. 1: 83–100.

Bowler, J. M., 1982. Aridity in the late Tertiary and Quaternary of Australia. In Barker, W. R. & P. J. M. Greenslade (eds), Evolution of the Flora and Fauna of Arid Australia. Peacock Publications, Frewville: 35–45.

Bowler, J. M., 1983. Lunettes as indices of hydrologic change: a review of Australian evidence. Proc. r. Soc. Vic. 95: 147–168.

Brinkhurst, R. O., 1971. The aquatic Oligochaeta known from Australia, New Zealand, Tasmania, and the adjacent islands. Univ. Queensland Pap. 3: 99–128.

Brinkhurst, R. O., 1982. Additional aquatic Oligochaeta from Australia and New Zealand. Rec. Queen Vic. Mus. 78: 1–13.

Brinkhurst, R. O., 1984. Two new species of Tubificidae (Oligochaeta) from the Northern Territory of Australia. Proc. Biol. Soc. Wash. 97: 142–147.

Brinkhurst, R. O. & W. Fulton, 1979. Some aquatic Oligochaeta from Tasmania. Rec. Queen Vic. Mus. 64: 1–13.

Brinkhurst, R. O. & W. Fulton, 1980. On *Haplotaxis ornamentus* sp. nov. (Oligochaeta, Haplotaxidae) from Tasmania. Rec. Queen Vic. Mus. 72: 1–8.

Brinkhurst, R. O. & M. R. Marchese, 1989. Guide to the Freshwater Aquatic Oligochaeta of South and Central America. Asociacion Ciencias Naturales del Litoral, Santo Tome, Argentina: 179 pp.

Campbell, I. C., 1981. Biogeography of some rheophilous aquatic insects in the Australian region. Aquat. Insects 3: 33–43.

Collado, R. & R. M. Schmelz, 2000. *Pristina silvicola* and *Pristina terrena* spp. nov., two new soil dwelling species of Naididae (Oligochaeta, Annelida) from the tropical rain forest near Manaus, Brazil, with comments on the genus *Pristinella*. J. Zool. Lond. 251: 509–516.

Cooling, M. P. & A. J. Boulton, 1993. Aspects of the hyporheic zone below the terminus of a South Australian arid-zone stream. Aust. J. mar. Freshwat. Res. 44: 411–426.

Cooling, M. P. & A. J. Boulton, 1993. Aspects of the hyporheic zone below the terminus of a South Australian arid-zone stream. Aust. J. mar. Freshwat. Res. 44: 411–426.

Davies, O. B., 1913. On two new species of *Chaetogaster*. Proc. r. Soc. Vic. 26: 88–98.

Davis, J. A., 1986. Revision of the Australian Psephenidae (Coleoptera): systematics, Phylogeny and historical biogeography. Aust. J. Zool. Suppl. 119: 1–97.

DeDekker, P., 1986. What happened to the Australian aquatic biota 18 000 years ago? In DeDekker, P. & W. D. Williams (eds), Limnology in Australia. CSIRO/Dr. W. Junk Publishers, Melbourne: 487–507.

Doeg, T.J., 1989. Colonization of experimentally disturbed patches by stream macroinvetebrates in the Acheron River. Aust. J. Ecol. 14: 207–220.

Dyne, G. R. & D. W. Walton (eds), 1987. Fauna of Australia. Volume 1a. General Articles. Australian Government Publishing Service, Canberra: 339 pp.

Erséus, C., 1993. Taxonomy of *Capilloventer* (Capilloventridae), a little-known group of aquatic Oligochaeta, with redescriptions of two new species. J. nat. Hist. 27: 1029–1040.

Erséus, C., 1997. The marine Tubificidae (Oligochaeta) of Darwin Harbour, Northern Territory, Australia, with descriptions of fifteen new species. In Hanley, R. H., G. Caswell, D. Megirian & H. K. Larson (eds), The Marine Flora and Fauna of Darwin Harbour, Northern Territory, Australia. Proceedings of the Sixth International Marine Biological Workshop, Darwin. Museums and Art Galleries Association of Australia and the Australian Marine Sciences Association, Darwin: 99–132.

Erséus, C. & R. Grimm, 1998. *Pristina proboscidea* and *Pristinella osborni* (Oligochaeta, Naididae) from a freshwater creek near Darwin, Northern Territory, Australia, with description of the genital organs of both species. The Beagle 14: 149–158.

Frakes, L. A., B. McGowran & J. M. Bowler, 1987. Evolution of Australian Environments. In Gyne, D. R. & D. W. Walton (eds), Fauna of Australia. Volume 1a. General Articles. Australian Government Publishing Service, Canberra: 1–16.

Fulton, W., 1983a. Macrobenthic fauna of Great Lake, Arthurs Lake and Lake Sorell, Tasmania. Aust. J. mar. Freshwat. Res. 34: 775–785.

Fulton, W., 1983b. Qualitative and quantitative variation in the macrobenthic fauna of the original lake and new lake areas of Great Lake and Arthurs Lake, Tasmania. Aust. J. mar. Freshwat. Res. 34: 787–803.

Gentilli, J., 1972. Australian Climate Patterns. Nelson, Melbourne: 285 pp.

Giani, N., P. Martin & J. Juget, 1995. A new species of Phreodrilidae (Oligochaeta), Astacopsidrilus naceri sp. nov., from Morocco (North Africa), with notes on the biogeography of the family. Can. J. Zool. 73: 2375–2381.

Goddard, E. J., 1909a. Contribution to a further knowledge of Australian Oligochaeta. Part I. Description of two species of a new genus of Phreodrilidae. Proc. linn. Soc. N. S. W. 33: 768–793.

Goddard, E. J., 1909b. Contribution to a further knowledge of Australian Oligochaeta. Part II. Description of a Tasmanian phreodrilid. Proc. linn. Soc. N. S. W. 33: 845–866.

Grimm, R., 1987. Contributions towards the taxonomy of the African Naididae (Oligochaeta). IV. Zoogeographical and taxonomical considerations of African Naididae. Hydrobiologia 155: 27–37.

Growns, J. E., J. A. Davis, F. Cheal, L. G. Schmidt, R. S. Rosich & S. J. Bradley, 1992. Multivariate pattern analysis of wetland invertebrate communities and environmental variables in Western Australia. Aust. J. Ecol. 17: 275–288.

Habermehl, M. A., 1980. The Great Artesian Basin, Australia. Bureau Min. Res. J. Aust. Geol. Geophys. 5: 9–38.

Halse, S. A., R. J. Shiel, A. W. Storey, D. H. D. Edward, I. Lansbury, D. J. Cale & M. S. Harvey, 2000. Aquatic invertebrates and waterbirds of wetlands and rivers of the southern Carnarvon Basin, Western Australia. Rec. W. Aust. Mus., Suppl. 61: 217–265.

Hopper, S. D., 1992. Patterns of plant diversity at the population and species levels in south-western Australian Mediterranean ecosystems. In Hobbs, R. J. (ed.), Biodiversity in Mediterranean Ecosystems in Australia. Surrey Beaty & Sons, Chipping Norton: 27–46.

Hopper, S. J., M. S. Harvey, J. A. Chappill, A. R. Main & B. Y. Main, 1996. The Western Australian biota as Gondwanan heritage. In Hopper, S. J., J. Chappill, M. S. Harvey & A. George (eds), Gondwanan Heritage: Past Present and Future of the Western Australian Biota. Surrey Beaty & Sons, Chipping Norton: 1–46.

Horwitz, P., 1990. A taxonomic revision of species in the freshwater crayfish genus *Engaeus* Erichson (Decapoda: Parastacidae). Invert. Tax. 4: 427–614.

Horwitz, P., 1997. Comparative endemism and richness of the aquatic invertebrate fauna in peatlands and shrublands of far south-western Australia. Mem. Mus. Vic. 56: 313–321.

Hynes, H. B. N., 1989. Tasmanian Plecoptera. Australian Society for Limnology Special Publication No. 8: 81 pp.

Jackson, A., 1931. The Oligochaeta of South-Western Australia. J. r. Soc. W. Aust. 17: 71–136.

Jackson, J., 1998. Preliminary guide to the identification of late instar larvae of Australian Calocidae, Helicophidae and Coenoesucidae (Insecta, Trichoptera). Identification Guide 16. Cooperative Research Centre for Freshwater Ecology, Albury: 81 pp.

Jamieson, B. G. M., 1978. *Rhyacodrilus arthingtonae*, a new species of freshwater oligochaete (Tubificidae) from North Stradbroke Island, Queensland. Proc. r. Soc. Qld 89: 39–43.

Jamieson, B. G. M., 1981. Historical biogeography of Australian Oligochaeta. In Keast, A. (ed.), Ecological Biogeography of Australia. Dr. W. Junk Publishers, The Hague: 885–921.

Kathman, R. D. & R. O. Brinkhurst, 1998. Guide to the Freshwater Oligochaetes of North America. Aquatic Resources Center, Tennessee: 264 pp.

Kay, W. R., M. J. Smith, A. M. Pinder, J. M. McRae, J. A. Davis & S. A. Halse, 1999. Patterns of distribution of macroinvertebrate families in rivers of north-western Australia. Freshwat. Biol. 41: 299–316.

Keast, A. (ed.), 1981. Ecological Biogeography of Australia. Dr. W. Junk Publishers, The Hague: 2142 pp.

Knott, B. & S. A. Halse, 1999. *Pilbaraophreatoicus platyarthricus* n. gen., n. sp. (Isopoda: Phreatoicidea: Amphisopodidae) from the Pilbara region of Western Australia. Rec. aust. Mus. 51: 33–42.

Learner, M. A., G. Lochhead & B. D. Hughes, 1978. A review of the biology of British Naididae (Oligochaeta) with emphasis on the lotic environment. Freshwat. Biol. 8: 357–375.

Main, B. Y. & A. Main, 1991. Report on the Southern Forest Region of Western Australia. Gondwanic Faunal Elements of the Southern Forest Region-Western Australia. The Australian Heritage Commission, Canberra: 20 pp.

Malipatil, M. B. & J.D. Blyth, 1982. A qualitative study of the macroinvertebrate fauna of the Thomson River and its major tributaries, Gippsland, Victoria. Rep. Nat. Mus. Vic. 1: 1–95.

Marchant, R. 1982. The distribution of macroinvertebrates along the lower reach of the LaTrobe River and its response to disturbance. LaTrobe Valley Water Resources Biological Studies, Volume 5. SEC Planning Investigations Department, Melbourne.

Martin, H. A., 1990. Tertiary climate and phytogeography in southeast Australia. Palaeobot., Palynol. 65: 47–55.

Martinez–Ansemil, E. & C. Giani, 1987. The distribution of aquatic oligochaetes in the south and eastern Mediterranean area. Hydrobiologia 155: 293–304.

Mesibov, R. 1994. Faunal breaks in Tasmania and their significance for invertebrate conservation. Mem. queensland Mus. 36: 133–136.

Mesibov, R., 1996. Invertebrate bioregions in Tasmania. Report to the Tasmanian RFA Environment and Heritage Technical Committee. Queen Victoria Museum and Art Gallery, Launceston: 50 pp.

Michaelsen, W., 1907. Die Fauna Sudwest-Australiens. Ergenb. Hamb. sudwest Austral. Forsch. 1: 1–68.

Morgan, D., H. Gill & I. C. Potter, 1996. The Distribution of Freshwater Fish in the South-Western Corner of Australia. Water and Rivers Commission, Perth: 26 pp.

Naidu, K. V. & K. A. Naidu, 1980a. *Nais pseudobtusa* Piguet, 1906 (Oligochaeta: Naididae) new to Australia. Hydrobiologia 68: 91–92.

Naidu, K. V. & K. A. Naidu, 1980b. Two species of Phreodrilidae (Oligochaeta) new to Australia and Tasmania. Hydrobiologia 75: 179–180.

Nicholls, G. E., 1921. On a new species of naidiform worm, *Dero roseola*. J. r. Soc. W. Aust. 7: 90–94.

Paoletti, A. de C. & B. Sambugar, 1996. Aquatic Oligochaeta in Italy, with special reference to the Naididae. Hydrobiologia 334: 37–49.

Pinder, A. M. & R. O. Brinkhurst, 1994. A Preliminary Guide to the Identification of the Microdrile Oligochaeta of Australian Inland Waters. Identification Guide No. 1. Cooperative Research Centre for Freshwater Ecology, Albury: 137 pp.

Pinder, A. M. & R. O. Brinkhurst, 1997a. A review of the Phreodrilidae (Annelida: Oligochaeta: Tubificida) of Australia. Inv. Tax. 11: 443–523.

Pinder, A. M. & R. O. Brinkhurst, 1997b. The family Capilloventridae (Annelida, Clitellata) in Australia with descriptions of two new species of *Capilloventer*. Zool. Scr. 26: 255–265.

Pinder, A. M. & R. O. Brinkhurst, 1998. First records of *Slavina* (Oligochaeta: Naididae) in Australia and description of a new species. Proc. r. Soc. Vic. 109: 149–155.

Pinder, A. M. & R. O. Brinkhurst, 2000. A review of the Tubificidae (Annelida: Oligochaeta) from Australian inland waters. Mem. Mus. Vic. 58: 39–75.

Pinder, A. M. & C. Erséus, 2000. New Phreodrilidae (Annelida: Clitellata) from Tasmanian estuaries. Pap. Proc. r. Soc. Tas. 134: 29–33.

Pinder, A. M., B. D. Hill & P.E. Green, 1998. A new species of *Dero* (*Allodero*) (Naididae: Oligochaeta) inhabiting the Wolffian Ducts of the Green Tree Frog (*Litoria caerula*) in Queensland. Mem. queensland Mus. 42: 559–564.

Pinder, A. M., S. A. Halse, R. J. Shiel & J. M. McRae, 2000. Granite outcrop pools in south-western Australia: foci for diversification and refugia for aquatic invertebrates. J. r. Soc. W. Aust. 83: 117–129.

Playford, P. E., A. E. Cockburn & G. H. Low, 1976. Geology of the Perth Basin Western Australia. Geological Survey of western Australia Bulletin 124: 1–311.

Poore, G. C. B. & W. F. Humphreys, 1998. First records of Spelaeogriphacea from Australasia: a new genus and species from an aquifer in the arid Pilbara of Western Australia. Crustaceana 71: 721–742.

Shiel, R. J., W. Koste & L. W. Tan, 1989. Tasmania revisited: rotifer communities and habitat heterogeneity. Hydrobiologia 186/187: 239–245.

Smith, J. M. B., 1982. A history of Australasian Vegetation. McGraw-Hill Book Company, Sydney: 168 pp.

Sperber, C., 1948. A taxonomical study of the Naididae. Zool. Bid. Fran Upps. 28: 1–296.

Thackway, R. & I. D. Cresswell, 1995. An Interim Biogeographic Regionalisation for Australia: A Framework for Setting Priorities in the National Reserves System Cooperative Program, Version 4.0. Australian Nature Conservation Agency, Canberra: 95 pp.

Timm, T., 1980. Distribution of aquatic oligochaetes. In Brinkhurst, R. O. & D. G. Cook (eds), Aquatic Oligochaete Biology. Plenum Press, New York: 55–77.

Timm, T., 1999. Distribution of freshwater oligochaetes in the west and east coastal regions of the North Pacific Ocean. Hydrobiologia 406: 67–81.

Timms, B. V., 1973. A limnological survey of freshwater coastal lakes of east Gippsland, Victoria. Aust. J. mar. Freshwat. Res. 24: 1–20.

Timms, B. V., 1977. A study of some coastal dune lakes in western Victoria. Proc. r. Soc. Vic. 89: 167–172.

Timms, B. V., 1978. The benthos of seven lakes in Tasmania. Arch. Hydrobiol. 81: 422–444.

Timms, B. V., 1981. Animal communities in three lakes of differing salinity. Hydrobiologia 81: 181–193.

Timms, B. V., 1983. A study of benthic communities in some shallow saline lakes of western Victoria, Australia. Hydrobiologia 105: 165–168.

Trayler, K. M., J. A. Davis, P. Horwitz & D. Morgan, 1996. Aquatic fauna of the Warren bioregion, south-west Western Australia: does reservation guarantee preservation. J. r. Soc. W. Aust. 79: 291–291.

Tyler, P. A., 1992. A lakeland from the dreamtime. The second Founder's Lecture. Br. Phycol. J. 27: 353–368.

Wardell–Johnson, G. & P. Horwitz, 1996. Conserving biodiversity and the recognition of heterogeneity in ancient landscapes: a case study from south-western Australia. For. Ecol. Man. 85: 219–238.

Wardell–Johnson, G. & J. D. Roberts, 1993. Biogeographic patterns in a subdued landscape: the distribution of the *Geocrinea rosea* (Anura: Myobatrchidae) complex in south-western Australia. Aust. J. Biogeog. 19: 95–108.

Wilson, G. D. F. & S. J. Keable, 1999. A new genus of phreatoicidean isopod (Crustacea) from the north Kimberley region, Western Australia. Zool. J. linn. Soc. 126: 51–79.

Wopfner, H. & C. R. Twidale, 1971. Geomorphological history of the Lake Eyre Basin. In Jennings, J. N. & J. Mabbutt (eds), Landform Studies from Australia and New Guinea. Australian National University Press, Canberra.

Zwick, P., 1981. Blephariceridae. In Keast, A. (ed.), Ecological Biogeography of Australia. Dr. W. Junk Publishers, The Hague: 1184–1193.

Hydrobiologia **463:** 65–74, 2001.
P. Rodriguez & P.F.M. Verdonschot (eds), Aquatic Oligochaete Biology VIII.

Oligochaetes in southern European groundwater: new records and an overview

Narcisse Giani[1], Beatrice Sambugar[2], Pilar Rodriguez[3] & Enrique Martínez-Ansemil[4]
[1]*CESAC, UMR CNRS 5576, Université Paul-Sabatier, 118 route de Narbonne, 31062 Toulouse Cedex 4, France*
[2]*Museo Civico di Storia naturale, Lungadige Porta Vittoria 9, 37129 Verona, Italy*
[3]*Departamento de Biología Animal y Genética, Universidad del País Vasco, Apdo 644, 48080 Bilbao, Spain*
[4]*Departamento de Bioloxía Animal, Bioloxía Vexetal e Ecoloxía, Universidade da Coruña, Campus de Zapateira s/n, 15071 A Coruña, Spain*

Key words: groundwater, aquatic oligochaetes, biodiversity, stygobiont, stygophile, southern Europe

Abstract

This work deals with an overview of the faunal composition and presents a general ecological approach to the groundwater oligochaetes in southern Europe. The species richness is high and similar to that of the superficial waters in the same area. Among the 150 species that we found, about 29 are stygobiont, so the total number of stygobiont species in the area studied has been increased to 58. Ten of these species are tubificids with marine phyletic affinities. A list of the most frequently found species and the relative abundance of the different families and genera collected in the main environments investigated are presented. Although enchytraeids often dominated the fauna, lumbriculids and tubificids exhibit a high level of diversity and comprised the majority of stygobiont and stygophilous taxa found. The discovery of members of the family Parvidrilidae in southern European caves is the first outside of North America. This indicates that the family is widely distributed in the Palearctic region and typical of underground waters. Here, we update the previous evaluation of the knowledge about oligochaetes inhabiting groundwater: 96 species of stygobiont freshwater oligochaetes (*sensu stricto*) are now known in the world and 81 of them have been recorded from the Palearctic region.

Introduction

The two most recent surveys concerning groundwater oligochaetes from Southern Europe are included in the large works edited by Botosaneanu (1986) and Juberthie & Decu (1994). In an extensive survey of the underground fauna of the Iberian Peninsula, Bellés (1987) reviewed the few studies concerning aquatic oligochaetes of the region (Hrabě, 1963a; Delay, 1973; Giani & Martínez-Ansemil, 1981). Although still relatively poor, knowledge of groundwater oligochaetes in this area has been greatly improved over the last few years (Rodriguez & Giani, 1987, 1989, 1994; Giani & Rodriguez, 1988, 1994). The first data concerning France are rather old (Moniez, 1888; Fauvel, 1903; Vandel, 1920; Černosvitov, 1931, 1936; Hrabě, 1958; Juget, 1959; Delay, 1972; and others) but in the last two decades several new studies have been reported (Juget, 1984, 1987; Rodriguez & Giani, 1989, 1994; Giani et al., 1990; Lafont & Durbec, 1990; Lafont et al., 1992; Erséus et al., 1992; Sambugar et al., 1999). As far as Austria is concerned, although the hyporheic interstitial environment has been the object of in-depth studies (Danielopol, 1991; Danielopol et al., 1992), oligochaetes were not taken into account. Thus, Juget & Dumnicka (1986) in their review only reported one species for Austrian subterranean waters. Similarly in Greece, only a few species have been reported (Juget & Dumnicka, 1986; Dumnicka, 1990). Again, a recent review of the subterranean fauna of Italy (Cobolli-Sbordoni et al., 1994), did not consider aquatic oligochaetes. The data concerning groundwater oligochaetes in Italy are based on the studies of Stammer (1932), Hrabě (1936, 1942, 1966), Dumnicka (1981, 1987, 1990), Erséus & Dumnicka (1988), Martínez-Ansemil et al. (1997) and Sambugar et al. (1999) with information on the interstitial hyporheic environment being reported by Ferrarese & Sambugar

(1976) and Centurioni & Sambugar (1980). In Slovenia and Croatia, where karst areas and hypogean habitats are widespread, knowledge of subterranean fauna as a whole is considerably greater than the sparse records on groundwater oligochaetes (Hrabě, 1963b; Juget & Dumnicka, 1986; Karaman, 1983; Sket et al., 1994; Martínez-Ansemil et al., 1997; Sambugar et al., 1999).

These studies from the literature together with our results (including some ecological notes) obtained over almost 20 years in the Iberian Peninsula, France, Italy, Slovenia, Croatia, Austria and Greece, will be considered in this paper. This review will collate the available knowledge on oligochaetes inhabiting the groundwaters of Southern Europe, and establish a basis for future research into subterranean oligochaete fauna.

Sites and area studied

Samples of groundwater oligochaetes from 303 localities were studied. Some of them belong to our own collections and others were placed at our disposal by the following colleagues: B. Lebreton, C. Bou and P. Leclerc (French localities), R. Rouch and J. Notemboom (Iberian Peninsula) and F. Stoch (Italy, Slovenia, Austria, Croatia and Greece). The samples came from 219 caves (watercourses, lakes, siphons, rimstone pools in the vadose zone), 40 springs, 16 wells, 15 hyporheal environments of rivers and 13 resurgences). The total number of localities sampled was 47 in the Iberian Peninsula in 10 northern provinces; 28 in France in 10 southern 'départements'; 174 in Italy in 10 provinces, mainly in the northern part; 45 in Slovenia; 3 in Austria; 1 in Croatia and 5 in Greece (Fig. 1). The list of all the localities with their communities will be published in a future paper as well as a description of new species mainly belonging to the genera *Trichodrilus* Claparède and *Cernosvitoviella* Nielsen & Christensen.

Materials and methods

Different methods were used to sample the various subterranean environments: Bou–Rouch or Karaman–Chappuis methods for hyporeal watercourses, phreatobiological net of Cvetkov for wells, hand net of 100 μm mesh size for caves, resurgences and springs and a little pump to collect fauna in the vadose zone and in the rimstone pools.

Samples were fixed in the field with 4% formaldehyde and stored in 70% ethanol. Before identification, some of the specimens, when necessary, were stained with Erlich's haematoxylin and dissected under a stereo-microscope and then mounted in Canada balsam.

Results

About 150 species were identified by the authors in the groundwaters of the studied area, and 36 of them deserve special attention either being found only in subterranean waters (i.e. stygobionts) or being biogeographically, phylogenetically or ecologically, significant. Most seem to be true stygobionts. They are the following:

Lumbriculidae

Cookidrilus speluncaeus Rodriguez & Giani, 1987. Only known from the type locality, La Bouiche Cave, Ariège, France.

Trichodrilus angelieri Giani & Rodriguez, 1994. Known from the hyporheic waters of a tributary of San Pietro River, Corsica, France and from the Llobregat River, Barcelona, Spain (Giani & Rodriguez, 1994).

Trichodrilus capilliformis Rodriguez & Giani, 1994. Only known from the type locality, hyporheic waters of Fluvia River, Girona, Spain.

Trichodrilus longipenis Rodriguez & Giani, 1994. Only known from the type locality, hyporheic waters of Muga River, Girona, Spain.

Trichodrilus rouchi Rodriguez & Giani, 1994. Known from hyporheic waters of Alcandre, Huesca, of a tributary of the Montsant River, Lleida, of Vero and Muga Rivers, Girona, Spain (Rodriguez & Giani, 1994).

Tubificidae

Rhyacodrilus amphigenus Juget, 1987. Described in the hyporheic environment of the alluvial plain of the River Rhône, France (Juget, 1987). We also found it in the Aven de la Cuisinière Cave (03.22.1987), Aveyron, France.

Rhyacodrilus dolcei Martínez-Ansemil et al., 1997. Only known from the type locality, Carbone Cave, Venezia Giulia, Italy.

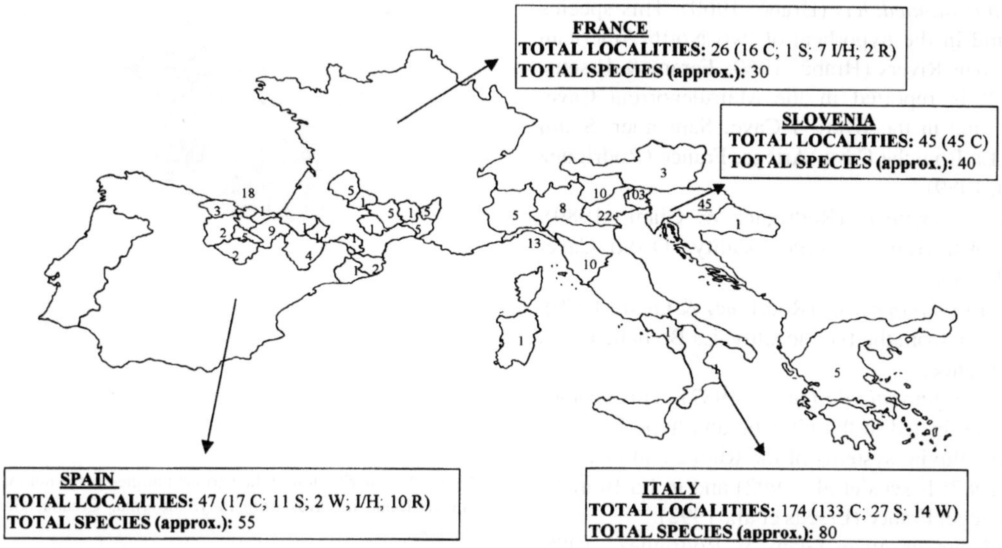

Figure 1. Number of sample sites in each region or country and total number of species found. C – caves; H – hyporheal water of rivers; R – resurgences; S – springs; W – wells.

Rhyacodrilus gasparoi Martínez-Ansemil et al., 1997. Only known from the type locality, Montefosca Cave, Friuli, Italy.

Rhyacodrilus gernikensis Giani & Rodriguez, 1988. Only known from the type locality, Rekalde karstic spring, Bizkaia, Spain.

Rhyacodrilus lindbergi Hrabě, 1963. Known from caves in Portugal (Hrabě, 1963a); in the present work it is reported from Seynes (03.03.1985) and Bedousse Cave (05.20.1983), Gard, France.

Rhyacodrilus okamikae Giani & Rodriguez, 1988. Only known from Laminak and Goiketxe Caves, where this species was described.

Rhyacodrilus omodeoi Martínez-Ansemil et al., 1997. Only known from the type locality, Krizna Cave, Slovenia.

Stochidrilus glandulosus Martínez-Ansemil et al., 1997. Only known from the type locality, Virsnica Cave, Radensko Polje, Slovenia.

Krenedrilus ibericus Giani et al., 1990. Only known from the type locality, the hyporheal flow of the small stream Porto do Cabo, Galicia, Spain.

Krenedrilus sergei Giani et al., 1990. Only known from the type locality, Dragonnières Cave, Aveyron, France.

Tubifex pescei (Dumnicka, 1981). Previously reported from wells in Marche, Italy (Dumnicka, 1981), this species was reported from a cave near Loz, Cernisko Jezero, Slovenia (Martínez-Ansemil et al., 1997).

Haber monfalconensis (Hrabě, 1966). Known from the type locality, Moschenizze and Sardos springs, Venezia Giulia, Italy, this species was found by us in Grotta Nuova di Villanova Cave (05.03.1987), Friuli and in Antro delle Ninfe Cave (09.10.1992), Venezia Giulia, Italy; in Kompoljska Cave (07.16.1994) and in Podpeska Cave (07.16.1994), Suha Krajina, Slovenia.

Tubificoides galarzai Giani & Rodriguez, 1988. Known from the type locality, the karstic spring of Argatxa and from the Goiketxe Cave, Bizkaia, Spain (Giani & Rodriguez, 1988).

Abyssidrilus cuspis (Erséus & Dumnicka, 1988). Known from Umbrian wells (Erséus & Dumnicka, 1988; Dumnicka, 1990), this species was found in Bocca Lupara Cave, Liguria, in Papipano Cave, Friuli, Italy, and in Aven de la Cuisinière Cave, Aveyron, France (Sambugar et al., 1999).

Abyssidrilus cf. *cuspis*. – Under this name, we ascribed some specimens differing slightly from the typical *A. cuspis* (Sambugar et al., 1999). They were found in Romana Cave, Venezia Giulia, Italy and in Skocjanske Cave, Slovenia.

Abyssidrilus subterraneus Rodriguez & Giani, 1989. Only known from the type locality, Jivero 1 Cave, Santander, Spain.

Gianius aquaedulcis (Hrabě, 1960). This species was found in the hyporheal of two North American and German Rivers (Hrabě, 1960; Farara & Erséus, 1991). It is reported in the Mairulegorreta Cave, Bizkaia, and in the Jivero 1 Cave, Santander, Spain and the La Bouiche Cave, Ariège, France (Rodriguez & Giani, 1989).

Gianius crypticus (Rodriguez & Giani, 1989). Only known from the type locality, Orillo Cave, Santander, Spain.

Gianius labouichensis (Rodriguez & Giani, 1989). Only known from the type locality, La Bouiche Cave, Ariège, France.

Spiridion phreaticola (Juget, 1987). This species has been collected in the phreatic and hyporheic waters of the fluvial systems of the Rhône and Garonne (Juget, 1987; Erséus et al., 1992) and in La Bouiche Cave, Ariège, France (Erséus et al., 1992).

Aktedrilus argatxae Giani & Rodriguez, 1988. Only known from the type locality, the karstic spring of Argatxa, Bizkaia, Spain.

Aktedrilus ruffoi Sambugar et al., 1999. Only known from the type locality, the spring of the Tione River, Veneto, Italy.

In addition to these species, two new species belonging to the recently described family Parvidrilidae Erséus, 1999 were found during this work in several localities (unpublished data) and can be considered as true stygobionts.

Following Juget & Dumnicka (1986), some species, despite the fact that they have also been collected in superficial waters, can be considered as stygobionts due to their abundance or their constancy in underground water communities. This is probably the case for the first two below, the two others being stygophiles:

Lumbriculidae

Trichodrilus campoyi Rodriguez, 1988. Known from many watercourses of Northeastern Spain (Rodriguez, 1988) and France (Rodriguez & Giani, 1994), this species is now reported in subterranean waters: from the Okamika, Goiketxe and Bolunta (resurgence of Omaerreka) Caves, from the spring of Omaerreka River, Bizkaia, from resurgence I of the Larrun River, Navarra, Spain and from the resurgence of Lantouy (03.22.1987), Aveyron, France.

Trichodrilus diversisetosus Rodriguez, 1986. Described from a Spanish River and further collected in the San Pedro de Busturia Cave, Bizkaia, Spain and

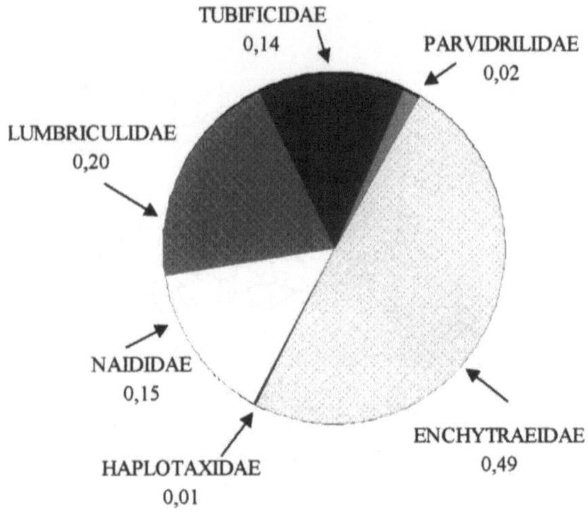

Figure 2. Contribution of the various families to the number of specimens found in caves from Italy, Slovenia, Croatia, Greece and Austria.

redescribed on the basis of this new material (Rodriguez & Giani, 1987). Found also by J. Juget in the hyporheic flow of the river Rhône, near Lyon, France (Rodriguez & Giani, 1994).

Naididae

Pristinella osborni (Walton, 1906). Rarely reported in Europe, known from hyporheic waters of the river Rhône, France (Lafont & Durbec, 1990; Lafont et al., 1992), this species is reported here for the first time in cave waters: Caverna Mainarda, Friuli; and Bocche del Timavo, Venezia Giulia, Italy.

Pristinella sima (Marcus, 1944). Rarely reported in Europe, this species is reported for the first time in subterranean waters of this continent: Laminak Cave, 05.04.1986, Bizkaia, Spain and hyporheic waters of the river Dourdou, 11.07.1986, Vabres, Aveyron, France. It was also reported in North America, in the subterranean waters of New York (Strayer & O'Donnell, 1988).

Five other species are worthy of mention because they mainly inhabit superficial water but are also found in underground water.

Naididae

Nais christinae Kasprzak, 1973. This species, widespread in surface waters of Europe and known from hyporheic waters of the river Rhône, France (Lafont & Durbec, 1990), is reported for the first time in cave water: Azerat Cave Dordogne, France.

Uncinais uncinata (Orsted, 1842). Found in the hyporheal water of the river Adige, Italy (Ferrarese & Sambugar, 1976) and reported in this work from the "Ruisseau souterrain de la Germaine", Goléjac, Dordogne, France.

Tubificidae

Peristodrilus montanus (Hrabě, 1962). This species is known from European surface waters, often in large populations, and from a Slovakian spring (Hrabě, 1962). This is the first report of its presence in a cave: Bolunta Cave (12.15.1984), Bizkaia, Spain.

Varichaetadrilus bizkaiensis Rodriguez & Giani, 1984. Previously known from a northwest Spanish river basin, we collected it in the Laminak Cave (01.05.85), Bizkaia, Spain, located in the same river basin.

Enchytraeidae

Cernosvitoviella palustris Healy, 1979. Known from surface waters of Southwestern Europe, this species is now reported for the first time in subterranean waters: Rekalde resurgence (12.30.1984), Aitziturri spring (02.20.1985), Goiketxe Cave (02.23.1985) and Laminak Cave (05.04.1986), Bizkaia, Spain; Cornejo Cave (10.25.1983), Burgos, Spain; Ariz Cave (11.03.1985), Ostolaza Cave (12.22.1985) and resurgence I of Larrun river (03.28.1986), Navarra, Spain; and Azerat Cave (02.03.2000), Dordogne, France.

About 30 other species, reported by other authors, complete the list of the most interesting findings in the groundwater of the area considered (see previous literature references).

Haplotaxidae

Delaya bureschi (Michaelsen, 1925), Greece; *D. cantabronensis* (Delay, 1973), Spain; *D. corbarensis* (Delay, 1972), France; *D. leruthi* (Hrabě, 1958), France; *D. navarrensis* (Delay, 1973), Spain

Dorydrilidae

Dorydrilus michaelseni Piguet, 1913 France

Lumbriculidae

Guestphalinus wiardi Michaelsen, 1933, Slovenia; *Trichodrilus cernosvitovi*, Hrabě, 1937, France; *T. claparedei* Hrabě, 1937, Italy; *T. intermedius* Fauvel, 1903, France; *T. leruthi* Hrabě, 1937, Italy; *T. macroporophorus* Hrabě, 1954, France; Austria, *T. pragensis* Vejdovsky, 1875, France, *T. ptujensis* Hrabě, 1963, Slovenia; *T. sketi* Hrabě, 1963, Slovenia; *T. stammeri* Hrabě, 1937, Slovenia; *T. strandi* Hrabě, 1936, France, Italy; *T. tacensis* Hrabě, 1963, Slovenia

Tubificidae

Peloscolex latens Dumnicka, 1983, Greece; *Rhyacodrilus balmensis* Juget, 1959, France; *R. carsticus* Kosel, 1980, France; *R. sketi* Karaman, 1974, Slovenia; *Epirodrilus slovenicus* Karaman, 1976, Slovenia; *Haber turquini* (Juget & Lafont, 1979), France; *H. zavreli* (Hrabě, 1942), Italy; *Tubifex flabellisetosus* Hrabě, 1966, Slovenia; *Psammoryctides hadzii* Karaman, 1974, Slovenia; *Potamothrix postojnae* Karaman, 1974, Slovenia; *Gianius riparius* (Giani & Martínez, 1981) Spain and France

Enchytraeidae

Enchytronia hellenica Dumnicka, 1980, Greece.

Even though most of the localities were sampled only once, various habitats are involved and represented differently in the study. Moreover, some species have not yet been identified; we give here a preliminary ranking of the most frequent species (Table 1), taking into account the large number of localities studied (303, of which 219 were caves – Fig. 1) and the lack of relevant data in the literature.

Owing to the large numbers of individuals in the different families and genera, we can now also present a picture of their relative abundance in the 189 sampled caves of Italy, Slovenia, Croatia, Greece and Austria (Figs 2 and 3). Composing about 50% of the individuals collected, enchytraeids are the dominant family; the other half of the oligochaete community is almost entirely represented by lumbriculids, naidids and tubificids, but the abundance of parvidrilids must be also emphasized. The enchytraeids *Enchytraeus*, *Cernosvitoviella* and *Marionina* and the lumbriculid *Trichodrilus* are the most abundant genera.

Taking into account the diversity of stations sampled in Italy (Fig. 1), we represent the percentage of individuals belonging to the different families found in caves, springs and wells (Fig. 4). We should point out that the results concerning springs and wells have an indicative value only as the number of localities sampled was only 27 and 14, respectively. Comparing the results is particularly interesting when considering the springs, due to the different types involved (12 sandstone, 11 karstic, 2 alluvial and 2 granite springs). In caves, the enchytraeids are clearly dominant (55%) as previously pointed out by Juget &

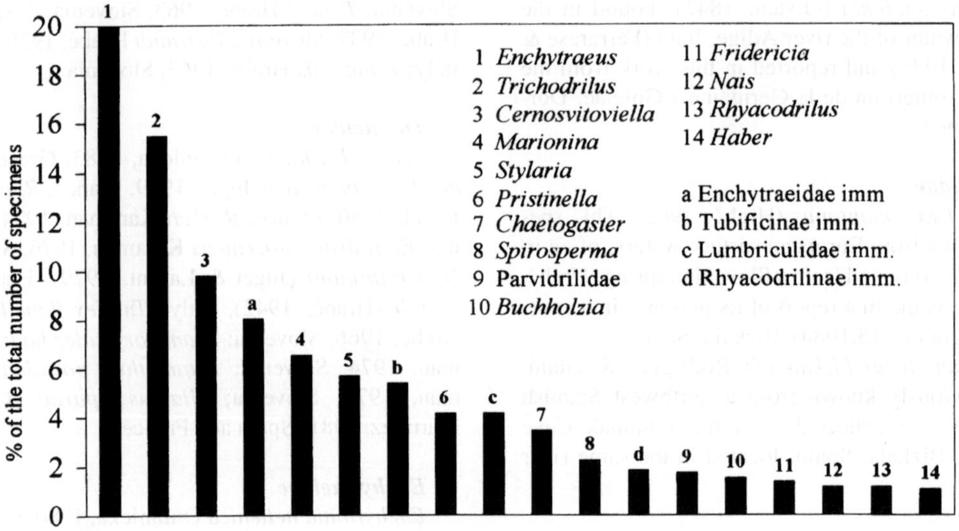

Figure 3. Contribution of the various genera to the number of specimens found in caves from Italy, Slovenia, Croatia, Greece and Austria (limited to genera contributing more than 1%).

Table 1. Ranking of the most frequent species for all sampled localities and for caves

Rank	All sampled localities	Caves
1	*Pristinella jenkinae* (Stephenson)	*Pristinella jenkinae* (Stephenson)
2	*Marionina argentea* (Michaelsen)	*Marionina argentea* (Michaelsen)
3	*Spirosperma velutinus* (Grube)	*Enchytraeus buchholzi* Vejdovský
4	*Pristina aequiseta* Bourne	*Pristina aequiseta* Bourne
5	*Cernosvitoviella palustris* Healy	Parvidrilidae gen. sp.
6	*Cernosvitoviella atrata* (Bretscher)	*Cernosvitoviella atrata* (Bretscher)
7	*Enchytraeus buchholzi* Vejdovský	*Spirosperma velutinus* (Grube)
8	*Eiseniella tetraedra* (Savigny)	*Cernosvitoviella palustris* Healy
9	*Buchholzia fallax* Michaelsen	*Nais variabilis* Piguet
10	*Nais variabilis* Piguet	*Eiseniella tetraedra* (Savigny)
11	Parvidrilidae gen. sp.	*Buchholzia fallax* Michaelsen
12	*Nais communis* Piguet	*Haplotaxis gordioides* (Hartman)
13	*Stylodrilus parvus* (Hrabě & Černosvitov)	*Trichodrilus* sp.
14	*Stylodrilus lemani* (Grube)	*Rhyacodrilus coccineus* (Vejdovský)
15	*Rhyacodrilus falciformis* Bretscher	*Haber monfalconensis* (Hrabě)
16	*Tubifex tubifex* (Müller)	*Psammoryctides barbatus* (Grube)

Dumnicka (1986), followed by lumbriculids (21%), naidids (13%) and tubificids (10%); some parvidrilids and haplotaxids are also present. In springs, naidids appear to be the most abundant family (45%), followed by enchytraeids (32%) and tubificids (20%); only a few lumbriculids and lumbricids have also been found. More than half of the oligochaetes collected in wells were lumbriculids (52%) and the only other families that we found in this habitat were tubificids (26%), enchytraeids (15%) and naidids (7%).

Discussion

The study of various collections of oligochaetes collected in groudwaters of southern Europe have resulted in the identification of 150 species. Thirty ad-

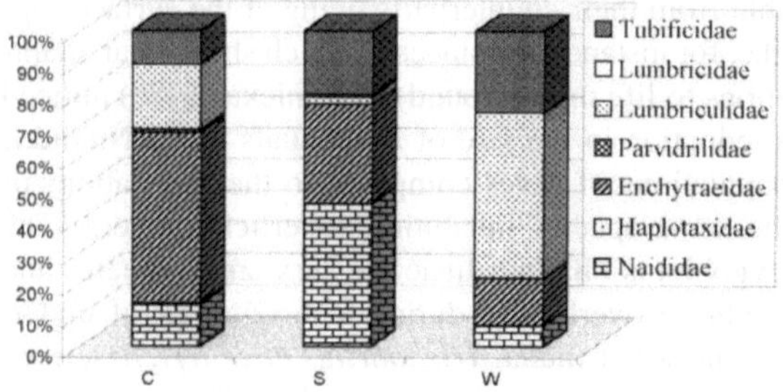

Figure 4. Percentage of families in caves (C), springs (S) and wells (W) from Italy.

ditional species were previously recorded in the literature (see above) bringing the total to 180 species recorded for the groundwaters of southern Europe. The specific richness and diversity of the underground freshwater oligochaete fauna are approximately comparable to those of superficial waters of the same area: Martínez-Ansemil (1993) estimated at 218 the number of species inhabiting superficial freshwater in Spain and France.

Most of the 150 species collected are ubiquitous, some of them being more or less stygophilous. Nevertheless, 31 (see list above) are stygobiont or markedly stygophile; we can add 26 species, not collected during the present work, but quoted in the literature for the area under study. Among these 57 species exclusively or preferentially inhabiting underground waters of southern Europe, 22 (of which two belong to two new genera: *Cookidrilus* and *Stochidrilus*) were described during the investigations carried out by the authors on the underground waters of southern Europe since the end of the eighties.

Černosvitov (1939) in his "*Catalogue des Oligochètes hypogés*" mentioned only 19 aquatic stygobiont oligochaetes. Juget & Dumnicka (1986) reported 65 species of oligochaetes (*Aphanoneura* excluded) characteristic of subterranean waters; among them 56 are only known from the Palearctic area. Since that work, new data have become available and we can now add:

- For North America: two species of *Stylodrilus*, Lumbriculidae (Rodriguez, 1996; Rodriguez & Coates, 1996) and a representative of a new family (Parvidrilidae) recently described by Erséus (1999).

- For Europe: *Abyssidrilus cuspis* (Erséus & Dumnicka, 1988) from Italy and *Phallodriloides macmasterae* (Erséus, 1983) from an anchialine cave (Bermuda), *Rhyacodrilus caudosetosus* Karaman, 1983 and *R. maculatus* Karaman, 1983 from Slovenia. We could also add 20 species of Lumbriculidae and Tubificidae described during our investigations and two new species belonging to the new family Parvidrilidae: one from Italy and Slovenia (unpublished data) and the other from France (J. Juget, unpublished data and pers. com.) extending the area of distribution of that new family to the whole Holarctic area.

- For Africa: one species of Phreodrilidae from Morocco (*Astacopsidrilus naceri* Martin et al., 1995) and two of Haplotaxidae from Guinea: *Haplotaxis villiersi* Omodeo, 1987 and *Villiersia guanivora* Omodeo, 1987.

Thus, 96 species of stygobiont freshwater oligochaete (*sensu stricto*) are actually known in the world; in this estimation we are not taking into account the new taxa described by Botea (1983) from Cuba because they are poorly described. Among these 97 species 82 are Palearctic, reflecting a better knowledge of European underground water fauna.

It is often difficult to make a clear separation between stygobiont and the stygophile oligochaetes as previously pointed out by Juget & Dumnicka (1986) and so, it is very difficult to determine the main faunistic characteristics of the underground oligochaete communities. In Europe, the genera *Cookidrilus*, *Dorydrilus*, *Guestphalinus* and *Stochidrilus* are only known from underground freshwater biota. The species of the small genus *Krenedrilus* and the larger genus *Trichodrilus* are almost entirely restricted to subterranean water, notably phreatic (Juget & Dumnicka, 1986); the genus *Rhyacodrilus* is well represented (in number of species and number of individuals) in such biota. These three genera, with the Haplotaxidae and the Parvidrilidae, appear to be the more characteristic faunistic elements of the underground freshwater oligochaete communities, even though these communities are numerically dominated by the Enchytraeidae. These investigations of the groundwater oligochaete fauna led to the discovery of tubificids of marine lineage (Rodriguez & Giani, 1989; Sambugar et al., 1999) among the Tubificinae (*Tubificoides*) and especially the Phallodrilinae (*Aktedrilus*, *Abyssidrilus*, *Gianius*, *Spiridion*, *Phallodriloides*); these species with marine affinities could now be added to the list of the more notable species of hypogean waters.

Many stygobiont taxa are known from one locality only so the stygobiont oligochaetes seem to exhibit a high degree of endemism (Juget & Dumnicka, 1986).

Intensive sampling, such as those presently carried out in southern Europe, has led to this apparent degree of endemism being lowered. This is the case for the family Parvidrilidae, the genus *Krenedrilus* and for species such as *Spiridion phreaticola*, *Abyssidrilus cuspis*, *Rhyacodrilus amphigenus*, *R. lindbergi* and various species of *Trichodrilus*, previously only known from their type-locality and now collected by us in other localities and even on other continents as is the case for the Parvidrilidae and *Krenedrilus* for instance.

It is very difficult to draw conclusions about the geographical distribution of the various taxa at a local level (countries), as the sampling effort was not the same in the different countries under study. However, according to the literature and to our work, the fauna of Slovenia seems very diversified. A possible explanation could be the geographical location of Slovenia, which is influenced by a number of very different biogeographical units and the aquatic troglobiont fauna is the richest in the world (Sket et al., 1991). Concerning the whole area studied, our data on troglobiont species reveal the great diversity and specific richness encountered in the groundwaters of southern Europe. The turbulent geological past, which enhanced possibilities for multiple immigrations and speciation, can explain the particularly high diversity in the wider Dinaric Region (S. Europe) and in Slovenia in particular as pointed out by Sket (1999).

In the sampling stations, the specific richness was usually high and similar to those observed in superficial waters, but the population densities were comparatively very low; the very small number of specimens collected for the stygobiont species is a handicap for taxonomic studies. The most abundant species in our samples are ubiquitous ones, usually widespread in superficial waters. Local abiotic factors or seasonal biotic parameters (e.g. asexual reproduction) could lead to a great development of some populations such as that observed for *Stylaria lacustris* in the well of Vermigliano, Italy. Such species pullulations are misleading in quantitative studies: in this way, *Stylaria lacustris* is ranked 5th in number of specimens of the species collected (Fig. 3), though it was only collected in a small number of stations among the 303 sampled during the present work. In addition, it is noteworthy that none or few species of Naididae are stygobionts: in the present work on southern Europe, only 2 species of *Pristinella* are considered as stygophilous.

As pointed out Dumnicka & Juberthie (1994), groundwater oligochaetes are not easily distinguishable from their counterparts living at the surface (unlike, for instance, crustaceans which show clear adaptations to life underground). Dumnicka (1986) noticed a reduction in the size of individuals in the Naididae populations of caves compared to the populations of the same species inhabiting superficial waters. The stygobionts Phallodrilinae are very small species and might be overlooked during investigations of underground water fauna. *Trichodrilus angelieri*, described during our investigations in southern Europe, present glandular pads in the epidermis of the posterior segments of the body; such tegumentary formations have been described in various taxa living in interstitial water, in particular the annelid *Rheomorpha neiswestnovae* (Lastočkin) (Aphanoneura). Sambugar et al. (1997) also suggest that the shifting or the asymmetrical bending of the spermathecae could be related to a reduction of body diameter on the gonadal region.

Acknowledgements

Many of the oligochaetes studied in this work were collected by many colleagues. Many thanks to all of them: F. Stoch, F. Gasparo, R. Rouch, B. Lebreton, P. Leclerc, C. Bou, J. Notemboom, F. Bacchia, J. Broder, M. Bude Dancevich, G. Caoduro, G. Crevatin, S. Dolce, E. Lana, E. Piva, E. Spadaro, L. Tomelleri, A. Camacho and E. Beruete. We wish to thank Stuart Gelder for reviewing the first part of this manuscript and Peter Winterton for checking the English language.

References

Bellés, X., 1987. Fauna cavernícola i intersticial de la Península ibérica i les illes Balears. Consell superior d'investigacions científiques, Moll Ed., Mallorca: 207 pp.

Botea, F., 1983. Oligochètes souterraines de Cuba. Résultats des expéditions biospéologiques cubano-roumaines à Cuba 4. Ed. Acad., Bucuresti: 19–38.

Botosaneanu, L., 1986. Stygofauna Mundi. A Faunistic, Distributional and Ecological Synthesis of the World Fauna inhabiting Suberranean Waters (including the Marine Interstitial). E. J. Brill/Dr. W. Backhuys, Leiden: 740 pp.

Centurioni, C. & B. Sambugar, 1980. I Macroinvertebrati del bacino idrografico dell'Adige. I. Gli Oligocheti. Mem. Mus. Civ. St. Verona, II ser., sez. biol. 6: 125–151.

Černosvitov, L., 1931. Note sur les Oligochètes des grottes de Sainte-Reine, près de Toul. Arch. Zool. exp. gén. 71: 62–66.

Černosvitov, L., 1936. Oligochètes des grottes artificielles du nord-est de la France. Arch. Zool. exp. gén. 78: 1–7.

Černosvitov, L., 1939. Études Biospéologiques 10 (1). Catalogue des Oligochètes hypogés. Bull. Mus. r. Hist. nat. Belg. 15: 1–92.

Cobolli-Sbordoni, M., M. Paoletti, V. Sbordoni & A. Vigna Taglianti, 1994. Italie. In Juberthie, C. & V. Decu (eds), *Encyclopaedia Biospeologica* I. Société de Biospéologie, Moulis, Bucuresti: 737–751.

Danielopol, D. L., 1991. Spatial distribution and dispersal of interstitial Crustacea in alluvial sediments of a backwater of the Danube at Vienna. Stygologia 6: 97–110.

Danielopol, D. L., J. Dreher, A. Gunatilaka, M. Kaiser, R. Niederreiter, P. Pospisil, M. Creuzé Des Chatelliers & A. Richter, 1992. Ecology of organisms living in a hypoxic groundwater environment at Vienna (Austria); methodological questions and preliminary results. Proc. First Int. Conf. Ground-water Ecology. U.S.E.P.A., A.W.R.A., Tampa: 79–90.

Delay, B., 1972. Un nouvel Oligochète Haplotaxidae souterrain des Pyrénées-Orientales (France): *Haplotaxis corbarensis* n. sp. Ann. Spéléol. 27: 329–340.

Delay, B., 1973. Deux nouveaux Oligochètes Haplotaxidae troglobies d'Espagne: *Haplotaxis navarrensis* n. sp. et *Haplotaxis cantabronensis* n. sp. Ann. Spéléol. 28: 405–411.

Dumnicka, E., 1981. *Peloscolex pescei* sp. n. (Oligochaeta, Tubificidae) from the subterranean waters of central Italy. Bull. Pol. Acad. Sci. Biol. 28: 663–666.

Dumnicka, E., 1986. Amsterdam Expeditions to the West Indian Islands, report 50. Naididae (Oligochaeta) from subterranean waters of West Indian Islands. Distribution, taxonomic remarks and description of a new species. Bijdr. Dierk. 56: 267–281.

Dumnicka, E., 1987. The revision and re-description of *Peloscolex pescei* Dumnicka (Oligochaeta, Tubificidae) with formation of a new genus. Bull. Pol. Acad. Sci., Biol. 35: 47–52.

Dumnicka, E., 1990. Oligochaetes from subterranean waters of Italy and Greece. Mém. Biospéol. 17: 163–168.

Dumnicka, E. & C. Juberthie, 1994. Aphanoneura et Oligochaeta. In Juberthie, C. & V. Decu (eds), *Encyclopaedia Biospeologica* I, Société de Biospéologie, Moulis, Bucuresti: 67–75.

Erséus, C., 1999. *Parvidrilus strayeri*, a new genus and species, an enigmatic interstitial clitellate from underground waters in Alabama. Proc. biol. Soc. Wash. 112: 327–337.

Erséus, C. & E. Dumnicka, 1988. A new *Phallodrilus* (Oligochaeta, Tubificidae) from subterranean waters in central Italy. Stygologia 4: 116–120.

Erséus, C., N. Giani & P. Rodriguez, 1992. The taxonomy of *Spiridion* Knöllner (Oligochaeta, Tubificidae), with a redescription of *Spiridion phreaticola* (Juget, 1987) comb. n. Stygologia 7: 77–83.

Fauvel, P., 1903. Un nouvel oligochète des puits (*Trichodriloïdes intermedius*, n. gen. n. sp.). C. r. Ass. fr. Avanc. Sci. 32: 221.

Farara, D. G. & C. Erséus, 1991. *Phallodrilus aquaedulcis* Hrabě, 1960, a meiobenthic freshwater oligochaete (Tubificidae) previously known only from Europe, recorded from the Niagara River, North America. Can. J. Zool. 69: 291–294.

Ferrarese, U. & B. Sambugar, 1976. Ricerche sulla fauna interstiziale iporreica dell'Adige in relazione allo stato di inquinamento del fiume. Rivista di Idrobiologia, Perugia 15: 47–127.

Giani, N. & E. Martínez-Ansemil, 1981. Observaciones acerca de algunos Tubificidae (Oligochaeta) de la Peninsula Iberica, con la descripcion de *Phallodrillus riparius* n. sp. Ann. Limnol. 17: 201–209.

Giani, N. & P. Rodriguez, 1988. Description de quelques espèces nouvelles de Tubificidae (Oligochaeta) de grottes et de sources karstiques de la péninsule ibérique. Stygologia 4: 121–137.

Giani, N. & P. Rodriguez, 1994. New species of the genus *Trichodrilus* (Oligochaeta, Lumbriculidae). Zool. Scr. 23: 33–41.

Giani, N., P. Martin & J. Juget, 1995. A new species of Phreodrilidae (Oligochaeta), *Astacopsidrilus naceri* sp. nov., from Morocco (North-Africa), with notes on the biogeography of the family. Can. J. Zool. 73: 2375–2381.

Giani, N., C. Erséus & E. Martínez-Ansemil, 1990. Redefinition of the subterranean genus *Krenedrilus* Dumnicka (Oligochaeta, Tubificidae) with a redescription of *K. papillatus* Dumnicka, 1983 and the description of two new species. Stygologia 5: 55–65.

Hrabě, S., 1936. *Trichodrilus strandi* n. sp., ein neuer vertreter der Höhlen-Lumbriculiden. Festschr. f. Prof. Dr Embrik Strand, Riga 1: 404–407.

Hrabě, S., 1942. Zur kenntnis der Brunnen- und quellenfauna aus der Slowakei. Sbor. Kl. prir. Brne 24: 23–30.

Hrabě, S., 1958. A new species of Oligochaeta from the southwest of France. Notes biospéologiques 13: 171–178.

Hrabě, S., 1962. *Rhizodrilus montanus* n. sp. from the glacial lake in the Perister Mountains in south Macedonia. Spisy prir Fac. Univ. Brne 435: 335–346.

Hrabě, S., 1963a. On *Rhyacodrilus lindbergi* n. sp., a new cavernicolous species of the family Tubificidae (Oligochaeta) from Portugal. Bol. Soc. Port. Ciênc. nat. 10, 2ª S: 52–56.

Hrabě, S., 1963b. Oligochaeta limicola from Slovenija. Biol. Vestnik 11: 67–75. Hrabě, S., 1966. New or insufficiently known species of the family Tubificidae. Publ. Fac. Sci. Univ. J.E. Purkyne Brne 470: 57–67.

Juberthie, C. & V. Decu, 1994. Encyclopædia Biospeologica I. Société de Biospéologie, Moulis, Bucuresti: 834 pp.

Juget, J., 1959. Recherches sur la faune aquatique de deux grottes du Jura méridional français: la grotte de La Balme (Isère) et la grotte de Corveissiat (Ain). Ann. Spéléol. 14: 391–401.

Juget, J., 1984. Oligochaeta of the epigean and underground fauna of the alluvial plain of the French upper Rhone. Hydrobiologia 115: 175–182.

Juget, J., 1987. Contribution to the study of the Rhyacodrilinae (Tubificidae, Oligochaeta), with description of two new stygobiont species from the alluvial plain of the French upper Rhône, *Rhyacodrilus amphigenus* sp. n. and *Rhizodriloides phreaticola* g. n., sp. n. Hydrobiologia 155: 107–118.

Juget, J. & E. Dumnicka, 1986. Oligochaeta (including Aphanoneura) des eaux souterraines continentales. In Botosaneanu, L. (ed.), Stygofauna Mundi. E. J. Brill/Dr W. Backhuys, Leiden: 234–244.

Karaman, S., 1983. Third contribution to the knowledge of freshwater Oligochaeta of Slovenia. Biol. Vestn. 31: 29–36.

Lafont, M. & A. Durbec, 1990. Essai de description biologique des interactions entre eau de surface et eau souterraine: vulnérabilité d'un aquifère à la pollution d'un fleuve. Ann. Limnol. 26: 119–129.

Lafont, M., A. Durbec & C. Ille, 1992. Oligochaete worms as biological describers of the interactions between surface and groundwaters: a first synthesis. Regul. River 7: 65–73.

Martínez-Ansemil, E., 1993. Etude sur les Oligochètes aquatiques des pays du pourtour de la Méditerranée: taxonomie, phylogénie, biogéographie et écologie; Part I. Doctoral Thesis, Univ. Toulouse, France: 197 pp.

Martínez-Ansemil, E., B. Sambugar & N. Giani, 1997. Groundwater Oligochaetes from southern-Europe. I. A new genus and three new species of Rhyacodrilinae (Tubificidae) with a redescription of *Tubifex pescei* (Dumnicka) comb. n. Ann. Limnol. 33: 33–44.

Moniez, R., 1888. Faune des eaux souterraines du département du Nord et en particulier de la ville de Lille. Revue biol. du Nord 1: 82–94.

Rodriguez, P., 1988. Sur certaines espèces de Lumbriculidae du Nord de la Péninsule Ibérique. Ann. Limnol. 24: 203–211.

Rodriguez, P., 1996. *Stylodrilus californianus* n. sp., a new lumbriculid (Annelida, Oligochaeta) from North America. Hydrobiologia 333: 161–164.

Rodriguez, P. & K. A. Coates, 1996. A new American *Stylodrilus* species (Lumbriculidae, Oligochaeta). Can. J. Zool. 74: 92–96.

Rodriguez, P. & N. Giani, 1987. Sur deux espèces de Lumbriculidae (Oligochaeta) cavernicoles: *Cookidrilus speluncaeus* n. gen., n. sp. et *Trichodrilus diversisetosus* Rodriguez, 1986. Bull. Soc. Hist. nat. Toulouse 123: 45–49.

Rodriguez, P. & N. Giani, 1989. New species of *Phallodrilus* (Oligochaeta, Tubificidae) from caves of northern Spain and southwestern France. Hydrobiologia 180: 57–63.

Rodriguez, P. & N. Giani, 1994. A preliminary review of the taxonomic characters used for the systematics of the genus *Trichodrilus* Claparède (Oligochaeta, Lumbriculidae). Hydrobiologia 278: 35–51.

Sambugar, B., N. Giani & E. Martínez-Ansemil, 1999. Groundwater oligochaetes from southern-Europe. Tubificidae with marine phyletic affinities: new data with description of a new species, review and consideration on their origine. Mém. Biospéol. 26: 107–116.

Sket B., 1999. The nature of biodiversity in hypogean waters and how it is endangered. Biodivers. Conserv. 8: 1319–1338.

Sket B., P. Habic & C. Juberthie, 1994. Slovenia. In Juberthie, C. & V. Decu (eds), Encyclopaedia Biospeologica I, Société de Biospéologie, Moulis, Bucuresti: 803–807.

Sket B., J. Bole, A. Benovic, A. Brancelj, J. Brglez, M. Cucek, B. Curcic, A. Jaklin, G. Karaman, I. Katavic, M. Kerovec, I. Kos, M. Legac, N. Mrsic, A. Malej, T. Novac, S. Petkovski, T. Petkovski, A. Polenec, F. Potocnik, V. Pujin, B. Radujkovic, Z. Stevcic, K. Tarman, A. Travizi, M. Velikonja, F. Velkovrh, J. Vidakovic & D. Zavodnik, 1991. Richness and state of knowledge of the fauna of Yugoslavia: lower invertebrates (Metazoa: Invertebrata, ex Insecta). Biol. Vestn. 39: 37–52.

Stammer, H. J., 1932. Die Fauna des Timavo. Ein Beitrag zur Kenntnis der Höhlengewässer, des Süss-und Brackwässers im Karst. Zool. J. Syst. 63: 521–656.

Strayer, D. L. & E. O'Donnell, 1988. Aquatic Microannelids (Oligochaeta and Aphanoneura) of Underground Waters of Southeastern New York. Am. midl. Nat. 119: 327–335.

Vandel, A., 1920. Sur la faune des sources. Bull. Soc. zool. Fr. 45: 177–183.

Hydrobiologia **463:** 75–81, 2001.
P. Rodriguez & P.F.M. Verdonschot (eds), Aquatic Oligochaete Biology VIII.

Oligochaete communities in the hyporheic zone of a glacial river, the Roseg River, Switzerland

Michel Lafont[1] & Florian Malard[2]
[1]*Cemagref, UR Biologie des Ecosystèmes Aquatiques, 3 bis Quai Chauveau, C.P. 220, F-69336 Lyon Cedex 09, France*
Tel: 33 4 78 47 87 21. E-mail: michel.lafont@cemagref.fr
[2]*UMR CNRS 5023, Ecologie des Hydrosystèmes Fluviaux, Université Lyon 1, 43 Bd 11 Novembre 1918, F-69622 Villeurbanne Cedex, France*

Key words: glacial river, hyporheic zone, Oligochaeta

Abstract

We examined the composition and distribution of oligochaete communities in the hyporheic zone of the Roseg River, a glacial river in southeastern Switzerland. Ten oligochaete species were collected from 11 sites distributed along an 11-km reach of the river, downstream of the Tschierva Glacier. The most frequently encountered species were *Dorydrilus michaelseni, Propappus volki, Cernosvitoviella atrata* and *Cernosvitoviella carpatica*. Six other species were relatively scarce and exhibited a discontinuous distribution pattern only in the downstream sites. At the most downstream site, a population of eyeless *Nais communis* was discovered with anatomical traits apparently related to a subterranean life-adaptation of this surface-living species. The species richness and abundance of oligochaete communities were relatively constant over time, but increased at sites where groundwater entered the stream. The spatial distribution of several oligochaete species was linked to the longitudinal arrangement of groundwater upwelling areas. This observation suggested that groundwater was an important upstream migration pathway for oligochaetes during the glacial retreat.

Introduction

Freshwater oligochaetes have long been recognized as common and permanent inhabitants of diverse aquatic habitats including lotic and lentic systems, surface waters, groundwater, and coarse as well as fine sediments (Brinkhurst & Jamieson, 1971). In particular, oligochaetes are recognized as permanent and abundant taxa in hyporheic sediments where they constitute a relevant tool for ecological studies (e.g. Ferrarese & Sambugar, 1976; Juget, 1984; Strayer & Bannon-O'Donnell, 1988; Lafont et al., 1992, 1996; Giere, 1993; Strayer et al., 1997). In addition, the importance of the hyporheic zone as a key component of running water systems has long been emphasized (Hynes, 1983; Danielopol, 1989; Brunke & Gonser, 1997; Jones & Mulholland, 2000).

Minimal substantial research focusing on the oligochaete communities of cold freshwater ecosystems in Europe is available. Past research has focussed on the study of oligochaetes in surface sediments of high elevation lakes or streams (Juget & Giani, 1974; Giani & Lavandier, 1977). Because of the fast actual retreat of most European glaciers, there has been a recent increase in the number of studies on biological diversity in glacial streams (Milner & Petts, 1994). However, most of these studies have focussed on the ecology of benthic invertebrates (Kownacki, 1991; Milner and Petts, 1994; Gislason et al., 1998; Füreder, 1999). Recent investigations carried out in the hyporheic zone of a glacial river in the Swiss Alps revealed the presence of a number of permanent aquatic taxa, such as cyclopoids, harpacticoids, and ostracods – and of particular interest – permanent oligochaete communities (Malard et al., unpublished).

This paper examines the species composition and distribution of oligochaete communities in the hyporheic zone of the Roseg River, a glacial river in

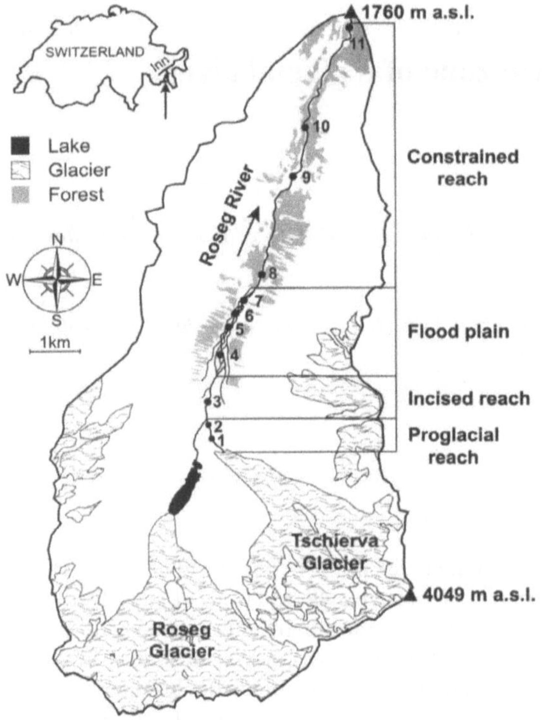

Figure 1. Map of the catchment showing the distribution of geomorphological reaches and location of sampling sites (1–11) along the Roseg River in southeastern Switzerland.

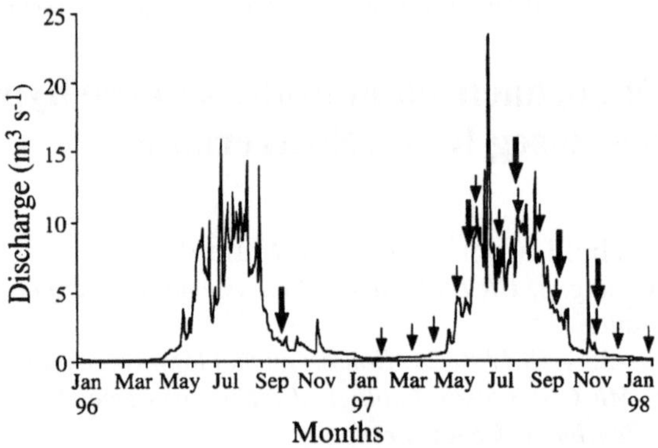

Figure 2. Hydrograph of the Roseg River in southeastern Switzerland, in 1996 and 1997 (daily discharge). Thick arrows indicate the sampling dates of oligochaetes. Thin arrows indicate dates when surface water was sampled for the analysis of silica concentrations.

southeastern Switzerland. We studied more specifically the influence of groundwater upwelling on the longitudinal distribution of oligochaete species.

Study site

The Roseg River, a glacial tributary of the River Inn (Danube catchment), is a 11.5 km-long second-order stream located in the Bernina Massif of the Swiss Alps in southeastern Switzerland (9° 53′ 57″ E, 46° 29′ 28″ N, Fig. 1). The main glaciers of the Roseg River catchment (Tschierva: 6.2 km^2; Roseg: 8.5 km^2) have continuously retreated (ca. 1100 m) during the last 120 yr (Maisch, 1988). From the terminus of the Tschierva glacier, four reaches can be distinguished along the valley floor: (a) an unstable pro-glacial braided reach (length: 900 m); (b) a single thread channel incised in glacial till (length: 600 m); (c) a complex flood plain comprising up to 21 km of channels (length: 2750 m); and (d) a long reach constrained by steep valley slopes (length: 7050 m).

Detailed information on the geomorphological, hydrological, and hydrochemical characteristics of this glacial river system were reported by Malard et al. (1999), Tockner et al. (1997) and Ward et al. (1999). The bedrock consists of granite and diorite belonging to the Bernina, Corvatsch and Sella series. Subalpine coniferous forests are restricted to the lateral sides of the valley (treeline: 2300 m a.s.l).

The mean annual river discharge, recorded by the Swiss Hydrological and Geological Survey for the period 1955–1997, was 2.76 m^3s^{-1}. The hydrograph is typical of a glacial-melt dominated river (Fig. 2). There is a distinct seasonal flow regime and marked daily fluctuations in flow during the ice-melt period. Daily mean discharge peaks during the ice-melt season range from 6 to 10 m^3s^{-1}, in July and August. Minimum discharge occurs in winter (ca. 0.2 m^3s^{-1} from December to March) when surface flow is sustained solely by groundwater inputs.

Materials and methods

Eleven sampling sites were located along the main channel of the Roseg River – from 400 m below the terminus of the Tschierva glacier (site 1), downstream to the river end (site 11), a reach of 11 km (Fig. 1). Streambed sediments at all sites were composed of an extremely heterogeneous mixture of cobble, pebble, gravel, and sand. From February 1997 to January 1998, surface water was collected monthly in 1-l polyethylene bottles and filtered (Whatman GF/F filters; 0.7 μm) within 1–4 h. Silica concentration was analyzed with the molybdate - heteropoly blue method

(APHA, 1989). Silica was considered as an effective tracer of groundwater input to the river because concentrations were distinctively higher in groundwater than in glacial water. Malard et al. (1999) showed that elevated silica concentrations in groundwater originated from soil contact and alteration of plagioclase during long subsurface flow paths in weathered rock materials of the valley slopes.

Faunal sampling was carried out on 27 September 1996 (decreasing discharge), 4 June (increasing discharge), 1 August (high discharge), 24 September (decreasing discharge) and 13 November 1997 (low discharge) (Fig. 2). Three faunal samples were collected at each site on each date. Oligochaetes were collected by driving a mobile pipe (25 mm internal diameter, with 5-mm holes at the base) to a depth of 30 cm below the streambed. Ten litres of interstitial water were immediately extracted using a Bou-Rouch pump (Bou & Rouch, 1967) and filtered through a 100 μm-mesh net. Samples were preserved in 4% formaldehyde. Oligochaetes were sorted using a dissecting microscope, mounted on slides in a solution of glycerin (50%) and lactic acid, and identified to species level using a compound microscope. For the identification of Enchytraeidae, we used the keys from Baroudi (1987) and Nielsen & Christensen (1959). For the other families, we referred to Brinkhurst & Jamieson (1971) and Coates (1986). Abundance and richness were expressed respectively as the number of oligochaetes and number of species collected in 10 l of pumped-water.

A two-way repeated measures analysis of variance (RMANOVA), followed by Tukey multiple comparison tests, were used to test for differences in species richness and total abundance of oligochaetes between sites and dates. The sampling date was introduced as a repeated measures factor in the ANOVA. The analysis was performed on data collected in 1997 (11 sites, 3 replicates per site, and 4 dates), because sites 4, 5, 6 and 11 were not sampled in September 1996. Abundance and richness were Ln $(x+1)$ transformed to meet the assumption of homogeneity of variances. Differences in species richness and abundance were considered significant when $P < 0.05$. Statistical tests were performed with the STATISTICA software package (StatSoft, 1994).

Table 1. Frequency of occurrence (FO%) and relative abundance (RA%) of oligochaete species collected in the hyporheic zone of the Roseg River. Frequency of occurrence was calculated as the number of samples containing a particular species divided by the total number of samples. Relative abundance is the total number of individuals for a particular species divided by the total number of oligochaetes

Taxa	FO%	RA%
Propappus volki Michaelsen 1916	47.5	41.2
Dorydrilus michaelseni Piguet 1913	38.8	3.9
Cernosvitoviella atrata (Bretscher, 1903)	27.3	52.2
Cernosvitoviella carpatica Nielsen & Christensen 1959	7.9	1.1
Haplotaxis gordioides (Hartmann, 1821)	5.8	0.2
Nais communis Piguet, 1906 (eyeless specimens)	<5	1.1
Chaetogaster diastrophus (Gruithuisen, 1828)	<5	0.2
Enchytraeidae g 1. sp. 1 (*Hemienchytraeus* ?)	<5	0.1
Cognettia glandulosa (Michaelsen, 1888)	<5	<0.1
Marionina argentea (Michaelsen, 1889)	<5	<0.1

Results

List of species

Ten species of aquatic oligochaetes were identified during this study (Table 1). Only 3 species – *Dorydrilus michaelseni, Propappus volki* and *Cernosvitoviella atrata* – occurred in more than 10% of the hyporheic samples. The two recorded species of the genus *Cernosvitoviella*, only identifiable when sexually mature, represented about 10% of the genus population. This suggested that other congeners may also be present at these sites. Immature specimens could be separated into two morphologically different groups. One homogeneous group, with specimens exhibiting a fairly developed prostomium and three pairs of septal glands, was considered as *Cernosvitoviella carpatica.* The second group did not exhibit any distinct morphological characters, suggesting that – in addition to *C. atrata* – one or more as yet undetermined taxa might also be present.

The mean oligochaete abundance was 29 individuals per 10 l of water and the maximum abundance reached 1527 individuals per 10 l of water at site 8 in September 1996. *Propappus volki* and *C. atrata* were the predominant species (Table 1). The collection of *D. michaelseni* and *P. volki* from sites 1 and 5 are thought to be the highest altitude records (2120 m and 2020 m, respectively) for these species in Europe.

At site 11, the most downstream site during the present study, 41 individuals of *Nais communis* were discovered (39 in August and 2 in September 1997).

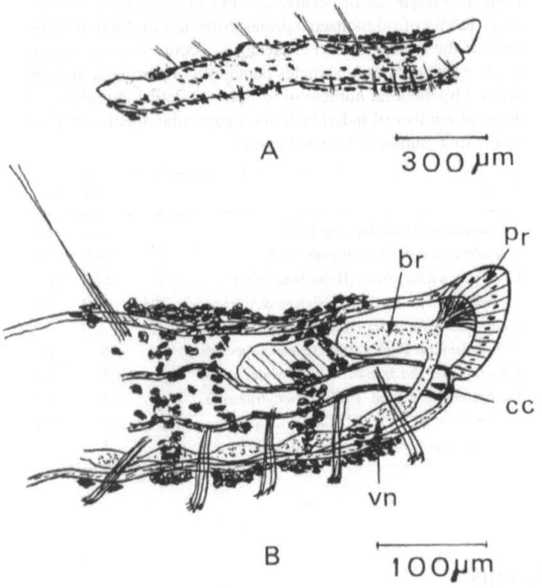

Figure 3. Schematic lateral view of *Nais communis* from the hyporheic zone of the Roseg River in southeastern Switzerland. (A) Whole specimen. (B) Details of the anterior end. Abbreviations: br – brain; cc – circum-oesophageal connectives; pr – prostomium; vn – ventral nerve cord.

Each *N. communis* specimen was observed to bear several particular anatomical traits (Fig. 3): (i) body densely covered with foreign material, suggesting a high mucus secretion; (ii) lack of eyes; (iii) strong development of the brain and circum-oesophagal connectives; (iv) presence of well developed prostomial cells, apparently densely connected to nerves originating from the brain. We suggest that these anatomical traits are related to a hyporheic life-adaptation of this surface-living species.

Groundwater upwelling and distribution of oligochaete species

Longitudinal changes in silica concentration showed that groundwater influx to the river occurred at the downstream end of the proglacial reach (sites 1–2) and more particularly from the downstream end of the floodplain reach (from site 6) (Fig. 4a). Indeed, silica concentration increased markedly from the downstream end of the flood plain.

Upstream sites 1 and 2 were colonized mainly by *D. michaelseni* (Fig. 4b). Oligochaetes were almost never collected at sites 3 and 4. Hyporheic assemblages at sites 6–11 were dominated by *P. volki*, *C. atrata* and *D. michaelseni*. The most downstream site during this study, site 11, was characterized by the presence of eyeless specimens of *N. communis* and the higher abundance of *C. carpatica*.

Species richness per site exhibited a major discontinuity along the longitudinal dimension (ANOVA site effect: $F = 34.24$, $P < 0.0001$). Multiple comparison tests indicated that sites located upstream of the strong groundwater upwelling area (sites 1 through 5) contained significantly fewer species than those located downstream (sites 6–11) ($P < 0.004$). Differences in species richness among dates were less pronounced (date effect: $F = 3.71$, $P = 0.0155$); samples contained significantly fewer species in June than in September 1997 ($P = 0.01$).

A similar distribution pattern was observed for oligochaete abundance. Tukey multiple comparison tests showed that sites located upstream of the strong groundwater upwelling area (sites 1–5) contained significantly fewer individuals than those located downstream (sites 6–11) ($P = 0.0002$). There were no significant differences in abundance among dates ($F = 0.98$, $P = 0.406$). However, the maximum average abundance was recorded in September 1996 (i.e. 115±330 individuals per 10 l of water; $n = 21$ samples), a date not considered in the analysis of variance.

Discussion

This study demonstrated that some oligochaete species, in particular *D. michaelseni*, *P. volki* and *C. atrata*, were able to maintain constant populations within the sediments of a harsh environment such as that of a glacial river. Tockner et al. (1997) measured concentrations of particulate organic carbon (POC) ranging from 0.40 to 0.93 mg g^{-1} of dry sediment in the bed sediment of the Roseg River. Despite this very low organic carbon content, we found a relatively high mean abundance of oligochaete assemblages (see for instance site 8, September 1996). Moreover, abundances were relatively constant over time. Oligochaete communities may be considered as an indirect measurement of the trophic potential of the hyporheic zone. Efficient trophic pathways probably exist in this extreme ecosystem so that recorded species are able to exploit limited nutrient resources. For example, the higher abundance of oligochaetes recorded in September 1996 is attributed to the rapid development in autumn of a thick cover of the chrysophyte *Hydrurus foetidus* in the surface stream (Uehlinger et al., 1998).

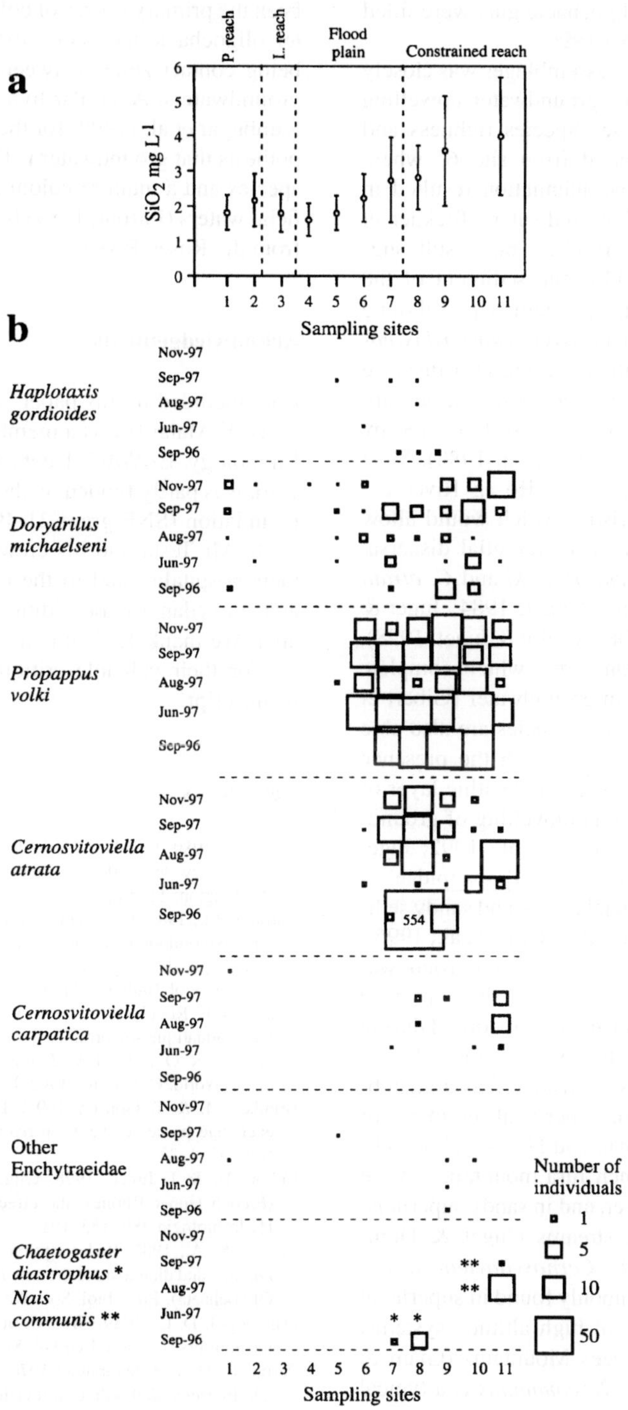

Figure 4. (a) Average silica concentrations in surface water (n = 12 dates) for each sampling site (except site 10). (b) Distribution and abundance of oligochaete species in the hyporheic zone of the Roseg River in southeastern Switzerland. Sides of squares are proportional to the Napierian logarithm of the average number of oligochaetes (n = 3 replicates per site and date). Other Enchytraeidae: *Marionina argentea*, *Cognettia glandulosa*, and *Hemienchytraeus* sp.

Indeed, we observed that oligochaete guts were filled with these algae in September 1996.

The spatial distribution of assemblages was closely related to the distribution of groundwater upwelling zones along the river course. Species richness and abundance markedly increased from site 6, where valley constriction and slope attenuation resulted in a large-scale upwelling of groundwater (Tockner et al., 1997; Malard et al., 1999). This result suggested that groundwater within the sediment of the valley floor was an important upstream dispersal route for oligochaetes in this glacial river. *Dorydrilus michaelseni* was present at site 1 located at a distance of only 400 m from the glacier terminus, an upwelling groundwater site which was still covered by the glacier about 30 years ago (Maisch, 1988). Most of the species collected from the Roseg River exhibited ecological characteristics which would allow them to use groundwater as a preferential dispersal route. *Dorydrilus michaelseni*, *P. volki* and *C. atrata* were considered as stygobites (Juget, 1984; Juget & Dumnicka, 1986; Lafont, 1989; Cellot & Juget, 1998), that is, obligatory hypogean forms which complete their life cycle exclusively in groundwater (Gibert et al., 1994). However, these three species are also able to develop abundant populations (with the presence of sexually mature specimens) in the benthic layer of rivers characterized by a strong upwelling of groundwater (Gaschignard, 1984; Lafont et al., 1992; Juget & Lafont, 1994). *Propappus volki* (Propappidae) is widely distributed in superficial coarse and sandy sediments (Coates, 1986; Timm, 1994; Timm et al., 1996). But this species evidently originated from groundwater in the Roseg River; it reproduced in the hyporheic zone because mature specimens were recorded during all sampling seasons (from June to November 1997). *Dorydrilus michaelseni*, also considered as a stygobite, is known to occur in fine superficial sediments of the deep zone of lakes Léman and Neuchâtel, in subterranean rivers of the French Jura mountains, in the underflow of the Rhône River, and in sandy superficial sediments of fast running streams (Juget & Dumnicka, 1986; Lafont, 1989). *Cernosvitoviella atrata* (Enchytraeidae) is also commonly found in superficial sediments, including those of high altitude systems, up to 2370 m in the Pyrenees Mountains (Giani & Lavandier, 1977). Although *N. communis* is a typical surface-living species, we observed hyporheic populations which expressed morphological characteristics different from those observed in surface populations. Lafont (1989) suggested that groundwater might have been the primary source of colonization of all freshwater oligochaete ancestors, the colonization pathways being contact zones between marine and freshwater groundwaters. A similar hypothesis was proposed by Sambugar et al. (1999) for the Phallodrilinae. The hypothesis that groundwater is likely to be a reservoir of species and a main re-colonization pathway of superficial waters (Lafont, 1989) is supported by our results from the Roseg River.

Acknowledgements

Data included in this paper were collected while one of us (F. Malard) was a member of the Department of Limnology, EAWAG, Duebendorf, Switzerland. This work was partly funded by the Swiss National Science Foundation (SNF grant 21-49243.96). We are grateful to Mr Testa and his crew at the Roseg Hotel for their hospitality and to the communes of Pontresina and Samedan for providing access to the sampling area. We thank T. Timm and two anonymous reviewers for their valuable critiques which improved the manuscript.

References

A.P.H.A., 1989. Standard Methods for the Examination of Water and Wastewater. 17th edn, American Public Health Association, Washington: 1134 pp.

Baroudi, M., 1987. Contribution à l'étude des oligochètes: I – Etude taxinomique de quelques genres d'Enchytraeida; II – Etude faunistique des oligochètes aquatiques du Maroc. PhD Thesis, University of Toulouse, France: 269 pp.

Bou, C. & R. Rouch, 1967. Un nouveau champ de recherches sur la faune aquatique souterraine. C. r. Acad. Sci., Paris 265: 369–370.

Brinkhurst, R. O. & B. G. M. Jamieson, 1971. Aquatic Oligochaeta of the World. Oliver and Boyd, Edinburgh: 860 pp.

Brunke, M. & T. Gonser, 1997. The ecological significance of exchange processes between rivers and groundwater. Freshwat. Biol. 37: 1–33.

Cellot, B. & J. Juget, 1998. Oligochaete drift in a large river (French Upper Rhône): the effect of life cycle and discharge. Hydrobiologia 389: 183–191.

Coates, K. A., 1986. Redescription of the oligochaete genus *Propappus*, and diagnosis of the new Family Propappidae (Annelida: Oligochaeta). Proc. biol. Soc. Wash. 99: 417–428.

Danielopol, D. L., 1989. Groundwater fauna associated with riverine aquifers. J. n. am. Benthol. Soc. 8: 18–35.

Ferrarese, U. & B. Sambugar, 1976. Ricerche sulla fauna interstiziale iporreica dell'Adige in relazione allo stato di inquinamento del Fiume. Riv. Idrobiol. 15: 47–127.

Füreder, L., 1999. High alpine streams: cold habitat for insect larvae. In Margesin, R. & F. Schinner (eds), Cold Adapted Organisms. Ecology, Physiology, Enzymology and Molecular Biology. Springer-Verlag, Berlin: 181–196.

Gaschignard, O., 1984. Impact d'une crue sur les invertébrés benthiques d'un bras mort du Rhône. Verh. int. Ver. Limnol. 22: 1997–2001.

Giani, N. & P. Lavandier, 1977. Les Oligochètes du torrent d'Estaragne (Pyrénées Centrales). Bull. Soc. Hist. Nat. Toulouse 113: 234–243.

Gibert, J., J. A. Stanford, M.-J. Dole-Olivier & J. V. Ward, 1994. Basic attributes of groundwater ecosystems and prospects for research. In Gibert, J., D. L. Danielopol & J. A. Stanford (eds), Groundwater Ecology. Academic Press, San Diego: 7–40.

Giere, O., 1993. Meiobenthology. The Microscopic Fauna in Aquatic Sediments. Springer-Verlag, Berlin: 328 pp.

Gislason, G. M., J. S. Olafsson & H. Adalsteinsson, 1998. Animal communities in Icelandic rivers in relation to catchment characteristics and water chemistry. Preliminary results. Nord. Hydrol. 29: 129–148.

Hynes, H. B. N., 1983. Groundwater and stream ecology. Hydrobiologia 100: 93–99.

Jones, J. B. & P. J. Mulholland, 2000. Streams and Ground Waters. Academic Press, San Diego: 425 pp.

Juget, J., 1984. Oligochaeta of the epigean and underground fauna of the alluvial plain of the French Upper Rhône (biotypological try). Hydrobiologia 115: 175–182.

Juget J. & N. Giani, 1974. Répartition des Oligochètes lacustres du massif de Néouvielle (Hautes-Pyrénées) avec la description de *Peloscolex pyrenaicus* n. sp. Annls Limnol. 10: 33–53.

Juget, J. & E. Dumnicka, 1986. Oligochaeta (incl. Aphanoneura) des eaux souterraines continentales. In Botosaneanu, L. (ed.), Stygofauna Mundi, E. J. Brill, Leiden: 234–244.

Juget, J. & M. Lafont, 1994. Theoretical habitat templets, species traits, and species richness: aquatic Oligochaetes in the Upper Rhône River and its floodplain. Freshwat. Biol. 31: 327–340.

Kownacki, A., 1991. Zonal distribution and classification of the invertebrate communities in high mountain streams in South Tyrol (Italy). Verh. int. Ver. Limnol. 24: 2010–2014.

Lafont, M., 1989. Contribution à la gestion des eaux continentales: utilisation des oligochètes comme descripteurs de l'état biologique et du degré de pollution des eaux et des sédiments. Thèse de Doctorates Sciences, Université Lyon I: 311 pp.

Lafont, M., A. Durbec & C. Ille, 1992. Oligochaete worms as biological describers of the interactions between surface and groundwater: a first synthesis. Regul. Rivers 7: 65–73.

Lafont M., J. C. Camus & A. Rosso, 1996. Superficial and hyporheic oligochaete communities as indicators of pollution and water exchange in river Moselle, France. Hydrobiologia 334: 147–155.

Maisch, M., 1988. Die Veränderungen der Gletscherflächen und Schneegrenzen seit dem Hochstand von 1850 im Kanton Graubünden (Schweiz). Zeitschrift fuer Geomorphologie Neue Folge Supplementband 70: 113–130.

Malard, F., K. Tockner & J. V. Ward, 1999. Shifting dominance of subcatchment water sources and flow paths in a glacial floodplain (Val Roseg, Switzerland). Arct. Antarct. Alp. Res. 3: 135–150.

Milner, A. M. & G. E. Petts, 1994. Glacial rivers: physical habitat and ecology. Freshwat. Biol. 32: 295–307.

Nielsen, C. O. & B. Christensen, 1959: The Enchytraeidae. Critical revision and taxonomy of European species. Natura Jutl. 8–9: 1–160.

Sambugar, B., N. Giani & E. Martínez-Ansemil, 1999. Groundwater oligochaetes from Southern-Europe. Tubificidae with marine phyletic affinities: new data with description of a new species, review and consideration on their origin. Mém. Biospéol. 26: 107–116.

StatSoft, 1994. STATISTICA for the Macintosh. Vol. 1. StatSoft, Tulsa, U.S.A.: 379 pp.

Strayer, D. & E. Bannon-O'Donnell, 1988. Aquatic Microannelids (Oligochaeta and Aphanoneura) of undergroundwaters of Southeastern New-York. Am. midl. Nat. 119: 327–335.

Strayer, D., S. E. May, P. Nielsen, W. Wolheim & S. Hausam, 1997. Oxygen, organic matter, and sediment granulometry as controls on hyporheic animal communities. Arch. Hydrobiol. 140: 131–144.

Timm, T., 1994. Propappidae and aquatic Enchytraeidae (Oligochaeta) from the farthest southeast of Russia. Hydrobiologia 278: 67–78.

Timm, T., C. Erséus & S. Lundberg, 1996. New and unusual records of freshwater Oligochaeta from the Scandinavian Peninsula. Nord. J. Freshwat. Res. 72: 15–29.

Tockner, K., F. Malard, P. Burgherr, C. T. Robinson, U. Uehlinger, R. Zah & J. V. Ward, 1997. Physico-chemical characterization of channel types in a glacial floodplain ecosystem (Val Roseg, Switzerland). Arch. Hydrobiol. 140: 433–463.

Uehlinger, U., R. Zah & H. Bürgi, 1998. The Val Roseg project: temporal and spatial patterns of benthic algae in an Alpine stream ecosystem influenced by glacier runoff. In Kovar, K., U. Tappeiner, N. E. Peters & R. G. Craig (eds), Hydrology, Water Resources and Ecology in Headwaters. IAHS Press, Wallingford, U.K.: 419–424.

Ward, J. V., F. Malard, K. Tockner & U. Uehlinger, 1999. Influence of ground water on surface water conditions in a glacial floodplain of the Swiss Alps. Hydrol. Process. 13: 277–293.

Hydrobiologia **463:** 83–92, 2001.
P. Rodriguez & P.F.M. Verdonschot (eds), Aquatic Oligochaete Biology VIII.

Oligochaetes in Lake Towada, Japan, an oligotrophic caldera

Akifumi Ohtaka
Department of Natural Science, Faculty of Education, Hirosaki University, Hirosaki 036-8560, Japan
Tel/Fax: 81 172 39 3369. E-mail: ohtaka@cc.hirosaki-u.ac.jp

Key words: Oligochaeta, Caldera Lake Towada, Japan, *Rhyacodrilus komarovi*, depth distribution, new distribution records

Abstract

Twenty-nine oligochaete taxa were recorded from the oligotrophic caldera Lake Towada, northern Japan. All the taxa were found at littoral bottoms shallower than 20 m, and 79% (23 taxa) occurred there exclusively. The profundal bottoms deeper than 40 m were inhabited by three tubificids (*Rhyacodrilus komarovi, Tubifex tubifex* and Tubificidae gen. sp.) and one lumbriculid (*Lumbriculus variegatus*). Among them, *R. komarovi* and Tubificidae gen. sp. were restricted to the profundal zone during the summer stratified season, while they were found at littoral bottoms in winter, suggesting that their bathymetric distributions are regulated by water temperature regime and change seasonally. *R. komarovi,* a primarily riverine tubificid, which was previously known only from Primor'e (Russia) and it is fully described and illustrated in this article, dominated at the profundal bottoms. The cool, oxygen-rich environment at the profundal bottoms of the lake allowed colonization by stenotherm and oxyphilic species. Descriptions and taxonomic remarks are provided for two naidids new to Japan, *Amphichaeta leydigii* and *Pristinella amphibiotica*, based on L. Towada material.

Introduction

There are many volcanic caldera lakes distributed along the north-western Pacific rim including the Japanese islands, the Kuriles and the Kamchatka Peninsula. Most of these calderas were formed in or after the Last Glacial period, and are characterized by deep and strongly isolated basins, and by oligotrophic and often acidic waters. The lake waters are generally rich in oxygen throughout the strata. Thus it is to be expected that many oxyphilic zoobenthos, including oligochaetes, are distributed at every depth of their bottoms. Some ecologists found that oligochaetes are one of the most dominant zoobenthos at the profundal bottoms of Japanese oligotrophic caldera lakes (Miyadi, 1932a, b; Kitagawa, 1974, 1975a, b). However, the species composition is almost entirely unknown at present. *Tubifex* (*Peloscolex*) *nomurai* Yamamoto & Okada, 1940, which was recorded from profundal bottoms deeper than 400 m in Lake Tazawa, is the only species identified solely from the profundal of oligotrophic caldera lakes. However, due to insufficient description and the lack of type material, the validity of this species has been questioned (Brinkhurst, 1981).

The current work is one part of a comprehensive review of oligochaete fauna in oligotrophic caldera lakes in northern Japan. As the first report of this study, the present paper describes the species composition and bathymetric distribution of oligochaetes in Lake Towada in northern Honshu, based on practical surveys carried out from 1997 to 2000.

Study site and methods

Lake Towada, located at the border of the Aomori/Akita Prefectures, northern Honshu, Japan (40° 24′–40° 30′ N, 140° 49′–140° 58′ E; Fig. 1). The lake is situated at an elevation of 400 m, with a surface area of 59.1 km^{2}and a 46.2 km shoreline. The lake basin is composed of a double caldera. The main basin was formed 15 000 years ago and has a broad and flat profundal bottom with a depth of 70–100 m. The secondary basin was formed by an eruption at the southern part of the main basin 6 300 years ago and has a deep bottom of 327 m at the maximum (Hayakawa, 1985; the ages are calibrated). The present lake water has continued without catastrophe for lake organ-

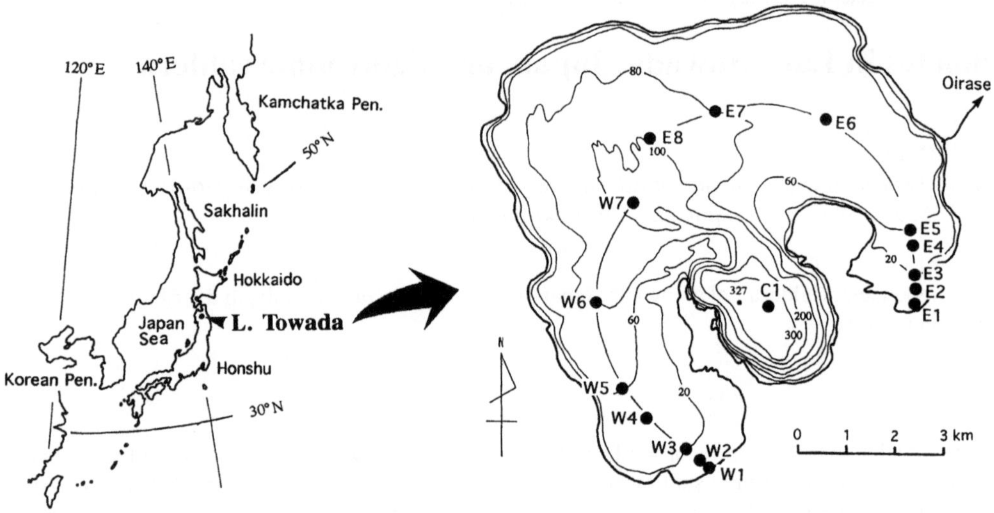

Figure 1. Geographic location of Lake Towada (left) and bathymetric outline of the basin, showing study sites for quantitative surveys (right). Numerals in right figure indicate water depth in m.

isms since the eruption forming the secondary basin (Hayakawa, pers. com.). The lake stratifies thermally during May–November (Takamura et al., 1999). This is an oligotrophic lake; thus, the annual means of TP and TN concentrations in the surface water ranged from 2 to 7 μg l^{-1} during 1981–1996, and from 32 to 82 μg l^{-1} during 1984–1996, respectively (Takamura et al., 1999). The pH of the lake water is almost neutral, and dissolved oxygen concentrations at a central point in the lake did not decrease below 8 mg l^{-1} throughout the water column in 1995–1997 (Takamura et al., 1999). The secchi disk transparency used to be over 20 m, but it has been decreasing markedly during 1985–1996. The recent decrease of transparency has been suggested to be related to a drastic change in fish species from kokanee to pod smelt (Takamura et al., 1999). Eleven species of aquatic macrophytes were recorded from littoral bottoms shallower than 20 m (Nohara et al., 1999). On the basis of functional and ecological difference in organic input, the bottom of the lake is bathymetrically divided into littoral (<20 m), sublittoral (20–40 m) and profundal (>40 m) zones. The lake is fed by 72 small tributaries, half of which dry up frequently, and it is drained by the Oirase River. Lake Towada is protected as a part of the Towada Hachimantai National Park, and is now used for multiple purposes, such as electric-power generation, sight-seeing, irrigation and drinking water, fisheries and sport fishing. Miyadi (1932b) and Kitagawa (1974) once recorded oligochaetes from the lake, but their specific composition has never been studied.

For taxonomic and faunal studies, oligochaete specimens used in the present study were collected from over a hundred sites of various areas and depths in the lake from 1996 to 2000, by the author and by the colleagues whose names are given in the text. The collecting was carried out with the aid of an Ekman-Birge bottom sampler (base 225 cm^2) at deep sites, or with dipnets at littoral sites. Specimens were fixed in 10% formalin solution, and whole mounted on microscope slides in CMCP10, or in Canada Balsam after being dehydrated in a graded series of ethanol and water solutions, and then cleared in methyl salicylate. They were observed and illustrated under a microscope using bright-field and Normarski illumination. Anatomical observation was made of dissected specimens and of specimens cut serially (sections 8–10 μm thick) and stained with haematoxylin and eosin. Specimens of *Rhyacodrilus komarovi* from other localities in Japan were also examined for comparison with Lake Towada form. In addition, to estimate worm densities and biomass, triplicate bottom samples were taken with the Ekman-Birge sampler mentioned above at selected 16 sites among the above sites for taxonomic studies (Fig. 1). The samples were passed through a screen with a 0.25 mm mesh, the retained material was fixed in a 10% formalin solution, and the worms were sorted and counted under a stereomicroscope. Wet weights of fixed oligochaetes were determined by an electronic balance. For identification in the quantitative analysis, temporal slide specimens mounted in water were examined under a microscope when needed. Dissolved oxygen concentrations of overlaying waters were determined at W7 (100 m depth) and C1 (320 m depth) in September 1998 and June 1999. All the specimens used in the present study are deposited in the Faculty of Education, Hirosaki University.

Results

Systematic account

In the present study, a total of 29 oligochaete taxa were found in the lake (Table 1). Descriptions and taxonomic remarks are provided for four notable species as follows.

Family Naididae

Subfamily Naidinae

Amphichaeta leydigii Tauber, 1879 (Fig. 2 A–D)

Amphichaeta leydigii Tauber. Brinkhurst & Jamieson, 1971, p. 315; Brinkhurst & Kathman, 1983, p. 2308, Figure 1.

Material: One immature specimen, off Namariyama, 15 m depth, sandy mud, H. Kato collected.

Description: Single zooid 1.3 mm long in fixed state, 13 segments. No eyes. Prostomium conical. Pharynx elongate. Dorsal chaetal bundles beginning in III, each consisting of 3–4 bifurcate crotchets, 40–52 μm in length (Fig. 2A). Ventral chaetae bifurcate; those in II 4–5 per bundle, 33–42 μm long (Fig. 2B); those in the following segments 2–4 per bundle, 45–55 μm long (Fig. 2 C,D). Upper tooth of ventral chaetae in II a little shorter and thinner than lower, whereas in the following segments both teeth are almost equally long but the upper one still thinner in both dorsal and ventral chaetae. Nodulus of chaetae in II and III situated almost medially, while those in the following segments situated a little distally.

Remarks: Each chaetal characteristic in the present specimen coincides well with that of the North American specimens described by Brinkhurst & Kathman (1983). This is the first recorded occurrence of this species in Japan.

Pristinella amphibiotica (Lastočkin, 1927) (Fig. 2 E–G)

Pristina amphibiotica (Lastočkin). Sperber, 1948, p. 217, Figures 23 A–C, pl. 19; Brinkhurst & Jamieson, 1971, p. 393, Figure 7.22 F–H.

Material: Two immature specimens, gravelly shore at Namariyama, 5 Dec. 1998, Y. Takaya and H. Nakata collected. One immature specimen, gravelly shore at Wainai, 31 July 1999.

Description: Single zooid 1.5–1.9 mm long, 22–24 segments. No proboscis. Dorsal chaetal bundles beginning in II, consisted of 1 smooth hair with 120–250 μm long and 1–2 needle chaetae with 52–61 μm long. The dorsal needle chaetae with nodulus distal and with parallel teeth in which upper tooth shorter and thinner than lower (Fig. 2E). Ventral chaetae in anterior segments 5–7 per bundle, 48–58 μm long, with nodulus distal and with upper tooth almost as long as but thinner than lower (Fig. 2F); those in posterior segment 3–5 per bundle, 45–55 μm long, with nodulus slightly distal and with upper tooth shorter and thinner than lower (Fig. 2G).

Remarks: Distal form of dorsal needle chaetae in the present species closely resembles that of *P. jenkinae* (Stephenson, 1931); Kathman (1985) suggested that *P. jenkinae* is potentially synonymous with *P. amphibiotica*. However, in this article *P. amphibiotica* is regarded as a distinct species because the upper teeth of its ventral chaetae have become shorter than the lower teeth in the posterior segments, unlike *P. jenkinae*, in which both upper and lower teeth are equally long in all segments. The present species was collected together with *P. jenkinae* in Lake Towada. This is the first recorded occurrence of this species in Japan.

Family Tubificidae

Subfamily Rhyacodrilinae

Rhyacodrilus komarovi Timm, 1990 (Figs 3 and 4)

Rhyacodrilus komarovi Timm, 1990, p. 62, Figures 2, 9–17; 1999. p. 3.

Material: Towada basin specimens: over 100 mature specimens, profundal bottoms from 40 to 320 m depth, mud, 6 June 1997, 16 June, 9 Sep. 1998, 15 June 1999. One mature specimen, gravelly shore at Wainai, water temperature 1.5 °C, 23 May 1999. Five mature specimens, a nameless brook on Mt. Shiroji at 775 m altitude and leading to L. Towada, water temperature 7.5 °C, 14 June 1999. Other Japanese specimens: Two mature specimens, lower Churui Stream near Lake Akan, Hokkaido, 6 Aug. 1990, R. B. Kuranishi collected. Two mature specimens, Ninosawa Stream entering Lake Akan-Panke, Hokkaido, 6 Aug. 1990, R. B. Kuranishi collected. 13 mature specimens, upper River Toyohira, Sapporo, Hokkaido, 8, 15 June 1983. 18 mature specimens, a phreatic stream on Mt.

Table 1. Batymetric distribution of oligochaetes in Lake Towada. All records from 1997 to 2000 are compiled. Parenthetic ones are recorded only in winter

Depth (m)	< 20	20–40	40–60	60–80	80–100	320
Haplotaxidae						
Haplotaxis sp.	+					
Lumbriculidae						
Lumbriculus variegatus(Müller, 1774)	+					+
Enchytraeidae						
Enchytraeidae non det.	+					
Naididae						
Chaetogaster diastrophus (Gruithuisen, 1828)	+					
Chaetogaster diaphanus (Gruithuisen, 1828)	+					
Amphichaeta leydigii Tauber, 1879	+					
Nais communis Piguet, 1906	+					
Nais variabilis Piguet, 1906	+					
Nais pardalis Piguet, 1906	+					
Nais bretscheri Michaelsen, 1899	+					
Nais barbata Müller, 1773	+					
Specaria josinae (Vejdovský, 1883)	+					
Dero digitata (Müller, 1773)	+					
Pristina longiseta Ehrenberg, 1828	+					
Pristina aequiseta Bourne, 1891	+					
Pristinella jenkinae (Stephenson, 1931)	+					
Pristinella amphibiotica (Lastočkin, 1927)	+					
Slavina appendiculata (d'Udekem, 1855)	+					
Tubificidae						
Limnodrilus udekemianus Claparède, 1862	+					
Limnodrilus claparedianus Ratzel, 1869	+					
Bothrioneurum vejdovskyanum Stolč, 1888	+					
Branchiura sowerbyi Beddard, 1892	+					
Aulodrilus japonicus Yamaguchi, 1953	+					
Aulodrilus pigueti Kowakewski, 1914	+					
Limnodrilus hoffmeisteri Claparède, 1862	+	+				
Rhyacodrilus coccineus (Vejdovský, 1875)	+	+				
Tubufex tubifex (Müller, 1774)	+	+	+	+		+
Rhyacodrilus komarovi Timm, 1990	(+)		+	+	+	+
Tubificidae gen. sp.	(+)		+	+	+	+

Maruyama, Sapporo, Hokkaido, water temperature 7.5–11.3 °C, 4 Oct. 1983, 23 Apr., 18 May 1984, 20 June 1985. Seven mature specimens, upper Izari River on Mt. Eniwa, Eniwa, Hokkaido, altitude 600 m, water temperature 4.2 °C, 28 Apr. 1984. Four mature specimens, upper Shiribetsu River, Hokkaido, 10 Apr. 1997, S. Urawa collected. Two mature specimens, upper Ukawa Stream on Mt. Iwaki, Hirosaki, Aomori Prefecture, 23 Sep. 1991. Eight mature and 2 immature specimens, upper Ôwasawa Stream, Hirosaki, Aomori Prefecture, 8 Aug. 1993, water temperature 5.6 °C. One mature specimen, upper Komine Stream on Mt. Shirakami, Iwasaki Village, Aomori Prefecture, altitude 435 m, water temperature 13.5 °C, 14 Aug. 1996.

Description: Body pinkish in living state, mature worms 7–13 mm and 4–9 mm long in living and fixed state respectively, 0.3–0.45 mm wide at middle of the body, segments 46–62. Prostomium conical and pointed. Dorsal and ventral chaetae uniform in shape, all bifurcate crotchets. Chaetae in anterior segments 3–7 per bundle, 76–120 μm long, with nodulus slightly

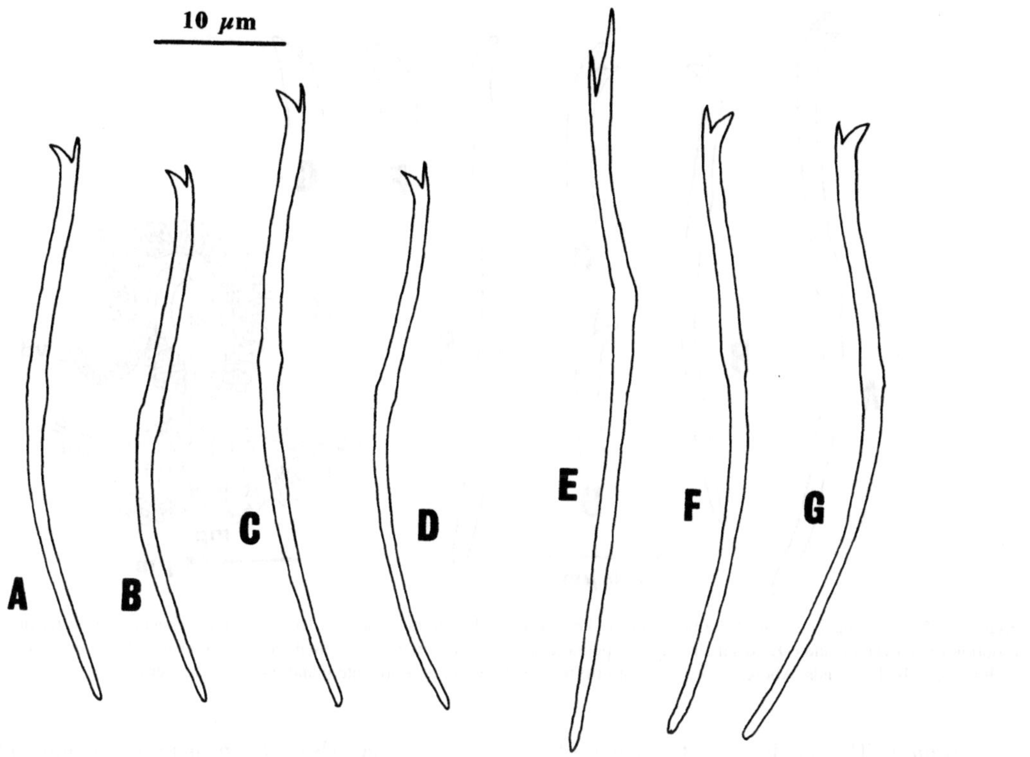

Figure 2. Amphichaeta leydigii and *Pristinella amphibiotica* from Lake Towada. (A–D) *Amphichaeta leydigii*: (A) dorsal chaeta in III; (B) ventral chaeta in II; (C) ventral chaeta in III; (D) ventral chaeta in VII. (E–G) *Pristinella amphibiotica*: (E) dorsal needle chaeta in VI; (F) ventral chaeta in III; (G) ventral chaeta in XV.

distal, and with upper tooth as long as or a little longer, and the lower one thinner (Fig. 3A); those in posterior segments 3–5 per bundle, 68–120 μm long, with nodulus distal and with upper tooth as long as or a little shorter and thinner than the lower one (Fig. 3B). One to several short and very fine intermediate teeth often found in dorsal as well as ventral chaetae (Fig. 3 C,D). Penial chaetae situated close to male pore medially in XI, 3–4 per bundle, 90–104 μm long, with the heads close together; nodulus situated at 1/4 from distal end (Fig. 3E) and distal end simply hooked or split into some to several rudimentary teeth (Fig. 3 F,G).

Transparent cutaneous glands abundant throughout the body. Coelomocytes numerous, spherical in shape with 8–14 μm in diameter. Pharynx in III, covered with dense pharyngeal glands dorsally. Chloragogen cells beginning in V and abundant from VI on. Intestine becoming wider from XII to XV on. Nephridia from VIII on, absent from X to XII in mature worms.

Clitellum from 3/4X to end of XII, swollen laterally and flattened ventrally. Testes and ovaries in X and XI, respectively. Male funnels 50–80 μm in diameter, small and flat. Vasa deferentia little convoluted, 10–24 μm thick, the lumen less than 4 μm thick. Ectal parts of vasa deferentia thickened and jointed with ental one-third of atria (Fig. 4A), the junctions of lumen with long cilia. Atria (Fig. 3H) elongate ovoid in shape, about 100 μm long by 50 μm in maximum width. Inner epithelium of atria 5–6 μm thick and the muscular covering 4–6 μm thick. Atria densely covered with diffuse prostate cells. Ectal part of atria becoming narrow and leading to ciliated protrusible pseudopenis. Male pores located laterally to ventral chaetal bundles in XI (Fig. 4B). Pseudopenes conical when protruded. Spermathecae in X, opening immediately behind septum 9/10 ventrolaterally (Fig. 4C). Spermathecal ampullae globular, 70–80 μm in diameter, the wall 6–8 μm thick, not connected with alimentary canal. Spermathecal ducts short and thick-walled, about 50 μm long. Some to several loose sperm masses in spermathecal ampullae. Sperm sac from IX to XIII and egg sac up to XI.

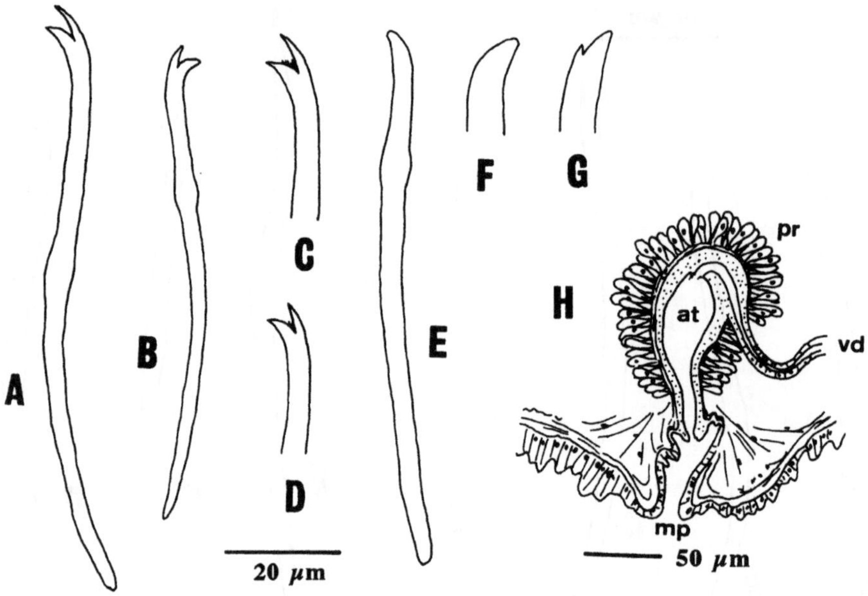

Figure 3. Rhyacodrilus komarovi from profundal Lake Towada. (A) ventral chaeta in III; (B) ventral chaeta in a posterior segment; (C, D) variation of (C) ventral and (D) dorsal chaetae; (E) penial chaeta; (F, G) variation of penial chaetae; (H) sagittal section of atrium. C, D, F, and G showing only distal ends. Abbreviations: at – atrium; mp – male pore; pr – prostate gland; vd – vas deferens.

Variation: The number and frequency of occurrence of intermediate teeth on chaetae varied considerably among specimens. They were less developed in Lake Towada material, in which a few short teeth were sometimes found in ventral or dorsal chaetae. On the other hand, they were greater in number and more distinct in other Japanese material, and were often found both in dorsal and ventral chaetae in almost all segments.

Remarks: *Rhyacodrilus komarovi* Timm, originally described and recorded from Primor'e (Timm, 1990, 1997, 1999), is characterized in lacking hair chaetae and having an elongate atrium joined to the ectal part of vas deferens. Although Timm (1990) suggested that *R. komarivi* is close to *R. levanidovae* Sokolskaya, 1973, he regarded the absence of intermediate teeth in chaetae, the chloragogen tissue starting from VI (V in *levanidovae*), and the position of the male pores located laterally from the penial chaetae as the characteristics distinguishing *R. komarovi* from *R. levanidovae*. Intermediate teeth were described in posterior ventral chaetae of *R. levanidovae* (Sokolskaya, 1973). On the other hand, in the present material they vary considerably and are often found both in dorsal and ventral chaetae throughout the segments. Thus, intermediate teeth cannot be regarded as a specific characteristic, as in many other tubificid and naidid species. In addition, chloragogen tissue may begin gradually, and the segmental position is often difficult to specify clearly in the present material and is not a reliable specific characteristic, either. The glandular cuticle described in the original description of *R. levanidovae* and also found in the present material was confirmed in *R. komarovi* (Timm, 1999). In the present material, the male pores were invariably located exactly lateral to the penial chaetal bundle, at a location identical with *R. komarovi*. On the other hand, in the original account of *R. levanidovae*, Sokolskaya (1973) described how the male pore is located on the line of chaetae (occasionally behind them) below the ventral chaetal bundle. The position of the male pores is the only essential difference between *R. komarovi* and *R. levanidovae*. In the present paper, *R. komarovi* and *R. levanidovae* are regarded as different species because the location of the male pores can be regarded as an important specific characteristic. However, it is also possible that the location of the male pores varies geographically and that both are synonyms.

R. komarovi has been recorded from Primor'e (Timm, 1990, 1997, 1999) and is now recorded from Hokkaido and northern Honshu. It is one of the most common tubificids in several streams in Primor'e. In

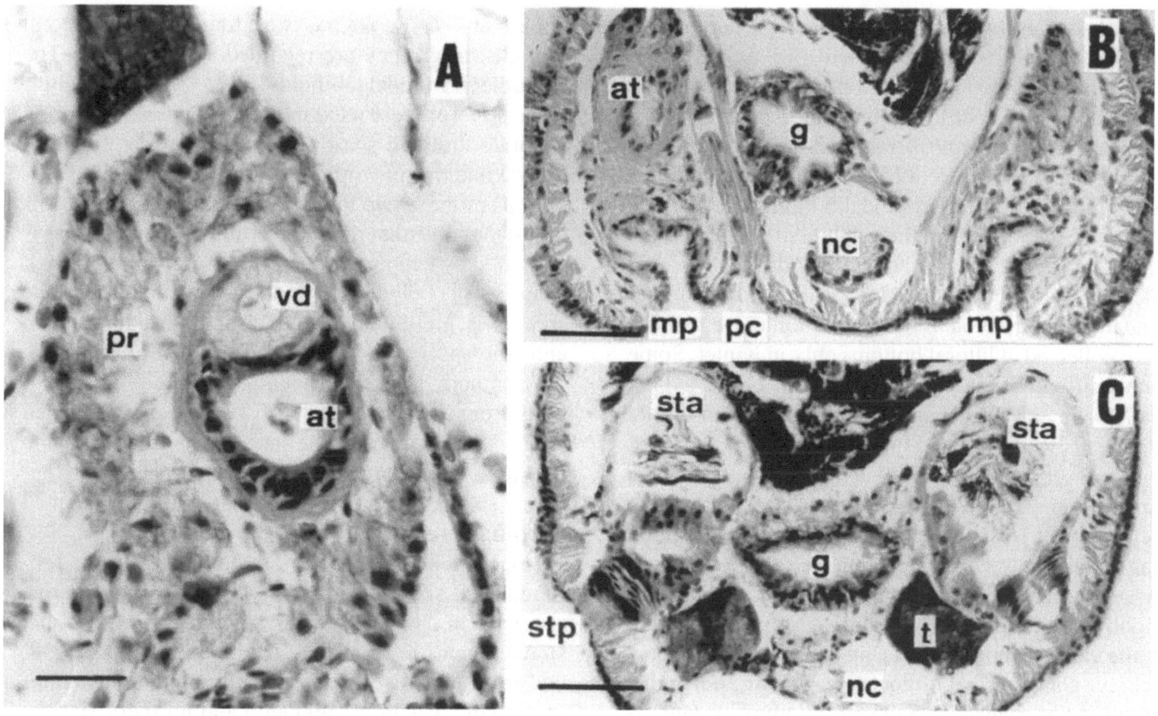

Figure 4. Genital organs in *Rhyacodrilus komarovi* from Japan. (A) ental part of atrial ampulla in cross section, Maruyama specimen; (B) cross section at line of male pore, Ôwasawa Stream specimen; (C) cross section at line of spermathecal pore, Maruyama specimen. Abbreviations: at – atrium; g – gut; mp – male pore; nc – nerve cord; pc – penial chaeta; pr – prostate gland; sta – spermathecal ampulla; stp – spermathecal pore; t – testis; vd – vas deferens. Scale bar: A – 20 μm; B and C – 50 μm.

Japan, it also dominates in many mountain streams and phreatic waters, and in the profundal of Lake Towada.

Tubificidae gen. sp.

Material: Eight mature and 18 immature individuals, profundal Lake Towada, with 40–320 m depth, in mud, 6 June 1997, 16 June, 9 Sep. 1998, 15 June 1999. Two immature specimens, gravelly shore at Wainai, Lake Towada, water temperature 1.5 °C, 23 May 1999.

Remarks: The present species is less than 10 mm long, and has long hair and bifid chaetae in dorsal bundles, and simple-pointed and bifurcate crotchets in anterior half of ventral bundles. This species provides two kinds of genital chaetae in XI. There is no coelomocyte. The atrium is oval to spindle-shaped with a large prostate gland apically, but the whole structure of the male duct has not been reconstructed, due to the limited amount of material. The compositions of dorsal and ventral somatic chaetae and the presence of two kinds of genital chaetae in XI in the present species suggest that this species belongs to the small and poorly known Tubificinae genus *Krenedrilus* Dumnicka, 1983, which was redefined by Giani et al. (1990). However, in the present paper, I reserve the generic status of this species, because the structure of the male duct, which is one of the most important subfamilial and generic criteria, is not reconstructed yet.

Bathymetric distribution of oligochaetes

In Lake Towada, the composition of bottom sediments is heterogeneous at littoral sites at depths from 0 to 20 m; they are gravelly, sandy or muddy, often with submerged vegetation. On the other hand, the sediments are uniformly fine-silty at profundal bottoms deeper than 40 m. Dissolved oxygen concentrations in overlaying waters are more than 90% saturated, even at the deepest bottom in summer stratified seasons.

The species composition of oligochaetes changed clearly along the water depth gradient (Table 1). All 29 taxa recorded from the lake were found at littoral bottoms shallower than 20 m, comprising one haplotaxid, one lumbriculid, one enchytraeid, 15 naidids, and 11 tubificids. Among them, 23 taxa were exclusively found at littoral zones and never at deeper bottoms. Two tubificids, *Limnodrilus hoffmeisteri* and *Rhyacodrilus coccineus*, were collected at sublittoral bottoms as deep as 40 m. The profundal bottoms deeper than 40 m were inhabited by three tubificids, *Rhyacorilus komarovi*, *Tubifex tubifex* and Tubificidae gen. sp. and one lumbriculid, *Lumbriculus variegatus*. Two profundal tubificids, *R. komarovi* and Tubificidae gen. sp. were collected at littoral bottoms only in winter. Some faunal differences are found between the broad and flat main basin and the deeper secondary basin, in that *L. variegatus* was not found in the former.

Total oligochaete density and biomass in the lake ranged from 222 to 8207 m^{-2} and from 0.07 to 4.07g wet wt m^{-2}, respectively (Table 2). Little relation was found between water depth and the total density or biomass, except that they tended to increase slightly at sublittoral zones and were markedly high at the deepest bottom (C1) in the secondary basin. Relative abundance of oligochaetes among zoobenthic communities ranged 12.8–96.5% by number, and increased clearly at deeper bottoms (Ohtaka et al., 1999). *T. tubifex* and/or *Limnodrilus* species dominated at littoral and sublittoral bottoms (W1–W3; E1–E3). On the other hand, *R. komarovi* dominated widely at the profundal bottoms (W4–W7; E4–E8; C1).

Discussion

Lake Towada stratifies thermally from May to November with a metaliminion around 20 m in depth (Shoji et al., 1985; Takamura et al., 1999). The water temperature attains 21–22 °C in the epiliminion in summer, while in deeper strata below 40 m it is constant about 4 °C throughout the year, except for the central basin in which unusual changes of temperature were detected (Shoji et al., 1985). The oligochaete composition in the present lake changed clearly along the water depth gradient. Among 29 taxa recorded from the lake, 23 taxa (79%) exclusively occurred at littoral bottoms shallower than 20 m depth. On the other hand, the profundal fauna at bottoms deeper than 40 m was composed of only four species, *Rhyacodrilus komarovi, Tubifex tubifex*, Tubificidae gen. sp. (Tubificidae) and *Lumbriculus variegatus* (Lumbriculidae). Among them, *T. tubifex* was collected at nearly every depth, and *L. variegatus* was found also at littoral bottoms; thus, they are regarded as eurybathic. The two other profundal tubificids, *R. komarovi* and Tubificidae sp., were restricted in the profundal zone during summer stratified season, while they were found at littoral bottoms in winter. It suggests that depth distributions of these two tubificids change seasonally. It is highly possible that their upper limit of distributions is regulated by temperatures and shifts upward in winter. Thus, it is concluded that no euprofundal oligochaete species is present in Lake Towada. In addition, no endemic species, nor any form adapted to profundal environment, has so far been found in any groups of biota from Lake Towada. This absence may be related to the short history of the lake and is different from that of so-called ancient lakes, which have many endemic species, including some that are highly adapted to profundal environments (Martin, 1996).

One tubificid, *R. komarovi,* dominated in the profundal of Lake Towada. This is primarily a riverine species that has been recorded from several cold and stony streams in Primor'e (Timm, 1990, 1997, 1999) and from many stony and gravelly mountain streams or phreatic waters in Hokkaido and northern Honshu (present study). Colonization of stenotherm, oxyphilic and primarily riverine species at profundal bottoms also applies to other groups of animals in Lake Towada. For example, Kawakatsu et al. (1976) found a stream-dwelling planarian, *Polyceris (Polyceris) sapporo* (Ijima & Kaburaki, 1916) from profundal bottoms deeper than 70 m. An amphipod, *Eogammarus kygi* (Derzhavin, 1923), and a copepod, *Eucyclops serrulatus* (Fischer, 1851), both of which are eurybathic but flourish at the profundal bottoms of the lake (Ohtaka et al., 1999), are also known from mountain streams or phreatic waters in Japan (Ishida, 1997; Morino & Totsuka, 1999). This suggests that the riverine species are not bound to current or stony bottoms, but to low temperature with a good oxygen supply. Caldera lake basins are generally strongly isolated with limited watersheds. In addition, many of them do not have histories long enough to allow remote populations to invade. Therefore, it can also be expected that the lacustrine fauna in caldera lakes is closely related to that within its watershed, rather than to show a faunal assemblage typical of profundal deep lakes. The present survey found *Rhyacodrilus komarovi* in the upper course of a stream leading into Lake Towada, and the profundal form could be directly derived from stream populations within the watershed.

Table 2. Densities and biomass of oligochates in Lake Towada, based on the quantitative surveys carried out in June 1997 (W1–W7), September 1997 (C1), and June 1998 (E1–E8)

Station	W1	W2	W3	W4	W5	W6	W7	E1	E2	E3	E4	E5	E6	E7	E8	C1
Depth (m)	5	10	20	40	60	80	100	5	10	20	40	60	70	80	100	320
Density (No m^{-2})																
Tubificidae gen. sp.	0	0	0	726	0	15	281	0	0	0	0	0	30	59	133	0
Rhyacodrilus komarovi	0	0	0	1156	385	1215	252	0	0	0	178	1496	711	711	1526	5970
Rhyacodrilus coccineus	0	0	30	30	0	0	0	44	15	0	0	0	0	0	0	0
Tubifex tubifex	0	89	178	0	0	0	0	252	193	548	0	178	0	0	0	2237
Aulodrilus pigueti	74	0	0	0	0	0	0	0	15	0	0	0	0	0	0	0
Limnodrilus hoffmeisteri	119	44	148	0	0	0	0	44	44	44	0	0	0	0	0	0
Limnodrilus immature	548	59	652	0	0	0	0	667	311	533	15	0	0	0	0	0
Other Tubificidae	74	0	0	0	0	0	0	15	0	0	15	0	0	0	0	0
Chaetogaster diastrophus	0	0	0	0	0	0	0	0	0	15	0	0	0	0	0	0
C. diaphanus	0	0	0	0	0	0	0	0	89	0	0	0	0	0	0	0
Nais pardalis	0	0	0	0	0	0	0	15	0	0	0	0	0	0	0	0
Specaria josinae	0	0	0	0	0	0	0	0	0	44	0	0	0	0	0	0
Slavina appendiculata	0	0	0	0	0	0	0	44	0	0	0	0	0	0	0	0
Dero digitata	0	30	0	0	0	0	0	0	15	0	0	0	0	0	0	0
Total density	815	222	1007	1911	385	1230	533	1081	681	1185	207	1674	741	770	1659	8207
Oligochaeta biomass (g wet wt. m^{-2})	1.04	0.12	2.04	1.41	0.18	0.90	0.10	1.23	0.47	2.73	0.07	0.58	0.24	0.15	0.50	4.07

Tubifex (*Peloscolex*) *nomurai* Yamamoto & Okada, 1940 was originally described from the profundal bottom deeper than 400 m of the oligotrophic caldera Lake Tazawa, adjacent to L. Towada. Brinkhurst & Jamieson (1971) once regarded it as a valid species, while Holmquist (1979) noted that the genital organs of this species are very suggestive of the condition in *Embolocephalus nikolskyi* (Lastočkin & Sokolskaya, 1953); Brinkhurst (1981) later dubiously synonymized it with *E. nikolskyi*. *Embolocephalus nikolskyi* has been recorded from several mountain streams and currents originating from phreatic waters in Hokkaido and northern Honshu, Japan (Ohtaka, unpubl.), and it is highly probable that this species invaded profundal Lake Tazawa. At present, we have too little information to compare and discuss oligochaete fauna among the oligotrophic caldera lakes. To understand the faunal characteristics of such kind of lakes and their formation processes, further faunal surveys are needed in other lakes as well as in their watersheds.

It is generally accepted that besides temperature and dissolved oxygen, a trophic condition is the most important factor determining the abundance of oligochaetes in lakes (Brinkhurst, 1974; Reynoldson, 1990). In the present study, oligochaetes were fewer than 2000 m^{-2} in density, and less than 3 g m^{-2} in wet weight in the main basin. The density compares with those in the lower dimictic or upper abyssal zones of the oligotrophic Siberian Lake Baikal (Martin et al., 1999), while it is 2–4 times lower than in the profundal zone of the mesotrophic north basin of Lake Biwa, Japan (Nishino & Ohtaka, unpubl.), and 10–50 times lower than in the hypertrophic Lake Suwa, Japan (Fukuhara et al., 1987). On the other hand, the oligochaete density and biomass were much higher at the deepest bottom of the secondary basin than in the main basin. The reason for the abundance is not clear, but the trophic condition may be better in the secondary basin because it has steep cliffs and the profundal zone is closer to shore than in the main basin, and a greater amount of nutritional matter may be supplied. The thermal condition may also affect the abundance. An unusual vertical distribution in water temperature has been detected in the secondary basin of Lake Towada, in that the temperatures gradually increased downward from 100 m depth and reached about 7.5 °C at the bottom at over 300 m depth (Shoji et al., 1985). Shoji et al. (1985) suspected that there are some heat resources at the bottom. The higher temperature in the secondary basin may be responsible for the accelerated secondary production for oligochaetes.

Acknowledgements

I am grateful to Dr Tarmo Timm (Võrtsjärv Limnological Station, Ranau, Estonia) for his invaluable taxonomic advice. Thanks are also due to Dr Yukio Hayakawa (Gunma University, Japan) for showing me his geologic history of the Towada volcano. I am also indebted to Dr Hideo Kato and Mr Ryuhei Ueno (National Institute for Environmental Studies, Tsukuba, Japan), Dr Shigehiko Urawa (National Salmon Resources Center, Sapporo, Japan), Messrs. Yosuke Takaya (Hokkaido University, Japan), Nobuhiro Nakata (Hirosaki University, Japan), and Mr Ryoichi B. Kuranishi (Chiba Natural History Museum and Institute, Japan), for providing me with oligochaete specimens for the present study.

References

Brinkhurst, R. O., 1974. The Benthos of Lakes. Macmillan Press, London: 190 pp.

Brinkhurst, R. O., 1981. Contribution to the taxonomy of the Tubificinae (Oligochaeta: Tubificidae). Proc. biol. Soc. Wash. 94: 1048–1067.

Brinkhurst, R. O. & B. G. M. Jamieson, 1971. Aquatic Oligochaeta of the World. Oliver & Boyd, Edinburgh: 860 pp.

Brinkhurst, R. O. & R. D. Kathman, 1983. A contribution to the taxonomy of the Naididae (Oligochaeta) of North America. Can. J. Zool. 61: 2307–2312.

Dumnicka, E., 1983. Tubificidae (Oligochaeta) from subterranean waters, with description of two new genera. Bijdr. Dierk. 53: 255–261.

Fukuhara, H., A. Ohtaka, Y. Isobe & M. Sakamoto, 1987. Seasonal changes in vertical distribution, biomass and faecal production of tubificids in the profundal region of a shallow Japanese lake. Holarct. Ecol. 10: 8–13.

Giani, N., C. Erséus & E. Martínez-Ansemil, 1990. Redefinition of the subterranean genus *Krenedrilus* Dumnicka (Oligochaeta, Tubificidae) with a redescription of *K. papillatus* Dumnicka, 1983 and the description of two new species. Stygologia 5: 55–65.

Hayakawa, Y., 1985. Pyroclastic geology of Towada Volcano. Bull. Earthquake Res. Inst. Univ. Tokyo 60: 507–592.

Holmquist, C., 1979. Revision of the genus *Peloscolex* (Oligochaeta, Tubificidae) 2. Scrutiny of the species. Zool. Scr. 8: 37–60.

Ishida, T., 1997. *Eucyclops roseus*, a new Eurasian copepod, and the *E. serrulatus-speratus* problem in Japan. Jap. J. Limnol. 58: 349–358.

Kathman, R. D., 1985. Synonymy of *Pristina jenkinae* (Oligochaeta: Naididae). Proc. biol. Soc. Wash. 98: 1022–1027.

Kawakatsu, M., W. Teshirogi & T. Tokui, 1976. Record of a freshwater planarian, *Polycelis sapporo* (Ijima et Kaburaki, 1916), from the bottom of Lake Towada-ko in north Japan, with a note on the southern limit of distribution of this species. Physiol. Ecol. Jap. 17: 477–483.

Kitagawa, N., 1974. Studies on the bottom fauna of seven lakes of north Japan. Jap. J. Limnol. 35: 162–172 (in Japanese with English abstract).

Kitagawa, N., 1975a. Studies on the bottom fauna of five lakes in southern Hokkaido (Lakes Shikotsu-ko, Kuttara-ko, Toya-ko, Hangetsu-ko and Oshima O-numa). Jap. J. Limnol. 36: 48–54 (in Japanese with English abstract).

Kitagawa, N., 1975b. Studies on the bottom fauna of lakes Mashu-ko, Akan-Panke-ko and Toyoni-ko, Hokkaido, Japan. Jap. J. Ecol. 25: 155–159.

Lastočkin, D. A. & N. L. Sokolskaya, 1953. New species of the genus *Peloscolex* (Tubificidae) from the Amur basin. Zool. Zh. 32: 409–412 (in Russian).

Martin, P., 1996. Oligochaeta and Aphanoneura in ancient lakes: a review. Hydrobiologia 334: 63–72.

Martin, P., K. Martens & B. Goddeeris, 1999. Oligochaeta from the abyssal zone of Lake Baikal (Siberia, Russia). Hydrobiologia 406: 165–174.

Miyadi, D., 1932a. Studies on the bottom fauna of Japanese lakes. VII. Lakes of Hokkaido. Jap. J. Zool. 4: 223–255.

Miyadi, D., 1932b. Studies on the bottom fauna of Japanese lakes. VIII. Lakes of North Japan. Jap. J. Zool. 4: 258–287.

Morino, H. & T. Totsuka, 1999. Taxonomy, distribution and breeding activity of *Eogammarus kygi* (Crustacea: Amphipoda) in Lake Towada. Res. Rep. Natl. Inst. Environ. Stud. Jap. 146: 87–94 (in Japanese).

Nohara, S., H. Kato, N. Takamura & H. Mikami, 1999. Aquatic higher plant vegetation studies in Lake Towada, results of a 1997 survey. Res. Rep. Natl. Inst. Environ. Stud. Jpn. 146: 49–54 (in Japanese).

Ohtaka, A., H. Kato, R. Ueno, T. Ishida, H. Abé, K. Ida & H. Morino, 1999. Fauna and bathymetric distribution of zoobenthos in Lake Towada, northern Japan. Res. Rep. Natl. Inst. Environ. Stud. Jap. 146: 55–71 (in Japanese).

Reynoldson, T. B., 1990. Distribution patterns of oligochaetes in the English Lake District. Arch. Hydrobiol. 118: 303–339.

Shoji, H., T. Nara, K. Oyamada, M. Noda, T. Hayakari, Y. Noro & N. Sato, 1985. Vertical distribution of water temperature in Lake Towada. Rep. Aomori Pref. Inst. Environ. Center 7: 90–94 (in Japanese).

Sperber, C., 1948. The taxonomical study of the Naididae. Zool. Bidr. Upps. 28: 1–296.

Sokolskaya, N. L., 1973. New Tubificidae species from Kamchatka and new findings of oligochaetes in the basin of the peninsula. Bull. Moscow Soc. Nat. Sect. Biol. 78 (5): 54–66 (in Russian).

Stephenson, J., 1931. Oligochaeta from Burma, Kenya, and other parts of the world. Proc. zool. Soc. Lond. 1–2: 33–91

Takamura, N., H. Mikami, H. Mizutani & K. Nagasaki, 1999. Did a drastic change in fish species from kokanee to pond smelt decrease the secchi disc transparency in the oligotrophic Lake Towada, Japan? Arch. Hydrobiol. 144: 283–304.

Timm, T., 1990. Aquatic Oligochaeta from the farthest south-east of the USSR. Proc. Estonian Acad. Sci. Biol. 39: 55–67.

Timm, T., 1997. Freshwater Oligochaeta of some urban water-courses in the Russian Far East. Int. Rev. ges. Hydrobiol. 82: 437–467.

Timm, T., 1999. Oligochaeta of the Edinka Stream, Russian Far East. Species Diversity 4: 1–7.

Yamamoto, G. & K. Okada, 1940. An anatomical study of *Tubifex* (*Peloscolex*) *nomurai* n. sp., obtained at the bottom of Lake Tazawa, the deepest lake of Japan. Sci. Rep. Tohoku Imp. Univ. Biol. 15: 427–455.

Hydrobiologia **463:** 93–102, 2001.
P. Rodriguez & P.F.M. Verdonschot (eds), Aquatic Oligochaete Biology VIII.

Distribution and dispersal capacity of the Ponto-Caspian tubificid oligochaete *Potamothrix moldaviensis* Vejdovský et Mrázek, 1903 in the Baltic Sea Region

Göran Milbrink[1] & Tarmo Timm[2]
[1]*Department of Animal Ecology, Evolutionary Biology Centre, Uppsala University, Norbyv. 18D, S-752 36 Uppsala, Sweden*
Fax: 46 18 4716484. E-mail: goran.milbrink@ebc.uu.se
[2]*Võrtsjärv Limnological Station, EE 61101 Rannu, Tartumaa, Estonia*
E-mail: ttimm@zbi.ee

Key words: Ponto-Caspian species, *Potamothrix*, Tubificidae, dispersal, ballast water, zoogeography

Abstract

Over the last few centuries, several Ponto-Caspian tubificid oligochaetes have gradually dispersed from the Black Sea – Caspian Sea region to the north-west and west over Europe. The present world distribution comprising also the Great Lakes of North America clearly demonstrates that anthropochorous vectors of dispersal are involved. Passive transportation in the ballast water of ships has radically changed the possibilities of dispersal for many invertebrate species and has even made dispersal between continents possible. The construction of navigable canals has furthermore facilitated the crossing of watersheds and continents. Other likely vectors of longway dispersal for oligochaetes, as well as for other small-size aquatic invertebrates, are birds and mammals. The dispersal of the *Potamothrix* species is likely to have taken place in successive waves (three) with front-lines still on the move from the east to the west over the Baltic States and Scandinavia. The rheophilous species *Potamothrix moldaviensis* has presently reached – apart from the large rivers of Russia and many Central-European water bodies – also the Baltic States and south-eastern Sweden. Trajectories of dispersal demonstrate routes across the Baltic Sea – via the ballast water of ships. In the largest rivers of the Eastern Baltic Region (Neva, Daugava, Nemunas), downstream dispersal is the most likely way of transportation. *P. moldaviensis* together with *P. heuscheri* (second wave) and *P. vejdovskyi* (third wave) are presently forming front-lines running obliquely from the north-east to the south-west over south Sweden. In mesotrophic-eutrophic basins of eastern Lake Mälaren, the abundance as well as the species diversity of oligochaetes is particularly high wherever Ponto-Caspian *Potamothrix* species – often several species together – are involved.

Introduction

There is a growing awareness in the world of human-aided dispersal of species beyond their natural distribution ranges. This is a major threat to biodiversity (Vitousek, 1997). Aquatic invertebrates are no exceptions in this respect (Havel et al., 1995). Today many euryoic invertebrate species of Ponto-Caspian origin (from the Black, Caspian and Azov Sea areas) are expanding their distribution ranges, some – such as the zebra mussel, *Dreissena polymorpha* – with obvious negative effects on the indigenous fauna (Mills et al., 1993; Dumont, 1998; Ricciardi & MacIsaac, 2000). This fauna native to fresh and brackish waters of the Ponto-Caspian region once evolved in different basins of that region, in reality the last remnants of the Tethys Sea (Dumont, 1998). A very complicated geological history leading to fluctuating water levels and salinities has promoted the evolution of tolerance to extreme salinity fluctuations in these species.

Ballast water is the main agent of dispersal of small aquatic animals and plants from one continent to the next (Hallegraeff & Bolch, 1991). Shipping routes around the world – along which numerous invading species are known to have dispersed particularly frequently – have been labelled 'conveyor belts'

(Carlton & Geller, 1993). The construction of navigable canals has furthermore facilitated the crossing of watersheds and continents after which rapid downstream migration could take place. We know that many Ponto-Caspian species have expanded their distribution ranges across Europe during the 19th (maybe even the 18th) century, much due to, of course, the construction of shipping canals and reservoirs, but also by new habits of stocking invertebrates as food to increase fish production.

Freshwater tubificid oligochaetes of Ponto-Caspian origin pertain to the fauna rapidly expanding over Europe, for instance to the north-west, reaching the Baltic Sea Region. Several species have even reached the Great Lakes of North America. Ballast water is the most likely means of transport in addition to ordinary downstream migration, rafting, cocoons spread by the clothes of sport-fishermen, equipment and boats, dispersal by bird and mammal vectors, etc. (Timm, 1980, 1987; Milbrink, 1999). Active dispersal by tubificid oligochaetes is in any case an immensely slow process and could here be reasonably ruled out (Timm, 1980). Ponto-Caspian tubificids are known to have dispersed over Europe in three more or less discrete waves (T. Timm, unpublished) and these fronts are fully discernible in Sweden. Figure 1 is an attempt to illustrate the early phases of the second wave, which comprises the species *P. moldaviensis*. In Milbrink (1999), the distribution pattern in the Baltic Sea area of another member of the second wave, *Potamothrix heuscheri*, was presented along with likely vectors of dispersal, i.e. ballast water and birds. Figure 2 shows the present positions of the three waves in Sweden. The most important species of the first wave is *P. hammoniensis*, which has not yet reached Norway, and of the third wave, *P. vejdovskyi*, which has reached the eastern and central parts of Lake Mälaren. In this work, we present the known distribution pattern of *P. moldaviensis* – mainly a rheophilous species, and discuss the associated likely modes and routes of dispersal in the Baltic Sea Region (Fig. 3 – world distribution). We make the realistic assumption that ships and the clothes, equipment and boats of sport-fishermen are the most important vectors of dispersal. We also hypothesize that birds, which appeared to be most realistic vectors of dispersal in the case of *P. heuscheri*, under certain circumstances also are likely vectors for *P. moldaviensis*. Circumstances of that kind are wading birds temporarily exploiting littoral areas of lakes, as well as rock-pools, especially along migratory routes (Milbrink, 1999).

World distribution of Potamothrix moldaviensis

Potamothrix moldaviensis has a world distribution (Fig. 3) which is rather similar to that of *P. vejdovskyi*, although the latter species is believed to belong to the third wave of dispersal. *P. moldaviensis* is today a common species in Central Europe – Switzerland and France (Juget, 1967; Giani, 1979; Lang, 1984; Lang & Reymond, 1996), and in Eastern Europe (Timm, 1987). However, this species has not yet been found in this respect, in more peripheral parts of Europe, such as Spain (Prat et al., 1992), Italy (Bonacina et al., 1992; Bazzanti et al., 1993), Norway and Finland (Särkkä, 1978). However, most species of the genus *Potamothrix* are presently expanding their distribution (Milbrink, 1980, 1999) and *P. moldaviensis* is likely to have reached some of the countries of southern Europe, as well, even though records to date are lacking.

P. moldaviensis is one of three species of the genus (*P. moldaviensis, P. vejdovskyi* and *P. bedoti*), that have reached the St. Lawrence River and the Great Lakes of North America. This species has also reached a tributary of the Enisej River and the vicinity of Lake Baikal in Siberia (Gold & Zapekina-Dulkejt, 1972; Popcenko, 1974).

Materials and methods

The present study is based on bottom fauna material from Sweden, Norway, the Baltic countries, and northeastern Russia, and covers analyses from the 1960s until today of the oligochaete fauna of numerous lakes and rivers, especially in Sweden (Milbrink, 1980; Milbrink, unpublished) and Estonia (Timm, 1987).

Extensive sampling networks exist since the early 1960s from the large lakes of southern Sweden – Mälaren, Vänern, Vättern and Hjälmaren. Since then, samples have been taken from these Swedish lakes at regular intervals from these networks by the Swedish Environmental Protection Agency (E.P.A.). On each locality and sampling occasion, 5 Ekman grab samples had been taken and sieved through a net with a 0.5 mm mesh-size. The samples were preserved in alcohol and the animals were mounted in glycerin to facilitate species identification.

Comparisons have also been made with earlier bottom fauna surveys carried out in Lake Mälaren. In the summers of 1915 and 1916, fishery biologists Oskar Nordqvist and Gunnar Alm (Institute of Freshwater

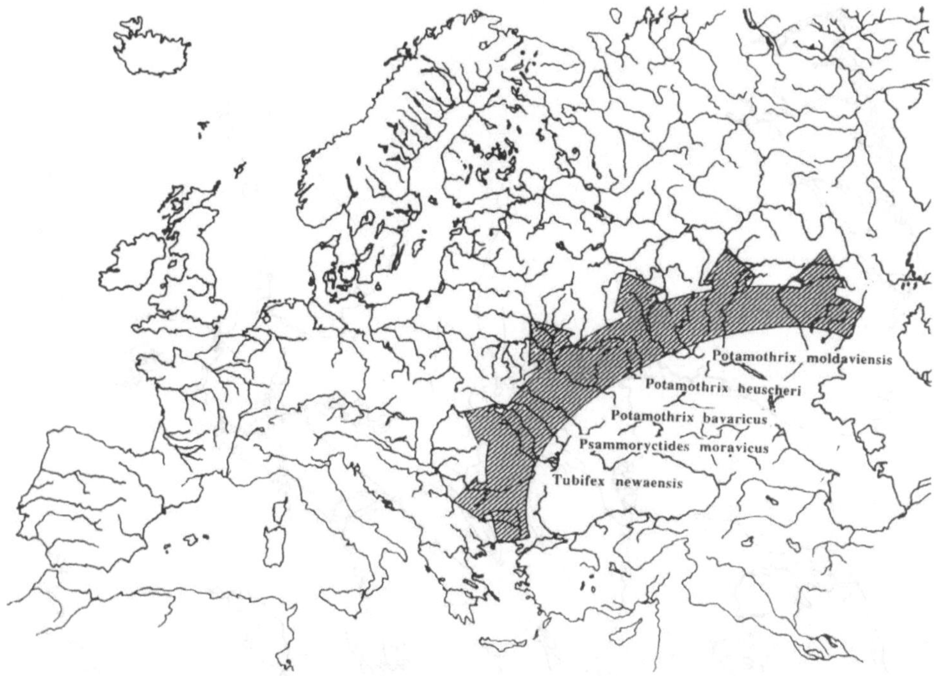

Figure 1. Ponto-Caspian tubificid species of the second wave dispersing from the Black, Azow and Caspian Sea regions (initial phases).

Research, Drottningholm) carried out bottom fauna surveys in the lake. In the first year (1915), 36 stations representing the various basins of the lake and in the second year 18 stations were sampled. These samples were obtained with a Petersen sampler (aperture 0.3 m^2), and the material was sieved through a net with a 1.0 mm mesh. Preservation was made in alcohol, and the material remained unidentified until the 1970s when the first author re-discovered the vials (G. Milbrink, unpublished).

Results

Distribution pattern of Potamothrix moldaviensis *in Scandinavia*

There are no records of *P. moldaviensis* from Scandinavian countries other than Sweden. In Sweden, the species has so far only been found in eastern Lake Mälaren, in Lake Roxen, and in Lake Algunnen in the River Alsterån catchment area in south-eastern Sweden – the latter locality only about 30 km inland from the Baltic Sea. All these localities have one thing in common; they are – or have been – accessible by boat from the Baltic Sea during the last century (Fig. 2).

Due to its rheophilous way of life, *P. moldaviensis* is not so often found in profundal samples from Lake Mälaren, but rather in samples from the littoral and from running water localities. Except for the Ekoln basin in the north, all other records from Lake Mälaren are from more or less narrow straits, from bottoms adjacent to moraine ridges or from other localities where there are large quantities of upwelling ground water. *P. moldaviensis* is thus particularly common all along the narrow waterway connecting the Ekoln basin in the north with the Görväln basin in the east and in the waters surrounding the large islands of the eastern parts of the lake (Fig. 4). However, one rather unexpected locality for the species is the relatively secluded basin of Lilla Ullevifjärden, a rather small lake with subterranean inlets surrounded by tall moraine ridges. There are no records of *P. moldaviensis* from nearby basins to the north, from the large central basins or from the western parts of the lake (see discussion below).

Records from Lake Roxen in 1985 were rather close to the inflow of River Svartån, the major tributary to the lake. This lake is located along the shipping

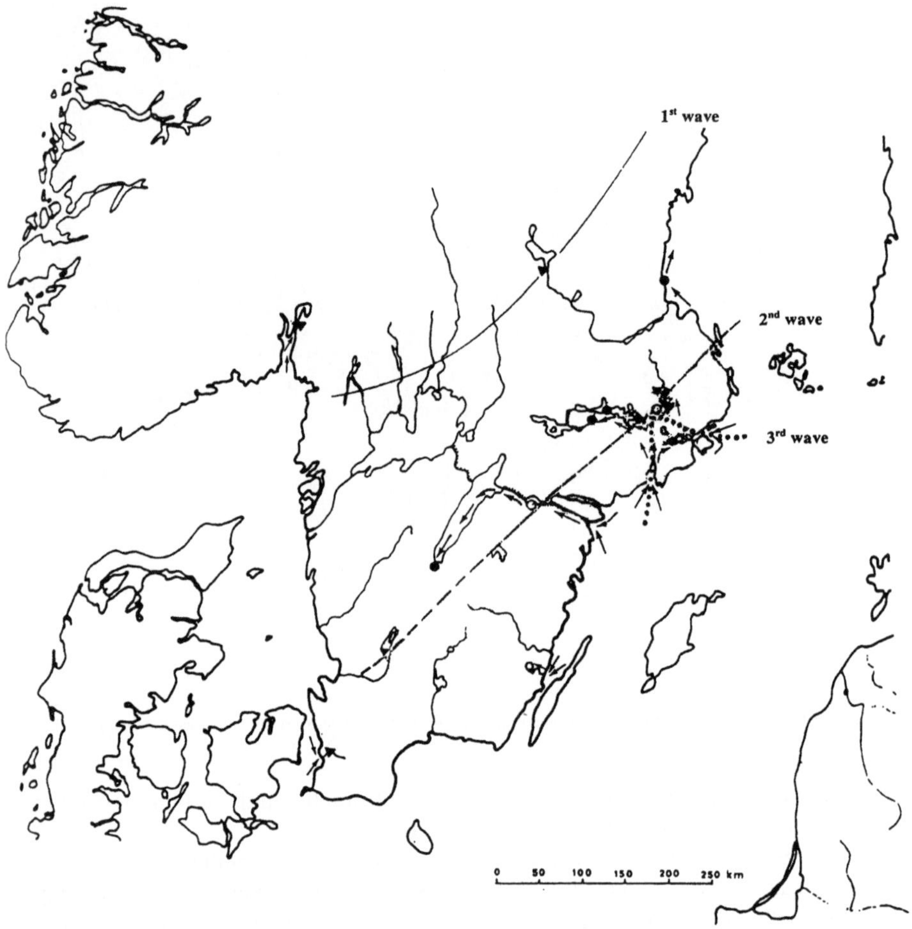

Figure 2. Fronts of dispersal of species of the genus *Potamothrix* of the first (continuous line), second (streaked line) and third wave (dotted line). The symbols are the same as in Figure 4. *P. hammoniensis* is considered to belong to the 1st wave, *P. moldaviensis* and *P. heuscheri* to the 2nd, and *P. vejdovskyi* to the 3rd (Timm, unpublished). For further details, see text. Likely trajectories of dispersal for the Ponto-Caspian species indicated by arrows (from Milbrink, 1999, in part).

route connecting the eastern and western coasts of Sweden, as well as Lakes Vättern (Milbrink, 1999) and Vänern, via a canal system – the Göta Canal (Fig. 2).

P. moldaviensis is mostly associated with the oligochaete fauna typically found in mesotrophic and eutrophic inland waters, and it clearly prefers running waters and the bottom water of lakes with persistent deep-water currents or subterranean inflows (Milbrink, 1980). In Lake Mälaren, the species diversity of oligochaetes is characteristically very high wherever *P. moldaviensis* and other *Potamothrix* species of the second wave occur (Milbrink, 1993). There is a clear positive relationship between oligochaete abundance and the trophic degree of each basin of Lake Mälaren (Fig. 5). However, what is even more interesting is primarily that in all basins of the lake, where Ponto-Caspian *Potamothrix* species have been recorded, abundance is significantly higher than normal (two-tailed t-test, $n = 5$, $P = 0.0021$), and secondly, wherever one such species has been recorded it is generally accompanied by another one or two Ponto-Caspian species, more or less irrespectively of the trophic degree. This is the situation represented in Figure 5.

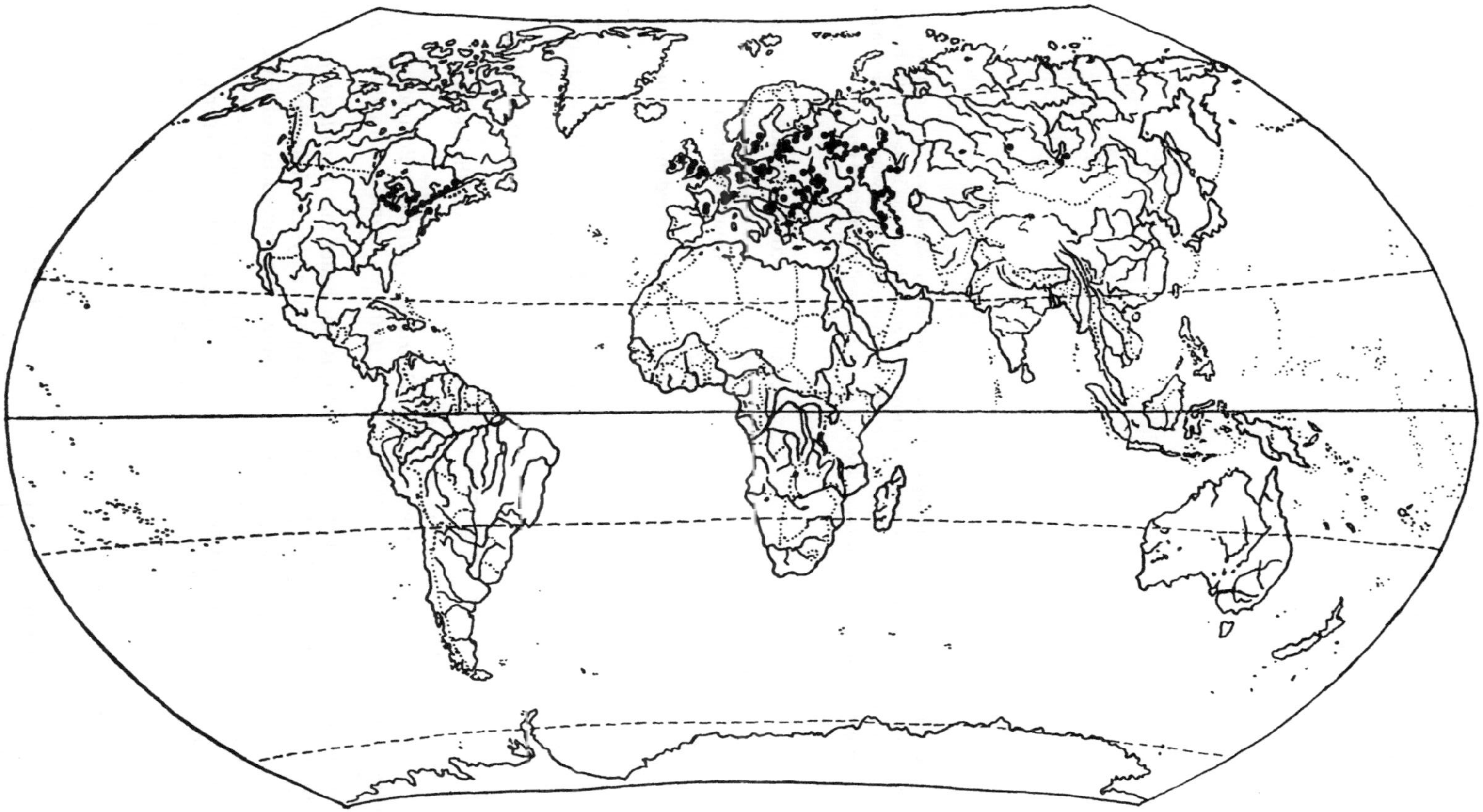

Figure 3. World distribution (year 2000) of the tubificid oligochaete *Potamothrix molodaviensis*.

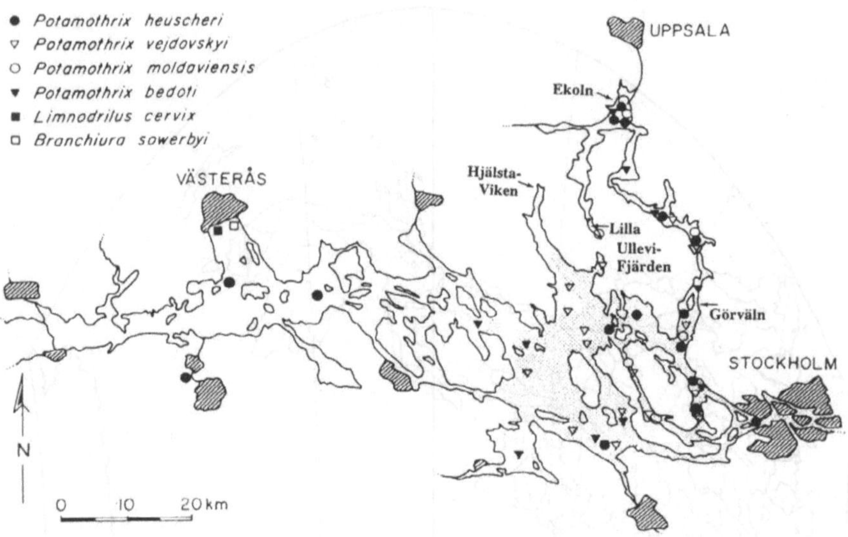

Figure 4. Present distribution pattern (by July 2000) of immigrating tubificid species in Lake Mälaren (partly redrawn from Milbrink, 1999). The occurrence of a fifth *Potamothrix* species, i.e. *P. bavaricus,* in south-eastern Lake Mälaren is not indicated in the Figure. The names of the various basins discussed in text are indicated in the figure. Maximum distribution of brackish bottom water is indicated in grey. (Partly redrawn from Milbrink, 1980). *P. hammoniensis* of the first wave is distributed all over the lake and not indicated here.

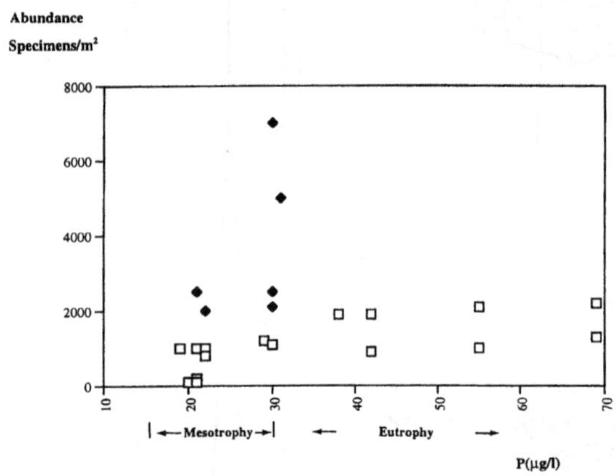

Figure 5. Total abundance of tubificid oligochaetes in a trophic gradient from east to west in Lake Mälaren in relation to mean total-phosphorus concentrations (in $\mu g\ l^{-1}$) in the surface water outgoing from the E.P.A. permanent stations net (Sept 1973 and Sept 1976, pooled data). Presence of Ponto-Caspian *Potamothrix* species in the samples (solid symbols) resulted in increased abundance levels in the mesotrophic section of the Figure in comparison with the eutrophic section representing the western basins of the lake (two-tailed *t*-test; $n = 5$, $P = 0.0021$). (Partly redrawn from Milbrink, 1993).

Patterns of distribution of Potamothrix *species in Lake Mälaren 1915–1916*

The very early bottom-fauna programs carried out in Lake Mälaren by Nordqvist and Alm in 1915 and 1916 clearly showed that besides *P. hammoniensis,* there were no less than three Ponto-Caspian *Potamothrix* species represented in the samples, i.e. *P. heuscheri, P. moldaviensis* and *P. bedoti.* The first two species were very well represented in all samples obtained from the basin of Görväln in the east – marking the beginning of the waterway leading to the northernmost basin of the lake – Ekoln (Fig. 4). Sexually immature specimens likely to be *P. heuscheri* were also found in the material from other basins of eastern Lake Mälaren. *P. bedoti* was found in samples from the south-central and eastern basins of the lake.

Distribution pattern of Potamothrix moldaviensis *in the Baltic States and north-eastern Russia*

It is our belief that *P. moldaviensis* has reached the Neva River System from River Volga with the help of ships. The construction of navigable canal systems connecting the Russian Rivers has greatly facilitated the dispersal of this and other species, e.g. the Vyšnij Voločok Canal System from Tver to the Msta – Volhov Rivers in the 18th century, and the Maria Canal System from the (later) Rybinsk Reservoir Region via Lake Beloe to Lake Ladoga in the 19th century. From the latter waterway, *P. moldaviensis* has evidently reached the Severnaja Dvina River System of the White Sea Region.

There is no doubt that *P. moldaviensis* has been carried by ships from River Neva over the brackish water barrier constituted by the Gulf of Finland into several river-mouths in the Baltic States, e.g. the rivers Narva, Pärnu, Salaca, Gauja, Daugava, Lielupe, Venta, Nemunas and Pregel, also including quite recently the small River Sötke to the west of River Narva. In Estonia, *P. moldaviensis* has mainly colonized the mouths of the two largest rivers and the sublittoral of eutrophic Lake Peipsi.

There is also another canal system located further south, i.e. between Rivers Dnepr and Nemunas, but so far there are no records of *P. moldaviensis* in the middle or upper reaches of River Nemunas. River Daugava, though very close to River Dnepr, has no connection with the latter river.

In Estonia, the navigable estuaries of the Rivers Narva and Pärnu were recorded as *P. moldaviensis* sites already in the 1960s. However, the upper reaches of those rivers have remained uneffected. From River Narva, the species is likely to have been carried by human activities into Lake Peipsi, where it is common on sandy bottoms. In our opinion, this introduction happened long ago, since *P. moldaviensis* has, in fact, occupied all suitable habitats in the lake, in contrast to the situation in Lake Võrtsjärv (see below). Since the species does not seem to exist in the lower reaches of the large River Velikaja entering Lake Peipsi from the south, this also means that invasions from the south are unlikely to have taken place. In Lake Võrtsjärv, *P. moldaviensis* has repeatedly been found on two of the numerous sampling stations located in the southern part of the lake but, so far, never in the northern part of the lake, which would seem to have quite suitable habitats for the species (sandy bottoms and lotic environments), or in the navigable River Emajõgi connecting the two lakes.

Discussion

Mass invasions of Ponto-Caspian invertebrates and the underlying mechanisms

Recent reviews of mass invasions into the North American Great Lakes by Ponto-Caspian invertebrate species (MacIsaac & Grigorovich, 1999), such as the zebra mussel, *D. polymorpha* (cf. Mills et al., 1993), the predatory cladoceran *Cercopagis pengoi* (cf. MacIsaac et al, 1999), and the amphipod *Echinogammarus ischnus* (cf. Dermott et al., 1998), have stressed the fact that invasions by alien aquatic invertebrates are often very effective and difficult to stop. Because of the latter, such invasions of continents are likely to increase in the future unless the problem of the dumping of ballast water in inland ports by transoceanic ships could be handled properly (Ricciardi & MacIsaac, 2000).

The main reasons for the accelerating rate of dispersal of Ponto-Caspian species into New World waters are both the euryhalinity of the Ponto-Caspian fauna and their increasing colonization rates of European ports (Morduchai–Boltovskoi, 1979; Dumont, 1998; Ricciardi & MacIsaac, 2000). Accordingly, Ponto-Caspian species residing in river deltas in coastal waters in Europe thereby easily end up in the ballast water of ships. Tolerant resting stages of bottom and planktonic invertebrates such as the cocoons of oligochaetes, ephippia of cladocerans, etc. come into the ballast water tanks along with small portions of sediment. All these species may even survive actions of precaution taken by the U.S. Government, for instance, in which the ballast water of ships from Europe by law has to be exchanged with salt water (exceeding about 13 ppt) before entering inland waters (Dumont, 1998; Ricciardi & MacIsaac, 2000).

Whereas cladoceran species such as *Daphnia lumholtzi* from Africa or south Asia and *Bythotrephes cederstroemi* from northern Europe are presently dispersing at a tremendous rate in the U.S. and feared to outcompete the indigenous cladoceran fauna (Havel et al., 1995) no such effects are known for invading tubificid oligochaetes. There are instead indications that the presence of Ponto-Caspian *Potamothrix* species in Scandinavian waters may rather stimulate the indigenous oligochaete fauna through interspecific mutualistic interactions with reference to bacterial food (Milbrink, 1993; Milbrink, unpublished).

Dispersal of Ponto-Caspian tubificids across Europe in three discrete waves

Ponto-Caspian tubificid oligochaete species – most of them of the genus *Potamothrix* – are known to have spread, or alternatively are just about to start dispersal, across Europe and Asia in three consecutive waves (Timm, unpublished).

The first wave actually started in early postglacial times utilizing natural but fluctuating water connections. Amongst species of the first wave may be mentioned *Psammoryctides barbatus* and *Potamothrix hammoniensis*. The latter species is today the species which is most characteristic of eutrophic lowland lakes all over the European continent (Milbrink, 1980). However, it has not yet colonized Norway and north-western Sweden. There is a fairly distinct front-line running from west to east across the county of Värmland to the north of Lake Vänern and then turning to the north-east (Fig. 2).

Several *Potamothrix* species of the second wave have reached the Baltic Sea Region including western Scandinavia in the 20th century and this wave is still on the move (Timm, 1980; Milbrink, 1980, 1999). Those species have presently reached another distinct front-line running obliquely across Lake Mälaren and further down to the south-west (Fig. 2). The distribution and dispersal capacity of one of these species in Scandinavia and the Baltic Sea area, *Potamothrix heuscheri*, was discussed in some detail in Milbrink (1999). Other species of the second wave are *P. moldaviensis*, *P. bavaricus*, *Psammoryctides moravicus* and *Tubifex newaensis* (Fig. 1). The second front-line would seem to be quite distinct and is obviously slowly moving to the west or north-west.

The third wave is a species pool most of them still waiting in the slightly brackish estuaries of the Black and Caspian Seas. Species of the third wave could be *Potamothrix alatus* and *Psammoryctides deserticola* and several more. As was mentioned above, a forerunner of the third wave, *P. vejdovskyi*, has reached Lake Mälaren at some time during the last 80 years (cf. below). Its curved front-line in Sweden is shown in Figure 2. It is today a common profundal species in the basins and connecting waterways in eastern Lake Mälaren and has just reached the central basins of the lake (Milbrink, 1980).

Another *Potamothrix* species, *P. bedoti*, is supposed to be an outsider in this respect, not necessarily fitting into the above framework. At least in Estonia *P. bedoti* seldom reproduces sexually – and only at low temperatures (Timm, 1987). Therefore it is often overlooked. Even though there are a few records of *P. bedoti* from western Scandinavia suggesting that it might belong to the first wave, the majority of the records, however, are from eastern Lake Mälaren (Milbrink, 1999).

The results from the early bottom-fauna surveys in Lake Mälaren, especially those carried out by Nordqvist and Alm in 1915 and 1916, would give us a minimum date of arrival for Ponto-Caspian species of the second wave, i.e. they had certainly reached eastern Sweden approximately 100 years ago. Of particular interest in this context is the fact that *P. vejdovskyi* was not found in any samples collected by Nordqvist and Alm in 1915 and 1916. It is therefore likely that this species has reached Sweden during the last 85 years or so. The majority of the inland canals in Russia were dug during the 19th century. *P. vejdovskyi* was first observed (i.e. had become abundant) in the Volga only in the 1960s after the Kujbyšev and Saratov Reservoirs had been built on the middle course of this river (Ljubin, 1971). Apparently its abundance there has promoted further dispersal towards the Baltic Sea along the canals. This would support the contention that the majority of the Ponto-Caspian species had reached the Baltic Sea Region via boat traffic on Russian inland waterways.

Could bird and mammal vectors of dispersal be involved and do we need to know which vectors are involved?

Most records of *P. moldaviensis* from Sweden are well concentrated to eastern Lake Mälaren and to the central basins of the lake, the bottoms of which as late as in 1955 were reached by an undertow of brackish water from the Baltic Sea. Boat traffic between ports in the Baltic and into Lake Mälaren is the most likely dispersing agent. However, there are three records mentioned above, which are well separated from the rest. The first record is from Lake Algunnen in south-eastern Sweden (Fig. 2), the second is from Lake Roxen (Fig. 2) and the third is from a small basin (Lilla Ullevifjärden) vaguely connected with the rest of the northern part of Lake Mälaren (Fig 4). Lake Algunnen is in the River Alsterån water system only about 30 km inland. Before the damming-up period, this lake could at least theoretically have been reached by boats from nearby sea ports and thereby also by oligochaetes in the ballast water. Lake Roxen could by the construction of the Göta Canal at the end of the 19th century, be reached by ships from distant harbours around the Baltic Sea, but also from more distant ports, again via Russian inland waterways, for instance.

However, in the third example, it is more doubtful whether boat transportation could at all have been involved. The small basin of Lilla Ullevifjärden is geographically and topographically much isolated from the rest of the lake, and the water quality is oligotrophic, i.e. more oligotrophic than the central basins of the lake. This basin is surrounded by steep moraine ridges and the bottom-sides are also steep. Utilizing a micro-stratification sampler technique (Milbrink, 1974), the first author found that this basin is fed by subterranean water of low salt contents. The basin is not readily accessible by boat. The only outlet – there is no inlet – is quite narrow and there has since long been restricted access to the area for the public due to military reasons. The bottom fauna of this lake is remarkable since this is the only locality in the region for some relict crustaceans from the ice age, i.e. *Pallasea quadrispinosa* and *Gammaracanthus loricatus*. *P. moldaviensis* is here a dominating oligochaete species from the upper littoral down to the largest depth – about 40 m. Its stronghold in Lake Mälaren – and so was already the case in the material from Nordquist and Alm in 1915 and 1916 – is the Görväln basin further to the east, at depths between 40 and 50 m (Fig. 4). This basin is characterized by high water velocity caused by open connections with adjacent basins but also by subterranean inflows.

It could be speculated that bird vectors could be involved in the case of the Lilla Ullevifjärden basin (cf. Milbrink, 1999). Not far from this basin is a parallel very shallow bay – Hjälstaviken – of one of the central basins of Lake Mälaren (Fig 4). This bay is famous for its seasonally large concentrations of wading birds and other migrating birds temporarily halting here before moving further either to the north or to the south. It is not unlikely that oligochaete cocoons could stick to the plumage and feathers of bathing birds – as was already suggested by Benham (1903) – and thereafter be shed when the bird is bathing or rinsing its downs and feathers again in another body of water near or far away. At least the cocoons of terrestrial earthworms could withstand draught for extended periods (Prof. de Pauw, pers. com.), which transportation in the open air would involve (cf. Milbrink, 1999). It is known that small landsnails (Dundee, 1967; Vagvolgyi, 1976), as well as the ephippia of *Daphnia* (Anders Hobaek, Norway, personal information), could readily be transported in this way. Hobaek has found many ephippia in the downy feathers of wading birds. Prof. Sven Ekman proposed that the dispersal of oligochaetes would be facilitated by the fact that the cocoons easily adhered to foreign objects (Ekman, 1955). *P. moldaviensis*, in particular, usually sticks its cocoons to hard substrates (Timm, 1987).

Another question which would seem to be relevant to ask is how specimens of *P. moldaviensis* could have reached the distant Enisei River and Lake Baikal region across Siberia (Fig. 2) in the first place, if bird vectors had not been involved. However, it is quite possible that boat traffic also here could have affected the present distribution pattern. Land and aerial transport of fish fry, adult fish and fish food organisms, which has been widely practiced in Russia, may otherwise at least theoretically be responsible for the transfer of tubificid oligochaetes over large distances.

We do not fully understand the mechanisms by means of which tubificid oligochaetes disperse. However, the dispersal of the Ponto-Caspian fauna and flora provides us with useful markers which could be traced at the present time, providing us with useful data for constructing the necessary trajectories of dispersal. As far as we know, invasions by oligochaetes including leeches have not yet caused great ecological problems nor even public concern, even though we know of cases of interactive segregation leading to replacements of similar kinds as those we recognize for some species of molluscs and cladoceran crustaceans. *Branchiura sowerbyi*, for instance, is a large tropical tubificid, the dispersal of which is little known. This species may totally exclude native oligochaetes near warm-water effluents from atomic power plants, greenhouses, etc. even at high latitudes, for instance in Lake Mälaren (see Fig. 4 and Milbrink, 1980). We have learnt that there is always a potential threat when exotic species massively enter new environments. We should be prepared for the worst scenario. Transportation in the ballast water of transoceanic ships could be restricted – if not stopped completely (Carlton & Geller, 1993; Ricciardi & MacIsaac, 2000). However, passive dispersal via migrating birds can hardly be avoided.

References

Bazzanti, M., M. Seminara & C. Tamorri, 1993. Eutrophication in a deep lake: depth distribution of profundal benthic communities as an indicator of environmental stress. Verh. int. Ver. Limnol. 25: 784–789.

Benham, W. B., 1903. A note on the Oligochaeta of the New Zealand lakes. Trans. Proc. N.Z. Inst. 36: 192–198.

Bonacina, C., G. Bonomi & A. Pasteris, 1992. Some remarks on the macrobenthos community of the profundal zone of the large Italian lakes. Mem. Ist. ital. Idobiol. 50: 79–106.

Carlton, J. T. & J. B. Geller, 1993. Ecological roulette: The global transport of nonindigenous marine organisms. Science 261: 78–82.

Dermott, R., J. Witt, M. Um–Young & M. Gonzales, 1998. Distribution of Ponto-Caspian amphipod *Echinogammarus ischnus* in the Great Lakes and replacement of native *Gammarus fasciatus*. J. Great Lakes Res. 24: 442–452.

Dumont, H. J., 1998. The Caspian lake: history, biota, structure and function. Limnol. Oceanogr. 43: 44–52.

Dundee, D. S., 1967. Snails on migratory birds. The Nautilus 80: 89–91.

Ekman, S., 1955. Djurens Liv (Brehm-Ekman). Band 7: Ryggradslösa Djur. Sohlmans Förlag, Stockholm: 688 pp. (in Swedish).

Giani, N., 1979. Les Oligochètes aquatiques du Sud-Ouest de la France. Bull. Soc. Hist. nat. Toulouse 115: 347–358.
Gold, Z. G. & J. I. Zapekina–Dulkejt, 1972. Kčizučeniju maloščetinkovyh červej (Oligochaeta) r. Many. Trudy Gosudarstvennogo zapovednika 'Stolby' 9: 180–186 (in Russian).
Hallegraeff, G. M. & C. J. Bolch, 1991. Transport of toxic dinoflagellate cysts via ships' ballast water. Mar. Poll. Bull. 22: 27–30.
Havel, J. E., W. R. Mabee & J. R. Jones, 1995. Invasion of the exotic cladoceran *Daphnia lumholtzi* into North American reservoirs. Can. J. Fish. aquat. Sci. 52: 151–160.
Juget, J., 1967. Quelques données novelles sur les Oligochètes du Leman: composition et origine du peuplement. Ann. Limnol. 3: 217–229.
Lang, C., 1984. Eutrophication of Lakes Léman and Neuchatel (Switzerland) indicated by oligochaete communities. Hydrobiologia 115: 131–138.
Lang, C. & O. Reymond, 1996. Reversal of eutrophication in four Swiss lakes: evidence from oligochaete communities. Hydrobiologia 334: 157–161.
Ljubin, V. A., 1971. O nahoždenii *Euilyodrilus vejdovskyi* (Oligochaeta, Tubificidae) v Volžskih vodohraniliščah. Zool. Zh. 50: 1579–1580 (in Russian).
MacIsaac, J. J. & I. A. Grigorovich, 1999. Ponto-Caspian invaders in the Great Lakes. J. Great Lakes Res. 25: 1–2.
MacIsaac, H. J., I. A. Grigorovich, J. A. Hoyle, N. D. Yan & V. E. Vadim, 1999. Invasion of Lake Ontario by the Ponto-Caspian predatory cladoceran *Cercopagis pengoi*. Can. J. Fish. aquat. Sci. 56: 1–5.
Milbrink, G., 1974. On the vertical distribution of oligochaetes in lake sediments. Rep. Inst. Freshw. Res. Drottningholm 53: 34–50.
Milbrink, G., 1980. Oligochaete communities in pollution biology. The European situation with special reference to lakes in Scandinavia. In Brinkhurst, R. O. & D. G. Cook (eds), Aquatic Oligochaete Biology. Plenum Press, New York: 433–455.
Milbrink, G., 1993. Evidence for mutualistic interactions structuring oligochaete communities. Oikos 68: 317–322.
Milbrink, G., 1999. Distribution and dispersal capacity of the Ponto-Caspian tubificid oligochaete *Potamothrix heuscheri* (Bretscher, 1900) in Scandinavia. Hydrobiologia 406: 133–142.
Mills, E. L., J. H. Leach, J. T. Carlton & C. L. Secor, 1993. Exotic species in the Great Lakes: a history of biotic crises and anthropogenic introductions. J. Great Lakes Res. 19: 1–54.
Morduchai–Boltovskoi, F. D., 1979. Composition and distribution of Caspian fauna in the light of modern data. Int. Rev. ges. Hydrobiol. 64: 1–38.
Popcenko, V. I., 1974. O faune maloščetinkovyh červej pribajkalskih vodoemov. In Produktivnost' Bajkala i antropogennye izmenenija ego prirody, Irkutsk: 277–287 (in Russian).
Prat, N., M. Real & M. Rieradevall, 1992. Benthos of Spanish lakes and reservoirs. Limnetica 8: 221–229.
Ricciardi, A. & H. J. MacIsaac, 2000. Recent mass invasion of the North American Great Lakes by Ponto-Caspian species. TREE 15: 62–65.
Särkkä, J., 1978. New records of profundal Oligochaeta from Finnish lakes, with ecological observations. Ann. zool. fenn. 15: 235–240.
Timm, T., 1980. Distribution of aquatic oligochaetes. In Brinkhurst, R. O. & D. G. Cook (eds), Aquatic Oligochaete Biology. Plenum Press, New York: 55–77.
Timm, T., 1987. Maloščetinkovye červi (Oligochaeta) vodoemov Severo-Zapada SSSR (Aquatic Oligochaeta of the northwestern part of the USSR). Valgus, Tallinn: 299 pp.
Vagvolgyi, J., 1976. Body size, aerial dispersal and origin of the Pacific land snail fauna. Syst. Zool. 24: 465–488.
Vitousek, P. M., 1997. Introduced species: a significant component of human-caused global change. New Zeal. J. Ecol. 21: 1–16.

Hydrobiologia **463:** 103–106, 2001.
P. Rodriguez & P.F.M. Verdonschot (eds), Aquatic Oligochaete Biology VIII.

Temporal pattern of the double sperm line production in *Tubifex tubifex* (Annelida, Oligochaeta)

Silvia Boi & Marco Ferraguti*
Dipartimento di Biologia, Università di Milano, via Celoria 26, 20133 Milano, Italy
Tel: 39 0226 604451. Fax: 39 0226 604462. E-mail: marco.ferraguti@unimi.it
*(*Author for correspondence)*

Key words: spermiogenesis, dichotomous sperm lines, life cycle, reproduction, cohort cultures

Abstract

Laboratory cohort cultures of the tubificid *Tubifex tubifex* were carried out at 21 °C in homogeneous conditions. Our aim was to study the temporal pattern of spermatogenesis of the two sperm lines which characterise this species. Starting from the second week after cocoon laying, we analysed the seminal vesicle content. We counted the cysts in the vesicles and arranged them into three classes: premeiotic, paraspermatids and euspermatids. Our results show that spermatogenesis begins early, at week 3, but cysts of spermatids are not found until week 5 for the parasperm line and week 6 for the eusperm line. These data are derived from a preliminary study of complex population dynamics. Nonetheless, they allow us to formulate some hypothesis regarding the mechanism which regulate the production of the two types of sperm.

Introduction

Two sperm types, differing in nearly all ultrastructural details, are produced by the oligochaete annelids of the subfamily Tubificinae. Only one of the two sperm types is able to fertilise ('eusperm' in Healy & Jamieson, 1981), whereas the other ('parasperm' in Healy & Jamieson, 1981) may aid, but not contribute genetically to, fertilisation (Hodgson, 1999).

In *Tubifex tubifex,* eusperm and parasperm differ in many features (Braidotti & Ferraguti, 1982; summarised in Ferraguti, 2000). The main differences concern the acrosome which in parasperm appears reduced in comparison to the eusperm acrosome; the nucleus which is 30 μm long and twisted column-shaped with fully condensed chromatin in eusperm whereas it is 3 μm long and comma-shaped with uncondensed areas in the chromatin in parasperm; the basal cylinder of the axoneme is absent or reduced in parasperm; the plasma membrane of the tail is close to the axoneme in eusperm, but widely separated from the axoneme in parasperm.

After mating, the two sperm types are grouped in the spermathecae in peculiar bundles called spermatozeugmata. These are regular arrangements of a core of eusperm and a cortex of parasperm. The parasperm tails are connected to one another for most of their length by a junctional complex (Braidotti et al., 1980). Only the tips of the tails are free and move forming a metachronal wave (Ferraguti et al., 1988). Thus, protection and progression of the eusperm are probably the main functions of the tubificine parasperm.

The presence of two different types of spermatozoa raises the problem of differentiation of the two lines. In *Tubifex tubifex,* spermatogonia are released in groups of four by the testes into the seminal vesicles, where the process of spermiogenesis occurs. The isogenic groups of cells, called cysts, undergo several divisions maintaining their connection. At the 32-cell stage, in some cysts a regular meiosis gives rise to 128 euspermatids for each cyst, whereas in other cysts the parasperm are produced through a peculiar fragmentation of the 32 nuclei (Boi et al., 2001).

Paraspermatid cysts are formed by a much greater and variable number of cells (between 550 and 3000: Ferraguti, 2000). Aside from this numerical difference, eu- and paraspermatid cysts can be easily distinguished; the paraspermatid cysts are larger (100–150

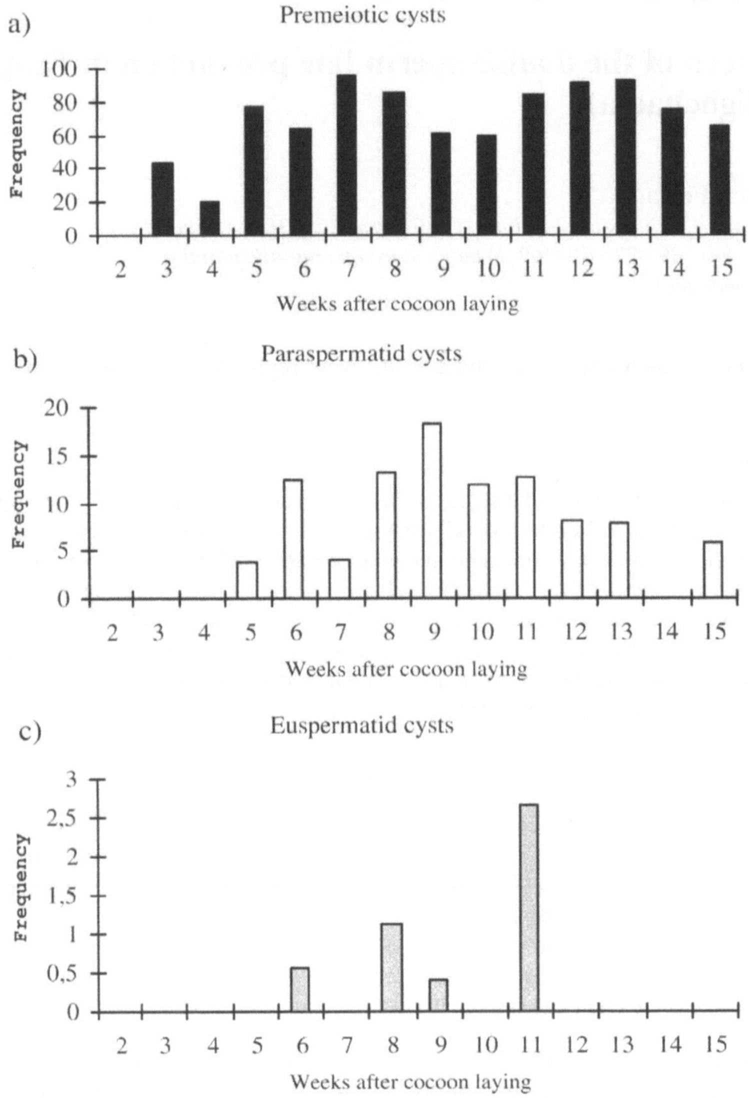

Figure 1. Frequency of occurrence of premeiotic cysts (a), paraspermatid cysts (b), and euspermatid cysts (c), produced from hatching up to the week 15 after cocoon laying.

μm compared to 70–80 μm of the euspermatid cysts), whereas the paraspermatid nuclei are much shorter than those of euspermatids.

Despite these differences between the two seminal elements at spermatid and sperm stage, it has been impossible to distinguish the earlier stages of dichotomous spermatogenesis. To contribute to the understanding of the mechanism underlying the production of the two sperm lines, in the present work we monitored the dichotomous sperm line production by means of cohort cultures. We observed the switch from one sperm line production to the other and hypothesised a possible regulating mechanism.

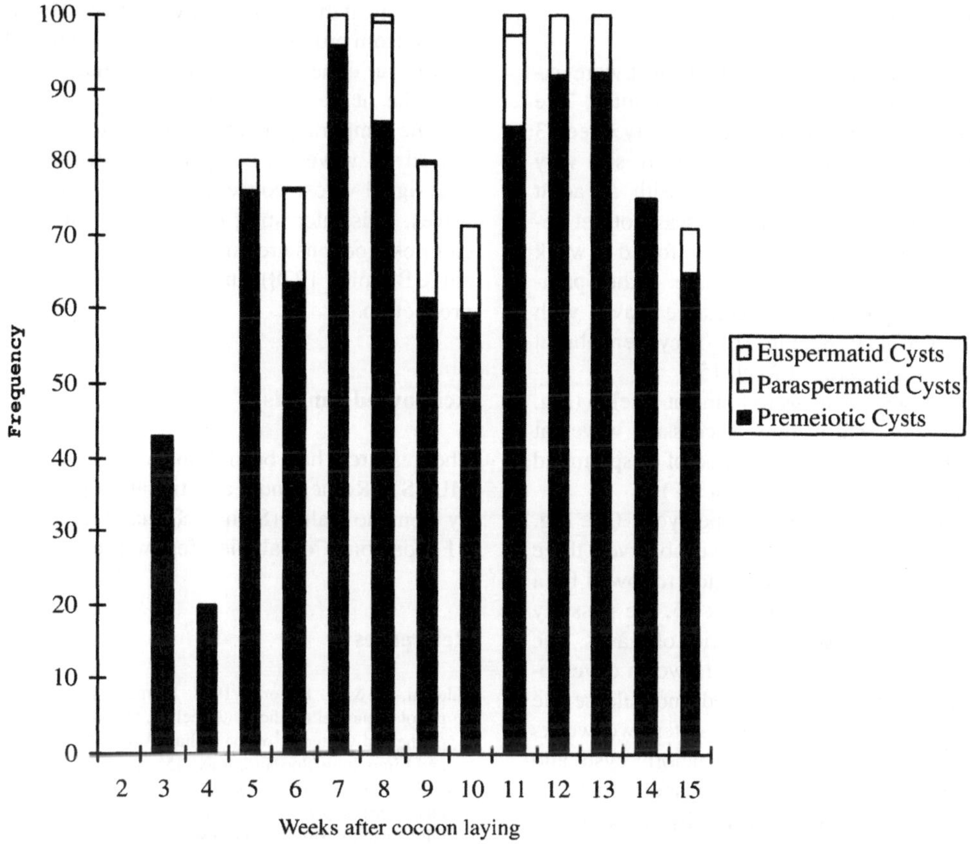

Figure 2. The same data shown in Figure 1, arranged in one single histogram to show the relative frequency in the seminal vesicles content, of cysts belonging to the three classes. Black for the premeiotic cysts, white for the paraspermatid cysts and grey for euspermatid cysts. The sum of the relative frequencies is not always 100% because the values are an average of all the samples including worms with empty seminal vesicles.

Materials and methods

Tubifex tubifex were collected from the Lambro River (province of Milano, Italy) and reared on a sand substrate covered with tap water at 21 °C. A cohort of cocoons was obtained by removing all the cocoons previously laid in the rearing vessels and collecting those laid during the following 48 h. Cocoons were placed in thirteen 250-ml beakers (five cocoons in each) which contained a 1–2 cm layer of sediment from Monate Lake (Province of Varese), mixed with 0.2 g Tetramin (food for aquarium fish, TetraWerke), and 180 ml tap water. The beakers were kept at 21 °C and aerated. The expected number of surviving worms per beaker was between 10 and 20. Starting from the second week, every week the contents of a beaker were filtered and the worms collected. Their seminal vesicles were opened on a slide in a drop of phosphate buffered saline (pH 7.6), left dry and stained for DNA with DAPI (4,6-diamidino-2-phenylindole, Sigma) 2 μg ml^{-1} for 5 min. The slides were mounted with glycerol and phosphate buffered saline 1:1 and examined under a fluorescence microscope (Zeiss III RS). The cysts were counted and assigned to one of three classes: premeiotic cysts, paraspermatid cysts and euspermatid cysts. The number of cysts per class was expressed as a share of the contribution of the cyst production by all members of a given cohort (Fig. 1). The sum of the relative frequencies is not always 100% because the null contribution of worms with empty seminal vesicles was included in the average.

Results and discussion

An average number of 15 worms per cohort were examined and roughly 13 000 cysts were counted. The first cysts were released from the testes early (week 3 after cocoon laying) when the worms were still very small (around 5 mm, in comparison with an adult which is 2–3 cm long) and clitellum was not yet developed. The first spermatid cysts were found at week 5 and belonged to the parasperm line. Other paraspermatid cysts appeared in successive waves with peaks at weeks 6 and 9 (see Fig. 1b). They were absent at week 14 but reappeared at week 15.

The first eusperm cysts were found at week 6 (Fig. 1c) and were produced in three successive waves at week 6, 8–9 and 11. The percentage of euspermatid cysts was always very low (less than 3%).

Sperm production appeared to be cyclic (see Fig. 2): in the 15-weeks time lapse, we observed three successive waves of sperm production followed by a resting period at week 14. At week 15, we possibly observed the beginning of a new series of waves. The spermatogenetic cycle begins early in worm development, as soon as testes are formed and release the first cysts in the seminal vesicles. After two weeks the seminal vesicles are full of premeiotic cysts and we found the first spermatids. The parasperm line production is active first and is followed by eusperm line production. Cysts undergoing nuclear fragmentation, i.e. the process that will give paraspermatid cysts, and cysts undergoing regular meiosis, which will give euspermatid cysts, were observed in the same slide in four cohorts. Hence, they are present at the same time in the seminal vesicles.

Conclusions

From our data, we see that euspermatid and paraspermatid cysts and their precursors (i.e. meiotic cysts and fragmenting cysts) are present in the seminal vesicles at the same time, implying that the two sperm line productions overlap. This leads us to exclude a sequential commitment due to hormone production during development, as is the case, for instance, of Lepidoptera (Friedländer, 1997).

Since we were not able to distinguish in our previous research and in the present study, the early precursors of the two lines, we can hypothesise that the switch from production of one seminal line to the other may occur either in the testis or in the spermatocyte I cysts, i.e. at the 32-cell stage.

The temporal pattern of sperm production characterised by waves is compatible with the hypothesis of a signal whose concentration oscillates, provoking waves. It is interesting to notice that in low density cultures, cocoons are laid in successive waves (Adreani & Bonomi, 1978) thus paralelling euspermatid cyst production.

Acknowledgements

The research has been finnanced by a grant from MURST (Rome) under a National Project coordinated by Romano Dallai (Siena). SB has been supported by a Fondazione Confalonieri fellowship.

References

Adreani, L. & G. Bonomi, 1978. Elementi per una dinamica di popolazione dei tubificidi (Annelida, Oligochaeta). Proceedings of the 3rd Congress of the Italian Association for Oceanology and Limnology. Sorrento, Italy: 355–366.

Boi, S., U. Fascio & M. Ferraguti, 2001. Nuclear fragmentation characterises paraspermiogenesis in *Tubifex tubifex* (Annelida, Oligochaeta). Mol. Reprod. Devl. 59: 442–450.

Braidotti, P., M. Ferraguti & T. Fleming, 1980. Cell junctions between spermatozoa flagella within the spermatozeugmata of*Tubifex tubifex* (Annelida: Oligochaeta). J. Ultrastruct. Res. 73: 299–309.

Braidotti, P. & M. Ferraguti, 1982. Two sperm types in the spermatozeugmata of *Tubifex tubifex* (Annelida, Oligochaeta). J. Morphol. 171: 123–136.

Ferraguti, M., G. Bernardini, G. Melone & R. Dallai, 1988. Structure and function of the metachronal wave in *Tubifex tubifex* spermatozeugmata (Annelida, Oligochaeta). J. Ultrastruct. Mol. Struct. Res. 99: 79–95.

Ferraguti, M., 2000. Clitellata. In Adiyodi, K. G. & R. G. Adiyodi (eds), Reproductive Biology of Invertebrates. Vol IX, part B. Progress in Male Gamete Ultrastructure and Phylogeny (volume Editor B. G. M. Jamieson). John Wiley & Sons, New York: 125–182.

Friedländer, M., 1997. Control of the eupyrene-apyrene sperm dimorphism in Lepidoptera. J. Insect Physiol. 43: 1085–1092.

Healy, J. M. & B. G. M. Jamieson, 1981. An ultrastructural examination of developing and mature spermatozoa in *Pyrazus ebeninus* (Mollusca, Gastropoda, Potamididae). Zoomorphology 98: 101–119.

Hodgson, A. N., 1999. Paraspermatozoa. In Knobil, E. & J. D. Neil (eds), Encyclopedia of Reproduction, Vol III. Academic Press, S. Diego: 656–668.

Hydrobiologia **463**: 107–113, 2001.
P. Rodriguez & P.F.M. Verdonschot (eds), Aquatic Oligochaete Biology VIII.

External structures used during attachment and sperm transfer in tubificids (Annelida, Oligochaeta)

Sergio Cuadrado & Enrique Martínez-Ansemil*
Departamento de Bioloxía Animal, Bioloxía Vexetal e Ecoloxía, Facultade de Ciencias, Universidade da Coruña, Campus da Zapateira s/n, 15071 A Coruña, Spain
Fax: 34 981 167065. E-mail: eansemil@udc.es
*(*Author for correspondence)*

Key words: Tubificidae, attachment structures, SEM, reproduction, sperm transfer

Abstract

The genital region of seven species of Tubificidae has been studied by SEM (Scanning Electron Microscopy). The form and the position of penial and spermathecal chaetae, male and spermathecal pores and other special structures have been examined. *Peristodrilus montanus* shows a special system to hold the partner: the penial chaetae anchor in an elaborated structure of the body wall formed between the spermathecal pores, the 'anchorage bridge'. *Protuberodrilus tourenqui* has a long glandular porophore with the male pores at the tip, allowing contact with the spermathecal pores which are located in deep, close to the mid-ventral line of the body. The grooved and strongly curved penial chaetae of *Rhyacodrilus falciformis* seem to be used both for attachment and for sperm transfer, entering into the lateral spermathecal pores. The embrace of the partners, as suggested by observations on *Psammoryctides barbatus*, *Potamothrix bavaricus*, *Potamothrix hammoniensis* and *Potamothrix heuscheri*, seems to be another important mechanism to fix contact between male and spermathecal pores and allow sperm transfer. The spermathecal chaetae could be interpreted as piercing chaetae with a chemical or mechanical stimulating role. Sensitive cilia near the penial chaetae seem to be used by the three rhyacodrilines studied to find the correct anchorage place. There is a great variety of structures which appear to be used for attachment and sperm transfer in tubificids, and consequently their role in the evolution of the whole family may be profound.

Introduction

Sexual reproduction, with mutual sperm transfer between two partners, constitutes the main reproductive mode for most oligochaetes. This process has been observed in detail only in some species of earthworms, where the role of glandular secretions and genital chaetae, and the embrace of the mating worms has been pointed out (Feldkamp, 1924; Grove, 1925; Oishi, 1930; Stephenson, 1930; Omodeo, 1998). For the aquatic families, the mechanisms involved in the attachment of the worms are unknown and can hardly be deduced from the genital structures which are visible with light microscopy. The small size of aquatic oligochaetes, as well as their habitat (particularly in tubificids), make the observation of hold systems difficult, and the use of SEM (Scanning Electron Microscopy) should be an important tool to facilitate a better understanding of such systems.

The aim of this work was to carefully study the genital region of some rhyacodrilines and tubificines (Tubificidae) by SEM, in order to characterise the form, location and role of the structures used by these worms for attachment and sperm transfer. In this regard, only very few data are known at present (e.g. Poddubnaya & Pataridze, 1989; Finogenova & Poddubnaya, 1990).

Materials and methods

Seven species, collected in several watercourses and reservoirs of the Iberian Peninsula, have been studied: the rhyacodrilines *Peristodrilus montanus* (Hrabě,

1962) – 11 individuals, *Protuberodrilus tourenqui* Giani & Martínez-Ansemil, 1979 – 18 individuals and *Rhyacodrilus falciformis* Bretscher, 1901 – 5 individuals, and the tubificines *Psammoryctides barbatus* (Grube, 1861) – 7 individuals, *Potamothrix bavaricus* (Öschmann, 1913) – 8 individuals, *Potamothrix hammoniensis* (Michaelsen, 1901) – 7 individuals and *Potamothrix heuscheri* (Bretscher, 1900) – 4 individuals.

All these specimens were observed by SEM, except the two *P. montanus* which were sectioned and observed by light microscopy.

All the *Psammoryctides* and *Potamothrix* specimens and some *P. tourenqui* and *R. falciformis* were fixed with 4% formalin and subsequently transferred to 70% ethanol. The other worms observed by SEM were anaesthetised with 7.5% magnesium chloride, added progressively to water, and then fixed with 5% glutaraldehyde in 0.1 M cacodylate buffer for 2 h (*P. montanus*) or with 2% glutaraldehyde and 1% paraformaldehyde for 2 h in 0.1 M cacodylate buffer (some *P. tourenqui* and *R. falciformis*). These worms were then washed three times in 0.2 M cacodylate buffer for 10 min and post-fixed with 1% osmium tetroxide for 2 h in cacodylate 0.1 M. Finally, all the above mentioned worms were dehydrated through an increasing acetone series, then critical point dried (except one *P. barbatus*, two *P. bavaricus*, two *P. hammoniensis* and one *P. heuscheri*), mounted on stubs and sputter coated with gold. Observations were carried out using a Jeol JSM-6400 and a Philips XL 30 SEM.

The two specimens of *P. montanus* used for light microscopy observations were anaesthetised with 7.5% magnesium chloride, fixed with 4% formalin and preserved in 70% ethanol. Eight-μm transverse sections were made, stained in hematoxylin and eosin and mounted in Canada balsam. The observations were performed using a Nikon Optiphot light microscope provided with Nomarski DIC.

Results

The external structures of the species studied that may be involved in attachment and sperm transfer can be observed in Figs 1–20. The following comments emphasise some data and complete the results of our observations.

Peristodrilus montanus (Figs 1–4)

Genital pores: Male and spermathecal pores small and aligned with somatic ventral chaetae. Male pores open in the transverse line of the penial chaetal bundles, outside of them. Spermathecal pores located anteriorly in X, half-way between the chaetae of IX and X (Fig. 1). The distance between the male pores (176–314 μm) and between the spermathecal pores (173–257 μm), although highly variable depending on the specimens, are similar for the same worm.

Genital chaetae: The two bundles of penial chaetae are separated by a distance of 93–145 μm. These chaetae are strongly curved towards the mid-ventral line of the body (Figs 1 and 2).

Other anatomical structures: A transverse elliptical elevation (87–132 μm long, 160–257 μm wide) is present between the two male pores in fully mature individuals (Fig. 1); there are numerous cilia on the prominent outer part of the tegument surrounding the penial chaetal bundles (Fig. 2); a transverse band of several bundles of cilia between these chaetal bundles was also observed. These worms have a very singular structure which we call an 'anchorage bridge', between the spermathecal pores. This structure consists of an elevation of the body wall (length: 67–104 μm; width: 50–71 μm at the middle part, 73–100 μm at the anterior and posterior ends) with a thin median longitudinal groove and two lateral 'pockets' (Figs 1,3 and 4). We observed that these two specialised areas develop after the formation of the male and spermathecal pores.

In the transverse sections for light microscopy, we observed the presence of two very powerful muscles attached to the ventral male elevation, close to each penial chaetal bundle, and crossing the coelomic cavity ventro-dorsally to branch and establish several contact points with the dorsal body wall.

Protuberodrilus tourenqui (Figs 5–10)

Genital pores: Male and spermathecal pores small and aligned near the mid-ventral line of the body (Fig. 7).

Genital chaetae: The characteristic set of genital chaetae of this species (one, rarely two chaetae in each ventral bundle of segments IX, X and XII) is very homogeneous; all these chaetae have a typical hypodermic needle form and emerge from prominent chaetophores (Figs 5–7). The distance between the two bundles of genital chaetae of each segment is three to four times that of the genital pores.

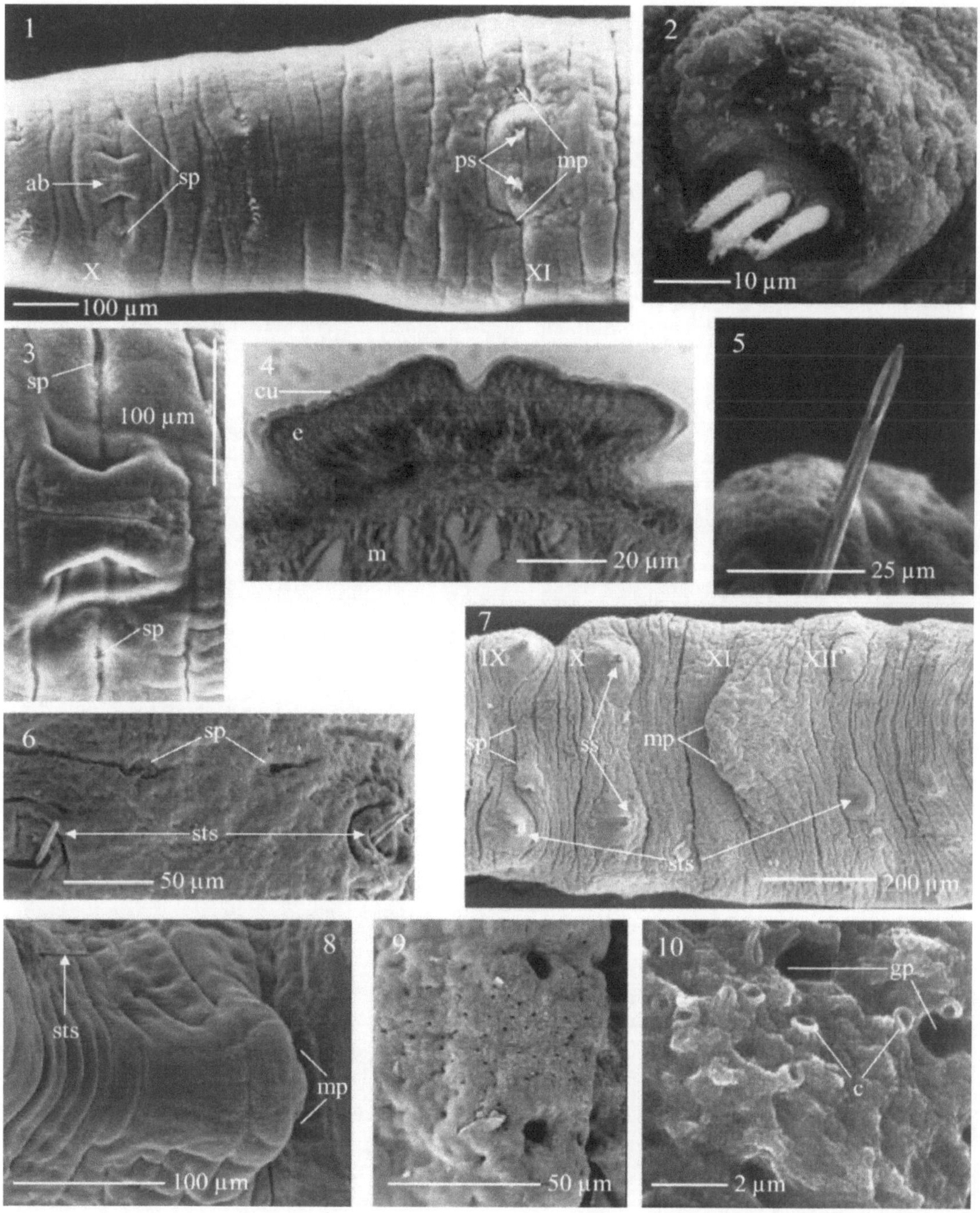

Figures 1–10. *Peristodrilus montanus* (1–4). (1) General ventral view of the genital region, (2) penial chaetae and ciliated epithelium surrounding them, (3) 'anchorage bridge' and spermathecal pores, (4) transversal section of the 'anchorage bridge'. *Protuberodrilus tourenqui* (5–10). (5) Spermathecal-type chaeta (segment XII), (6) spermathecal-type chaetae of segment IX and spermathecal pores, (7) general ventral view of genital region, (8) male porophore, (9) male pores and small glandular pores at the tip of the porophore, (10) cilia (?) and glandular pores between male pores. Abbreviations: ab – anchorage bridge, c – cilium, cu – cuticle, e – epithelium, gp – glandular pore, m – muscle, mp – male pore, ps – penial chaeta, sts – spermathecal –type chaeta, sp – spermathecal pore, ss – spermathecal chaeta. IX, X, XI, XII: segment number.

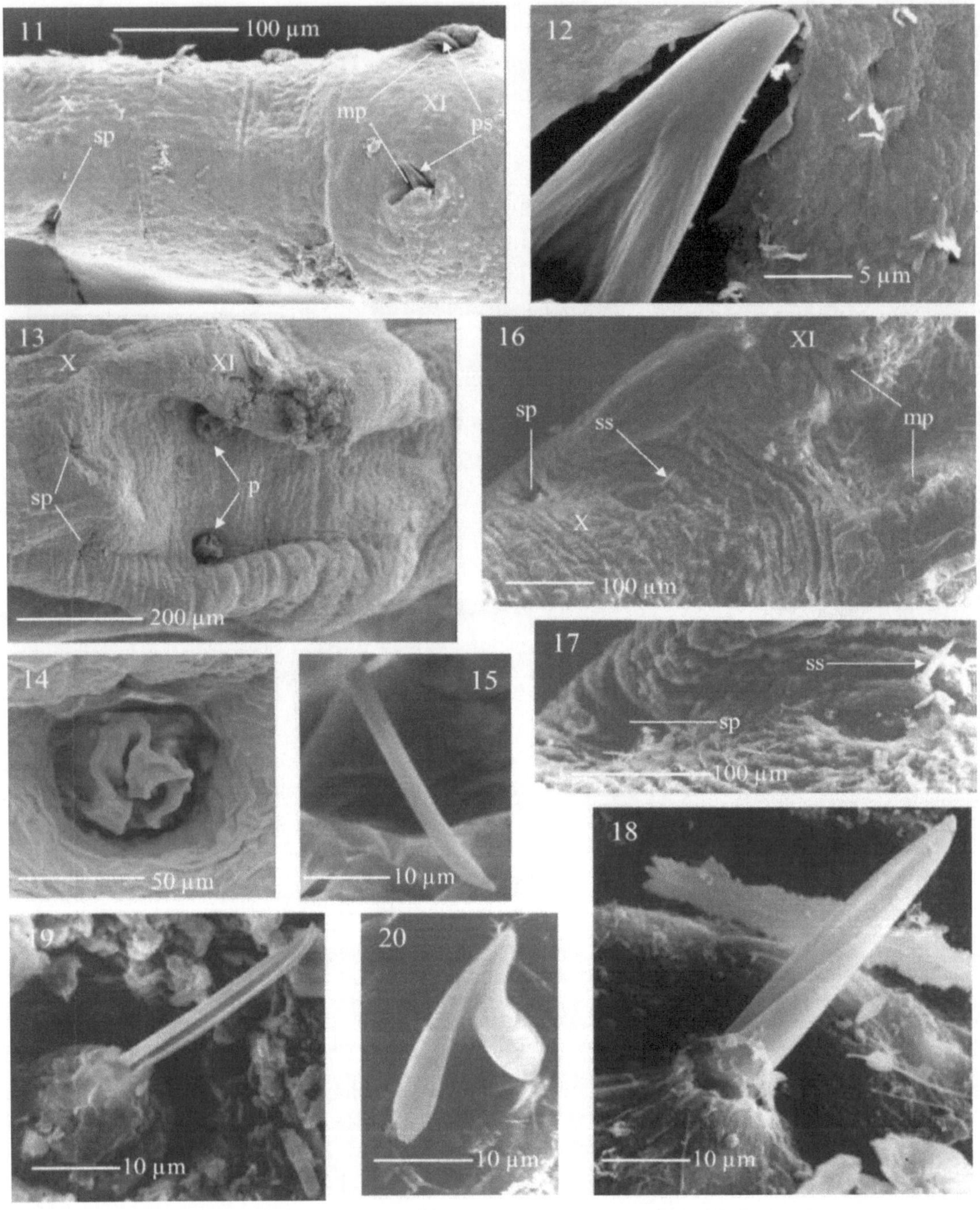

Figures 11–20. Rhyacodrilus falciformis (11–12). (11) General ventral view of the genital region, (12) penial chaeta and cilia. *Psammoryctides barbatus* (13–15). (13) General ventral view of the genital region, showing the protruded spermathecal region and the retracted male region, (14) penis, (15) spermathecal chaeta. *Potamothrix hammoniensis* (16–18). (16) General ventral view of the genital region, showing the retracted region between the male pores, (17) location of spermathecal chaeta and spermathecal pore, (18) spermathecal chaeta. *Potamothrix heuscheri* (19). Spermathecal chaeta. *Potamothrix bavaricus* (20). Spermathecal chaeta. Abbreviations: mp – male pore, p – penis, ps – penial chaeta, sp – spermathecal pore, ss – spermathecal chaeta. X, XI: segment number.

Other anatomical structures: The male pores are located at the tip of a prominent porophore, and the spermathecal pores are located deeply just behind the chaetophores of segment IX (Figs 6–9); porophore and chaetophores can be easily observed in all live mature worms.

At the tip of the male porophore there are many glandular pores and scattered cilia (?) (to be confirmed by TEM,Transmitted Electron Microscopy) (Figs 9 and 10).

Rhyacodrilus falciformis (Figs 11 and 12)

Genital pores: Male pores large and aligned with ventral chaetae. Spermathecal pores large and lateral. (Fig. 11)

Genital chaetae: The typical falciform penial chaetae of this species appear to be deeply grooved and oriented toward the mid-ventral line of the body and slightly pointed to the anterior region (Figs 11 and 12).

Other anatomical structures: There are several bundles of cilia near the male pores (Fig. 12), distributed in a half-moon area in front of each pore. These cilia were not observed in the specimens fixed with formalin.

Psammoryctides barbatus, Potamothrix hammoniensis, Potamothrix heuscheri** and **Potamothrix bavaricus (Figs 13–20)

All these species display a similar pattern in terms of the different structures involved in attachment and sperm transfer.

Genital pores: The flexibility of the body, observed in all individuals as a protrusion between the spermathecal pores and a retraction in the male pores region (Figs 13 and 16), allows the alignment of male and spermathecal pores, even if the latter have a lateral position. The distance between genital pores changes depending on the degree of contraction of the worms.

Genital chaetae: The spermathecal chaetae of these species are located far from the spermathecal pores (posterior to the pores and closer to the mid-ventral line) (Figs 16 and 17). All these chaetae are clearly gutter-shaped, although with different length and shape (Figs 15,18–20).

Other anatomical structures: No other distinctive anatomical structures were observed, but the presence of cilia cannot be discarded because all the individuals studied were fixed with formalin.

Discussion

It is generally stated that during copulation, two oligochaete worms are apposed with their heads pointing in opposite directions, and a mutual transfer of sperm occurs by contact between male and spermathecal pores or by passing through seminal grooves (Lumbricidae) (Stephenson, 1930; Omodeo, 1998). Only in some taxa, such as the glossoscolecid *Criodrilus* Hoffmeister or the tubificids *Bothrioneurum* Stolč and *Bacescuella* Hrabě, spermatozoids are placed in spermatophores on the body wall of the couple. To judge by the location of genital pores, direct sperm transfer with simple apposition of male and spermathecal pores, or with penetration of differentiated structures such as penes or pseudopenes, appears to be the mechanism used by most tubificids.

Our observations of the genital region of *Potamothrix* and *Psammoryctides* suggest that the embrace of the partners, due to the flexibility of their bodies allowing the formation of a ventral protrusion between spermathecal pores and a retraction of the male pores area, could allow contact between male and spermathecal pores, even in species where the latter are clearly lateral. This 'embrace' has been observed in some megadriles, where the clitellar region can practically enclose the body of the partner (Grove, 1925; Grove & Cowley, 1926).

When male and spermathecal pores are aligned, an embrace is not essential to keep them in contact, and highly elaborate attachment systems appear as in *P. tourenqui* and *P. montanus*.

The ciliate (?) and glandular porophore of *P. tourenqui* could allow the approach of male and spermathecal pores and the adherence of the two worms during sperm transfer. The presence of glands, whose secretions could facilitate the adherence of the worms during copulation (mucous clitellar glands, glands associated with genital markings, chaetal glands, etc.) has been described many years ago (Ditlevsen, 1904; Stephenson, 1930; Righi & Kanner, 1979).

As regard to *P. montanus*, it is likely that the cilia close to the penial chaetal bundles will facilitate the finding of the correct anchorage place. The combined action of the penial chaetal muscles and the strong dorso-ventral muscles could allow the penial chaetae to enter into the anchorage bridge pockets, to firmly hold the couple while the pseudopenes enter into the spermathecal pores. The small epidermal papilla present in the mid-ventral line of the spermathecal segment of the four species of *Krenedrilus*

Dumnicka so far known (Giani et al., 1990; Martínez-Ansemil & Collado, 1996), combined with their penial chaetae, probably represent a quite similar anchorage system. The so-called 'X-shaped mid-ventral slit' of *Bathydrilus rohdei* (Jamieson, 1977) has been also interpreted by Erséus (1981) as a structure that apparently facilitates the mutual holding of copulating worms. A great number of species belonging to several genera from all the tubificid subfamilies have similar penial chaetae that, based on form, location and orientation, may also have an anchorage function. Certainly they could anchor directly on the body wall, but perhaps new anchorage structures will be described when other species are carefully observed by SEM.

R. falciformis has a particular system to facilitate attachment and probably also to assist sperm transfer. The large and deeply grooved falciform penial chaetae, assisted by several bundles of cilia, may enter into the spermathecal pores, clasping the couple and assisting in the sperm transfer.

Five of the seven species studied in this work have spermathecal chaetae. *Protuberodrilus tourenqui* also has spermathecal-type chaetae in two non-spermathecal segments (IX and XII). All these chaetae are gutter-shaped, as are most tubificid spermathecal chaetae. Several reasons lead us to think that they act as piercing chaetae, whose principal role is perhaps stimulation rather than attachment of the concopulatants or the sperm transfer: (a) they are generally long enough to penetrate deeply into the coelomic cavity; (b) some species from several genera have spermathecal chaetae together with typical anchorage penial chaetae (e.g. *Krenedrilus*, *Rhizodrilus* Smith, *Limnodriloides* Pierantoni); (c) in many species, the spermathecal chaetae are not located near the spermathecal pores, and there are even some species with supplementary sets of spermathecal-type chaetae in non-spermathecal segments (e.g. *Protuberodrilus* Giani & Martínez-Ansemil, *Krenedrilus*). If a mechanical stimulation really takes place, the secretions of the large glands associated with spermathecal chaetae, present in many tubificids, could serve to firmly attach the partners while spermathecal chaetae protract and retract alternatively to stimulate the worms. It has been suggested that spermathecal chaetae could serve to inoculate secretions into the blood stream, acting as a sexual stimulant (Feldkamp, 1924). In our opinion, the possibility of a chemical stimulation should not be discarded and it would be interesting to characterise the composition of the gland secretions in order to verify their possible stimulating role. Either chemical or mechanical stimulation could help to contract the atrium or to contract and expand the coelomic cavity, producing embrace and release movements which help in sperm transfer. Stephenson (1930) did note that a negative pressure induced in spermathecal segments could produce the dilation of the spermathecae and aspiration of the spermatozoids.

There are other structures and mechanisms, not present in the species studied in this work, that probably represent other important ways of holding partners during sperm transfer in tubificids. This could be the case of the common median chamber (*bursa*) into which male pores open in some species (e.g. some *Rhizodrilus*, some *Monopylephorus* Levinsen), the development of a cuticular penial sheath of various lengths and forms (e.g. *Limnodrilus* Claparède, *Aktedrilus* Knöller) and the development of special penial chaetae (e.g. *Adelodrilus* Cook, *Inanidrilus* Erséus). The long and horn-shaped penis sheath of several *Aktedrilus* species could be interpreted as a way to embrace the partner and to bring the sperm into the unique dorsal spermatheca (*Aktedrilus* is devoid of penial chaetae). Erséus (1979) and Erséus & Baker (1982) interpreted the giant penial chaetae of *Adelodrilus* and *Inanidrilus* as structures used to transfer sperm into the spermathecae of the mate; the small penial chaetae would be used for attachment.

As we have seen, in spite of the small number of species studied, there appears to be a large variety of external structures used by tubificids during sperm transfer. Consequently their role in the evolution of the whole family must be emphasised.

Acknowledgements

We are indebted to our colleagues Julian Yañez, Carlos Caramelo and Bernardo Fernández for their technical support. We would also like to express our appreciation of the constructive comments on our paper by P. Martin and K. A. Coates. Financial support was provided by the Xunta de Galicia (XUGA 10305B98).

References

Ditlevsen, A., 1904. Studien an Oligochäten. Z. wiss. Zool. 77: 398–480.

Erséus, C., 1979. *Bermudrilus peniatus* n. g., n. sp. (Oligochaeta, Tubificidae) and two new species of *Adelodrilus* from the Northwest Atlantic. Trans. am. Micros. Soc. 98: 418–427.

Erséus, C., 1981. Taxonomic Studies of Phallodrilinae (Oligochaeta, Tubificidae) from the Great Barrier Reef and the Comoro Islands with Descriptions of Ten New Species and One New Genus. Zool. Scr. 10: 15–31.

Erséus, C. & H. R. Baker, 1982. New species of the gutless marine genus *Inanidrilus* (Oligochaeta, Tubificidae) from the Gulf of Mexico and Barbados. Can. J. Zool. 60: 3063–3067.

Feldkamp, J., 1924. Untersuchungen über die Geschlechtsmerkmale und die Begattung der Regenwürmer. Zool. Jahrb. Anat. 46: 609–632.

Finogenova, N. P. & T. L. Poddubnaja, 1990. One more revision of the genus *Potamothrix* Vejdovský and Mrázek, 1902 (Oligochaeta, Tubificidae). Zool. Jb. Syst. 117: 55–83.

Giani, N., C. Erséus & E. Martínez-Ansemil, 1990. Redefinition of the subterranean genus *Krenedrilus* Dumnicka (Oligochaeta, Tubificidae) with a redescription of *K. papillatus* Dumnicka, 1983 and the description of two new species. Stygologia 5: 55–65.

Grove, A. J., 1925. On the reproductive processes of the earthworm, *Lumbricus terrestris*. Quart. J. Micr. Sci. 69: 245–290.

Grove, A. J. & L. F. Cowley, 1926. On the reproductive processes of the brandling worm, *Eisenia foetida* (Sav.). Quart. J. Micr. Sci. 70: 559–581.

Martínez-Ansemil, E. & R. Collado, 1996. Two new species of freshwater Oligochaeta from the North-west Iberian Peninsula: *Krenedrilus realis* sp. nov. (Tubificidae) and *Cernosvitoviella bulboducta* sp. nov. (Enchytraeidae). J. Zool., Lond. 240: 363–370.

Oishi, M., 1930. On the reproductive processes of the earthworm *Pheretima communissima*. Sci. Rep. Tohoku Imp. Univ., Ser. 4, 5: 509–524.

Omodeo, P., 1998. History of *Clitellata*. Ital. J. Zool. 65: 51–73.

Poddubnaya, T. L. & A. I. Pataridge, 1989. A new species of the genus *Potamothrix* (Oligochaeta, Tubificidae) from the high mountain lakes of the Caucasus. Zool. Zh. 68: 153–156.

Righi, G. & E. Kanner, 1979. Marine Oligochaeta (Tubificidae and Enchytraeidae) from the Caribbean sea. Studies on the fauna of Curaçao and other Caribbean islands 58: 44–68.

Stephenson, J., 1930. The Oligochaeta. Clarendon Press, Oxford: 978 pp.

Hydrobiologia **463:** 115–122, 2001.
P. Rodriguez & P.F.M. Verdonschot (eds), Aquatic Oligochaete Biology VIII.

Mysterious 'crystals': found attached to the epidermal peritoneum of marine tubificid (Annelida, Clitellata) species

Kathryn A. Coates, Debra A. Sorocco, Nicole Maturen, Jaret Bilewitch & Aaron Thompson
Bermuda Biological Station for Research, Inc., St George's, Bermuda
Fax: 441 297 8143. E-mail: kcoates@bbsr.edu

Key words: marine tubificids, oligochaetes, annelids, biomineralization

Abstract

Marine tubificids are abundant and diverse in the carbonate sediments of Bermuda, as well as in many other tropical and subtropical locations. Recently, during microscopic observations of living specimens, crystal-like structures were observed attached to the coelomic peritoneum and in the coelomic cavity of some Bermuda species, including phallodrilines of the genera *Aktedrilus* and *Pectinodrilus,* and a rhyacodriline of the genus *Heterodrilus*. Similar structures were not seen in tubificid species of *Thallasodrilides* and other limnodriloidines, a second species of *Heterodrilus*, a tubificine of the genus*Tubificoides*, a phallodriline of the genus *Bathydrilus,* nor in a number of marine enchytraeid genera and species found in Bermuda. The crystal-like structures have two needle arms, each about 5–10 μm long and about 0.5 μm in diameter, meeting at an obtuse angle. At the junction of the arms, there is a small membrane-bound 'knob', about 1 μm in diameter, which may be continuous with the coelomic peritoneum. The numbers of 'crystals' per individual worm are estimated at 100–400 per body segment, or well over 2×10^3 in an adult worm. 'Crystals' are found: throughout the length of the worms, in all individuals of species in which 'crystals' occur, and over the range of environmental conditions where these species are found in Bermuda. Simple digestions with hypochlorite, weak and dilute acids, and staining with nuclear and cytoplasmic stains indicate that the composition of the knob is organic and the arms inorganic. The fluorescent tracer Calcein (Sigma) was not incorporated into any structures during a 24-h bath incubation of living worms, and the 'crystals' do not show birefringence when viewed between crossed polarizing filters. These last two results do not support an hypothesis that these are calcium carbonate 'crystals'. Geographically, the crystal-like structures are widespread, and have also been observed in a species of immature (unidentified) marine tubificid from Rottnest Island, Western Australia.

Introduction

Although oligochaete clitellates are most familiar as terrestrial or freshwater animals, there are well over 500 marine species. Most of these species have been discovered and described in the past 25 years, so that, not surprisingly, little is known about the details of their biology. The family Tubificidae is recognized as the most diverse clitellate family found in marine habitats.

In Bermuda, the family Tubificidae is represented in marine habitats by about 35 described species (Erséus, 1990). Specimens of many of these, and of marine enchytraeid oligochaetes, are easily found in shallow water and intertidal sites around Bermuda, and, since 1993, we have collected many marine clitellate specimens for studies of DNA sequences.

Making detailed observations of live worms is not a common practice in taxonomic studies of tubificids, in contrast to enchytraeids. Also, the structural details currently considered taxonomically significant are setal and reproductive organ forms, not structures of the coelomic lining. Finally, standard techniques of preparing marine tubificids for taxonomic studies involve the use of weak acids (including the fixative formalin, if not buffered) and other substances that we know can degrade or otherwise modify tissues or structural materials. Thus, it is not wholly surprising that we are reporting, for the first time, a very abundant and prevalent, small (subcellular sized), extracellular, structure from some common and abundant subtropical species.

In order to answer some very basic questions about the crystal-like structures, we undertook a study of marine tubificids from a number of sites in Bermuda. Some of the questions we hoped to answer were: what is the size and shape of the 'crystals' and how variable is their structure? What is the distribution in the body of the worms, abundance in an individual, frequency in a population at a given time? What is the local habitat distribution of the 'crystals'? What is the taxonomic distribution? What are the 'crystals' made of and how are they formed?

Methods and materials

Specimen collection

Collections of intertidal and shallow, subtidal, sediments were made during the years 1993–1999 at a number of sites throughout Bermuda (Fig. 1). Worms were extracted from the sediment samples by a simple wash-flotation and decanting method. The wash waters were sieved on screens with pore size 250–300 μm. The specimens retained on the screen were placed in seawater and sorted to family under a dissecting microscope. Living worms were then examined and identified on a compound microscope. In 1993, both enchytraeids and tubificids were identified live and examined for 'crystals', but in later years only tubificids were regularly examined for 'crystals'.

General observations and species identification

Live worms were placed in a small drop of seawater on a microscope slide and then covered by a coverslip (usually #1 weight, 18 mm squares). The movements of the worms were slowed by gradual withdrawal of fluid from under the coverslip, as needed. They were observed on Olympus BH2 and AX70 microscopes, both with DIC optics and video image capture capabilities. Videotapes were made of worms observed on the BH2. An image analysis system on the AX70 was used to capture images and make measurements of the 'crystals'. Mature worms were identified to species based on morphology of their setae and reproductive organs (Milligan, 1996), and with reference to the most recent list of tubificid species found in Bermuda (Erséus, 1990). Immature worms were also identified to species, if they had setal characteristics unique for mature specimens of species that were found in the same sample. The presence, abundance, location, and form of the unusual crystal-like structures were noted for each worm that was identified to a species rank.

Studies of 'crystals' in standard preparations

Some specimens with 'crystals', of each identified species, were fixed in ethanol and then prepared as if for taxonomic studies. The preparation of marine tubificid specimens for taxonomic studies used was: fixation in 70% ethanol, staining in alcoholic borax carmine, differentiation or destaining with dilute HCl (less than 10% solution), dehydration through an alcohol series, clearing with xylene, followed by mounting in Canada balsam. Some specimens were not stained, but were otherwise processed as noted above. Specimens were checked for the presence of 'crystals' after mounting in Canada balsam.

Chemical testing, staining of 'crystals', and crossed-polarizer observations

Smears of individual specimens of tubificid species with 'crystals' were made by rinsing the whole worm in freshwater (to prevent formation of salt crystals) and then spreading the specimen on a microscope slide using the edge of a coverslip. After drying, slides were rinsed with methanol. 'Crystals' were then located on the slide and their coordinates recorded for post-treatment observations. Two to six smear preparations each were treated with one of the following reagents or stains: hypochlorite (Chlorox), 100% ethanol, dilute HCl (1 drop of concentrated acid in 10 ml) in 70% ethanol or in distilled water, Giemsa (Fischer Scientific staining kit) – a nuclear and cytoplasmic stain, DAPI (4′,6′-diamidino-2-phenylindole; 5 μg ml^{-1}) – a fluorescent nuclear stain, or Grenacher's alcoholic borax carmine – a nuclear stain. Hypochlorite, HCl, and ethanol treated slides were left overnight, then rinsed with methanol or distilled water prior to reexamination. Giemsa staining was performed by Dr Gene Burreson (Virginia Institute of Marine Science), and both he and we examined these preparations. Borax carmine treatments were rinsed after five minutes. DAPI was applied for 30 min to both smears and whole, live, worms. The DAPI treated smears and living worms were observed on the AX70 microscope, epi-illuminated using a mercury lamp light source, with UV excitation (Chroma Technology Corp., filter cube UM536, UV wide band excitation cube, excitation filter 330–385 nm, dichroic mirror 400 nm, barrier filter 420 nm).

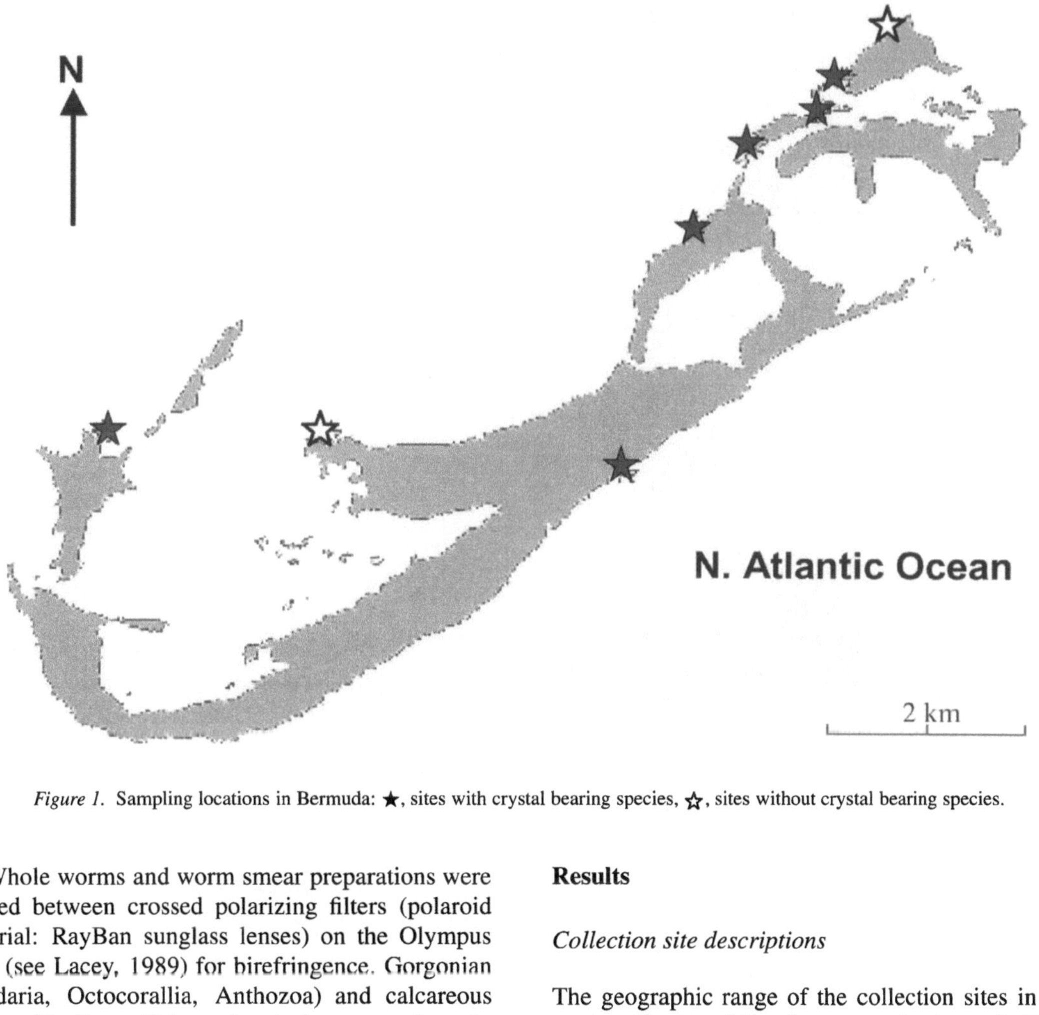

Figure 1. Sampling locations in Bermuda: ★, sites with crystal bearing species, ☆, sites without crystal bearing species.

Whole worms and worm smear preparations were viewed between crossed polarizing filters (polaroid material: RayBan sunglass lenses) on the Olympus BH2 (see Lacey, 1989) for birefringence. Gorgonian (Cnidaria, Octocorallia, Anthozoa) and calcareous sponge (Porifera, Calcarea) spicules were also observed for birefringence.

Twenty whole worms with 'crystals' were incubated in a Calcein bath ($C_{30}H_{26}N_{20}O_{13}$; Sigma Chemicals; Chem. Abstracts Id# [1461-15-0]; 500 mg l^{-1}) for 24 h. This is a fluorescent tracer of biological uptake and deposition of calcium (Rowley & MacKinnon, 1995). Calcareous green algae (Chlorophyta) and calcareous sponges were treated similarly, for comparison. Specimens were examined for yellow-green fluorescence (emissions about 514 nm), on the AX70 microscope when epi-illuminated with blue light (Chroma Technology Corp., filter cube UM41001, excitation 465–505 [480/40] nm, emission 510–560 [535/50] nm).

Results

Collection site descriptions

The geographic range of the collection sites in Bermuda with crystal-bearing worms is shown in Figure 1. The habitat condition ranges found at these sites were: pH of pooled interstitial water: 7.81–8.34; salinity of pooled interstitial water: 34–40 ppt; grain size characteristics: mixed sands (medium coarse to fine sand particles); and, organic carbon content: 2.4–4.7%.

Taxonomic distribution, frequency, and abundance of 'crystals'

Chevron 'crystals' (Fig. 2) were found in all identifiable specimens of the following species of Tubificidae, at all dates and locations where they were collected: *Pectinodrilus molestus* (Erséus, 1988), *Aktedrilus parvithecatus* (Erséus, 1978), and *Heterodrilus* sp. 1 (no fully mature specimens).

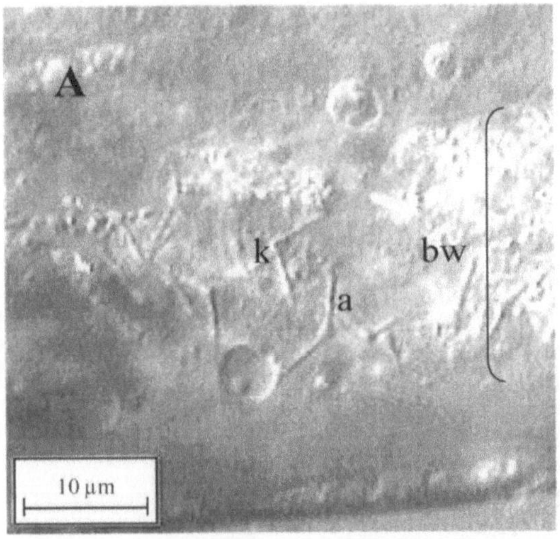

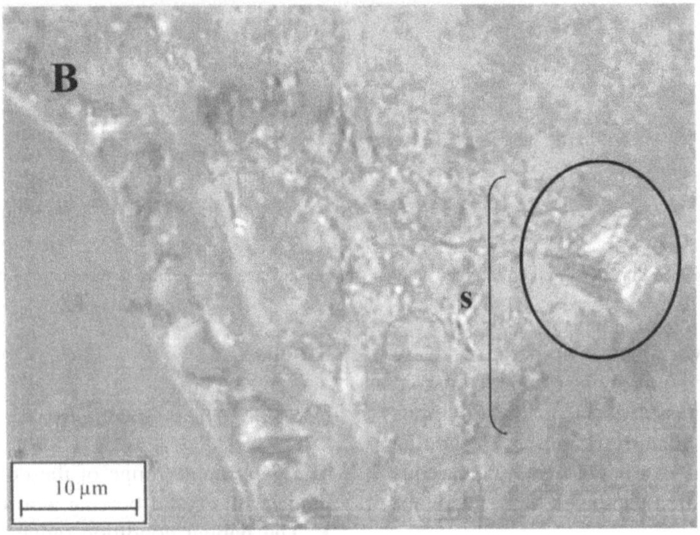

Figure 2. Chevron 'crystals'. (A) on the body wall peritoneum in *Heterodrilus* sp.; (B) in stacks on the septal membrane of *Aktedrilus parvithecatus*. Abbreviations: a – crystal arms; k – organic knob (and attachment point to peritoneum); bw – body wall peritoneum; s – septal peritoneum.

'Crystals' were not found in any of the following species: *Heterodrilus* sp. 2 (no fully mature specimens), *Limnodriloides monothecus* Cook, 1974 *Smithsonidrilus minusculus* (Erséus, 1983) *Thallasodrilides gurwitschi* (Hrabe, 1971), *Thallasodrilides belli* (Cook, 1974), and *Tubificoides bermudae* Råsmark & Erséus, 1986. Nor were they found in another two species of limnodriloidine, either *Limnodriloides* or *Smithsonidrilus* species, of which no fully mature specimens were examined. These were distinguished from each other by their different forms of somatic setae. 'Crystals' were not found in any species, which were examined alive, of the following genera of Enchytraeidae: *Grania, Marionina, Enchytraeus, Stephensoniella, Achaeta,* and *Randidrilus* (from Australia, KAC, pers. obs.).

A different form of 'crystal' (Fig. 3) was seen in *Bathydrilus adriaticus* (Hrabe, 1971). We have called this the 'starburst', but these have not been investigated to the same extent as the chevron 'crystals'. They

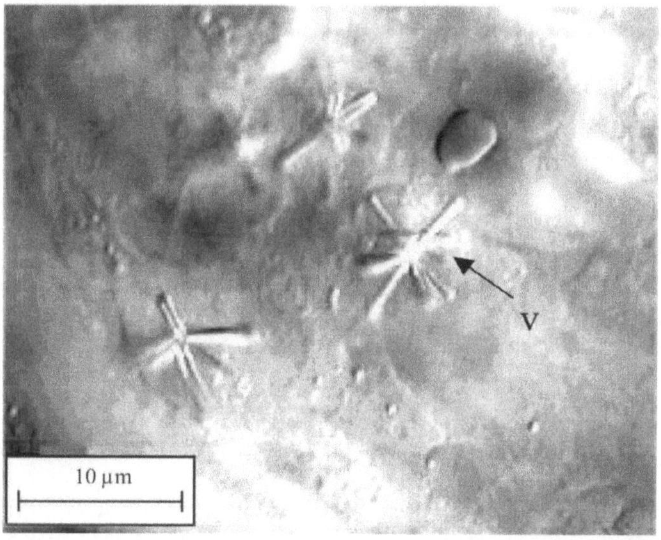

Figure 3. Starburst 'crystals' in membrane bound vesicles of *Bathydrilus adriaticus*. Abbreviation: v – vesicle of crystal.

are most often enclosed in an extracellular vesicle, unlike the chevron 'crystals', and are not as abundant.

Chevron 'crystals' (Fig. 2) were found throughout the body of a worm, from the prostomium to the pygidium, attached to the exposed surfaces of peritonea lining the body cavity, including the septa, and floating in the coelom. The attached 'crystals' usually appeared to be irregularly distributed (Fig. 2A) but sometimes were found aligned in obvious stacks (Fig. 2B). Segmental estimates (based on counts of subvolumes of segments) gave numbers of 100–400 'crystals' per segment in a mature worm, suggesting a total number of 10^3–10^4 per worm of an average-sized mature specimen of *A. parvithecatus*. Specimens of *A. parvithecatus* had the most 'crystals' per worm, of the species observed.

Description of chevron 'crystals'

The chevron 'crystals' (Fig. 2A,B) have arms of about equal lengths. In *A. parvithecatus*, these range from 5 to 10 μm, and were observed to be of a similar size in the other species in which they were found. The arms look glassy and have a diameter of about 0.4–0.5 μm. At the apex of the two arms is a transparent, membrane-bound knob, approximately 1 μm in diameter.

Fates of 'crystals' in treatments

The fates of the 'crystals' in the various treatments were:

1. Washing with ethanol – no change in appearance.
2. Hypochlorite digestion – central knob disappears, arms remain.
3. Acid (dilute hydrochloric acid and acetic acid) digestion, in either ethanol or water – arms disappear, knob remains.
4. Nuclear stains (borax carmine, DAPI, Giemsa) – no staining demonstrated of any of the crystal components.
5. Cytoplasmic stain (Giemsa) – deep blue staining of the 'knob' (pers. com. G. Burreson and authors' observations).
6. Calcein incubation – *Aktedrilus* incubation, no yellow-green fluorescence, but results are ambiguous as similar results were obtained in incubations of calcareous sponges and calcareous algae.
7. Cross-polarization – worm 'crystals', either isolated or *in situ*, no birefringence; whereas, gorgonian and calcareous sponge spicules produced brilliant birefringence.
8. Observations after standard taxonomic preparations, with and without staining – no 'crystals' found.

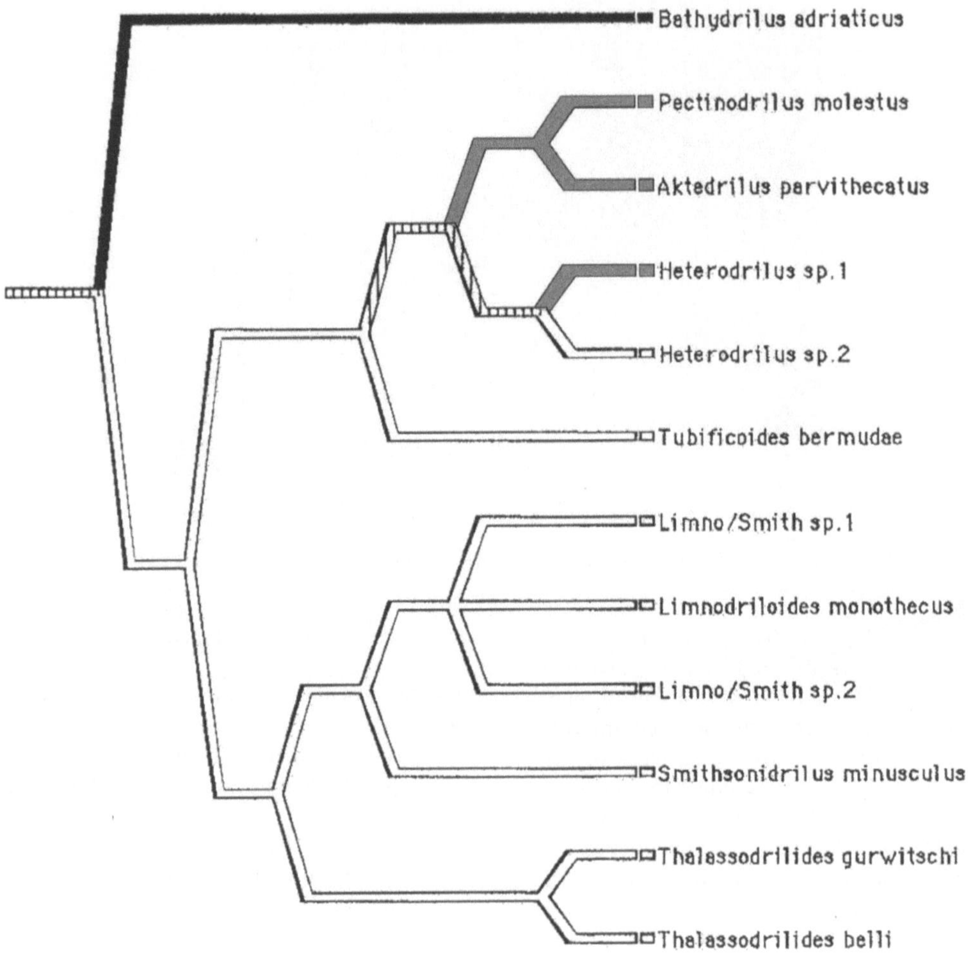

Figure 4. Distribution of taxa with 'crystals', plotted on reduced phylogeny after Erséus et al. (2000, Fig. 1). Genera that were not examined for 'crystals' have been excluded and the positions of genera studied by Erséus et al. (2000) have been named as the species we examined. Species with starbursts, ■ ; species with chevrons, ■; species without 'crystals' □ ; hypothetical taxa or lineages with unresolved 'crystal' character ◫.

Discussion

The knob at the junction of the chevron crystal arms is organic and appears to be a part of the peritoneal membrane. It may be involved in the production (?secretion) of the crystal arms. The arms of the 'crystals' are not organic and are not calcium carbonate, neither calcite nor aragonite, the most commonly produced biomineral (Lowenstam, 1981; Lowenstam & Weiner, 1989; Pearse et al., 1994).

'Crystals' are present in at least three species of tubificid found in Bermuda, and also are known from geographically distant locations. They were observed previously (KAC, pers. obs.) in unidentified, immature, tubificids from south-western Western Australia. The presence of 'crystals' is constant across the range of habitats and locations in which the three species are found in Bermuda and the specimens collected did not show obvious signs of disease, such as malformed reproductive structures, epidermal lesions, or unusual form of the gut. Thus, the presence of 'crystals' does not seem to be a pathological condition nor a response to a particular environmental condition of a subset of the habitats occupied by these species. Their prevalence and abundance along with staining characteristics confirm that the 'crystals' are not a cyst-form of a parasite.

When the distribution among taxa of the chevron 'crystals' is compared to a recent phylogeny (Erséus et al., 2000) of the same tubificid genera (Fig. 4), the degree of congruence of their presence with the phylogeny indicates a likely evolutionary 'context' for the crystals; they are not found randomly and widely distributed across many clades. They are found in phallodrilines, excluding *Bathydrilus*, and in at least some rhyacodriline species, *Heterodrilus* sp. 1, whereas they were not found in another species of *Heterodrilus*, one tubificine or a number of limnodriloidine tubificids. The analyses of Erséus et al. (2000) separated *Bathydrilus* from the other phallodriline genera, and they noted this was 'an unexpected result', compared to previous ideas about the relationships among these tubificid genera. Thus, it is of some additional interest that the crystal distribution is congruent with their result.

In this study, specimens were recognized as belonging to *Heterodrilus* based on the presence of trifid setae, with a distinctly reflected ectal end, and the absence of hair setae (Milligan, 1996). The two species distinguished here were separated by the shape and number of their setae, one species had single setae throughout, the other had single setae only in posterior segments but pairs in anterior segments. Both species had bifid setae in posterior segments. As far as the difference we found between our two *Heterodrilus* species in the presence of 'crystals', it is of note that the genus is itself speciose and diverse, and now encompasses all the species of *Heterodriloides* Erséus, 1981 and *Giereidrilus* Erséus, 1981 (Erséus, 1990; Milligan, 1996). Perhaps the setal characteristics commonly relied on to determine that species belong to *Heterodrilus* are not distinctive of a single lineage, and, perhaps, this taxonomic grouping is not monophyletic.

Congruence of crystal presence with the results of a multi-character (nucleotides) phylogenetic analysis (Erséus et al., 2000), suggest that the 'crystals' may have a single adaptive origin and a distinct function. However, any obvious structural function or any conditions which control or limit the production of 'crystals', and which might imply a physiological role, are unknown. Their scattered distribution and flexible connections to the coelomic membranes are not suggestive of a skeletal function, in contrast to *Nephtys* spp (Polychaeta), which have calcium phosphate granules in the core of muscle cells of the body wall (Gibbs & Bryan, 1984).

From the smear preparations, we know that the 'crystals' are not affected by 100% ethanol, whereas, they do dissolve in hydrochloric acid of a concentration similar to that used in destaining. Thus, the 'crystals' are probably not present in whole mounts that have been stained, rather than just impossible to see. This may also be true for specimens that have been fixed in unbuffered formalin, a common practice. However, this may not be the case for unstained whole mounts made from ethanol fixed and preserved specimens, in which we were also unable to see 'crystals'.

Plans for future studies include isolation of pure 'crystals' for other types of investigations of their composition, use of other preservation and observation protocols, such as glutaraldehyde, thin-sectioning and transmission electron microscopy, manipulations of microhabitats to determine whether environmental conditions affect their formation, and additional geographic and taxonomic records.

Acknowledgements

We thank Gene Burreson for guiding our understanding of parasite infestations and, in particular, our distinction of myxozoans from the crystals. The Canadian Associates of the Bermuda Biological Station provided funding to KAC, JB, and AT. The BBSR NSF-REU award #OCE 9619718 funded thc studics of NM and DS, and NSF DBI-9724335 to KAC funded the Olympus AX70 microscope facility. This is contribution # 1585 of the Bermuda Biological Station for Research, Inc. KAC thanks all of the co-authors for the pleasure of working with such fine, young scientists.

References

Erséus, C., 1990. The marine Tubificidae (Oligochaeta) of the barrier reef ecosystems at Carrie Bow Cay, Belize, and other parts of the Caribbean Sea, with descriptions of twenty-seven new species and revision of *Heterodrilus*, *Thalassodrilides* and *Smithsonidrilus*. Zool. Scr. 19: 243–303.

Erséus, C., T. Prestegaard & M. Kallersjo, 2000. Phylogenetic analysis of Tubificidae (Annelida, Clitellata) based on 18S rDNA sequences. Mol. Phylo. Evol. 15: 381–389.

Gibbs, B. E. & G. W. Bryan, 1984. Calcium phosphate granules in muscle cells of *Nephtys* (Annelida, Polychaeta) – a novel skeleton? Nature 310: 494–495.

Lacey, A. J., 1989. Rendering transparent specimens visible by inducing contrast. In Lacey, A. J. (ed.), Light Microscopy, a Practical Approach. IRL Press, Oxford: 25–59.

Lowenstam, H. A., 1981. Minerals formed by organisms. Science 211: 1126–1131.

Lowenstam, H. A. & S. Weiner, 1989. On Biomineralization. Oxford University Press, New York: 324 pp.

Milligan, M., 1996. Identification manual for the aquatic Oligochaeta of Florida. Vol. II. Estuarine and nearshore marine oligochaetes. Florida Department of environmental Protection, Tallahassee, Florida: 239 pp.

Pearse, V. B., T. Uehara & R. L. Miller, 1994. Birefringent granules in Placozoans (*Trichoplax adhaerens*). Trans. am. microsc. Soc. 113: 385–389.

Rowley, R. J. & D. I. MacKinnon, 1995. Use of the fluorescent marker calcein in biomineralization studies of brachiopods and other marine organisms. Bull. Inst. ocean. Monaco, 2 (special publication #14): 111–120.

Hydrobiologia **463:** 123–131, 2001.
P. Rodriguez & P.F.M. Verdonschot (eds), Aquatic Oligochaete Biology VIII.

Pattern formation in embryos of the oligochaete annelid *Tubifex*: cellular basis for segmentation and specification of segmental identity

Takashi Shimizu, Kaoru Kitamura, Asuna Arai & Ayaki Nakamoto
Division of Biological Sciences, Graduate School of Science, Hokkaido University, Sapporo 060-0810, Japan
Fax: 81 11 706 4851. E-mail: stak@sci.hokudai.ac.jp

Key words: Tubifex, embryogenesis, segmentation, segmental identity, teloblast

Abstract

The embryonic origin of metameric segmentation was examined in the oligochaete *Tubifex* using lineage tracers. Segments in *Tubifex* embryos arise from five bilateral pairs of longitudinal coherent columns (bandlets) of primary blast cells which are generated by five bilateral pairs of embryonic stem cells called teloblasts (M, N, O, P and Q). As development proceeds, an initially linear array of blast cells in each ectodermal bandlet gradually changes its shape in a lineage-specific manner. These morphogenetic changes result in the formation of distinct cell clumps, which are separated from the bandlet to serve as segmental elements (SEs). SEs in the N and Q lineages are each comprised of clones of two consecutive primary blast cells. In contrast, in the O and P lineages, individual blast cell clones are distributed across SE boundaries; each SE is a mixture of a part of the preceding anterior clone and a part of the next posterior clone. Morphogenetic events, including segmentation, in an ectodermal bandlet proceed normally in the absence of neighboring ectodermal bandlets. Without the underlying mesoderm, separated SEs fail to space themselves at regular intervals along the anteroposterior axis. It is suggested that ectodermal segmentation in *Tubifex* consists of two stages; autonomous morphogenesis of each bandlet leading to generation of SEs, and the ensuing mesoderm-dependent alignment of separated SEs. In contrast, metameric segmentation in the mesoderm (M lineage) is a one-step process in that it arises from an initially simple organization (i.e. a linear series) of primary m-blast cells, which individually serve as a founder cell of each segment. The boundary between mesodermal segments is determined autonomously. The results of a set of cell ablation and transplantation experiments, using alkaline phosphatase activity as a biochemical marker for segments VII and VIII suggest that segmental identities in primary m-blast cells are determined according to the genealogical position in the M lineage and that the M teloblast possesses a developmental program through which the sequence of blast cell identities is determined.

Introduction

Oligochaete annelids are segmented animals that display a high degree of metamerism in their body plan. The trunk portion of the body comprises multiple segments, each of which contains a similar complement of ectodermal and mesodermal tissues. The head, represented by the prostomium, is not a segment, nor is the terminal part of the body in which the anus is located. Trunk segmentation is visible externally as rings and is reflected internally by the serial arrangement of coelomic compartments separated from one another by intersegmental septa (Brusca & Brusca, 1990). Additionally, it should be noted that segments are regionally differentiated along the longitudinal body axis. There is no doubt that individual segments acquire specific identities during embryonic development of oligochaetes. Thus, the fundamental aspects of pattern formation during oligochaete embryogenesis include not only the generation of metameric segments but also the specification of their identities.

In oligochaete annelid embryos, five bilateral pairs of embryonic stem cells called teloblasts are formed early in development (Fig. 1C). Each teloblast divides repeatedly to give rise to primary blast cells, which are arranged into a coherent column or bandlet (see Fig. 1D). Four of the five bandlets on each side of the embryo join together to form an ecto-

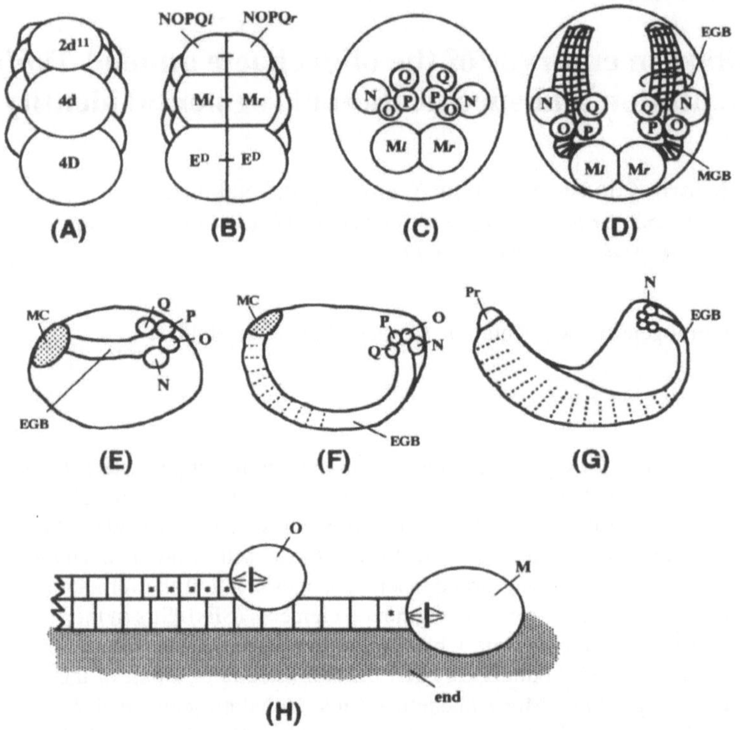

Figure 1. Summary of *Tubifex* development. (A and B) Posterior view with dorsal to the top; (C and D) dorsal view with anterior to the top; (E–G) side view with anterior to the left and dorsal to the top. (A) A 22-cell stage embryo. Cells $2d^{11}$, 4d and 4D all come to lie in the future midline. (B) 4d divides bilaterally into left and right mesoteloblasts (M*l* and M*r*); $2d^{111}$ derived from $2d^{11}$ devides into a bilateral pair of ectoteloblast precursors (NOPQ*l* and NOPQ*r*), and 4D devides into a pair of endodermal precursor cells E^D. (C) An embryo at about 30 h after the bilateral division of 4d. Only teloblasts are depicted. NOPQ on each side of the embryo has produced ectoteloblasts N, O, P and Q. (D) A two-day-old embryo following the bilateral division of 4d. Only teloblasts and associated structures are depicted. At this stage, a short ectodermal germ band (EGB) extending from the ectoteloblasts N, O, P and Q is seen on either side of the embryo. A mesodermal germ band (MGB) extending from the M teloblast is located under the ectodermal germ band. (E–G) Morphogenesis of the ectodermal germ band. Embryos are shown at (E) 2.5, (F) 4, and (G) 6 days after the 4d cell division. (E) The germ band (EGB) is associated, at its anterior end, with an anteriorly located cluster of micromeres (called a micromere cap; MC), and it is initially located at the dorsal side of the embryo. (F) The germ bands (EGB) on both sides of the embryo elongate and gradually curve round toward the ventral midline and finally coalesce with each other along the ventral midline. (G) The coalescence is soon followed by dorsalward expansion of the edge of the germ band, Pr, prostomium. (H) Longitudinal section showing the relative positions of the endoderm (end) and bandlets extending from teloblasts M and O. Anterior is to the left and posterior is to the right. The bandlet (germ band) derived from the m teloblast is overlain by the o-bandlet and is underlain by the endoderm (end). Asterisks indicate the presence of a single primary blast cell in each block of the bandlet. The remaining blocks individually represent a cell cluster, which is derived from a single primary blast cell.

dermal germ band (GB), while the remaining bandlet becomes a mesodermal GB that underlies the ectodermal GB (Fig. 1H). The results of previous descriptive and cell ablation studies (Wilson, 1889; Penners, 1924a,b, 1926; Devries, 1973a,b) have suggested that morphogenetic events such as body elongation and segmentation depend solely on the presence of teloblasts and their progenies. In fact, teloblasts are the main source of ectodermal and mesodermal segmental tissues; no non-teloblastic cells can replace missing teloblasts in this respect. However, the process by which the initially continuous GB is converted into an array of distinct cell clusters (i.e. segments) has not been studied. Thus, presently available information about metameric segmentation in oligochaetes is fragmentary and uncertain.

In this paper, we describe the process of ectodermal and mesodermal segmentation in embryos of *Tubifex hattai*, reviewing recent findings from our laboratory (Goto et al., 1999a,b; Kitamura & Shimizu, 2000a,b; Nakamoto et al., 2000; Arai et al., 2001).

We also discuss the mechanisms for specification of segmental identities in this animal.

Re-examination of cell lineages in the *Tubifex* embryo by intracellular injection of lineage tracers

Many tissues comprising body segments have been assigned to the progeny of the teloblasts. However, the developmental fate of each teloblast has been inferred only from its initial position in the embryo (Penners, 1922, 1924a). Therefore, we have followed the fate of progeny of each teloblast by means of intracellular injection of a tracer enzyme, horseradish peroxidase (HRP), and we have constructed cell fate maps for the teloblasts in the *Tubifex* embryo (Goto et al., 1999b).

M teloblasts give rise to nearly all of the mesodermal tissues, which include circular and longitudinal muscles, coelomic walls, nephridia (in segments VII and VIII) and primordial germ cells (in segments X and XI). Although few in number, M teloblasts also contribute cells to the ventral ganglion.

Similarly, each of the ectoteloblasts, N, O, P and Q, makes a topographically characteristic contribution to the ectodermal tissues such as the nervous system (i.e. ganglionic cells and peripheral neurons) and epidermis, all of which exhibit a segmentally repeated distribution pattern. P and Q teloblasts uniquely give rise to additional ectodermal tissues, namely ventral and dorsal setal sacs, respectively. Furthermore, O teloblasts make a contribution to nephridiopores in segments VII and VIII as well.

These results confirm the previously held views that ectodermal and mesodermal tissues comprising body segments are mainly derived from ectoteloblasts and mesoteloblasts, respectively (Penners, 1922, 1924a). However, it should be mentioned that there are several types of tissue whose embryonic origins have been incorrectly or insufficiently assigned. First, we have shown that circular muscles are derived from M teloblasts, not from ectoteloblasts. This is in contrast to Penners' conclusion that circular muscles originate from O, P and Q teloblasts (which are designated M^1, M^2 and M^3, respectively, in Penners' papers). Second, in contrast to the traditionally held view that N teloblasts are the exclusive source of ganglionic cells, the present study has shown that all of the five teloblasts (M–Q) make a topographically characteristic contribution to the central nervous system. Third, setal sacs, which were assumed to originate from N or O teloblasts, have been shown to be derived from P and Q teloblasts. Furthermore, it has also been suggested that ectoteloblasts contribute peripheral neurons to the body wall, which had not been mentioned in previous studies. Taken together, the present results suggest that all of the teloblasts in the *Tubifex* embryo generate more types of tissue than have been previously thought.

These results also confirm the previous notion that the prostomium and the supraesophageal ganglion (brain) are both derived from cells other than progenies of teloblasts (Penners, 1924a). Even when NOPQ or $2d^{111}$ is injected with HRP (see Fig. 1 A,B), no labeled progeny cells are found in either of these two regions (data not shown). Apparently, teloblasts and their precursors do not make any contribution to these anterior structures.

Segmentation in the ectoderm

Ectodermal segmentation in *Tubifex* is a process of separation of 50 μm-wide blocks of cells from the initially continuous ectodermal GB, a cell sheet consisting of four bandlets of blast cells derived from ectoteloblasts (Fig. 2). Using intracellular lineage tracers, we have characterized the morphogenetic processes that give rise to formation of these ectodermal segments (Nakamoto et al., 2000).

The formation of ectodermal segments begins with formation of fissures, first on the ventral side and then on the dorsal side of the GB. The unification of these fissures gives rise to separation of a 50 μm-wide block of cells from the ectodermal GB. As development proceeds, an initially linear array of blast cells in each ectodermal bandlet gradually changes its shape and its contour becomes indented in a lineage-specific manner (Fig. 2 C,D). These morphogenetic changes result in the formation of distinct cell clumps, which are separated from the bandlet to serve as segmental elements (SEs; Fig. 3A). SEs in the N and Q lineages are each comprised of clones of two consecutive primary blast cells. In contrast, in the O and P lineages, individual blast cell clones are distributed across SE boundaries; each SE is a mixture of a part of the preceding anterior clone and a part of the next posterior clone (Fig. 3B).

Autonomy of bandlet morphogenesis

When an ectoteloblast is forced to be "solitary" by removal of all of its (ipsilateral) sister teloblasts, it continues dividing at a rate comparable to that in intact embryos. All but o-bandlets are very similar to

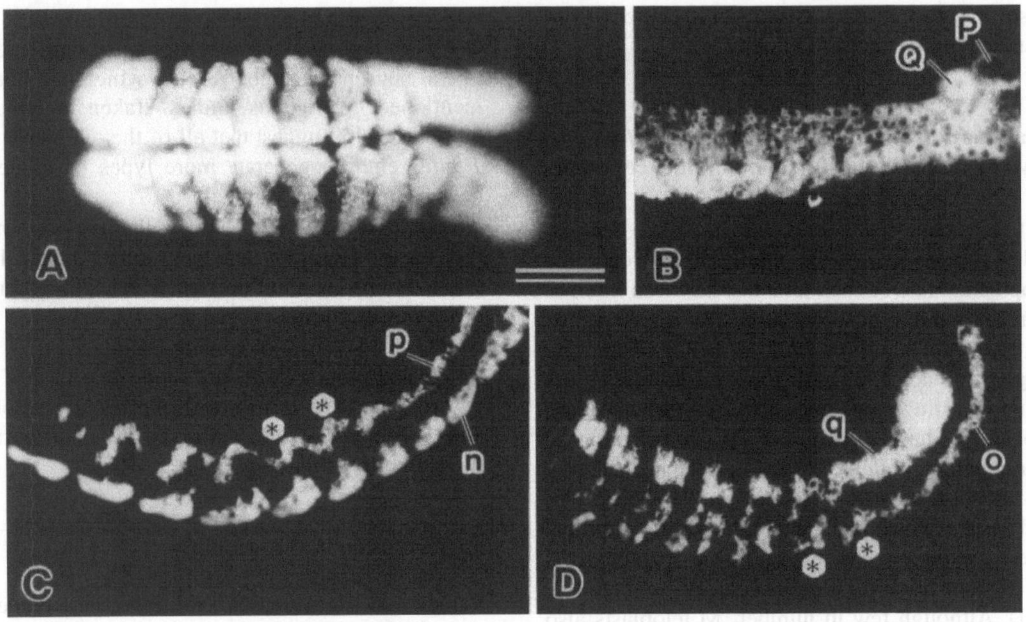

Figure 2. Fluorescence micrographs showing segmentation in ectodermal GBs. $2d^{111}$ cell (A), left NOPQ (B) or individual teloblasts (C, D) were injected with a lipophylic fluorescent dye DiI and allowed to develop for 3 days before fixation. Whole-mount preparations were viewed from the ventral side (A) or left side (B–D). In all panels, anterior is to the left; in B–D, dorsal is to the top. Bar: 100 μm (A, B); 80 μm (C, D). (A) Both the left and right germ bands (GBs) are labelled with DiI. Note that GBs are divided into 50 μm-wide blocks of labelled cells by intersegmental furrows, which are recognised as non-fluorescent transverse stripes. (B) The posterior portion of the left GB is shown. P and Q teloblasts are seen, but N and O teloblasts are outside the field. Note the formation of fissures in bandlets. (C) Fluorescent n- and p-bandlets in the left GB. These bandlets were derived from left N and P teloblasts which had been injected simultaneously with DiI shortly after the birth of the P teloblast. Asterisks indicate S-shaped segmental elements (SEs) in the P lineage. (D) Fluorescent o- and q-bandlets in the left GB. These bandlets were derived from O and Q teloblasts which had been injected simultaneously with DiI shortly after the birth of the O teloblast. Asterisks indicate W-shaped SEs of the O lineage. From Nakamoto et al. (2000).

the respective bandlets in intact embryos, not only in shape but also in periodicity of separated SEs. This suggests that lineage-specific bandlet transformation in the N, P and Q lineages occurs independently of adjacent bandlets (Nakamoto et al., 2000).

In contrast, 'solitary' o-bandlets exhibit features characteristic of the P lineage rather than the O lineage. This finding suggests that 'solitary' o-bandlets adopt the P fate rather than the O fate. The O-to-P fate conversion of 'solitary' o-bandlets has been shown to be maintained in the terminally differentiated progeny of the primary blast cells. Recently, it has also been shown that o-bandlets are induced to assume the O fate by signals from the p-bandlets (Arai et al., 2001).

Mesodermal control

During *Tubifex* embryogenesis, the ectodermal GB is normally underlain by the mesodermal GB (see Fig. 1H). There is a possibility that the mesodermal GB plays a role in ectodermal segmentation. The results of recent cell ablation and transplantation experiments have suggested that the mesodermal GB plays an important role in two aspects of ectodermal morphogenesis, viz. spatial arrangement of SEs along the anteroposterior axis and dorsalward expansion of the ectodermal GB (Nakamoto et al., 2000).

In embryos from which left M teloblasts have been ablated, ectodermal GBs on the mesoderm-deficient (left) side fail to generate any segmental organization. These GBs do not undergo the dorsalward expansion that normally accompanies ectodermal segmentation. As a result, they appear as a long rod running along the ventral midline. If a cell layer derived from M teloblasts is transplanted to the space between the mesoderm-deficient ectodermal GB and the endoderm, the ectodermal GB undergoes segmentation and dorsalward expansion only in the portion underlain by the transplanted (mesodermal) cell layer (A. Nakamoto, unpublished data).

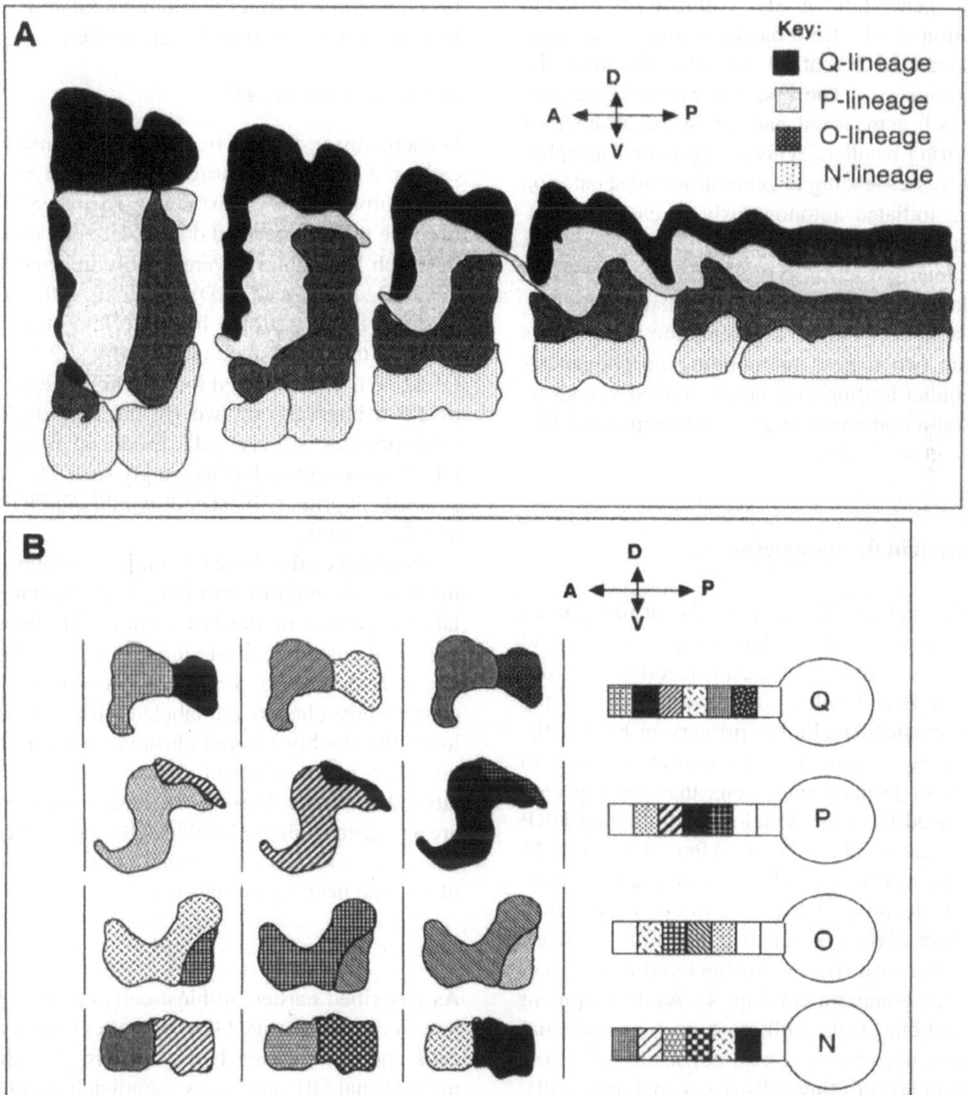

Figure 3. Schematic summary of morphogenetic events leading to ectodermal segmentation. (A) Formation of SEs is followed by their separation from bandlets. This separation occurs first in the N lineage, followed by the O, Q and P lineage in this order. Upon their integration into a discrete segment, segmental units further change their shape to intermingle with each other within the segment. (B) Segmental contribution of clones of primary blast cells. Three segments are shown; each pattern represents an individual clone. In each lineage, the order of primary blast cells is shown to the right of the figure, together with their parent teloblasts. In the N and Q lineages, two consecutive primary blast cells give rise to one segmental complement of progeny. In the O and P lineages, each of the serially homologous primary blast cell clones is divided into two parts, which are inherited separately by two consecutive segments. A – anterior; D – dorsal; P – posterior; V – ventral. Adapted from Nakamoto et al. (2000).

Two-step process

Ectodermal segmentation in *Tubifex* involves three key events: (a) generation of SEs within each bandlet, (b) separation of SEs from bandlets, and (c) arrangement of separated SEs at 50 μm intervals along the anteroposterior axis. The first two events occur normally in each ectodermal bandlet in the absence of its neighboring bandlets. This suggests that morphogenetic processes leading to generation and separation of SEs are initiated autonomously in each bandlet. In contrast, the distribution pattern of separated SEs along the anteroposterior axis apparently depends on the germ layers underlying ectodermal bandlets. Thus, we suggest that ectodermal segmentation in *Tubifex* is divided into two stages: autonomous morphogenesis of each bandlet leading to generation of SEs, and the ensuing non-autonomous alignment of separated SEs (Nakamoto et al., 2000).

Segmentation in the mesoderm

Mesodermal segmentation is a process of reorganization of the mesodermal GB into a linear array of 50 μm-wide clusters of cells. Careful stage-by-stage observations of HRP-labeled GBs have shown that the number of clusters (including primary m-blast cells, each of which is counted as one cluster) is equal to the number of primary blast cells that are expected to be produced from the M teloblast following HRP injection (Goto et al., 1999a). After birth from M teloblasts, each primary m-blast cell undergoes a spatiotemporally stereotyped sequence of cell divisions to generate three classes of cells (in terms of cell size), which together give rise to a distinct cell cluster (i.e. a mesodermal compartment; Fig. 4). As development proceeds and blast cells proliferate, each mesodermal compartment becomes a vesicle consisting of a surrounding thin layer of tiny cells (i.e. a coelomic wall), inner cells at one end, and the coelomic cavity.

Judging from the sizes of cells involved, two steps are discernible in the process of coelomic wall formation. First, tiny cells, which emerge during the third and fourth divisions, form a thin layer at the boundary between adjacent cell clusters. This is soon followed by formation of dorsal, ventral and lateral coelomic walls, all of which are made of smaller cells that are produced later than the fifth division. Around this time, cells of a similar small size spread along the thin layer of tiny cells at the cluster boundary, and they establish the septum; apparently, the thin layer of tiny cells serves as a template for the septum. As development proceeds, cells forming the coelomic walls individually become flattened, and the coelomic cavity becomes more distinct through cumulation of spaces between inner cells (Goto et al., 1999a).

Segmental founder cells

Is each cluster derived from a single primary m-blast cell or a mixture of progeny cells of two or more consecutive primary blast cells? To answer this question, we have performed double-labeling experiments in which M teloblasts were doubly injected first with Texas Red dextran (TRD) and then with fluorescein dextran (FLD) 2.5 or 5 h later (Goto et al., 1999a). Since M teloblasts divide repeatedly at 2.5 h intervals (at 22 °C), it is expected that the first one (in the case of 2.5 h intervals) or two (in the case of 5 h intervals) primary m-blast cells produced from these M teloblasts would be labeled singly with TRD, whereas subsequent primary blast cells would inherit both TRD and FLD labels.

In embryos that have been injected doubly at 2.5 h intervals, an anteriormost cluster of the fluorescently labeled portion of the GB exhibits TR fluorescence only, and ensuing clusters inherit both TRD and FLD labels. In the case of 5 h interval double labeling, two anteriormost clusters are labeled with TRD only, followed by doubly labeled clusters. In both cases, the boundary of the adjacent clusters that are differently labeled is sharp. There is no case in which clusters are a mosaic of singly labeled cells and doubly labeled cells. These results suggest that each cluster is a clone of a single primary m-blast cell (Goto et al., 1999a).

Absence of ectodermal control

As described earlier, m-blast cells come to be overlain by the ectodermal GBs soon after primary m-blast cells undergo the first two cell divisions. Thereafter, mesodermal GBs are always sandwiched between the overlying ectodermal GB and the underlying endoderm, at least up to the completion of their coalescence along the ventral midline. There is a possibility that segmentation in the mesodermal GB is regulated by the overlying cell layers. However, this is unlikely since even after deletion of 2d cells (precursors of ectoteloblasts), cellular events that lead to segmentation occur in a manner comparable to that in intact embryos (Goto et al., 1999a). Blast cells are organized in clusters surrounded by tiny cells. Furthermore, as demonstrated by double-labeling experiments, each cluster in ectoderm-deficient embryos is comprised

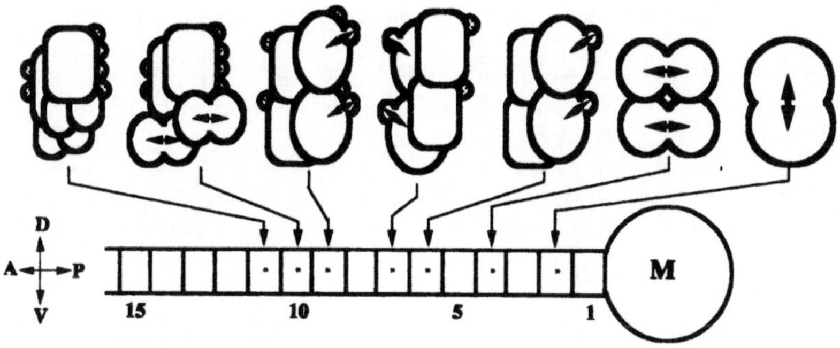

Figure 4. Schematic summary of pattern and sequence of divisions in mesodermal blast cells. Inequality and direction of divisions are reflected by position and orientation of mitotic spindles in dividing cells. The mesodermal GB extending from the M cell is illustrated in the lower part of the figure; each block in the GB represents a cell cluster. Arrows indicate the approximate position, along the GB, where each division occurs. A – anterior; D – dorsal; P – posterior; V – ventral. Adapted from Goto et al. (1999a).

of progeny cells of a single primary m-blast cell. These results suggest that segmentation in the mesodermal GB proceeds normally in the absence of the ectodermal GB.

Taken together, it is reasonable to assume that each primary m-blast cell serves as a founder cell of each mesodermal compartment (i.e. segment) and that the boundary between compartments is determined autonomously. It is concluded that the metameric body plan of *Tubifex* arises from an initially simple organization (i.e. a linear series) of segmental founder cells (Goto et al., 1999a).

Segment-specific expression of alkaline phosphatase (ALP) activity in the mesoderm

During normal *Tubifex* embryogenesis, more than 30 primary m-blast cells are produced on either side of the embryo (Meyer, 1929; Penners, 1934). There is no evidence that these cells are different from one another in either spatial or temporal aspects of their proliferation. It appears that primary m-blast cells are all identical in their ability to execute mitotic events that lead to segmentation (Goto et al., 1999a).

On the other hand, as is well known, segments are regionally differentiated along the longitudinal body axis, though they are homologous. Among mesodermal organs, for instance, nephridia and genital primordia are localized in a subset of segments, in a species-specific manner (Brusca & Brusca, 1990). These are morphological features that give each of the segments their own identity. It would be of great interest to learn when and how individual segments acquire specific identities.

Recently, we have found that segments VII and VIII specifically exhibit mesodermal alkaline phosphatase (ALP) activity during *Tubifex* embryogenesis (Kitamura & Shimizu, 2000a). At 4.5 days after 4d cell division, two bilateral pairs of ALP-positive tiny spots become detectable in the ventrolateral region in the midzone of the anterior half of the embryo (Fig. 5). The two pairs are located in the mesoderm of segments VII and VIII, respectively. As development proceeds, ALP-positive cells in each spot increase in number and become organized into a loop (i.e. a nephridium) that assumes a hairpin shape. Apparently, the ventrolateral ALP spots are detectable long before nephridia (which are morphological features specific to segments VII and VIII during embryogenesis) are morphologically brought into being. We consider that this mesodermal ALP activity serves as a very early marker of segments VII and VIII (Kitamura & Shimizu, 2000a).

To address the mechanism for specification of segmental identity in the mesoderm of *Tubifex*, we have examined whether this segment-specific expression of ALP activity depends on external cues (Kitamura & Shimizu, 2000b). Cell lineage analyses show that the ALP-expressing cells originate from M teloblasts. Furthermore, a set of teloblast-ablation experiments demonstrated that the 7th and 8th primary m-blast cells (m7 and m8) produced from M teloblasts give rise to ALP-expressing cells in segments VII and VIII, respectively, and that primary m-blast cells other than m7 and m8 lack the ability to generate ALP-expressing progeny cells. The results of another set of blastomere-ablation experiments suggest that ALP-expressing cells emerge independently of interactions with surrounding tissues. Teloblast-transplantation experiments have demonstrated that m8 can generate ALP-expressing cells in an ectopical position, sug-

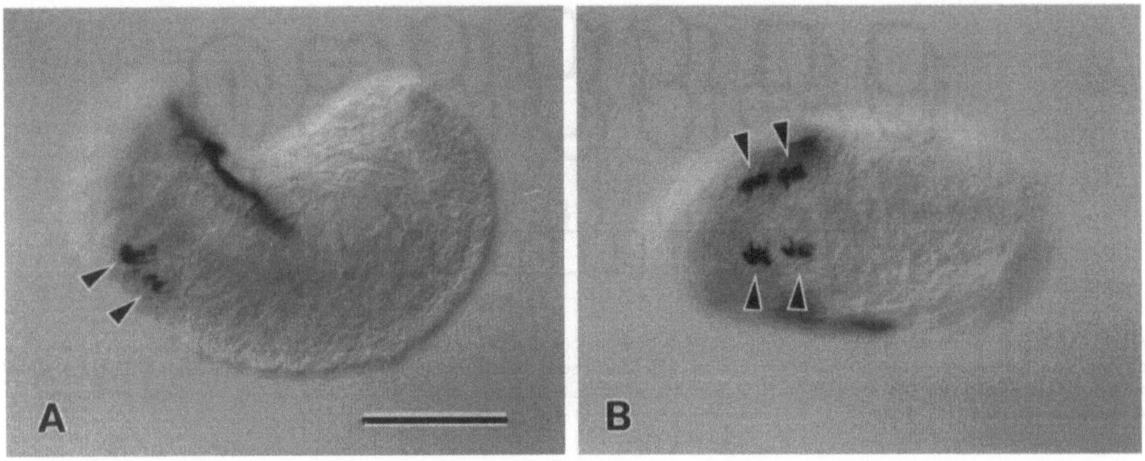

Figure 5. Histochemical localization of ALP activity in *Tubifex* embryos. Lateral (A) and ventral (B) views of a 5-day-old embryo following the 4d cell division. The anterior part is on the left. Two bilateral pairs of ALP-positive spots (arrowheads) are seen in the ventrolateral region. ALP-positive linear structures are also seen in the dorsoanterior region. Bar: 200 μm. From Kitamura & Shimizu (2000b).

gesting that it is unlikely that ALP activity emerges in response to the positional cues residing in the embryo (Kitamura & Shimizu, 2000b). These results suggest that m7 and m8 are exclusively specified as precursors of ALP-expressing cells at the time of their birth from M teloblasts.

Implications for specification of segmental identity

As described earlier, ALP spots in segments VII and VIII develop into nephridia, which are detected as ALP-positive, hairpin-shaped structures. During *Tubifex* embryogenesis, ALP spots and nephridia are present in these two segments but not in other segments; therefore, they are morphological features that give each of these two segments their own identity. The emergence of these features in segments VII and VIII is not because m-blast cells that have been situated in these segments are committed to take on segment-specific fates but because m-blast cells that have already been committed to these fates come to take up the positions corresponding to segments VII and VIII. Furthermore, it is at the time of their birth that primary m-blast cells acquire specific identities. Thus, our results strongly suggest that segmental identity is already specified at the level of the primary m-blast cell. That is to say, primary m-blast cells that are produced sequentially from the M teloblast may individually have an intrinsic segmental identity. Given that segmental identities in primary m-blast cells are determined according to the rank order of their birth, it is probable that the M teloblast possesses a developmental program through which the sequence of blast cell identities is determined.

Comparison with other annelids

Ectodermal and mesodermal segmentation has also been examined in leeches (*Helobdella triserialis* and *Theromyzon rude*). These animals exhibit many developmental features that appear to be homologous to those of *Tubifex*: e.g. generation of five bilateral pairs of teloblasts, ensuing formation of GBs, and exclusive origin of segmental tissues and organs from the GBs (Fernandez & Olea, 1982). Furthermore, segmental distribution patterns of blast cell progenies in leeches are very similar to the respective teloblast lineages in *Tubifex* (Weisblat & Shankland, 1985; Nakamoto et al., 2000). In both *Helobdella* and *Tubifex*, each mesodermal segment is comprised of descendants of a single primary blast cell produced from the m teloblast (Zackson, 1982; Goto et al., 1999a).

In contrast to the aforementioned common framework of pattern formation, cellular events leading to metameric segmentation appear to be diverse among clitellate annelids. First, although ectoteloblasts are produced through almost identical sequences of cell division (Fernandez & Olea, 1982; Shimizu, 1982), the mechanisms for cell fate decisions of teloblasts in *Helobdella* and *Tubifex* are distinct (Huang & Weisblat, 1996; Arai et al., 2001). Second, there are significant differences between oligochaetes and leeches in the details of fate maps of teloblasts (Weisblat et al., 1980, 1984; Torrence & Stuart, 1986; Goto et al., 1999b). Third, unlike autonomous mesodermal segmentation in *Tubifex*, mesodermal segmentation (or segmental organization) in *Helobdella* appears to depend on the presence of ectodermal GBs (Blair, 1982). Fourth, mesoteloblasts in leeches and *Tubifex* are provided with distinct developmental properties. As in *Tubifex*, leech M teloblasts serve as stem cells, from which nearly all mesodermal tissues and organs are derived. However, during *Tubifex* embryogenesis, nephridia and primordial germ cells, both of which are derived from M teloblasts, occur in segments VII–VIII and X–XI, respectively (Meyer, 1929; Goto et al., 1999a), whereas in leeches, nephridia are seen in segments II–XVIII (except for segments VI and VII) and primordial germ cells in segments VIII–XVIII (Weisblat & Shankland, 1985). In view of the notion that the developmental fate of each primary m-blast cell is specified at its birth (Gleizer & Stent, 1993; Kitamura & Shimizu, 2000b), this may indicate that developmental programs, residing in M teloblasts, for specification of primary m-blast cells are different in *Tubifex* and leeches.

These abovementioned differences seem to be only some of the differences existing between *Tubifex* and leeches, but they are sufficient to suggest that not only developmental fates of teloblasts but also pathways leading to metameric segmentation have been modified during evolutionary isolation of oligochaetes and leeches. The accumulation of such small, but distinct, modifications in teloblast lineages, together with changes in other cell lineages, may have given rise to diversity in the body plan of clitellate annelids.

References

Arai, A., A. Nakamoto & T. Shimizu, 2001. Specification of cell fates of ectoteloblast lineages in embryos of the oligochaete annelid *Tubifex*: involvement of novel cell–cell interactions. Development 128: 1211–1219.

Blair, S. S., 1982. Interactions between mesoderm and ectoderm in segment formation in the embryo of a glossiphoniid leech. Dev. Biol. 89: 389–396.

Brusca, R. C & G. J. Brusca, 1990. Invertebrates. Sinauer, Sunderland.

Devries, J., 1973a. La formation et la destinée des feuillets embryonnaires chez le lombricien *Eisenia foetida* (Annélide Oligochète). Arch. Anat. Microsc. 62: 15–38.

Devries, J., 1973b. Détermination précoce du développement embryonnaire chez le lombricien *Eisenia foetida*. Bull. Soc. Zool. Fr. 98: 405–417.

Fernandez, F. & N. Olea, 1982. Embryonic development of glossiphoniid leeches. In Harrison, F. W. & R. R. Cowden (eds), Developmental Biology of Freshwater Invertebrates. Alan R. Liss, New York: 317–361.

Gleizer, L. & G. S. Stent, 1993. Developmental origin of segmental identity in the leech mesoderm. Development 117: 177–189.

Goto, A., K. Kitamura & T. Shimizu, 1999a. Cell lineage analysis of pattern formation in the *Tubifex* embryo. I. Segmentation in the mesoderm. Int. J. Dev. Biol. 43: 317–327.

Goto, A., K. Kitamura, A. Arai & T. Shimizu, 1999b. Cell fate analysis of teloblasts in the *Tubifex* embryo by intracellular injection of HRP. Dev. Growth Differ. 41: 703–713.

Huang, F. Z. & D. A. Weisblat, 1996. Cell fate determination in an annelid equivalence group. Development 122: 1839–1847.

Kitamura, K. & T. Shimizu, 2000a. Embryonic expression of alkaline phosphatase activity in the oligochaete annelid *Tubifex*. Invert. Reprod. Dev. 37: 69–73.

Kitamura, K. & T. Shimizu, 2000b. Analyses of segment-specific expression of alkaline phosphatase activity in the mesoderm of the oligochaete annelid *Tubifex*: implications for specification of segmental identity. Dev. Biol. 219: 214–228.

Meyer, A., 1929. Die Entwicklung der Nephridien und Gonoblasten bei *Tubifex rivulorum* Lam. nebst Bemerkungen zum natürlichen System der Oligochäten. Z. wiss. Zool. 133: 517–562.

Nakamoto, A., A. Arai & T. Shimizu, 2000. Cell lineage analysis of pattern formation in the *Tubifex* embryo. II. Segmentation in the ectoderm. Int. J. Dev. Biol. 44: 797–805.

Penners, A., 1922. Die Furchung von *Tubifex rivulorum* Lam. Zool. Jb. Abt. Anat. Ontog. Tiere 43: 323–367.

Penners, A., 1924a. Die Entwicklung des Keimstreifs und die Organbildung bei *Tubifex rivulorum* Lam. Zool. Jb. Abt. Anat. Ontog. Tiere 45: 251–308.

Penners, A., 1924b. Über die Entwicklung teilweise abgetöteter Eier von *Tubifex rivulorum*. Verh. dt. zool. Ges. 29: 69–73.

Penners, A., 1926. Experimentelle Untersuchungen zum Determinationsproblem am Keim von *Tubifex rivulorum* Lam. II. Die Entwicklung teilweise abgetöteter Keime. Z. wiss. Zool. 127: 1–140.

Penners, A., 1934. Experimentelle Untersuchungen zum Determinationsproblem am Keim von *Tubifex rivulorum* Lam. III. Abtötung der Teloblasten auf verschiedenen Entwicklungsstadien des Keimstreifs. Z. wiss. Zool. 145: 220–260.

Shimizu, T., 1982. Development in the freshwater oligochaete *Tubifex*. In Harrison, F. W. & R. R. Cowden (eds), Developmental Biology of Freshwater Invertebrates. Alan R. Liss, New York: 283–316.

Torrence, S. A. & D. K. Stuart, 1986. Gangliogenesis in leech embryos: Migration of neural precursor cells. J. Neurosci. 6: 2736–2746.

Weisblat, D. A. & M. Shankland, 1985. Cell lineage and segmentation in the leech. Phil. Trans. r. Soc., Lond. B312: 39–56.

Weisblat, D. A., S. Y. Kim & G. S. Stent, 1984. Embryonic origin of cells in the leech *Helobdella triserialis*. Dev. Biol. 104: 65–85.

Weisblat, D. A., G. Harper, G. S. Stent & R. T. Sawyer, 1980. Embryonic cell lineages in the nervous system of the glossiphoniid leech *Helobdella triserialis*. Dev. Biol. 76: 58–78.

Wilson, E. B., 1889. The embryology of the earthworm. J. Morph. 3: 387–462.

Zackson, S. L., 1982. Cell clones and segmentation in leech development. Cell 31: 761–770.

Hydrobiologia **463:** 133–140, 2001.
P. Rodriguez & P.F.M. Verdonschot (eds), Aquatic Oligochaete Biology VIII.

Selective feeding by the aquatic oligochaete *Tubifex tubifex* (Tubificidae, Clitellata)

Pilar Rodriguez, Maite Martinez-Madrid, Jesús Angel Arrate & Enrique Navarro
Departamento de Biología Animal y Genética, Facultad de Ciencias, Universidad del País Vasco, Apdo. 644, E-48080 Bilbao, Spain
Tel: 34 94 601 2712. Fax: 34 94 464 8500. E-mail: ggprorop@lg.ehu.es

Key words: Feeding biology, feeding selection, tubificids, *Tubifex*, faeces

Abstract

The particle size distribution of faecal pellets produced by the tubificid worm *Tubifex tubifex* in laboratory culture, was measured with a Coulter® Multisizer. The faecal material from worms cultured in a range of sediments was composed of particles with a mean diameter of less than 63 μm, and only a few isolated larger particles were found by microscopic analysis. This suggests that this species actively selects the silt-clay fraction, avoiding larger sand particles. A more detailed analysis of faeces revealed that about 75%, by volume, was composed of particles with a mean diameter < 25μm, and the mode was < 10μm. *T. tubifex* fed selectively on the organic rich particles of the sediment, and this feeding was independent of particle size. Measurement of the organic content of faeces (measured as % loss on ignition) showed that they had a consistently higher organic content than the sediment, considered as whole sediments or the <63 μm sieved fraction. On the basis of these results, we hypothesise that this species exhibits two levels of selectivity in its feeding behaviour. Thus selection is primarily based on particle size, avoiding the ingestion of sand particles and also, on the preferential selection of particles associated with organic material, within the fine (silt-clay) fraction of the sediment.

Introduction

The idea that tubificids feed selectively on the fine fraction of sediments was initially pointed out by Wagner (1968). The study of soil bioturbation by oligochaetes led Kikuchi & Kurihara (1977) to conclude the existence of selective feeding. These authors showed that the particle size distribution of sediment was altered by tubificids feeding on the fine fraction, resulting in larger particles in the deeper layers. Similarly, Davis (1974) and Tevesz et al. (1980) after measuring particle size distribution in a vertical section of the sediment layer, concluded that tubificids selected silt-clay sized particles in sediment and transported them to the superficial layer via faecal egestion. Direct measurement of particle size in the gut of oligochaete worms has been performed by Juget (1979). This author measured the maximum diameter of the ingested particles in the gut of 13 different oligochaete species, although no information on the particle size distribution of the ingested sediment was provided.

Other studies have shown selective feeding by oligochaetes on the organic component (living and non-living) of sediment. Bacterial flora has been shown to represent a primary food source for tubificids (Brinkhurst & Chua, 1969; Wavre & Brinkhurst, 1971). Brinkhurst et al. (1972) and Brinkhurst & Austin (1979) studied selective feeding by tubificids and showed that the organic content in the faeces of *T. tubifex* and *L. hoffmeisteri* was significantly higher than that found in sediment, concluding that worms fed mainly on the organic rich fraction of the sediment. Poddubnaya (1961) measured a similar or higher content of nitrogen, organic matter and calories in faeces compared with sediment (in Brinkhurst et al., 1972). A more recent work by Matisoff et al. (1999) also concluded the existence of selective feeding on organic-rich, fine-grained particles by tubificids, as well as a positive dependence of selection on population density, using a radiotracer layer microcosm technique.

The objectives of this study were to determine the fraction of the sediment which was preferentially ingested by worms using direct granulometric analysis of faecal pellets, and then to establish if selective feeding depends on the organic content of the sediment. These factors may be important in bioaccumulation studies (Leppänen, 1995) and should be taken into account when relating chemical concentration in tissue with that measured in the sediment.

Materials and methods

Tubifex tubifex oligochaete worms used in our experiments were maintained in laboratory culture in a natural sediment from a reference site (Barazar) for over 9 years. For more details about culture of the animals, see Martinez et al. (1999). Three natural sediments from (1) the reference site at Barazar, (2) a stream at Ubidea (Natural Park of Gorbeia, Bizkaia, Spain), and (3) a water reservoir at Ordunte (Northern Burgos, Spain), as well as one artificial sediment, were examined. Natural sediments were sieved through a 250 μm mesh to eliminate resident benthic fauna (Reynoldson et al., 1994). The artificial sediment consisted (by weight) of 76% quartz sand (nominal size: 150–210 μm), 22% kaolinite clay and 2% peat (Egeler et al., 1997). The coarse granulometric characteristics of the four study sediments are summarised in Table 1. The artificial sediment has the highest percentage of sand (76%). Barazar has the lowest proportion of sand (10.7%), and is dominated by the silt-clay fraction (89.3%). Ubidea and Ordunte are intermediate in their granulometric characteristics with about 50% sand and 50% silt-clay, by weight. However, the former has a higher proportion of medium-sized sand particles and the latter is richer in the fine sand fraction. Two batches of Barazar sediment (split from the same sample) and the artificial sediment were studied and their general granulometric characteristics were assumed to be the same for both experiments. Results related to selective feeding are expressed for those sediments as Barazar 1/Barazar 2 and Artificial 1/Artificial 2, respectively.

Worms were kept in each test sediment during four weeks for acclimation, at a temperature of 22.5 ± 1 °C, in the dark. The worms were maintained at a density of 50–100 worms per 225 cm^2, and the overlying water was gently aerated. Tetramin® (1 g) was added to each culture baker. At the end of the acclimation period, 30 worms were washed in dechlorinated tap water and separated into three replicates of ten worms each, using Petri dishes containing dechlorinated water. In this water, worms were allowed to defecate for 4 h for future analysis of the faecal pellets. During this period, faeces were removed from the Petri dishes every 30 min using a micropipette, to avoid the reingestion of faeces by the worms. Triplicate samples from the sediments were also taken (avoiding the superficial layer containing faeces) for analysis of the sediment particle size distribution and organic content.

Sediment samples sieved through a 63 μm mesh and collected faeces were disaggregated by adding a basic solution of NH_4OH adjusted to pH 8. Sediment and faeces were dispersed with a vortex and subsequently in an ultrasonic treatment (5 min). Aliquots (1–2 ml) of these suspensions were then resuspended into Isoton® (100 ml) and analysed for particle size distribution in the range of 2–63 μm using a Coulter® Multisizer, fitted with a 100 μm aperture tube.

The organic content and density were calculated as % loss on ignition (% LOI) from other aliquots of a known volume of sediment and faeces suspensions. Samples were filtered over pre-weighed glass fiber filters, dried at 80 °C and ashed in a Mufla Dinko-Electric oven at 450 °C for 6 h. After drying and ashing, samples were allowed to cool at room temperature in a desiccator, before weighing in a precision Sartorius Micro M3P electrobalance (detection limit 1 μg). The determination of the organic content of sediments was performed on both whole sediment and on the < 63 μm fraction. The organic content in both sediment and faeces from Barazar and Artificial sediments was measured only for Barazar 2 and Artificial 2 samples. The packed particulate volume, as determined by the Coulter® Multisizer, and the dry weight of faecal and sediment particles were used to calculate mean particle density. Faecal samples were also analysed by microscopy to measure the maximum particle diameter.

The accumulated particle volume (%) of different diameters (channels of the Coulter® Multisizer) was fitted with an asymptotic function using a non-linear regression analysis with the SYSTAT 5.2 software. The model is based on the function:

$$y = 100\,[1 - e^{-k(x-x_o)}],$$

where y is the mean accumulated volume (%) and x is the particle size diameter (in μm). In this equation,

Table 1. Granulometric characteristics of the studied sediments calculated in dry weight percentage

Sediment source	Medium sand (100–250 μm)	Fine sand (63–100 μm)	Total sand (63–250 μm)	Silt and clay (< 63 μm)
Artificial sediment[a]	76	0	76	22
Barazar	3.08 ± 1.73	7.58 ± 1.53	10.66 ± 3.08	89.34 ± 3.08
Ubidea	34.41 ± 4.11	19.87 ± 0.94	54.28 ± 4.06	45.72 ± 4.07
Ordunte	18.79 ± 1.09	28.65 ± 0.43	47.44 ± 1.43	52.56 ± 1.44

[a]Nominal percentage of the components.

Table 2. Organic matter calculated as % LOI in different sediments and oligochaete faeces (not measured for Barazar 1 and Artificial 1 sediments). Asterisks indicate those cases with significant differences in organic content (expressed as % LOI) ($P < 0.05$, Mann–Whitney *U*-test), between worm faeces and the sediment fraction below 63 μm

Sediment	% in the culture sediment	% in <63 μm sediment fraction	% in oligochaete faeces
Artificial 2	2.47 ± 0.70	13.91 ± 0.22	15.66 ± 0.34*
Barazar 2	6.32 ± 0.45	7.56 ± 0.29	8.94 ± 1.47*
Ubidea	4.43 ± 0.36	7.27 ± 0.20	10.44 ± 0.97*
Ordunte	3.73 ± 0.30	5.45 ± 0.15	11.54 ± 3.22*

x_0 is the intercept with the abscissa, representing the minimum particle diameter (in μm) predicted by the model, and k is the coefficient relating total packed volume (100%) and particle size. Differences between k coefficients were evaluated using the Student t test (Zar, 1996). For comparative purposes, this model has been used to estimate the upper range of particle size comprising 50% and 75% of the accumulated volume in both sediment and faeces. The Student t-test, the Mann–Whitney U-test and the Spearman correlation (r_s) were calculated using StatView SE+ Graphics™ software.

Results

In all cases, the organic content (measured as % LOI) was lower in the whole sediment than in the <63 μm fraction, and in the latter, lower than in the faeces. These differences were found to be statistically significant (Mann–Whitney U-test, $P < 0.05$, Table 2).

The granulometric composition of faeces as measured by the Coulter® Multisizer showed that the maximum particle size of faecal pellets ranged in mean diameter from 36.7 to 51.2 μm and 26.3 to 41.01 μm in Barazar 1 and Barazar 2, respectively; 51.9 to 59.1 μm in the Ordunte sediment, and 28.0 to 43.9 μm in the Ubidea sediment (n = 3 for all determinations of mean values). The maximum faecal particle size from worms cultured in the artificial sediment ranged from 26.3 to 41.01 and from 32.8 to 36.7 μm in Artificial 1 and Artificial 2, respectively (n = 3 for both determinations). Microscopic examination of the faeces corroborated the data obtained by the Coulter® Multisizer in terms of the dominance of <63 μm particles, although a few faecal particles showed diameters in the lower limit of classification of the sand class. The maximum particle size recorded was 101 × 63 μm (maximum × minimum diameter). These observations suggested the existence of selective feeding by the oligochaetes, based primarily on particle size, which commonly involves avoidance of sand grains.

The volume percentage of the clay fraction (<4 μm) was significantly higher in faeces than in the <63 μm sediment fraction, in four out of six study cases. Non-significant differences were observed only in 2 cases (Barazar 2 and Ordunte; Table 3). This fraction in the faecal pellets is relatively homogeneous in all the study cases. It comprises around 20–23% of the total volume, with the exception of the Ubidea sediment, where it reaches a mean value of 42.54%. The mode of the particulate volume in faeces is <10 μm, whereas it is very variable in the sediments. In fact, there is no correlation between the modal particle size in sediments and in faeces, which is very similar in some cases and very different in others. In general, granulometric characteristics are clearly less variable in faeces than in sediments, the latter being dependent on the geological nature of the catchment basin. The modal size of faecal particles is lower than that of the <63 μm sediment fraction in the artificial sediment and in Ubidea sediment, and markedly lower in Barazar 1 (Table 3). However, in Barazar 2 modal

Table 3. Granulometric characteristics of sediment and oligochaete faecal pellets. Clay percentage in volume (< 4 μm fraction), mode of particle size in the sieved sediment (< 63 μm) and oligochaete faeces, together with the upper particle size comprising 50% and 75% of the total accumulated volume of particles are represented. Asterisks indicate those cases with significant differences between sediment and faecal composition (Mann–Whitney *U*-test, $P < 0.05$, $n = 3$)

Sediment	Granulometric characteristics	Sediment (<63 μm fraction)	Oligochaete faeces
Artificial 1	clay (<4 μm)%	14.79 ± 1.48	23.06 ± 1.12*
	Modal size range (μm)	11.98 – 12.10	5.99 – 6.22
	50% Particle size (μm)	10.17 ± 0.14	7.69 ± 0.64*
	75% Particle size (μm)	18.26 ± 0.30	13.54 ± 1.39*
Artificial 2	clay (<4 μm)%	18.59 ± 1.87	21.88 ± 1.20*
	Modal size range (μm)	7.99 – 8.68	6.22 – 6.95
	50% Particle size (μm)	8.12 ± 0.39	7.94 ± 0.28
	75% Particle size (μm)	14.11 ± 0.83	14.03 ± 0.60
Barazar 1	clay (<4 μm)%	11.52 ± 0.65	22.37 ± 0.903*
	Modal size range (μm)	35.94 – 36.70	4.99 – 7.57
	50% Particle size (μm)	14.68 ± 0.66	7.99 ± 0.39*
	75% Particle size (μm)	27.49 ± 1.61	14.17 ± 0.97*
Barazar 2	clay (<4 μm)%	19.06 ± 2.27	20.85 ± 0.83
	Modal size range (μm)	6.22 – 6.95	8.68 – 9.70
	50% Particle size (μm)	7.01 ± 0.64	7.91 ± 0.45
	75% Particle size (μm)	11.76 ± 1.28	13.88 ± 1.00
Ubidea	clay (<4 μm)%	27.29 ± 2.52	42.54 ± 5.457*
	Modal size range (μm)	2.00 – 4.00	< 2.00
	50% Particle size (μm)	7.79 ± 1.27	5.14 ± 0.95*
	75% Particle size (μm)	15.03 ± 2.63	10.26 ± 1.98*
Ordunte	clay (<4 μm)%	22.52 ± 3.74	22.60 ± 0.57
	Modal size range (μm)	2.00 – 4.00	2.00 – 4.00
	50% Particle size (μm)	9.15 ± 1.46	10.65 ± 0.75
	75% Particle size (μm)	17.62 ± 2.74	21.14 ± 1.53

particle size is slightly higher in faeces than in the sediment, but about the same range size in Ordunte sediment.

Granulometric curves for each sediment (<63 μm fraction) and faecal material are shown in Figure 1. In most cases, x_0 approximates the value of the lower detection limit of the 100 μm aperture tube in the Coulter® Multisizer (i.e. 2 μm). Faecal granulometric curves were very similar and comparisons of the *k* coefficients by the Student *t*- test revealed significant differences only for Ordunte with Artificial 1 and with Ubidea ($P < 0.05$, $n = 3$). Comparisons of faecal curves and those of their respective sediments showed that *k* coefficients for faecal curves are either higher or lower, being significantly higher ($P < 0.05$) only for Barazar 1 and Artificial 1 (Fig. 1). For comparative purposes, the particle size that comprises 50% and 75% of the total volume in both sediment (<63 μm fraction) and faecal samples has been estimated using asymptotic equations (see Fig. 1), and compared using the Mann–Whitney *U*-test. This analysis revealed that in most cases there were significant differences for both variables, except in Ordunte and Barazar 2 sediments (Table 3).

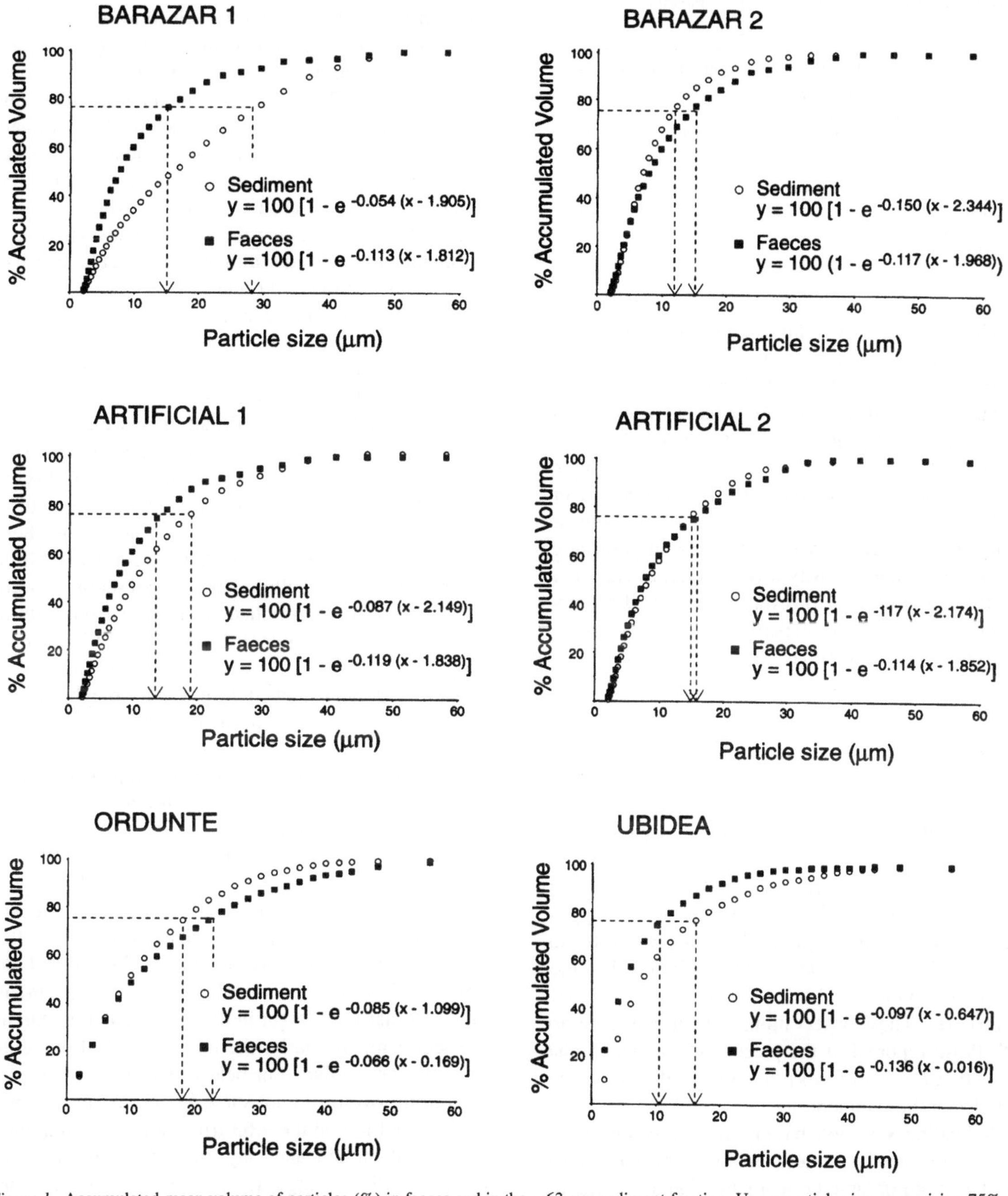

Figure 1. Accumulated mean volume of particles (%) in faeces and in the <63 μm sediment fraction. Upper particle sizes comprising 75% of the total volume are indicated by arrows with doted lines. The function is $y = 100\ [1 - e^{-k(x-x_0)}]$, where y is the mean accumulated volume (%), x is the particle size diameter (in μm), x_0 is the intercept with the abscissa and k is the coefficient relating total packed volume (100%) and particle size.

Table 4. Mean density (mg mm^{-3}) of particles in the sediments and in oligochaete faeces (n = 3). P represents the probability values according to the Student t-test (2-tails) comparing mean density values of sediment and faeces

Sediment	Sediment (<63 μm fraction)	Oligochaete faeces	t-test P
Artificial 2	4.15	3.85	0.246
Barazar 2	4.21	4.19	0.589
Ubidea	4.61	4.00	0.031
Ordunte	5.23	4.14	0.089

The mean density values of faecal material were always lower than those of their respective sediments (Table 4), although these differences are only significant for Ubidea sediment (Student t-test, $P < 0.05$). The density of faecal material was less variable (variation coefficient = 5.9%) than the density of the <63 μm sediment fraction (variation coefficient = 14.5%). Therefore, differences in density between sediment and faeces were more apparent in those cases where the sediment density was higher, for instance, at Ubidea and Ordunte. However, the expected negative correlation between faecal particle density and organic content was not observed ($r_s = -0.80$, $P > 0.05$).

Discussion

Tubifex tubifex avoids sand sized particles and feeds mainly in the silt-clay fraction. Thus, it can be said that there is a primary selection of the sediment based on particle size. By microscopic analysis, Juget (1979) reported that the maximum diameter of the biggest particle observed in the gut of *T. tubifex* was 130 μm, although the mean value of the maximum diameter for biggest particles observed in several individuals was 67 μm. These observations are not contradictory with those reported in the present study, which also shows that *T. tubifex* feeds most commonly on silt-clay particles of the sediment.

A feeding selection based exclusively on avoidance of sand sized particles would be expected to have resulted in an organic enrichment of ingested matter and hence in the faeces, since organic matter is mainly associated with the fine particle fraction of sediment below 63 μm (Mudroch & Azcue, 1995). However, we found that organic content was not only consistently higher in faeces than in whole sediment but also higher than in the fraction below 63 μm. In this respect, it is important to point out that in the present experiments, our evidence of selection is limited by the fact that organic content was measured in the faeces instead of in the ingested sediment which worms actively select. Therefore, a certain quantity of organic matter has been subtracted from the ingested sediment by gut absorption. Assimilation efficiency values (as a percentage of organic ingestion) in tubificids has been reported to be around 4.1% (Brinkhurst & Austin, 1979), implying that differences between culture sediment and ingested sediment are in fact higher than those measured in the present work.

Organic matter in faeces from worms fed in natural sediments was calculated to be around 10%, which is consistent with the data reported by Brinkhurst et al. (1972) and Brinkhurst & Austin (1979), at 20 °C. The former authors showed that there were no significant differences in the organic content of faeces in an experiment designed to test the effect of varying the particle size of mud used for the culture of tubificids. The present results are consistent with these observations as regards our natural sediments, while faeces from worms cultured in artificial sediment were significantly more enriched than those from natural sediments (Mann–Whitney U-test, $P < 0.05$). Microscopic analysis of the faeces from worms cultured in artificial sediment showed the existence of a high content of vegetal fibres from the peat component which is neither digested nor assimilated.

However, these results do not completely clarify the extent to which the organic enrichment of faeces is due to the selection of the organic matter itself or the mere consequence of the selection of the particle size fraction finer than 63 μm, which would be presumably richer in organic matter. A detailed analysis and comparison of the particle size distribution in the <63 μm sediment fraction and faeces showed that the volume of particles in the clay fraction (<4 μm) is either similar to or larger in the faecal material than in the sediment. Similarly, the mode of faecal particles may also be similar to or below that of the sediment. Therefore, worms do not show a constant feeding behaviour and fed on the <63 μm fraction or on a finer fraction of the silt and clay, probably depending on the distribution of organic matter in the fine fractions of sediment. On the other hand, granulometric curves showed a higher constancy of particle size distribution in faeces than in <63 μm sediment fraction, with fine particles being similar or more abundant in faeces with respect to sediment. Thus, after a primary selection based on particle size, avoiding sand particles, worms do not appear to effect a secondary selection based on particle size. The enrichment of the ingested sediment is most probably a consequence of the active selection of organic matter associated with a more or less fine fraction, which is variable for each sediment.

It can not be ignored that a proportion of the organic matter can also be ingested with few or no mineral particles, which would explain high enrichment of organic content and the slight modifications of the particle size frequency distribution of the faeces compared with the sediment (e.g. Barazar 2 and Ordunte). The selective feeding on one fraction or another in the silt/clay portion of the sediment most probably depends on both the quantity and composition of the bacterial community and of the detrital matter either free or associated with mineral sediment particles. These characteristics vary widely from sediment to sediment, and represent a second level of selective feeding of the worms. Selective feeding on bacteria was first demonstrated by Coler et al. (1968) and can also be inferred from the data by Brinkhurst (1974) and Milbrink (1993). Indeed, it may probably be an important factor to understand the ingestion of certain particles and the rejection of others by the worms. The existence of selective feeding on certain bacterial communities associated with sediment particles or with organic matter in the sediment may be another interesting factor to be investigated in the future.

The difference in density of particles packed in the faeces can also be interpreted as a result of selection of non-mineralised particles by worms. Tevesz et al. (1980) considered faeces to be of lower density than sediment, although the present study shows that they can also be of about the same density. The absence of this type of data in other studies precludes further discussion or comparison of density results.

In ecotoxicological studies, it is important to know the sediment fraction that is available to benthic organisms. The fact that *Tubifex tubifex* ingests particles mainly <63 μm in diameter is interesting from a toxicological point of view, since the principal portion of many metals is incorporated into the silt and clay fractions of the sediment (Mudroch & Azcue, 1995). On a more general scale, it is important to understand the feeding biology of aquatic oligochaetes in order to interpret correctly the various biological processes underlying growth, reproduction or bioaccumulation. Brinkhurst (1999) has recently commented on the requirement of a method for identifying factors which are directly related to worm biology rather than relying solely on physical and chemical factors related to the water column.

Acknowledgements

The authors are in debt to Dr J. I. Perez-Iglesias (UPV/EHU, Spain) and Dr G. Matisoff (Case Western Reserve University, U.S.A.) for their important contributions to the improvement of this paper. We would also like to express our thanks to Dr T. B. Reynoldson (NWRI, Canada) with whom we fruitfully discussed some of the methods employed in this paper. This work was financed by the University of the Basque Country (research project UPV/EHU 154.310-EA071/99).

References

Brinkhurst, R. O., 1974. Factors mediating interspecific aggregation of tubificid Oligochaeta. J. Fish. Res. Bd Can. 31: 460–462.

Brinkhurst, R. O., 1999. Retrospect and prospects: reflexions on forty years of study of aquatic oligochaetes. In Healy, B. M., T. B. Reynoldson & K. A. Coates (eds), Aquatic Oligochaetes. Proceedings of the 7th International Symposium on Aquatic Oligochaetes. Presque isle, 18–22 August 1997. Hydrobiologia 406: 9–19.

Brinkhurst, R. O. & M. J. Austin, 1979. Assimilation by Aquatic Oligochaeta. Int. Rev. ges. Hydrobiol. 64: 245–250.

Brinkhurst, R. O. & K. E. Chua, 1969. Preliminary investigation of the exploitation of some potential nutritional resources by three sympatric tubificid oligochaetes. J. Fish. Res. Bd Can. 26: 2659–2668.

Brinkhurst, R. O., K. E. Chua & N. K. Kaushik, 1972. Interspecific interactions and selective feeding by tubificid oligochaetes. Limnol. Oceanogr. 17: 122–133.

Coler, R. A., H. B. Gunner & B. M. Zuckermann, 1968. Selective feeding of tubificids on bacteria. Nature 216: 1143–1144.

Davis, R. B., 1974. Tubificids alter profiles of redox potential and pH in profundal lake sediments. Limnol. Oceanogr. 19: 466–488.

Egeler Ph., J. Römbke, M. Meller, T. Knacker, C. Franke, G. Studinger & R. Nagel, 1997. Bioaccumulation of Lindane and Hexachlorobenzene by tubificid sludgeworms (Oligochaeta) under standardized laboratory conditions. Chemosphere 35: 835–852.

Juget, J., 1979. La texture granulometrique des sediments et le regime alimentaire des oligochètes limnicoles. Hydrobiologia 65: 145–154.

Kikuchi, E. & Y. Kurihara, 1977. *In vitro* studies on the effects of tubificids on the biological, chemical and physical characteristics of submerged ricefield soil and overlying water. Oikos 29: 348–357.

Leppänen, M., 1995. The role of feeding behaviour in bioaccumulation of organic chemicals in benthic organisms. Ann. Zool. Fenn. 32: 247–255.

Martinez-Madrid, M., P. Rodriguez, J. I. Perez-Iglesias & E. Navarro, 1999. Sediment toxicity Bioassays for assessment of contaminated sites in the Nervion River (Northern Spain). 2. *Tubifex tubifex* reproduction sediment bioassay. Ecotoxicology 8: 111–124.

Matisoff G., X. S. Wang & P. L. McCall, 1999. Biological redistribution of lake sediments by tubificid oligochaetes: *Branchiura sowerbyi* and *Limnodrilus hoffmeisteri / Tubifex tubifex*. J. Great Lakes Res. 25: 205–219.

Milbrink, G., 1993. Evidence for mutualistic interactions in freshwater oligochaetes communities. Oikos 68: 317–322.

Mudroch, A. & J. M. Azcue, 1995. Manual of Aquatic Sediment Sampling. CRC Press, Boca Raton: 219 pp.

Reynoldson, T. B., K. E. Day, C. Clarke & D. Milani, 1994. The effect of indigenous animals on chronic endpoints in freshwater sediment toxicity tests. Envir. Toxicol. Chem. 13: 973–977.

Tevesz M. J. S., F. M. Soster & P. L. McCall, 1980. The effects of size-selective feeding by oligochaetes on the physical properties of river sediments. Journal of Sedimentary Petrology 50: 561–568.

Wagner, G., 1968. Zur Beziehung zwischen der Besiedlungsdischte von Tubificiden und dem Nahrungsangebot im Sediment. Int. Rev. ges. Hydrobiol. 53: 715–721.

Wavre, H. & R. O. Brinkhurst, 1971. Interactions between some tubificid oligochaete and bacteria found in the sediments of Toronto Harbour, Ontario. J. Fish. Res. Bd Can. 28: 335–341.

Zar, J. H., 1996. Biostatistical Analysis. Prentice Hall, New Jersey: 121 pp.

Hydrobiologia **463:** 141–148, 2001.
P. Rodriguez & P.F.M. Verdonschot (eds), Aquatic Oligochaete Biology VIII.

Cocoon deposition of *Rhyacodrilus hiemalis* Ohtaka (Tubificidae) in Lake Biwa, Japan

Tetsuya Narita
Center for Ecological Research, Kyoto University, 509 Ohtsuka, Kamitanakami-Hirano, Otsu, Shiga, 520-2113, Japan
Fax: 81 77 5498201. E-mail: narita@ecology.kyoto-u.ac.jp

Key words: Rhyacodrilus hiemalis, cocoon, breeding ecology, reproductive strategy, Tubificidae, Lake Biwa

Abstract

The breeding ecology of the tubificid worm *Rhyacodrilus hiemalis* was studied in Lake Biwa with special reference to timing and duration of cocoon deposition, number of eggs in a cocoon, and the substrate type and the position used for cocoon attachment. Cocoon deposition lasted from the beginning of January to the end of March, and peaked between mid-January and mid-February. Laboratory experiments showed that most *R. hiemalis* breed only once during their lifetime. In its natural habitat, most cocoons contained from 5 to 15 eggs (mean = 10.82, *SD* = 3.55; range = 0–33, *n* = 378), and the mean number of eggs per cocoon decreased as the deposition season progressed. The average number of eggs per cocoon of *R. hiemalis* is the greatest known among the tubificids. In Lake Biwa, *R. hiemalis* positioned its cocoons beneath the sediment, attached to the shell surface of bivalves and gastropods and to rock substrates; laboratory experiments showed that most cocoons were located less than 2 cm beneath the sediment surface. Under laboratory conditions, *R. hiemalis* produces a single cocoon during its lifetime, with a large number of eggs in midwinter, a period when potential predation on the eggs is supposed to be at its lowest during the year, in Lake Biwa. The site of cocoon attachment near the sediment surface guarantees a supply of well-oxygenated water. This may enhance embryonic development and shorten the embryonic stages, thereby achieving rapid preparing growth prior to summer aestivation and minimizing predation risk during the vulnerable early instar stages.

Introduction

Although aquatic oligochaetes are a dominant component of the infauna of muddy bottoms of lentic and lotic freshwaters, ecological studies of their breeding ecology and reproductive strategies are few (Newkla & Mutayoba, 1987), especially in natural habitats (Aston, 1968, 1973; Thorhauge, 1975). This is partly because most tubificids deposit small, free cocoons in muddy sediment, which makes it difficult to specify the day of deposition and to collect enough cocoon samples for ecological studies.

Rhyacodrilus hiemalis (Oligochaeta, Tubificidae) is endemic to Japanese freshwaters (Timm, 1997) and is common in lakes in Japan (Ohtaka, 1995). This species deposits its cocoons on hard substrata, a feature which facilitates the study of its breeding ecology in nature. *R. hiemalis* is an annual species which breeds during the cold season, grows in spring, and aestivates deep in the sediment during summer (T. Narita, unpublished data). This paper documents the breeding season, number of eggs in a cocoon and the usage of deposition substrata of *R. hiemalis* in Lake Biwa, Japan.

Materials and methods

To study the season of *R. hiemalis* cocoon deposition, two study sites were chosen off Shimosakamoto in the southern basin of Lake Biwa. Site A was about 400 m offshore and site B was about 200 m offshore. Both sites had a muddy bottom and were about 3 m deep.

At each site, two concrete bricks were placed on the bottom, to act as substrates for cocoon deposition.

The field experiment started on 16 October 1989 and finished on 30 May 1990. Every 7–10 days, the bricks at each site were lifted and inspected, and all of the attached cocoons were collected and preserved in a 5% formalin solution. After removal of cocoons, bricks were replaced on the lake bottom.

The length and width of each cocoon was measured with an eyepiece to the nearest 0.05 mm, and the volume of each cocoon was then calculated from these measurements by approximating its shape as being conical. Each cocoon was then dissected, the number of eggs counted, and the developmental stage of the eggs was identified.

Eggs in each cocoon were observed with a dissecting microscope without staining, and classified into the following six developmental stages: stage 1, eggs in which the fertilized membrane was not separated from the yolk; stage 2, eggs where the fertilized membrane was separated from the yolk; stage 3, eggs in which embryonic division had started; stage 4, embryos at the morula stage; stage 5, embryos of dog-legged shape; and stage 6, worms ready to emerge. State 1 eggs were about 0.4 mm in diameter.

Using clawed dredges, bivalves and gastropods were collected from areas adjacent to sites A and B in the southern basin, in order to study the kind of substratum of cocoon deposition. Collections were also made in the littoral zone, from 5 to 15 m deep, off Shinkai-hama, at the eastern part of the northern basin of the lake.

The water content, loss on ignition and carbon and nitrogen contents of the sediment of the uppermost 3 cm from site A and B were measured. Loss on ignition was determined by combusting the dried and ground sediment at 550 °C for 3 h. Carbon and nitrogen contents of the sediments were analyzed by means of a CHN Corder (Yanaco MT-2). Two sediment samples were collected at each site by using a K-K type core sampler once every month from October 1989 to January 1990.

To examine the number of times of cocoon deposition per individual, two or three mature and clitellate *R. hiemalis* specimens were maintained in glass containers (7 cm deep and 8 cm in diameter) containing a 4–5 cm thick layer of autoclaved lake sediment. The containers were soaked in aquaria at 7 °C in the laboratory, and the aquaria water was aerated. Two or three individuals (in 8 and 2 containers, respectively) were placed in each container, together with one individual of *Unio biwae*, the commonest bivalve, to serve as a substratum for cocoon deposition.

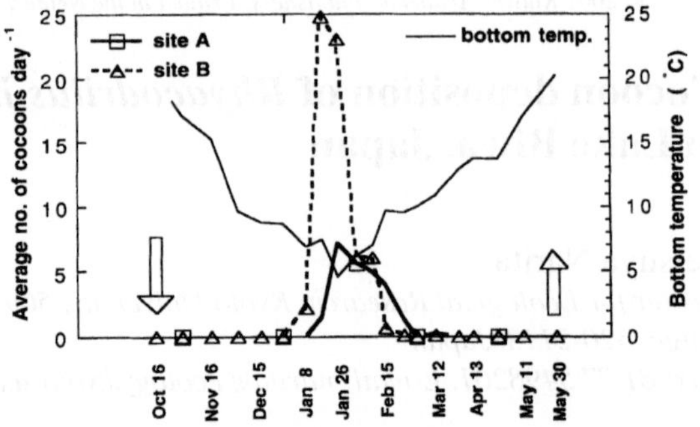

Figure 1. Rhyacodrilus hiemalis cocoon deposition. The ordinates indicate the average number of cocoons attached to the experimental concrete bricks per day and the bottom temperature at study sites A and B in the southern basin of Lake Biwa. The bottom temperature at site A is shown by a fine solid line. The bricks were first set on October 16, 1989 and removed finally on May 30, 1990.

To examine the depth at which cocoons were attached, 6 glass jars of 10 mature and clitellate worms with one *U. biwae*, and 5 deep glass jars of 15–20 mature and clitellate worms with two PVC sticks (3.3 cm wide and 16 cm long) each were prepared. The jars were soaked in aquaria, with aerated lake water and placed in a dark cold-room at 7 °C. The experiment was run for 45 days, from mid-February to the end of March, 1990. Cocoons on substrata were inspected at the end of the experiment and the depth from the sediment surface was measured.

Statistical analyses for correlation, regression and *t*-test were made using StatView-J5.0 software.

Results

Cocoon deposition and number of eggs in cocoons

Cocoons attached to the bricks were first found on January 17 at site A and January 8 at site B, and the last cocoons were observed on March 23 at both sites. The total number of cocoons attached to the bricks throughout the breeding season at sites A and B were 227 and 229, respectively. The average number of cocoons attached per day peaked on January 26 at site A and January 15 at site B. These peaks coincided with the minimum bottom temperature (4.6 °C) recorded at site A (Fig. 1).

Most cocoons contained between 5 and 15 eggs (range = 0–33 eggs) (Fig. 2). Broken cocoons, most of which were probably broken when they were detached

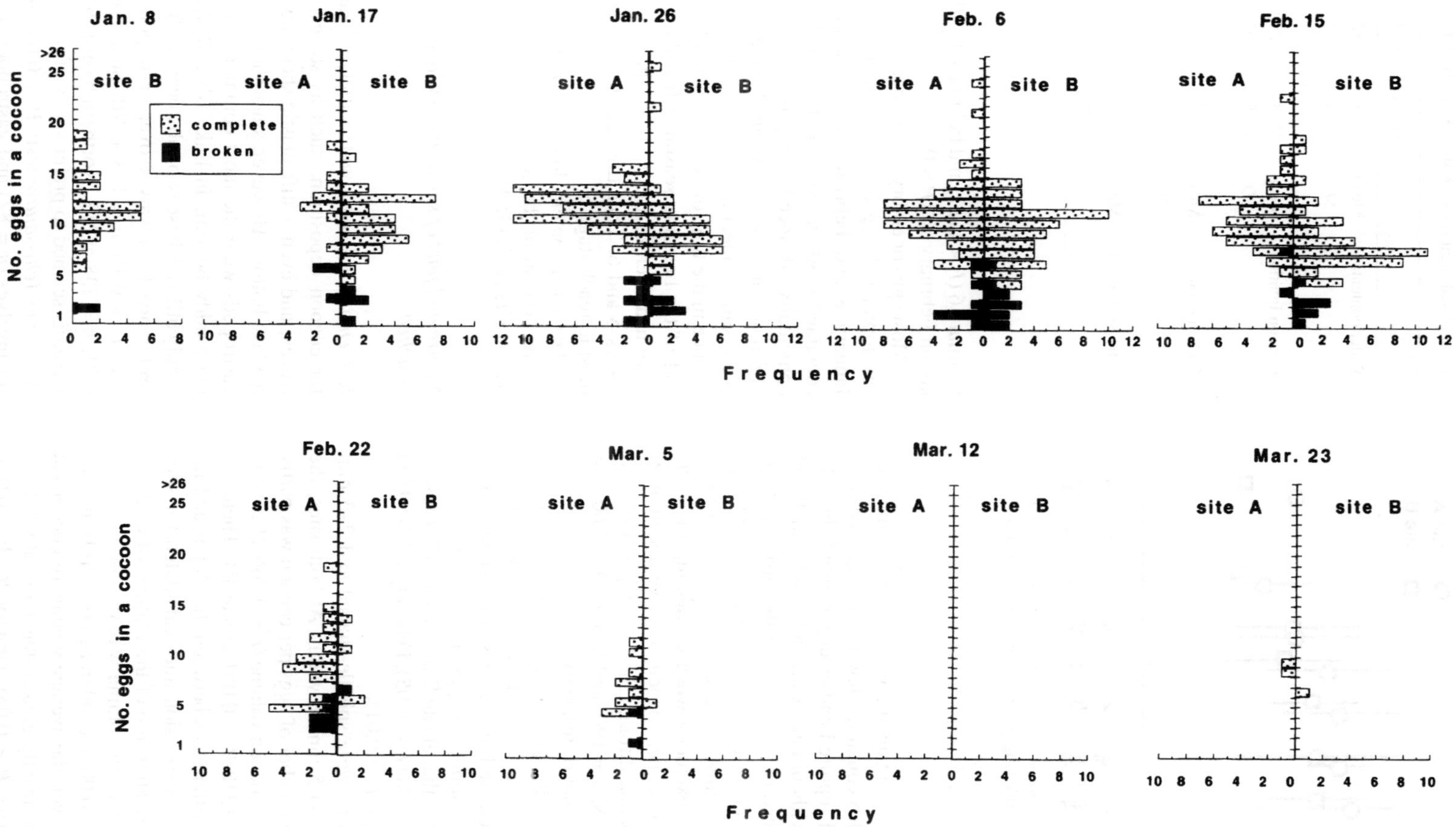

Figure 2. Frequency distribution of the number of eggs per cocoon produced by *R. hiemalis* at study sites A and B. Broken cocoons and those containing degenerated eggs are shown together by solid bars.

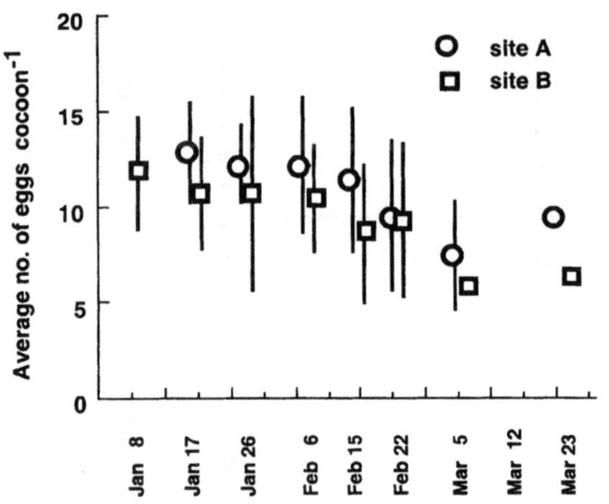

Figure 3. Average number of eggs per cocoon (with 1 *SD*) of *R. hiemalis* during the collecting period at study sites A and B.

Table 1. Water content, loss on ignition, carbon and nitrogen content of sediment (in%) at sites A and B, in Lake Biwa

		Site A	Site B	*t*-test
Water content	Mean	72.63	67.82	$P < 0.01$
	SD	1.45	3.93	
Loss on Ignition	Mean	9.98	8.28	$P < 0.001$
	SD	0.15	1.02	
Carbon content	Mean	2.84	2.63	$P = 0.095$
	SD	0.05	0.34	
Nitrogen content	Mean	0.31	0.27	$P < 0.05$
	SD	0.01	0.04	

from the bricks, contained few or no eggs. Some intact cocoons also contained few viable eggs, because some eggs had ruptured and/or degenerated (Fig. 2). The cocoons without eggs were not those which had hatched and become vacant since the inspection intervals of concrete bricks were short enough for eggs to complete hatching. All cocoons in the above two categories were excluded when calculating the average number of eggs per cocoon. The average number of eggs per cocoon through the breeding season was 11.48 (SD= 3.38, n = 190) and 8.76 (SD = 3.60, n = 188) for sites A and B, respectively.

At both sites, the average number of eggs per cocoon was highest at the beginning of the breeding season and decreased as the season progressed (Fig. 3). The average number of eggs at site A was significantly greater than that at site B on January 17 (*t*-test, $P < 0.05$), January 26 ($P < 0.05$), February 6 ($P < 0.01$) and February 15 ($P < 0.01$).

Most cocoons were spindle-shaped and 2–3 mm in length and 1–1.5 mm in width. At both sites, the variance in the number of eggs per cocoon was partly explained by the cocoon volume ($r^2 = 0.266$, $P < 0.001$ at site A; $r^2 = 0.318$, $P < 0.001$ at site B). There are no significant differences between the slopes and intercept values of regression lines calculated for sites A and B. The relationship of the number of eggs per cocoon to the cocoon volume was $y = 3.44 + 27.46\,x$ ($r^2 = 0.317$, $P < 0.001$), combining cocoons from sites A and B. However, the average volume of cocoons at site A was significantly greater than those at site B on January 17 (*t*-test, $P < 0.05$), January 26 ($P < 0.01$), February 6 ($P < 0.01$) and February 15 ($P < 0.01$), but not on February 22 ($P = 0.126$).

Most eggs or embryos in a cocoon were at the same developmental stage on each sampling date. Between January 8 and February 22, most of the cocoons contained eggs in stage 3, though the most developmentally advanced embryos were in stage 4. The difference in the developmental stage among cocoons at the same collection date may reflect partly the difference in the time of deposition. On March 23, at the end of the breeding season, when the water temperature had risen to about 11 °C, the eggs had reached stages 5 and 6. Thus the eggs took about 11 days at most to reach stage 6.

The water content, loss on ignition, carbon and nitrogen content of the sediment were higher at site A than at site B (Table 1).

Nature and part of substrate to which cocoons were attached

R. hiemalis utilized a wide variety of hard substrates for cocoon deposition, such as stones, wood, reed leaves, and even a stiff plastic sheet that had sunk to the lake bottom. However, since shells of bivalves and gastropods were the most common hard material on the muddy bottom, in Lake Biwa *R. hiemalis* commonly utilized these two substrates, with no significant preference for bivalves over gastropods (*chi-square* test: $P > 0.10$, n = 113 at SB, Southern basin; $P < 0.01$, n = 238 at NB, Northern basin). In addition, *R. hiemalis* seemed to prefer shells of living molluscs to dead ones (*chi-square* test: $P < 0.05$, and $P > 0.10$ for bivalves (n = 82) and gastropods (n = 31) at SB;

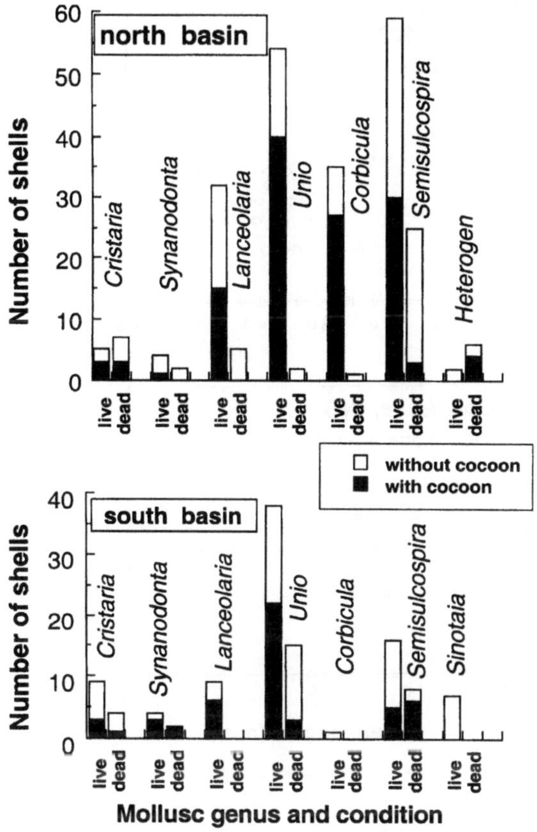

Figure 4. Number of living and dead shells with and without *R. hiemalis* attached cocoons for bivalve and gastropod genera in the northern (upper) and the southern basin (lower).

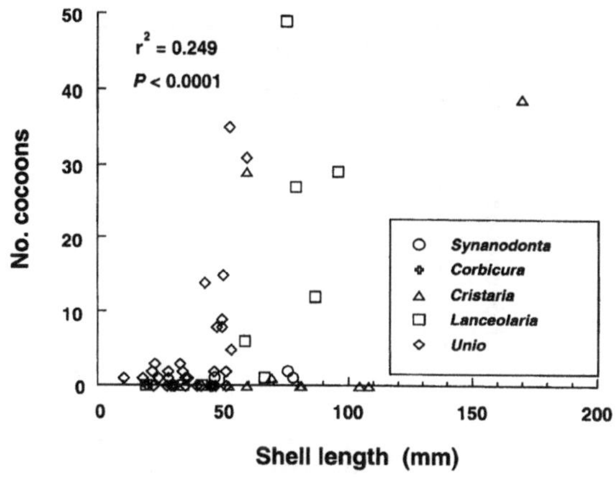

Figure 5. Number of *R. hiemalis* cocoons in relation to shell length in all living bivalve species in the southern basin of Lake Biwa.

$P < 0.01$ and $P < 0.05$ for bivalves ($n = 146$) and gastropods ($n = 92$) at NB) (Fig. 4).

R. hiemalis showed no selective preference among living bivalve species (chi-square test: $P > 0.10$ at SB, and $P > 0.10$ at NB), though it utilized most frequently*Unio biwae*, the most common bivalve in both basins of the lake.

A linear relationship ($y = 0.205\ x - 4.328$, $n = 61$, $r^2 = 0.249$, $P < 0.0001$) was found between the number of cocoons attached and the shell length, as an index of shell surface area, of all living bivalves of the southern basin considered as a group, but not for those of the northern basin (Fig. 5). Considering the species separately, the number of cocoons grows exponentially with the shell length for *Unio biwae* ($n = 38$, $r^2 = 0.223$, $P < 0.01$ at SB, and $n = 54$, $r^2 = 0.159$, $P < 0.01$ at NB), *Lanceolaria oxyrhyncha* ($n = 9$, $r^2 = 0.738$, $P < 0.01$ at SB), and *Corbicula sandai* ($n = 34$, $r^2 = 0.519$, $P < 0.001$ at NB) (Fig. 6), where linear regression analysis has been performed after transformation of number of cocoons to Ln (y+1). Among the gastropods, *R. hiemalis* utilized *Semisulcospira* spp. more frequently in both basins, but no relationship was found between the number of cocoons attached and the width of the first body whorl for any gastropod species.

To ascertain to which part of the mollusc shell *R. hiemalis* cocoons were attached, the bivalve shell surface was distinguished into four parts: I, postero-dorsal part; II, postero-ventral part; III, hinge part; and IV, anterior part. Since bivalves position themselves in sediment with the anterior part directing downwards, parts I and II will be located near the sediment surface. For gastropods which crawl on sediment, the whorl numbers on which cocoons were attached were recorded.

Most cocoons attached to bivalve shells were found on parts I and II, the shell parts positioned closer to the sediment surface. For all bivalve species in both basins of the lake, only a small proportion of cocoons was attached to parts III and IV (Fig. 7), which were deeper in the sediment. For gastropods, most of the cocoons were attached to whorl I, which was also near the sediment surface (Fig. 7). Thus, *R. hiemalis* attached its cocoons on the shells of molluscs at a position near the sediment surface, whether located above (gastropods) or beneath (bivalves) the sediment surface.

Cocoon proximity to substrate surface and number of times of cocoon deposition

In laboratory experiments, a total of 59 cocoons were laid in experimental glass containers. Contrary to ex-

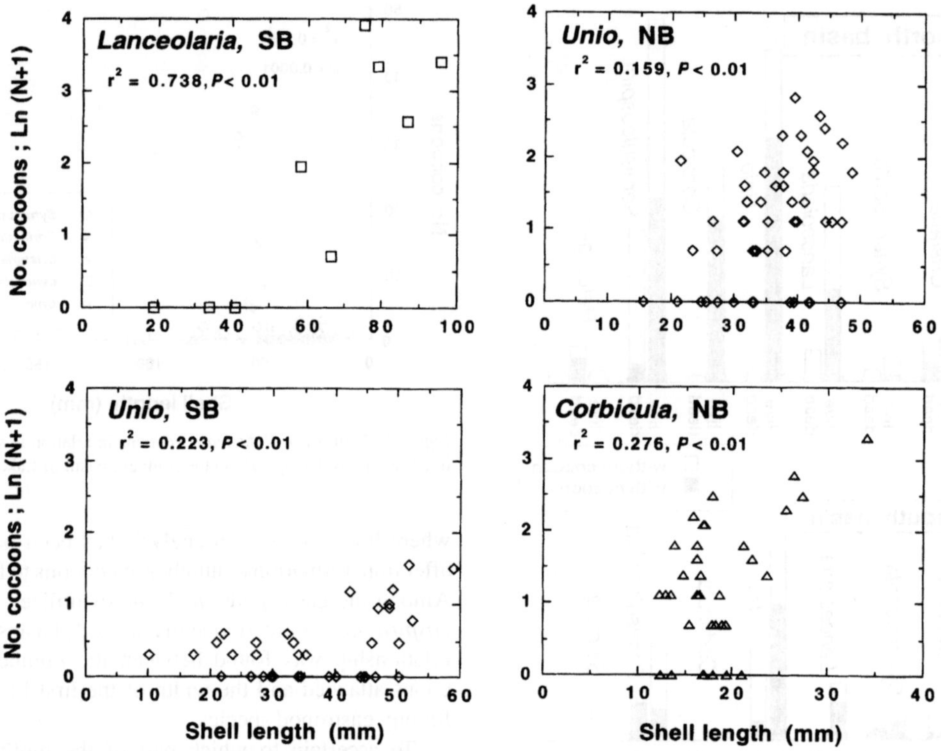

Figure 6. Number of *R. hiemalis* cocoons in relation with shell length in three living bivalve species in Lake Biwa; *Lanceolaria oxyrhyncha*, *Unio biwae* and *Corbicula sandai*. SB – southern basin, NB – northern basin.

pectation, the majority (34 cocoons) were attached to the walls of the container, and relatively few to the experimental substrates which were placed in the container; only 21 on *U. biwae* and 4 on PVC sticks.

Regarding the depth beneath sediment surface at which cocoons were attached to the substrate, 48 cocoons (81%) were attached within 2 cm of the sediment surface (Fig. 8). Eleven cocoons (19%: 9 on the container's wall and 2 on *U. biwae*) were found attached deeper than 2 cm. The deepest cocoon was located 4.7 cm beneath the sediment surface.

Of the nine cocoons attached to the container wall deeper than 2 cm, four took longer to develop than did the cocoons attached at a depth of less than 2 cm. The five other cocoons located deeper than 2 cm became black, seemed to decay and failed to hatch.

Regarding the frequency of cocoon deposition per individual within a breeding season, two individuals deposited two cocoons on four occasions, and on one occasion three individuals deposited four cocoons and two individuals deposited three cocoons. Excepting these instances, the laboratory experiments showed clearly that most individuals (11 of 13 individuals) deposit only one cocoon during a breeding season. All worms studied had resorpted genitalia at the end of the laboratory study.

Discussion

In most tubificids, breeding occurs in spring to summer and continues for several months (Brinkhurst, 1964; Thorhauge, 1975; Lazim & Learner, 1986a). However, *R. hiemalis* breeds during midwinter, when water temperatures are the lowest during the year and when most of the animals inhabiting subtropical Lake Biwa are inactive. In addition, the breeding period is concentrated over a rather short period of about one month from the beginning of January to mid-February. This timing and short duration of reproduction enables *R. hiemalis* to minimize potential predation on the eggs, such as by warm-water cyprinid fishes (Miura, 1966) during the period of development.

Cocoons produced in the earlier periods of the breeding season were larger, which may indicate that larger-sized mature individuals of *R. hiemalis* start to breed earlier than the smaller-sized individuals (Thorhauge, 1975; Bonomi & Di Cola, 1980). It may

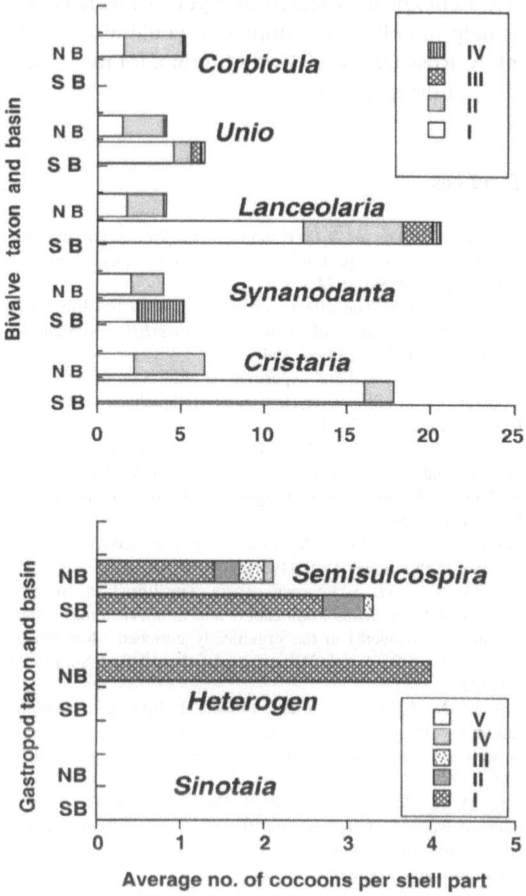

Figure 7. Average number of *R. hiemalis* cocoons attached to the different shell parts of bivalve (upper) and gastropod (lower) genera, in the southern (SB) and the northern basin (NB) of Lake Biwa. For bivalves, I, postero-dorsal part; II, postero- ventral part; III, hinge part; and IV, anterior part. For gastropods, roman numerals indicate shell whorl number.

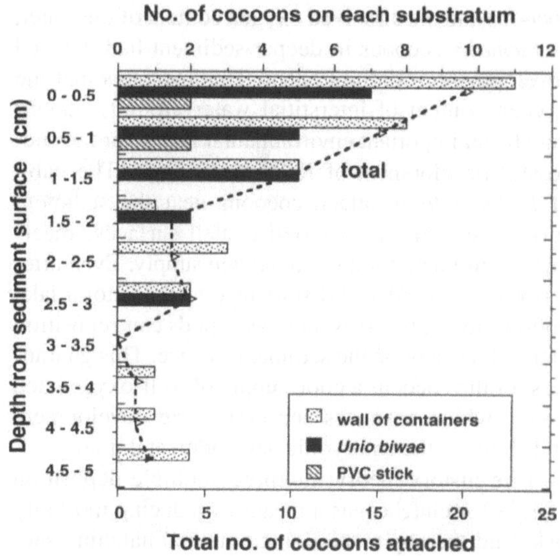

Figure 8. Number of *R. hiemalis* cocoons on different substrata and the distance from the sediment surface in laboratory experiments. The broken line shows the total number of cocoons on the three substrata: container walls, *Unio biwae* and PVC sticks.

also explain the observation that the cocoons laid during the earlier dates of the breeding season had more eggs. Within a given species of animal, larger individuals generally produce more eggs than do smaller individuals. Thorhauge (1975) showed that in *Potammothrix hammoniensis* the average number of eggs per cocoon increased with individual wet weight. Small, slow-growing individuals of *R. hiemalis* would probably have matured at a small size in the breeding season, since their failure to mature in this period would result in death without reproduction due to their annual cycle.

The average number of eggs per cocoon was greater at site A than at site B, probably because the organic content, and hence nutritious potential of the sediment would be higher at site A than at site B.

Among tubificids, *R. hiemalis* has the highest known number of eggs (9–11 eggs) per cocoon, e.g. *Tubifex tubifex*, 2–4 eggs (Adreani et al., 1981; Lazim & Learner, 1986b); *Limnodrilus hoffmeisteri*, 1–4 eggs (Adreani et al., 1981); *L. udekemianus*, 2 eggs (Lazim & Learner, 1986a); *Potammothrix hammoniensis*, 4–8 eggs (Thorhauge, 1975). *R. hiemalis* is an annual species (T. Narita, unpublished data) and most individuals are semelparous, depositing only one cocoon in their lifetime. However, most other tubificids deposit cocoons several times in a breeding season and the incidence of cocoon deposition does not change with individual weight (for instance, *P. hammoniensis*, see Thorhauge, 1975).

Most *R. hiemalis* cocoons were attached less than 2 cm beneath the sediment surface. In comparison, *T. tubifex* mainly deposits its cocoons at a depth greater than 2 cm (1–12 cm), and *L. udekemianus* usually deposits cocoons deeper than 5 cm (3–20 cm) beneath the sediment surface (Lazim et al., 1989). Newkla & Mutayoba (1987) suggested that *T. tubifex* hides its cocoons deeper than 2 cm beneath the sediment surface as a means of avoiding fish predation. Contrarily, *R. hiemalis* does not hide its cocoons.

Although Thorhauge (1975) noted no relationship between the developmental time of *P. hammoniensis* cocoons and the dissolved oxygen content of the water, *R. hiemalis* cocoons in deeper sediment had delayed development or did not hatch. This indicates that the oxygen content of interstitial water around cocoons may be an important environmental factor for the successful development of *R. hiemalis* eggs. The habit of *R. hiemalis* to attach cocoons near the sediment surface, especially on mollusc shell surfaces, meets this requirement for a good oxygen supply. Even after serious storm-induced disturbance to the littoral lake bottom, living bivalves and gastropods can reposition themselves in or on the sediment surface. This guarantees for the cocoons a good supply of well-oxygenated water, which enhances successful egg development and shortens the embryonic developmental time.

Life history theory interprets multiple deposition of eggs (or clutches) as a means of reducing mortality risks and thereby maximizing potential hatching success. Upon first examination, the breeding strategy of *R. hiemalis*, which breeds only once and produces a large clutch, would not seem to fit this general theory. However, *R. hiemalis* seems to compensate for the potential loss of all eggs in a single cocoon by breeding in midwinter, when predation is at its lowest, and by having embryonic development even at the lowest mid winter water temperatures. It achieves the latter by attaching cocoons near the sediment surface where they are always in contact with well-oxygenated lake water.

This midwinter breeding is also an advantageous life history trait from the viewpoint of the 6-month aestivation undertaken by *R. hiemalis* during summer; the timing of reproduction allows newly hatched individuals to feed and grow early in the year, thereby accumulating and reserving energy stores before burrowing into the sediment in May or June at the latest.

Acknowledgements

The author thanks A. Kawabata and T. Ueda, Center for Ecological Research of Kyoto University, for their help in collecting samples. I would also like to thank A. Rossiter, Lake Biwa Museum, for his critical reading of the manuscript.

References

Adreani, L., C. Bonacina & G. Bonomi, 1981. Production and population dynamics in profundal lacustrine oligochaeta. Verh. int. Ver. Limnol. 21: 967–974.

Aston, R. J., 1968. The effect of temperature on the life cycle, growth and fecundity of *Branchiura sowerbyi* (Oligochaeta: Tubificidae). J. Zool., Lond. 154: 29–40.

Aston, R. J., 1973. Field and experimental studies on the effects of a power station effluent on Tubificidae (Oligochaeta, Annelida). Hydrobiologia 42: 225–242.

Bonomi, G. & G. Di Cola, 1980. Population dynamics of *Tubifex tubifex*, studied by means of a new model. In Brinkhurst, R. O. & D. G. Cook (eds), Aquatic Oligochaete Biology. Plenum Press, New York: 185–203.

Brinkhurst, R. O., 1964. Observation on the biology of lake-dwelling Tubificidae. Arch. Hydrobiol. 60: 385–418.

Lazim, M. N. & M. A. Learner, 1986a. The life-cycle and production of *Limnodrilus hoffmeisteri* and *L. udekemianus* (Tubificidae; Oligochaeta) in the organically enriched Moat-Feeder Stream, Cardiff, South Wales. Arch. Hydrobiol., Suppl. 74: 200–225.

Lazim, M. N. & M. A. Learner, 1986b. The life-cycle and productivity of *Tubifex tubifex* (Oligochaeta; Tubificidae) in the Moat-Feeder Stream, Cardiff, South Wales. Holarct. Ecol. 9: 185–192.

Lazim, M. N., M. A. Learner & S. Cooper, 1989. The importance of worm identity and life history in determining the vertical distribution of tubificids (Oligochaeta) in a riverine mud. Hydrobiologia 178: 81–92.

Miura, T., 1966. Ecological notes of the fishes and the interspecific relation among them in Lake Biwa. Jap. J. Limnol. 27: 1–24.

Newrkla, P. & S. Mutayoba, 1987. Why and where do oligochaetes hide their cocoons? Hydrobiologia 155: 171–178.

Ohtaka, A., 1995. A new species of the genus *Rhyacodrilus* Bretscher (Oligochaeta, Tubificidae) from Japanese lakes. Zool. Sci. 12: 491–498.

Timm, T., 1997. Freshwater oligochaeta of some urban watercourses in the Russian Far East. Int. Rev. ges. Hydrobiol. 82: 437–467.

Thorhauge, F., 1975. Reproduction of *Potamothrix hammoniensis* (Tubificidae, Oligochaeta) in Lake Esrom, Denmark. A field and laboratory study. Arch. Hydrobiol. 76: 449–474.

Hydrobiologia **463:** 149–169, 2001.
P. Rodriguez & P.F.M. Verdonschot (eds), Aquatic Oligochaete Biology VIII.

Utility and relevance of aquatic oligochaetes in Ecological Risk Assessment

Peter M. Chapman
EVS Environment Consultants, 195 Pemberton Avenue, North Vancouver, B.C., Canada V7P 2R4
Tel: 604 986 4331. Fax: 604 662 8548. E-mail: pchapman@attglobal.net

Key words: oligochaetes, risk assessment, sediments

Abstract

Ecological risk assessment (EcoRA) provides both a process and a framework to evaluate the potential for adverse ecological effects occurring as a result of exposure to contaminants or other stressors. EcoRA begins with problem formulation/hazard identification, progresses to effects and exposure assessment, and culminates with risk characterization (an estimate of the incidence and severity of any adverse effects likely to occur). Key components of EcoRA include determining: stressors/contaminants of concern; sensitive, exposed biota; and, appropriate tests and organisms for evaluating effects. Aquatic oligochaetes are not generally used directly in EcoRA because of three major perceptions. First, EcoRA personnel are generally not familiar with or comfortable using this group of organisms. Second, there is believed to be a paucity of widely accepted toxicity tests with these organisms. Third, their taxonomy is considered difficult and uncertain. In fact, aquatic oligochaetes potentially have great utility and relevance to EcoRAs because of factors including: their importance in the aquatic food chain (e.g. prey to fauna including fish and waterfowl; as a vector for contaminant movement through the food chain from bacteria); many species are widely distributed and well studied; representatives include fresh, estuarine and marine species; as a group, they range from sensitive to insensitive over a wide range of environmental insults; they have a long history of use in pollution monitoring and assessment; and, relevant toxicity and biaccumulation tests exist. Toxicity testing under defined conditions is appropriate for problem formulation while more realistic testing for effects assessment (e.g. microcosms) is logistically easier with this group of organisms than with others due to their relatively small size. The importance of aquatic oligochaetes for EcoRA, in particular of sediments, is particularly compelling.

Introduction

Aquatic oligochaetes have an extremely long history of use for pollution assessments. They, or chironomids, may have been the 'bloodworms' collected by Aristotle, from freshwater muds downstream of where Athenians discharged their wastes (Hynes, 1960). If so, they are also the first organisms for which we have a record of toxicity testing: Aristotle conducted rudimentary bioassays by placing these organisms in salt water and observing their responses (Hynes, 1960).

Certain species of aquatic oligochaetes (e.g. *Tubifex tubifex* and *Limnodrilus hoffmeisteri*) are abundant in organically polluted waters because of lack of competition and an abundant food supply coupled with a tolerance to reduced oxygen conditions (Brinkhurst & Jamieson, 1971; Brinkhurst et al., 1983). Historically, a prevalence of aquatic oligochaetes has been linked to organic pollution (Hynes, 1960; Brinkhurst & Jamieson, 1971) such that the whole taxon has been and continues to be generally considered 'pollution tolerant'. This perception of 'tolerance' has been incorrectly extended to chemical pollution. In the last few decades, organic pollution issues resulting in widespread acute effects have been largely addressed in developed countries and the focus has moved to chemical pollution and to chronic effects. The regulatory approach to chemical pollution has historically been to reduce or eliminate it, thus the basic premise of both field studies and toxicity testing has been founded on the 'most sensitive species'. Largely because of their perceived 'tolerance' to pollution, aquatic oligochaetes have generally fallen out of favor with many researchers for modern-day pollution assessments, including ecological risk assessment (EcoRA). This is unfortunate, as aquatic oligochaetes can and should be extremely useful in EcoRA studies. In fact, a few (too few) species of aquatic oligochaetes are proving useful in various EcoRA components as discussed herein.

This paper fulfills three purposes: (1) a review of aquatic oligochaete usage in EcoRA-related studies; (2) a discussion of the importance of aquatic oligochaetes for EcoRA; and (3) recommendations for initiatives including research that will increase such usage.

Ecological Risk Assessment (EcoRA)

EcoRA is a process and a framework that, although it can be complex, basically evaluates the potential for adverse ecological effects occurring as a result of exposure to contaminants or other stressors. In doing so it provides a means for determining key uncertainties, i.e. areas where additional studies will make the most difference to the assessment and to the decisions which will subsequently be made.

EcoRA begins with a determination of potential hazard, i.e. the *possibility* of a stressor causing adverse effects. In the U.S., this first step is called problem formulation; in Canada and the European Union, it is called hazard identification. In this step, the EcoRA purpose and the problem are defined, potential stressors and biological receptors are identified, and a plan for subsequent analysis and characterization is defined. The next steps involve analysis/characterization of both exposure (emissions, pathways and rates of movement of stressors in the environment) and ecological effects (information on the relationship between stressor concentrations and the incidence and (or) severity of adverse effects). The ecosystem, which has been characterized to some extent in the first step, is more definitively characterized and a conceptual diagram of the ecosystem and potential stressor pathways and intensities either developed or refined. The final step (Fig. 1), risk characterization, determines the likelihood or *probability* that the potential hazard identified in the first step could be realized. In other words, a determination is made as to the likelihood and extent of adverse effects from particular stressors occurring to biological receptors. Adverse effects occur only when a receptor is exposed (e.g. via respiration, food or contact) to a stressor at a sufficient concentration and for a sufficient duration of time. Risk characterization can include evidence from laboratory studies, field observations, model predictions, and/or professional judgment. It also includes an estimate of uncertainty, both overall and for specific EcoRA components.

EcoRAs can be either retrospective (evaluating historic stressor-receptor relationships) or prospective (evaluating possible future stressor-receptor relationships). They typically involve a comparison, for chemicals, of the predicted environmental concentration (PEC) with the predicted no effect concentration (PNEC). If the PEC is larger than the PNEC, then there may be a risk. Such simple comparisons resulting in a value above, below or equal to unity can be derived from simply comparing measured or predicted chemical concentrations with one or more environmental quality values to determine a hazard quotient (HQ). However, HQs provide no information on risk; determining probability and intensity of effects requires a great deal more information (Hill et al., 2000).

Typically, EcoRAs are conducted on a site-specific basis. Regional and global EcoRAs are also possible, but necessarily have a much greater level of uncertainty than site-specific EcoRAs. Typically and historically, EcoRAs have been focused on chemical stressors; however, they are being broadened to include biotic (e.g. species introductions) and abiotic (e.g. habitat loss) stressors.

EcoRA can be done at either a screening level or a detailed level. Hill et al. (2000) summarize the differences between these two levels of detail.

Perceptions and facts regarding the use of oligochaetes in EcoRA

Aquatic oligochaetes have, in general, only been used to a limited extent in EcoRA or assessments that could comprise one or more components of an EcoRA related to chemical contaminants. A few individual species have been used extensively; however, the limited usage of the taxon as a whole is due to a number of mainly perceptual barriers, typified by a widely cited paper by Giesy & Hoke (1989) regarding appropriate freshwater species for sediment toxicity testing. These authors concluded that aquatic oligochaetes were not appropriate for reasons outlined in Table 1, where those reasons are compared with and invalidated by the contrasting reality.

Giesy & Hoke (1989) also stated that oligochaete toxicity tests "are not simple or standardized and the sensitivity and discriminatory power of such assays has been little studied". They concluded that "it is not likely that oligochaetes will prove to be any more sens-

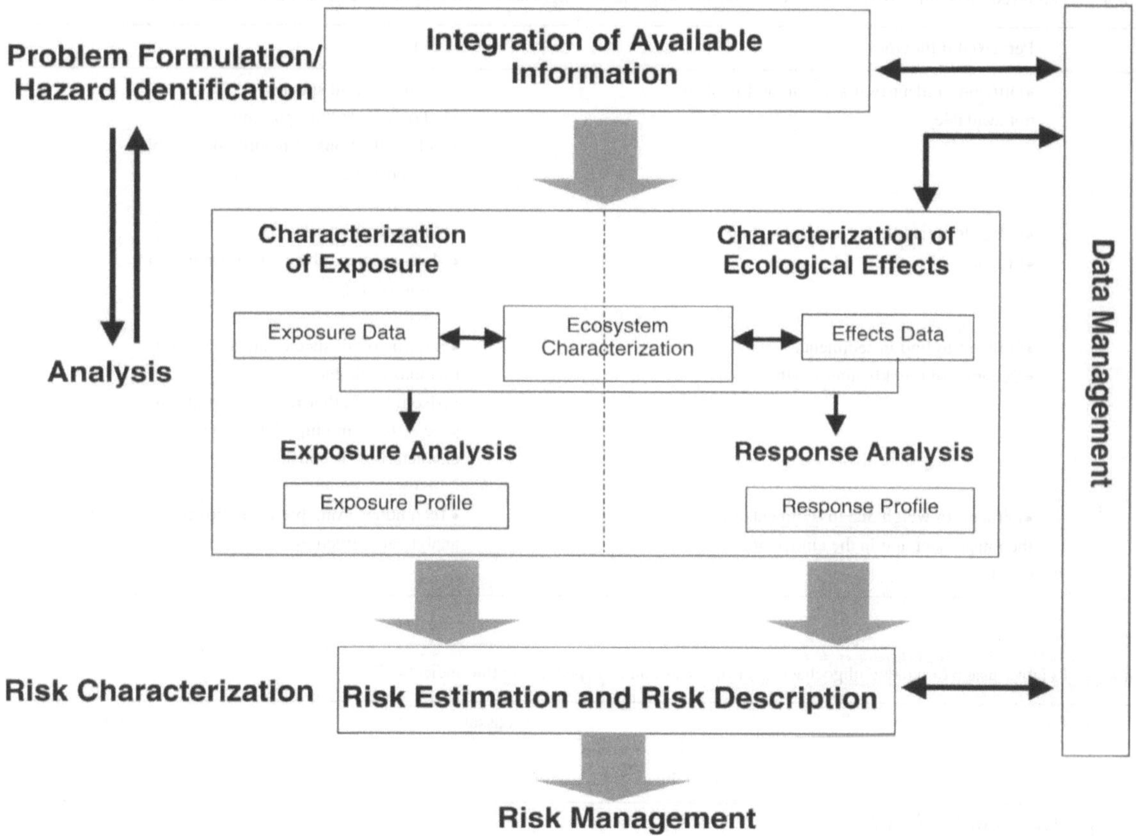

Figure 1. Risk assessment framework. Adapted from U.S. EPA (1992).

itive to in-place pollutants or add additional information not obtained from other more easily performed tests…"

These are important criticisms that, if true, would indeed invalidate wider use of aquatic oligochaetes, at least in toxicity testing. To examine the first part of this criticism ("not simple or standardized"), the readily available literature was reviewed. Specific aquatic oligochaete use in bioassay testing is summarized in Table 2 together with an evaluation of these species as benthic indicators. Such usage is significant. Further, *Standard Methods for the Examination of Water and Wastewater* (Eaton et al., 1995) has a section on aquatic oligochaete toxicity tests, and deals specifically with several species listed in Table 2. U.S. EPA (2000) describes a method for 28-d sediment bioaccumulation testing using *Lumbriculus variegatus*, including preliminary screening for toxicity. ASTM (1996) describes a method for 28-d survival and reproduction testing with *Tubifex tubifex*.

Six oligochaete species have had reasonably wide use in acute toxicity tests (Table 2). Five of these species are also widely used as benthic indicators. Eighteen other species have been used occasionally in acute toxicity tests. However, oligochaete use in chronic toxicity tests is much less extensive (Table 3), and is focused mainly on *T. tubifex*. Bioaccumulation testing has been focused primarily on *L. variegatus*, though testing has also been done with *T. tubifex* and *L. hoffmeisteri*. Also apparent from Table 2 (and Table 3) is the fact that bioaccumulation testing to date has focused on freshwater species (which tend to be larger than saltwater species).

Since Bailey & Liu (1980) first advocated its use in toxicity tests, *Lumbriculus variegatus* has been the most widely used oligochaete for acute toxicity, and for bioaccumulation tests involving a wide range

Table 1. Perceived difficulties (Giesy & Hoke, 1989) in using aquatic oligochaetes for bioassays, contrasted with the reality

Perceived difficulties	Reality
•Difficult to identify if appropriate life-stage not available	•Not true for all species •Laboratory culture possible •Field collections of readily identifiable adults possible
•Difficult to handle for toxicity testing •Fragile	•Not generally true •Many species robust and amenable to laboratory usage
•Difficult to find in sediments •Decompose quickly after death	•Not true of all species and collection methods available •Also true of other routinely used test species (e.g. amphipods), for which this characteristic is considered irrelevant
•Difficult to weigh due to sediments on the integument and in the alimentary canal	•Techniques available for removal or analytical correction

Table 2. Relative usage of aquatic oligochaetes for bioassays and as specific benthic indicators[a]

Species	Bioassays			Benthic Indicators[b]
	Acute Toxicity	Chronic Toxicity	Bioaccumulation	
Lumbriculus variegatus (L)[c]	****	**	****	
Tubifex tubifex (T)	****	****	***	****
Limnodrilus hoffmeisteri (T)	***	*	**	****
Stylodrilus heringianus (L)	***			****
Branchiura sowerbyi (T)	***	*	*	****
Tubificoides benedii (T, S)[d]	***		*	***
Other species used for acute toxicity tests				
Ilyodrilus frantzi (T, S)	*Pristina leidyi* (N)			
I. templetoni (T)	*Quistadrilus multisetosus* (T)			
Limnodrilus udekemianus (T)	*Rhyacodrilus montana* (T)			
L. claparedeanus (T)	*Spirosperma ferox* (T)			
Monopylephorus cuticulatus (T, S)	*S. nikolskyi* (T)			
Nais communis (N)	*Tubificoides fraseri* (N, S)			
Paranais frici (N, S)	*Tectidrilus diversus* (T, S)			
P. litoralis (N, S)	*Varichaetadrilus pacificus* (T)			
Potamothrix hammoniensis (T)				

[a] Stars indicate number of published references readily available: no star = 0; * = 1–3; ** = 3–5; *** = 5–10; **** = >10.
[b] Used individually to indicate pollution status.
[c] L = Lumbriculidae; T = Tubificidae; N = Naididae.
[d] (S) = saltwater species.

Table 3. Chronic (growth and reproduction) tests with oligochaetes

Species	Growth	Reproduction
L. variegatus (L)[a]	Leppänen & Kukkonen (1998c)	Nebeker et al. (1989); Hickey & Martin (1995) Leppänen & Kukkonen (1998b,c); Chapman et al. (1999)
T. tubifex (T)	Wiederholm et al. (1987); Milbrink (1987); Reynoldson (1987); Wiederholm & Dave; (1989) Reynoldson et al. (1991); Marchese & Brinkhurst (1996)	Wiederholm et al. (1987); Milbrink (1987); Wiederholm & Dave (1989); Reynoldson et al. (1991); Bailey et al. (1995); Day et al. (1995a,b); Marchese & Brinkhurst (1996); Milani et al. (1998); Martinez–Madrid et al. (1999); Chapman et al. (1999)
L. hoffmeisteri (T)	Wiederholm et al. (1987); Reynoldson (1987)	Wiederholm et al. (1987); Lotufo & Fleeger (1996)
B. sowerbyi (T)	Marchese & Brinkhurst (1996)	Marchese & Brinkhurst (1996); Casellato et al. (1992)
L. udekemianus (T)	Wiederholm et al. (1987)	Wiederholm et al. (1987)
L. claparedeanus (T)	Wiederholm et al. (1987)	Wiederholm et al. (1987)
Potamothrix hammoniensis (T)	Wiederholm et al. (1987)	Wiederholm et al. (1987)
I. templetoni (T)	Reynoldson (1987)	
P. leidyi (N)	Smith et al. (1991)	

L = Lumbriculidae; T = Tubificidae; N = Naididae.

of inorganic and organic contaminants (Ewell et al., 1986; Schuytema et al., 1988; Nebeker et al., 1989; Ankley et al., 1991a,b, 1992, 1993, 1994a,b; Dermott & Munawar, 1992; West et al., 1993; Phipps et al., 1993, 1995; Kukkonen & Landrum, 1994; Schubauer-Berigan et al., 1995; Hansen et al., 1996; Whiteman et al., 1996; West et al., 1997; Kosian et al., 1998; Brunson et al., 1998; Leppänen & Kukkonen, 1998a,b,c; Chapman et al., 1999; Mount et al., 1999; Fisher et al., 1999). The doubling rate for *L. variegatus* under standardized culture conditions is about 15 d (Phipps et al., 1993), rendering it readily amenable to measurements of fecundity even though this has not commonly been measured.

L. variegatus has also been used for assessing sublethal neurophysiological endpoints following exposure to various chemicals (Rogge & Drewes, 1993). These authors noted the following advantages to such testing with this oligochaete: cosmopolitan distribution in holarctic regions and introduction to Africa, Australia and New Zealand; sediment burrowing; ease of laboratory rearing, maintenance and handling; and presence of giant nerve fibers that make them amenable to noninvasive monitoring of electrical activity. The third advantage above is in direct contrast to Giesy & Hoke's (1989) contention that aquatic oligochaetes are generally difficult to handle and fragile.

Other reasonably (but less commonly) used species, primarily for toxicity testing, are *Tubifex tubifex* (Jones, 1938; Whitley, 1967; Brokovic-Popovic & Popovic, 1977a,b; Chapman et al., 1982a,b; Amon, 1985; Wiederholm et al., 1987; Milbrink, 1987; Wiederholm & Dave, 1989; Khangarot, 1991; Reynoldson et al., 1991; Fargasová, 1994; Marchese & Brinkhurst, 1996; Egeler et al., 1997; Meller et al., 1998; Chapman et al., 1999; Vecchi et al., 1999; Martinez–Madrid et al., 1999) and *Limnodrilus hoffmeisteri* (Chapman et al., 1982a,b; Wiederholm et al., 1987; Lotufo & Fleeger, 1996; Egeler et al., 1997; Meller et al., 1998; Keilty et al., 1998a,b). *Stylodrilus heringianus* is readily collectable and identifiable from areas of the Great Lakes but has been less widely used

(Chapman et al., 1982a,b; Keilty et al., 1998a,b,c). *Branchiura sowerbyi* is a large, readily identifiable (due to the gills on its posterior end) warm-water species originally from Asia, which has invaded North America. It has been used, but not extensively, for both toxicity and bioaccumulation tests (Chapman et al., 1982a,b; Casellato & Negrisolo, 1989; Casellato et al., 1992; Marchese & Brinkhurst, 1996; Bhunia et al., 2000). It is surprising that *B. sowerbyi* has not been more extensively used for bioaccumulation testing due to its large size; bioaccumulation tests with most oligochaete species are more difficult due to their small size relative to tissue requirements for chemical analyses (Chapman et al., 1980). The marine oligochaete *Tubificoides benedii* has been used to study adaptation to sulfidic sediments (Dubilier et al., 1994, 1995, 1997; Giere et al., 1999).

Toxicity studies by a few researchers add a further 17 aquatic oligochaete species to the list (Table 2: Chapman et al., 1982a,b; Chapman & Mitchell, 1986; Wiederholm et al., 1987; Smith et al., 1991). Detailed culture and test conditions have been developed for the naidid *Pristina leidyi* (Smith et al., 1991).

Oligochaetes have also been used to determine modes of action and entry of toxicants. For instance, mechanistic studies have been conducted to understand mechanisms regulating bioavailability, uptake and localization of various metals and chlorinated organics for oligochaetes including *Monopylephorus cuticulatus*, *L. variegatus*, *Limnodrilus udekemianus*, *L. hoffmeisteri* and *Tubifex tubifex*. (D'Angelo & Signorile, 1974; Dean, 1974; Fischer & Horvàth, 1977; Thompson et al., 1982; Oliver, 1984, 1987; Connell et al., 1988; Bauer-Hilty et al., 1989; Gunn et al., 1989; Back, 1990). Critical body residues have been proposed for PCBs in *L. variegatus* (Fisher et al., 1999).

Ingersoll et al. (1995) reviewed freshwater toxicity and bioaccumulation methods and applications. Ten separate toxicity test organisms were reviewed including *L. variegatus* and *T. tubifex*. They recommended that toxicity evaluations involve survival of *L. variegatus*, survival and growth of the chironomid *Chironomus tentans*, and survival of the amphipod *Hyalella azteca*. They further recommended that bioaccumulation evaluations be done using *L. variegatus*. *T. tubifex* was not recommended because of a lack of round robin testing and of a relative sensitivity toxicity data base. Both oligochaete species were considered to be amenable to laboratory culture, easy to identify, of ecological importance, having a wide geographic distribution and tolerance to different sediments, and to have had laboratory toxicity responses confirmed by benthic population studies.

Thus, in contrast to Giesy & Hoke's (1989) dismissal of aquatic oligochaetes for sediment toxicity tests, these organisms are in fact appropriate for both sediment and water column tests for the following reasons: they are ecologically important components of aquatic (fresh, estuarine, marine) ecosystems; many species are widely distributed; several species are suitable or have been used for long-term testing including chronic toxicity end-points (growth, reproduction); they are generally exposed to contaminants via both water and sediment; many have sufficient biomass for bioaccumulation testing; many are easily cultured and handled; in several cases standardized test procedures are available; and, many (e.g. *L. variegatus*) tolerate a wide range of sediment physical and chemical characteristics (e.g. grain-size, organic carbon content).

Chapman et al. (1999) have specifically noted that tests with the oligochaetes *L. variegatus* and *T. tubifex* meet the majority of Giesy & Hoke's (1989) characteristics for an ideal sediment toxicity test: speed, simplicity, reproducibility, reasonable expense, standardization, sensitivity, discriminatory power, ecological relevance, relationship to field effects and use in developing regulatory criteria. As they note, culturing these species in the laboratory 'is easy'. Similar comments apply to many other oligochaete species. Hickey & Martin (1995) compared the relative sensitivity of five New Zealand benthic species for use in sediment toxicity tests and found that the oligochaete *L. variegatus* was one of "the most promising species for sublethal test development".

Potential uses of oligochaetes in EcoRA

There are several distinct ways in which any organism can be used in EcoRA, as discussed in more detail below relative to aquatic oligochaetes: single-species acute toxicity tests, chronic and sublethal toxicity tests; mixed species toxicity tests; bioaccumulation tests; ecological assessment; food chain studies; and, integrative assessments.

Toxicity tests: tolerance, predictions, causality, criteria development and culturing

L. hoffmeisteri, collected from contaminated sites, has been shown to be less sensitive to contaminants than oligochaetes from uncontaminated sites (Klerks & Levinton, 1989; Klerks & Bartholomew, 1991). This may be due to increased detoxification in response to elevated Cd concentrations (Klerks & Levinton, 1989). As noted later (Food Chain Studies), this may have important implications not just for toxicity testing but also for food chain contaminant accumulation.

Klok & de Roos (1996) predicted the effects of copper-induced changes in feeding and metabolism for the earthworm *Lumbricus rubellus*. Predictions as to the copper concentrations at which population growth rate would equal zero corresponded well to an observed steep decline in field populations at those copper concentrations. Such predictions should also be possible for aquatic oligochaetes.

Aquatic oligochaetes have also been used to determine causes of observed toxicity. Kosian et al. (1998) used *L. variegatus* in acute lethality tests to assess reasons for phototoxicity of sediment pore waters from sediments contaminated with PAHs. A toxicity identification evaluation (TIE) was conducted using the worm to identify potential causative agents. Similarly, *L. variegatus* was used to assess ammonia bioavailability and toxicity from pore water, using spiked sediment and water-only exposures (Whiteman et al., 1996).

L. variegatus has also been an important part of the U.S. EPA's development of sediment quality criteria (Ankley et al., 1991a; Hansen et al., 1996). Acute lethality has been used to assess the predictability of metals toxicity based on interstitial water concentrations of metals and acid-volatile sulfide normalization.

It is generally not difficult to culture oligochaetes for toxicity testing. Lumbriculids (*L. variegatus*), tubificids and naidids have all been successfully cultured. Naidids are particularly easy to culture due to their largely asexual mode of reproduction (budding). Levinton & Stewart (1982) report no difficulty in culturing *Paranais litoralis*, which they used for predation studies, rather than for toxicity studies. Smith et al. (1991) have successfully cultured *P. leidyi* for toxicity testing. It is surprising that, despite their small size, more naidids have not been cultured and used for toxicity tests, particularly related to rates of population growth under different environmental conditions and contaminant concentrations.

Toxicity tests: acute endpoints

Single-species acute toxicity tests, which are conducted under defined conditions and which are not designed to be environmentally realistic, are primarily useful for the problem formulation stage of an EcoRA (Hill et al., 2000). This initial stage of an EcoRA involves developing assessment (ecosystem values to be protected) and measurement endpoints (related to those ecosystem values), sources and stressors of potential concern, and a conceptual model. The latter provides a visual representation of potential causal relationships among sources, stressors, receptors and endpoints.

Because *L. variegatus* is almost always only used for acute toxicity testing, it is often paired with other species for chronic toxicity testing. Such pairings are typified by Whiteman et al. (1996), who used the acute worm response together with *Hyalella azteca* survival and *Chironomus tentans* survival and growth endpoints. *C. tentans* is routinely used for chronic sediment toxicity tests and, because it occupies a similar ecological niche, its use appears to have obviated development of chronic tests with *L. variegatus* or the consistent use of other aquatic oligochaetes for which chronic toxicity tests are available.

Although enchytraeid oligochaetes are commonly found in fresh, estuarine and marine waters, they have not been used to date for bioassay testing. However, terrestrial enchytraeids (*Enchytraeus albidus* and *E.* cf. *buchholzi*) have been used for acute toxicity tests in freshwater (Römbke & Knacker, 1989). Tests were conducted with eight chemicals and the enchytraeids were considered suitable for acute tolerance tests in both soil and water. The same method has been used for studying the effects of soil extracts (Graefe, 1991) as well as of PAHs and PCBs in water-only tests (Achazi et al., 1995).

Toxicity tests: chronic and sublethal endpoints, sensitivity

Chronic toxicity tests generally involve a significant portion of an organism's life-cycle and typically involve determinations such as reproductive success (i.e. fecundity) or growth. Measures of growth and reproductive success have been developed and used with a variety of aquatic oligochaetes, primarily tubificids, as shown in Table 3.

Comparative tests with different species indicate that *L. variegatus* is more sensitive to metals in sediments than *T. tubifex* (Chapman et al., 1999). Among tubificid oligochaetes, *T. tubifex* appears to be the best candidate toxicity test organism because it is easiest to culture, reproduces consistently (Milbrink, 1987) and is easier to handle than *L. hoffmeisteri* (Meller et al., 1998). However, in both acute (Chapman & Brinkhurst, 1984) and chronic tests (Wiederholm et al., 1987), relative sensitivities of different oligochaetes are contaminant-specific. Similar specific differences exist within other taxa, including widely used sediment toxicity test species such as infaunal amphipods (McPherson & Chapman, 2000). For any and all taxa, uncertainty is reduced when the ultimate choice of test organism is specific to the situation (stressor(s) and environmental conditions).

Sublethal behavioral toxicity tests have been proposed for oligochaetes, including retraction of the tail of head-downward buried worms in the presence of potentially injurous water-column contaminants (Leynen et al., 1999), changes in burrowing/sediment reworking (Coler et al., 1988; Meller et al., 1998), avoidance of contaminated sediments (McMurtry, 1984; White & Keilty, 1988; Keilty et al., 1988a,b,c; Kukkonen & Landrum, 1994; West & Ankley, 1998; Meller et al., 1998), autotomy (Meller et al., 1998) and changes in escape reflexes (Drewes, 1997). Meller et al. (1998) argue that such behavioral responses are particularly useful and directly relevant to the overall health of populations of worms. However, to date such behavioral tests have been restricted to a few species: *L. variegatus*, *S. heringianus*, *T. tubifex*, *L. hoffmeisteri* and *L. udekemianus*.

Changes in aquatic oligochaete respiration have also been used to measure the sublethal effects of individual contaminants (Chapman et al., 1982c; Brinkhurst et al., 1983). They have also been used to categorize contaminated marine sediments (Chapman, 1987).

Mixed species toxicity tests

Toxicity and/or bioaccumulation studies can be conducted in the laboratory or in the field, and can be done in simple single-organism systems or in complex multiple-organism systems. Mesocosm or microcosm experiments by definition involve more than a single organism and attempt, to varying degrees, to test responses at the ecosystem level. The major difference between mesocosms and microcosms is size – the former are much larger than the latter and thus tend to comprise more ecosystem components and/or larger organisms, while the latter tend to be counter-top laboratory experiments. Odum (1984) defines mesocosms as "bounded and partially enclosed outdoor experimental setups" and suggests that the use of these systems in ecotoxicology represents an important shift from reductionist to holistic approaches.

Two mesocosms (which could be considered microcosms in comparison to, for instance, whole lake experiments) and a microcosm are shown in Figure 2. The mesocosms are 2–5 m in size while the microcosm is two orders of magnitude smaller. As is apparent from Figure 2, mesocosms and microcosms can be placed in water bodies to contain water, sediment and organisms. They can also be placed into water bodies with sediment added. That sediment can be from another area, or from that area but sterilized to allow for recolonization, and can be spiked with chemical(s). Mesocosms and microcosms can also involve structures placed on land beside waterbodies with water pumped in (static, partial replacement, or flow through), and sediment added. Again, that sediment can contain all organisms obtained during collection, can be sterilized to allow for recolonization, and/or can be spiked with chemical(s).

The microcosm design shown in Figure 2C was used in an experiment demonstrating the sensitivity of oligochaetes to a specific contaminant (Warren et al., 1998). Sediments were collected from the bottom of a shallow lake, and the top 10 cm removed. The remaining sediment was mixed and divided into six treatments. Each treatment received different doses of Cd spiked into the sediment, and was then distributed into individual microcosms (eight per treatment, n = 48). Lids were placed on the microcosms, they were then placed back in the lake bottom so their tops were flush with the sediment surface, and the lids removed. They were left in place for 11 months to allow for recolonization, then lids were put on them and they were removed to assess metals concentrations (overlying, pore water and sediments), organisms present and body burdens. Surprisingly, most of the benthic organisms that recolonized the sediments, including those constructing sediment burrows, were not affected by Cd in the sediment or pore waters, but by Cd in the overlying waters. The two exceptions were the two organisms that actually feed on the sediments, in this case a chironomid species and 'tubificid oligochaetes' (specific identifications of the oligochaetes were not undertaken). Based on this experiment, tubificid oligochaetes and chironomids are clearly key taxa for studying 'worst case' effects of sediment contamination since their exposure routes to chemicals occur not just through the water column, but also through sediment ingestion.

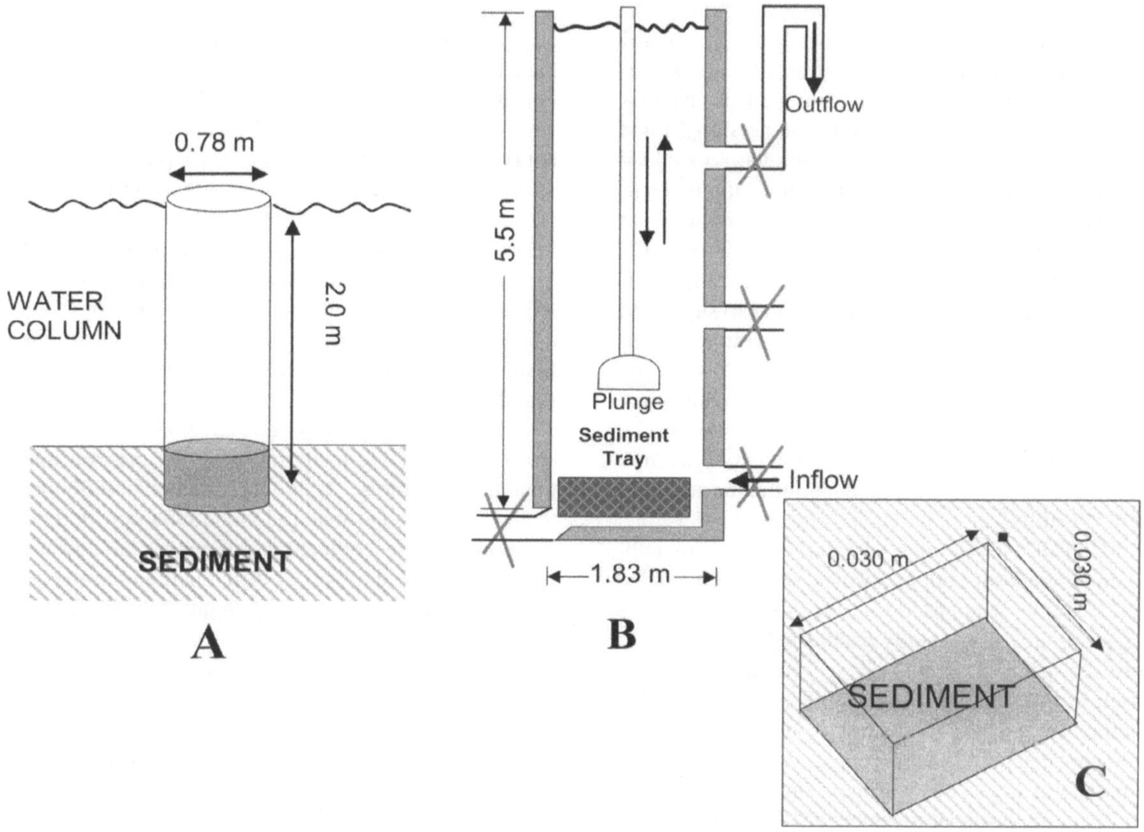

Figure 2. Different mesocosms/microcosms. (A) (modified from Diamond et al., 1990): a flexible, open-ended plastic tube allowing for vertical mixing inserted into the sediments. (B) (modified from Adler et al., 1980): from the Mesocosm Experimental Research Laboratory (MERL) at the University of Rhode Island, field-collected sediment overlain by static or flowing water that can be mixed using a plunger. (C) (modified from Warren et al., 1998): square plastic container which can be filled with sediment and placed into natural sediment at the bottom of a waterbody.

Verdonschot & Ter Braak (1994) experimentally manipulated oligochaete communities in constructed ditch mesocosms by applying chlorpyrifos or nutrients. Different ditches were used to study the effects of insecticide application *vs* the effects of eutrophication. Cylinders of substrate (6 cm depth, 10.4 cm diameter) were deployed so that colonization could occur, and were retrieved after 20-week periods (the mesocosms themselves were maintained for several years). The rate at which the oligochaete community colonized these cylinders was measured and compared between treatments and to the outside substratum. In addition to developing information on the effects of chlorpyrifos and eutrophication, this study indicated that such large-scale experiments or mesocosms require years to develop and stabilize before they can be effectively used. Such long time frames are almost certainly not as important for microcosms as for mesocosms. And, because of their small size, aquatic oligochaetes are particularly amenable to use in microcosms, though such use to date has been limited.

While megadrile earthworms have been used in studies comparing laboratory tests, mesocosms and field tests (Booth et al., 2000), no such detailed comparisons have been done with aquatic oligochaetes. In fact, most mesocosm and microcosm studies do not explicitly include aquatic oligochaetes. For instance, a major European research initiative comparing laboratory toxicity tests with microcosms, stream and pond mesocosms (Pascoe et al., 2000) did not include aquatic oligochaetes among the test species. Test taxa were Protozoa, Alga, Coelenterata, Rotifera, Crustacea and Insecta (*Chironomus tentans*). As previously noted, *C. tentans* is being used by many researchers for toxicity tests instead of aquatic oligochaetes. Stream mesocosms consisted of channels 5 m long divided into riffle and pool sections. Pond enclosures consisted of 1.25 m cylinders embedded in natural pond sediments and laboratory microcosms (1-l and 300-l vessels containing natural pond water and sediments). The results of this very large study (48 laboratory toxicity test methods and 58 different mesocosm/microcosm response criteria) suggested that "appropriately designed, relatively simple and inexpensive laboratory toxicity tests with a selection of test species are generally adequate, with small application factors, for predicting the environmental risk of polluting chemicals to freshwater ecosystems". Chemicals used were Cu, lindane, atrazine and 3,4-dichloroaniline.

ECETOC (1997) in a very large review of published microcosm and mesocosm studies concluded that a single standard design or set of designs is inappropriate. "Each study should be tailored to address the specific issues or data requirements that have arisen from earlier stages of testing". They found that microcosm and mesocosm test systems fell into the following categories: containing or using a natural assemblage of organisms and their environment (e.g. a whole pond); artificial assemblages in the laboratory; and, a mixture of natural and artificial assemblages, in the laboratory or field.

ECETOC (1997) suggest that sufficiently complex microcosms and mesocosms may in fact be "considered as realistic for the real world situation" and conclude that single-species toxicity tests with an application factor of about 8 are environmentally protective. Similarly, Maltby (1999) has suggested that "the effects of stressors on populations can be predicted from a knowledge of the effects of stressors on individual energy budgets".

In contrast, Landis et al. (2000) used microcosms (3-l laboratory containers with 16 species not including oligochaetes – algae, invertebrates, bacteria) to demonstrate the persistence of contaminant-induced effects over time. They note that patterns of impact can be detected "at concentrations up to 15 times lower than the experimentally derived single-species EC_{50}". Thus, they would not agree with the conclusions of ECETOC (1997) or Pascoe et al. (2000).

It is unfortunate that aquatic oligochaetes have not been used more extensively in microcosms and mesocosms. These and other bioturbating organisms will affect the bioavailability of sediment contaminants to other organisms as well as the movement of contaminants between the water and sediment compartments. For instance, Ciarelli et al. (2000) found that bioturbation by a marine amphipod increased fluoranthene uptake from sediments by a marine polychaete. Bioturbation by oligochaetes is a dynamic and powerful process (Karickhoff & Morris, 1985; Peterson et al., 1996; Wang & Matisoff, 1997; Mohanty et al., 1998).

Aquatic oligochaetes also demonstrate apparently unique interspecific interactions that have not been a focus of microcosm or mesocosm testing to date. Brinkhurst et al. (1972) and Chua & Brinkhurst (1973) found that mixed species cultures of *L. hoffmeisteri*, *T. tubifex* and *Peloscolex* (now *Quistadrilus*) *multisetosus* had better growth and lower respiration rates than pure cultures. They attributed this difference to feeding on fecal material containing concentrated bacterial cultures reflecting differences in bacterial removal and metabolism by different oligochaete species. Chapman et al. (1982c) found that respiration rates of *L. hoffmeisteri* and *T. tubifex* were reduced in mixed culture. They also found that toxicity (as measured by 96-h LC_{50} values) was also significantly reduced, sometimes several-fold. 96-h LC_{50} values for individual vs mixed species ranged as follows (all values in mg l^{-1}): NaPCP, 0.33 & 0. 38 vs 0.58; Hg, 0.14 & 0.18 vs 0.23; Cd, 0.17 & 0.32 vs 0.58. These reductions in toxicity have obvious implications for extrapolations between single-organism toxicity tests and the real world, which comprises mixed species.

The presence of aquatic oligochaetes can also affect the responses of other taxa. Reynoldson et al. (1994) showed that the presence of *T. tubifex* at concentrations simulating natural populations, reduced growth of *Chironomus riparius*, *Hyalella azteca* and *Hexagenia limbata* (Ephemeroptera) in sediment toxicity tests. They recommended removal of indigenous organisms before conducting toxicity testing because their presence "can confound the interpretation of toxicity test results". The reverse interpretation is also possible: that removing indigenous organisms reduces the ecological realism of the tests. Hence, the value of microcosms and mesocosms, particularly for reducing uncertainty in EcoRA.

Bioaccumulation tests

As many oligochaetes ingest sediment with associated contaminants, this makes them particularly appropriate for bioaccumulation studies (Klump et al., 1987).

Three freshwater species, a lumbriculid and two tubificids, have been used more than occasionally for bioaccumulation tests: *L. variegatus* (Schuytema et al., 1988; Nebeker et al., 1989; Carlson et al., 1991; Ankley et al., 1992, 1994a; Phipps et al., 1993; Harkey et al., 1994; Kukkonen & Landrum, 1994, 1995; Bremle & Ewald, 1995; West et al., 1997; Standley, 1997; Brunson et al., 1998; Mount et al., 1999; Fisher et al., 1999; Chapman et al., 1999); *T. tubifex* (Chapman et al., 1980; Oliver, 1984, 1987; Markwell et al., 1989; Egeler et al., 1997; Chapman et al., 1999); *L. hoffmeisteri* (Chapman et al., 1980; Oliver, 1984, 1987; Markwell et al., 1989; Egeler et al., 1997). In Europe, a draft guideline for bioaccumulation testing using *T. tubifex* (and other species: *L. variegatus* & *B. sowerbyi*) is being developed (Ph. Egeler, pers. com.).

The lumbriculid *L. variegatus* is easier to use for bioaccumulation tests compared to most tubificids because it is larger (though *T. tubifex* can reach similar individual biomass under optimal culture conditions) and thus more readily provides sufficient biomass for chemical analyses. Laboratory tests also appear to be reasonably predictive. Brunson et al. (1998) found a reasonable concordance between bioccumulation by *L. variegatus* in 28-d laboratory exposures with concentrations of the same contaminants (PAHs) in field-collected aquatic oligochaetes. Similar results have been reported by Ankley et al. (1992) for PCBs. These authors also found *L. variegatus* to be more useful and predictive of indigenous benthic invertebrate bioaccumulation (including primarily *Limnodrilus* spp. oligochaetes) than standardized tests using the fathead minnow, *Pimephales promelas*.

Interestingly, terrestrial oligochaetes have also been used to measure bioaccumulation in waterbodies. Mac et al. (1990) exposed *Lumbricus terrestris* to Detroit River sediments both in the laboratory and in cages in the field. They found it a useful surrogate, particularly because of the relatively large amount of tissue available for chemical analyses.

A major advantage to using aquatic oligochaetes for bioaccumulation measurements is the fact that many are sediment-ingesters that feed in a head-down position with their tails in the water column. These oligochaetes are exposed to contaminants through all possible exposure routes: directly from pore water, from ingestion of contaminated sediment, and from overlying waters. Ram & Gillett (1993) suggest that oligochaetes primarily take up chlorinated hydrocarbons from the overlying water and ingested sediments with minimal contributions from the pore water; however, this may not be true of all situations (e.g. all contaminants, all oligochaetes). Certainly, ingestion of contaminated sediment has been shown to be the major source of strongly particle-sorbed contaminants to other deposit-feeding organisms (Voparil & Mayer, 2000; Weston et al., 2000; Lee et al., 2000).

Ecological assessment

Aquatic oligochaete communities are affected by contaminant concentrations (Rosso et al., 1994; Bervoets et al., 1997; Lang, 1997; Prygiel et al., 2000) and environmental factors (e.g. temperature, oxygen, substrate – Reynoldson, 1987). Not all oligochaete species are equally tolerant to chemical contaminants (Chapman et al., 1982a,b). However, historic attempts to develop benthic community classification systems have consistently classified 'Oligochaeta' as tolerant (e.g. Schaeffer et al., 1985). For instance, Wildhaber & Schmitt (1996, 1998) used Lenat's (1993) tolerance values for aquatic oligochaetes (and other taxa) in their analyses of benthic community structure.

When aquatic oligochaetes are assessed as part of the benthos, without preconceived, generally incorrect ideas regarding their 'tolerance', they generally provide the most useful information for sediment classification. For example, Chapman et al. (1987), in one of the earliest Sediment Quality Triad studies, found that *Tubificoides brownae* was one of the most abundant taxa at their study stations and used this information as part of their benthic assessment. Other studies have used aquatic oligochaetes similarly (Watzin et al., 1997; Leslie et al., 1999). Reynoldson & Rodriguez (1999) point out the importance of aquatic oligochaetes, as an integral part of most sediment infaunal communities, and as a tool for the assessment and management of contaminated sediments. Numerous studies have used specific aquatic oligochaete worms as key parameters in benthic community assessments; in many cases, aquatic oligochaetes have, either with chironomids or alone, comprised the majority of the benthos (Canfield et al., 1994, 1996, 1998; Bailey et al., 1995; Day et al., 1995a,b; Besser et al., 1996; Swift et al., 1996). Other studies either have not identified oligochaetes beyond 'Oligochaeta' or possibly 'worms with hairs' and 'worms without hairs'. These studies have lost important benthic community structure information as a result.

Beyond the question of identifying aquatic oligochaetes to species, there is the problem of adequately sampling these relatively small organisms.

Many benthic studies use too large a sieve size when sorting benthic samples: ≥500 μm mesh sieves in freshwater (e.g. Malueg et al., 1984) and up to 1 mm sieves in marine waters and estuaries. These sieve sizes will not allow adequate representation of aquatic oligochaete communities (Barber & Kevern, 1974). Benthic studies need, at a minimum, to sieve samples using a 500 μm mesh, then a 250 μm mesh sieve to ensure adequate retention of aquatic oligochaetes.

Food chain studies

Tubificid oligochaetes have long been recognized as the preferred food of various fish species (Phillips & Buhler, 1978; Buddington & Doroshov, 1984). Efficient culturing and harvesting techniques have been developed for the fish food industry (Marian & Pandian, 1984; MacMichael et al., 1988). Most pet stores have live tubificids ('*Tubifex* worms') for sale as fish food, although most of these are not even tubificids but lumbriculids.

In the natural environment, oligochaetes (tubificids and naidids) are prey to leeches (Cross, 1976; Kaster, 1989), chironomids and other insect larvae (Loden, 1974; Thompson, 1978; Wisniewski, 1978), a wide variety of fish species in fresh, estuarine and marine waters (Strangenberg, 1956; Poddubnaya, 1962; Kennedy, 1969; Popchenko, 1971; Aarefjord et al., 1973; Arntz, 1978; Yaroshenko et al., 1980), diving ducks and probably other bottom-feeding waterfowl in freshwaters (Rofritz, 1977), snails (Levinton & Stewart, 1982), crabs and shorebirds on intertidal mudflats (Quammen, 1984), and shrimp (Wallace et al., 1998). Predation on bottom-dwelling tubificids which often feed in a head downward position with their tails waving in the water column to facilitate oxygen exchange (Leynen et al., 1999), can often involve cropping of those tails which subsequently regenerate. This may be an effective and unique means of contaminant passage to higher trophic levels. Lucan-Bouché et al. (1999) have shown that *T. tubifex* in areas contaminated by heavy metals preferentially accumulates heavy metals in its tail and may protect itself from metal toxicity by autotomy of the tail in a novel decontamination process. Cropping of oligochaete tails containing elevated contaminant concentrations may thus increase food chain transfer (while also decontaminating the worms).

Aquatic oligochaetes feed primarily (but not solely, e.g. some are carnivorous) on bacteria (Brinkhurst & Jamieson, 1971) and algae (Bowker et al., 1985; Kairesalo & Koskimis, 1987). The transfer of metals from bacteria to oligochaetes to fish in a simplified food chain has been demonstrated by Patrick & Loutit (1976, 1978). Dabrowska et al. (1996) built on work done by Rinderhagen & Butte (1995) and found that overall hexachlorobiphenyl tissue concentrations in channel catfish (*Ictalurus punctatus*) were significantly higher when the fish were fed *Tubifex* worms contaminated with hexachlorobiphenyl than when exposure was only via the water column. Egeler et al. (2000) demonstrated similar findings for sticklebak (*Gasterosterus aculeatus*) fed aquatic *T. tubifex* contaminated with hexachlorobenzene.

The efficiency of food chain transfer varies depending on whether the oligochaetes have evolved resistance to contaminants or not. Wallace et al. (1998) demonstrated that *L. hoffmeisteri* which has evolved resistance to Cd, transferred less Cd to shrimp eating them than was the case for non-resistant worms. Since the evolution of resistance to metals in sediments by oligochaetes has been demonstrated (Klerks & Levinton, 1989; Klerks & Bartholomew, 1991), this work has obvious implications to metals bioavailability and cycling within aquatic systems. Whether oligochaetes living in sediments contaminated with mixtures of organic and inorganic contaminants will show similar acclimation and food chain transfer of contaminants is, however, unknown. Recent work by Klerks (1999) with grass shrimp suggests that acclimation, at least, may become less likely as the number of contaminants increases. This may not be the case for evolved resistance, but this remains to be shown.

Although oligochaetes have great potential for toxicity testing, it is important to address the specific study objectives. As previously noted, chironomid larvae have been used by many researchers instead of aquatic oligochaetes for toxicity testing. Choices between chironomids and aquatic oligochaetes for food chain studies need to be made on the basis of the complete food chain; in other words, by determining which is the best choice relative to the questions being asked. There will be circumstances where, because chironomids have an adult winged form, they may be a more appropriate test organism. For instance, Reinhold et al. (1999) assessed the food chain risk from contaminated sediments to pond bats by studying food chain passage of contaminants from sediments to chironomids to bats (chironomid adults are the most important potential food source for pond bats).

Integrative assessments

Integrative assessments are a critical part of risk characterization, and involve a weight of evidence approach, often, for sediments, combining components such as chemistry, toxicity and resident community structure in the Sediment Quality Triad (Malueg et al., 1984; Chapman et al., 1987; Canfield et al., 1994, 1996, 1998; Besser et al., 1996; Wildhaber & Schmitt, 1996, 1998; Watzin et al., 1997; Ingersoll et al., 1997; Leslie et al., 1999).

Rodriguez & Reynoldson (1999) review single-species toxicity tests, microcosms and mesocosms for sediment bioassessment. Their key recommendations include: a battery of sensitive test species representing different trophic levels; combining acute and chronic endpoints, varying from organism-level to multi-species; and, integrating field benthic community data along with physical, chemical, and toxicity data.

Bailey et al. (1995) found that, in the North American Great Lakes, the 'best' taxa for assessing pollution were aquatic oligochaetes, and the 'best' toxicity test was *T. tubifex* reproduction (production of cocoons and young). Similarly, Day et al. (1995b) found that reproduction in *T. tubifex* was "strongly correlated with a degradation of the benthic community structure." However, these studies were conducted on the Canadian side of the Great Lakes. Many U.S. researchers in the Great Lakes, including the U.S. EPA's Assessment and Remediation of Contaminated Sediments (ARCS) Project, still do not include oligochaete toxicity tests in their test batteries (e.g. Canfield et al., 1996; Burton et al., 1996; Besser et al., 1996), even though aquatic oligochaete distributions are used in their benthic assessments.

A good example of the importance, and under-use, of aquatic oligochaetes for integrative assessments is provided by estuarine ecosystems, as reviewed by Chapman & Wang (2001). There are relatively few truly estuarine species of any taxa (i.e. species that live between about 5 and 20 g l^{-1} salinity), compared to freshwater and marine species. As a result, estuarine toxicity has generally been tested with freshwater or marine species, not estuarine species. This has meant that exposures have occurred not at the appropriate salinities but rather at adjusted salinities, which will affect contaminant bioavailability. Benthic assessments in estuaries have not considered seasonal movements up- and down-stream by benthic infauna responding to interstitial, not overlying water salinities (Chapman & Brinkhurst, 1981; Seys et al., 1999). Aquatic oligochaetes are abundant, and in some cases, dominate the benthic community in estuaries, and are therefore appropriate (though not generally used) for toxicity testing, as well as for infaunal analyses. The importance of aquatic oligochaetes in estuarine ecosystems has been well documented (Birtwell & Arthur, 1980; Diaz, 1980, 1989; McLusky et al., 1993; Seys et al., 1999). Brackish-water aquatic oligochaetes include but are certainly not restricted to: *Paranais* spp., *Tubificoides* spp. and *Heterochaeta* spp. (especially *H. costata*, which is widespread in Europe). Limited acute toxicity testing with *Paranais* spp. has been conducted (Table 2); much more is merited. In addition, chronic tests should be developed and integrated with predictive population models and community structure determinations. Such studies are also required for marine systems where few toxicity and no bioaccumulation studies have been conducted to date, using aquatic oligochaetes (Tables 2 and 3).

Summary – utility in EcoRA

The incorrect perception that aquatic oligochaetes are 'tolerant' to pollution, and thus not appropriately sensitive for use in EcoRA is still prevalent. Although the acute tolerances of many species are relatively high, tolerances are species-specific. Sublethal endpoints (behavior, growth, reproduction) are in fact very sensitive measures of pollution effects, and have been shown to affect not just individuals but also populations (Meller et al., 1998). Aquatic oligochaetes have a variety of uses in risk assessment as shown in Figure 3 and detailed below.

Problem formulation/hazard identification

This initial stage of an EcoRA involves the integration of available information. It may be triggered by an observed or possible ecological effect, for instance alteration of community structure or by the possibility of such alteration. Initial studies conducted as part of this stage can involve toxicity tests and/or benthic community studies. Aquatic oligochaetes can serve all of these functions. They have also served in the development of environmental quality values used in the problem formulation phase (e.g. *L. variegatus* has been used as part of sediment quality value development by the U.S. EPA – Ankley et al., 1991a; Hansen et al., 1996).

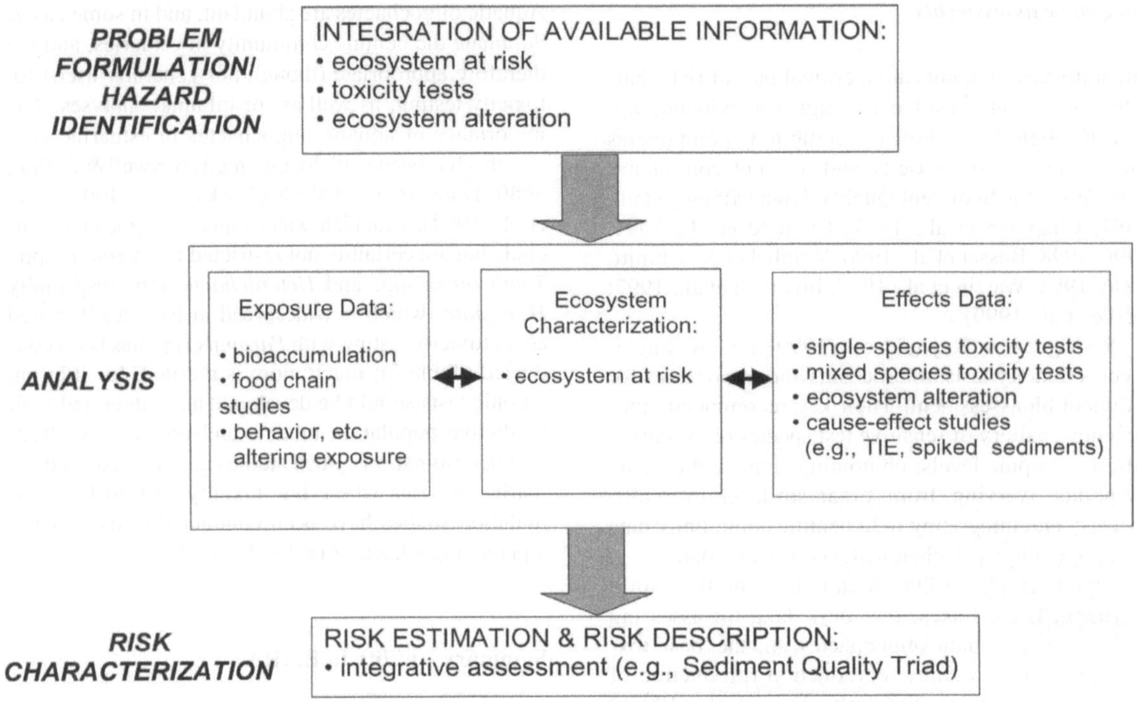

Figure 3. Roles of aquatic oligochaetes within the risk assessment framework.

Analysis

The analysis phase consists of exposure and effects characterization with concomitant ecosystem characterization. Exposure characterization can, in the specific case of aquatic oligochaetes, involve: bioaccumulation testing or analyses of resident species, food chain studies, and/or behaviors or other factors altering exposure of organisms to stressors (e.g. bioturbation, resistance to contaminants affecting predator uptake, concentration of contaminants in oligochaete tails which are cropped). Effects characterization can involve single-species toxicity tests, mixed species toxicity tests, community structure studies (to assess alteration), and/or cause-and-effect studies (e.g. toxicity identification evaluations [TIEs], spiked sediment studies). And, in many cases, aquatic oligochaetes will be a dominant feature of aquatic ecosystems.

Risk characterization

The final stage of an EcoRA involves a synthesis of all previous stages in an integrative assessment. Often, for sediments, this stage involves the Sediment Quality Triad where, as noted above, aquatic oligochaetes have often figured prominently.

Recommended future initiatives

The following future initiatives would render aquatic oligochaetes even more useful, and much more obviously so, in EcoRA.

Single species toxicity tests

Acute toxicity tests with *L. variegatus* are clearly useful and widely used; standardized chronic toxicity tests (reproduction, possibly also growth) with this species should be developed. Acute and chronic toxicity tests are also widely used with *T. tubifex*. Whether additional aquatic oligochaete toxicity tests are useful or necessary for freshwaters is uncertain; if *C. tentans* is a reasonable analogue there is no point in developing such tests for their own sake. However, if there are important freshwater ecosystems where aquatic oligochaete species would serve better as new test species, then it makes sense to develop such tests.

There presently is a paucity of truly estuarine toxicity test species; there is a real need and potential for developing estuarine oligochaete toxicity tests. A lesser but still real potential also exists in marine waters. It would be ideal to have the potential to conduct aquatic oligochaete acute and chronic tests from freshwater, through estuaries, and into marine waters. This has been done for amphipods and has resulted in their widespread usage (e.g. *Hyalella azteca* in freshwater; *Eohaustorius estuarius* in estuarine waters; *Rhepoxynius abronius* or *Ampelisca abdita* in marine waters).

Naidid species have not been generally used in toxicity tests; given their mainly asexual method of reproduction it is likely that useful population growth rate tests could be developed using these organisms. Similarly, the potential of behavioral tests for aquatic oligochaetes in general merits further investigation – the emphasis must of course be on ecological relevance (i.e. population-level effects).

Use as test species includes not only ecological assessment, but also investigations into causality (e.g. TIEs, spiked sediment tests). Where a test is to be developed, it is imperative that sound culture methods also be developed to supply test organisms.

Development of any new or modified aquatic oligochaete single-species test should conform with the guidelines set out by Giesy & Hoke (1989) and Ingersoll et al. (1995). In particular, the sensitivity, accuracy, reliability and ecological relevance of such tests need to be measured against existing tests to ensure that there are real advantages to the new tests (i.e. that new tests are not simply developed for their own sake). Such an approach has been reasonably well followed for earthworms (Motalib et al., 1997).

Mixed species toxicity tests

Given their relatively small size, aquatic oligochaetes are ideal organisms for use in microcosms or small mesocosms. Their use in such mixed species tests should be encouraged (e.g. Verdonschot & Ter Braak, 1994).

But in addition, and to maximize the value of such studies, research is needed to determine: how oligochaete bioturbation changes the exposure of other organisms, related to contaminant movement between environmental compartments (the latter of which has been well studied); how extensive and significant interspecific interactions are both between different aquatic oligochaete species, and between these worms and other taxa; and, the relationship between laboratory tests, microcosms/mesocosms and field populations of aquatic oligochaetes.

Bioaccumulation tests

The utility and relevance of *B. sowerbyi* for bioaccumulation testing (this is a large worm with relatively large biomass for analysis) should be determined. An alternative in freshwater systems is the use of caged terrestrial earthworms. It is not certain whether aquatic oligochaetes will be of use in estuarine and marine waters for bioaccumulation testing due to their relatively small size (and small biomass), but this is worth investigating. Certainly where aquatic oligochaetes are exposed to contaminants from sediments, pore waters and overlying waters, the advantage of these multiple exposure routes may outweigh the disadvantage of their small size.

Ecological assessment

Sediment, pore and overlying water exposure routes have been established for lake dwelling 'Oligochaeta' but specific determinations have not been confirmed, and this work has not been repeated with aquatic oligochaetes in estuarine or marine waters. Exposure routes for estuarine and marine oligochaetes remain to be determined (e.g. sediment vs pore water vs overlying water).

Investigators need to be convinced to use sufficiently small sieve sizes to representatively retain aquatic oligochaetes from sediments. They then need to be convinced to have such collections identified to species. In the latter regard, this could perhaps be addressed by the production of simple, up-to-date keys that allow technicians who are not oligochaete specialists to perform competent identifications of at least the dominant species.

Food chain studies

Studies should be conducted with aquatic oligochaetes to: more fully assess the effects of contaminant tolerance on food chain cycling including the occurrence of tolerance (e.g. is it also a feature of environments containing multiple, different contaminants?); develop and publish detailed conceptual diagrams indicating both contaminant exposure routes and predator–prey relationships; assess whether cropping of oligochaete tails which have concentrated contaminants is a meaningful food chain issue; and conduct food chain contaminant transfer studies (e.g. build on the work done several decades ago by Patrick & Loutit, 1976, 1978).

Integrated assessments

As modified from recommendations made by Rodriguez & Reynoldson (1999), integrated assessments should be built upon: a battery of sensitive toxicity tests representing different tropic levels and including aquatic oligochaetes where this is meaningful and realistic (such tests should also include acute and chronic end-points, and both single- and mixed-species tests); relevant bioaccumulation and food chain studies/data; and, the integration of field community data along with physical and chemical data and the above other two components. The ultimate goal is to develop useful and realistic predictive population models that are relevant to measured community structure changes in response to stressors.

Acknowledgements

I thank Pilar Rodriguez for inviting me to write this paper. Useful comments were provided by her, Phillip Egeler, Deedee Kathman, and Trefor Reynoldson.

References

Aarefjord, F., R. Borgstrom, L. Lien & G. Milbrink, 1973. Oligochaetes in the bottom fauna and stomach contents of trout, *Salmo trutta* (L.). Norw. J. Zool. 21: 281–288.

Achazi, R. K., G. Chrozcz, C. Düker, M. Henneken, B. Rothe, K. Schaub & I. Steudel, 1995. The effect of fluoranthene (Fla), benzo(a)pyrene (BAP) and cadmium (Cd) upon survival rate and life cycle parameters of two terrestrial annelids in laboratory test systems. Newslett. Enchytraeidae 4: 7–14.

Adler, D., M. Amdurer & P. H. Santschi, 1980. Metal tracers in two marine microcosms: sensitivity to scale and configuration. In Giesy, J. P. (ed.), Microcosms in Ecological Research. Conf-78-1101. U.S. Dept. of Energy, Washington, D.C.

Amon, H. U., 1985. Worm toxicity tests using *Tubifex tubifex*. FAO Workshop, Versailles. Colloq. INRA 31: 303–317.

Ankley, G. T., G. L. Phipps, E. N. Leonard, D. A. Benoit, V. R. Mattson, P. A. Kosian, A. M. Cotter, J. R. Dierkes, D. J. Hansen & J. D. Mahony, 1991a. Acid-volatile sulfide as a factor mediating cadmium and nickel bioavailability in contaminated sediments. Environ. Toxicol. Chem. 10: 1299–1307.

Ankley, G. T., M. K. Schubauer-Berigan & J. R. Dierkes, 1991b. Predicting the toxicity of bulk sediments to aquatic organisms using aqueous test fractions: pore water *versus* elutriate. Environ. Toxicol. Chem. 10: 1359–1366.

Ankley, G. T., P. M. Cook, A. R. Carlson, D. J. Call, J. A. Swenson, H. F. Corcoran & R. A. Hoke, 1992. Bioaccumulation of PCBs from sediment by oligochaetes and fishes: comparison of laboratory and field studies. Can. J. Fish. aquat. Sci. 49: 2080–2085.

Ankley, G. T., D. A. Benoit, R. A. Hoke, E. N. Leonard, C. W. West, G. L. Phipps, V. R. Mattson & L. A. Anderson, 1993. Development and evaluation of test methods for benthic invertebrates and sediments: effects of flow rate and feeding on water quality and exposure concentrations. Arch. envir. Toxicol. Chem. 25: 12–19.

Ankley, G. T., E. N. Leonard & V. R. Mattson, 1994a. Prediction of bioaccumulation of metals from contaminated sediments by the oligochaete, *Lumbriculus variegatus*. Wat. Res. 28: 1071–1076.

Ankley, G. T., D. A. Benoit, J. C. Balogh, T. B. Reynoldson, K. E. Day & R. A. Hoke, 1994b. Evaluation of potential confounding factors in sediment toxicity tests with three freshwater benthic invertebrates. Environ. Toxicol. Chem. 13: 627–635.

Arntz, W. E., 1978. Predation on benthos by flounder *Platichthus fleus* L. in the deeper parts of Keel Bay. Meeresforsching 26: 70–78.

ASTM, 1996. Standard test methods for measuring the toxicity of sediment-associated contaminants to fresh water invertebrates. ASTM Method E1706-95b. 1996 Annual Book of ASTM Standards, Volume 11.05. American Society for Testing and Materials, Philadelphia, PA.

Back, H., 1990. Edipermal uptake of Pb, Cd, and Zn in tubificid worms. Oecologia 85: 226–232.

Bailey, H. C. & D. H. W. Liu, 1980. *Lumbriculus variegatus*, a benthic oligochaete, as a bioassay organism. In Eaton, J. C., P. R. Parrish & A. C. Hendricks (eds), Aquatic Toxicology. ASTM STP 707, American Society for Testing and Materials, Philadelphia: 205–215.

Bailey, R. C., K. E. Day, R. H. Norris & T. B. Reynoldson, 1995. Macroinvertebrate community structure and sediment bioassay results from nearshore areas of North American Great Lakes. J. Great Lakes Res. 21: 42–52.

Barber, W. E. & N. R. Kevern, 1974. Seasonal variation of sieving efficiency in a lotic lake. Freshwat. Biol. 4: 293–300.

Bauer-Hilty, A., R. Dallinger & B. Berger, 1989. Isolation and partial characterization of a cadmium-binding protein from *Lumbriculus variegatus* (Oligochaeta, Annelida). Comp. Biochem. Physiol. 94C: 373–379.

Bervoets, L., R. Blust, M. de Wit & R. Verheyen, 1997. Relationships between river sediment characteristics and trace metal concentrations in tubificid worms and chironomid larvae. Envir. Pollut. 95: 345–356.

Besser, J. M., J. P. Giesy, J. A. Kubitz, D. A. Verbrugge, T. G. Coon & W. E. Braselton, 1996. Assessment of sediment quality in dredged and undredged areas of the Trenton Channel of the Detroit River, Michigan, U.S.A., using the Sediment Quality Triad. J. Great Lakes Res. 22: 683–696.

Bhunia, F., N. C. Saha & A. Kaviraj, 2000. Toxicity of thiocyanate to fish, plankton, worm and aquatic ecosystem. Bull. envir. Contam. Toxicol. 64: 197–204.

Birtwell, I. K. & D. R. Arthur, 1980. The ecology of tubificids in the Thames estuary with particular reference to *T. costatus* (Claparède). In Brinkhurst, R. O. & D. G. Cook (eds), Aquatic Oligochaete Biology. Plenum Press, New York: 175–184.

Booth, L. H., S. Hodge & K. O'Halloran, 2000. Use of cholinesterase in *Apporrectodea caliginosa* (Oligochaeta: Lumbricidae) to detect organophosphate contamination: comparison of laboratory tests, mesocosms, and field studies. Environ. Toxicol. Chem. 19: 417–422.

Bowker, D. W., M. T. Wareham & M. A. Learner, 1985. A choice chamber experiment on the selection of algae as food and substrate by *Nais elinguis* (Oligochaeta:Naididae). Freshwat. Biol. 15: 547–557.

Bremle, G. & G. Ewald, 1995. Bioconcentration of polychlorinated biphenyls (PCBs) in chironomid larvae, oligochaete worms and fish from contaminated lake sediment. Mar. Feshwat. Res. 46: 267–273.

Brinkhurst, R. O. & B. G. M. Jamieson, 1971. Aquatic Oligochaeta of the World. Oliver & Boyd, Edinburgh: 860 pp.

Brinkhurst, R. O., K. E. Chua & N. Kaushik, 1972. Interspecific interactions and selective feeding by tubificid oligochaetes. Limnol. Oceanogr. 17: 122–133.

Brinkhurst, R. O., P. M. Chapman & M. A. Farrell, 1983. A comparative study of respiration rates of some aquatic oligochaetes in relation to sublethal stress. Int. Rev. ges. Hydrobiol. 68: 683–699.

Brokovic-Popovic, I. & M. Popovic, 1977a. Effects of heavy metals on survival and respiration of tubificid worms: Part I – Effects on survival. Envir. Pollut. 13: 65–72.

Brokovic-Popovic, I. & M. Popovic, 1977b. Effects of heavy metals on survival and respiration of tubificid worms: Part I – Effects on respiration rate. Envir. Pollut. 13: 93–98.

Brunson, E. L., T. J. Canfield, C. J. Ingersoll & N. E. Kemble, 1998. Assessing the bioaccumulation of contaminants from sediments of the Upper Mississippi River using field-collected oligochaetes and laboratory-exposed *Lumbriculus variegatus*. Arch. Envir. Contam. Toxicol. 35: 191–201.

Buddington, R. K. & S. I. Doroshov, 1984. Feeding trials with hatchery produced white sturgeon (*Acipenser transmontanus*). Aquaculture 36: 237–243.

Burton, G. A. Jr., C. G. Ingersoll, L. C. Burnett, M. Henry, M. L. Hinman, S. J. Klaine, P. F. Landrum, P. Ross & M. Tuchman, 1996. A comparison of sediment toxicity methods at three Great Lake Areas of Concern. J. Great Lakes Res. 22: 495–511.

Canfield, T. J., N. E, Kemble, W. G. Brumbaugh, T. J. Dwyer, C. G. Ingersoll & J. F. Fairchild, 1994. Use of invertebrate community structure and the Sediment Quality Triad to evaluate metal-contaminated sediment in the Upper Clark Fork River, Montana. Environ. Toxicol. Chem. 13: 1999–2012.

Canfield, T. J., F. J. Dwyer, J. F. Fairchild, P. S. Haverland, C. G. Ingersoll, N. E. Kemble, D. R. Mount, T. W. La Point, G. A. Burton & M. C. Swift, 1996. Assessing contamination in Great Lakes sediments using benthic invertebrate communities and the Sediment Quality Triad approach. J. Great Lakes Res. 22: 565–583.

Canfield, T. J., E. L. Brunson, F. J. Dwyer, C. G. Ingersoll & N. E. Kemble, 1998. Assessing sediments from Upper Mississippi River navigational pools using a benthic invertebrate community evaluation and the Sediment Quality Triad approach. Arch. Envir. Contam. Toxicol. 35: 202–212.

Carlson, A. R., G. L. Phipps, V. R. Mattson, P. A. Kosian & A. M. Cotter, 1991. The role of acid-volatile sulfide in determining cadmium bioavailability and toxicity in freshwater sediments. Environ. Toxicol. Chem. 14: 1309–1319.

Casellato, S. & P. A. Negrisolo, 1989. Acute and chronic effects of an anionic surfactant on some freshwater tubificid species. Hydrobiologia 180: 243–252.

Casellato, S., R. Aiello, P. A. Negrisolo & M. Seno, 1992. Long-term experiment on *Branchiura sowerbyi* Beddard (Oligochaeta, Tubificidae) using sediment treated with LAS (Linear Alkylbenzene Sulphonate). Hydrobiologia 232: 169–173.

Chapman, P. M., 1987. Oligochaete respiration as a measure of sediment toxicity in Puget Sound, Washington. Hydrobiologia 155: 249–258.

Chapman, P. M. & R. O. Brinkhurst, 1981. Seasonal changes in interstitial salinities and seasonal movements of subtidal benthic invertebrates in the Fraser River, B.C. Estuar. coast. mar. Sci. 12: 49–66.

Chapman, P. M. & R. O. Brinkhurst, 1984. Lethal and sublethal tolerances of aquatic oligochaetes with reference to their use as a biotic index of pollution. Hydrobiologia 115: 139–144.

Chapman, P. M. & D. G. Mitchell, 1986. Acute tolerance tests with the oligochaete *Nais communis* (Naididae) and *Ilyodrilus frantzi* (Tubificidae). Hydrobiologia 155: 61–64.

Chapman, P. M. & F. Wang, 2001. Assessing sediment contamination in estuaries. Environ. Toxicol. Chem. 20: 3–22.

Chapman, P. M., L. M. Churchland, P. Thompson & E. Michnowsky, 1980. Heavy metal studies with oligochaetes. In Brinkhurst, R. O. & D. G. Cook (eds), Aquatic Oligochaete Biology. Plenum Press, New York: 477–502.

Chapman, P. M., M. A. Farrell & R. O. Brinkhurst, 1982a. Relative tolerances of selected aquatic oligochaetes to individual pollutants and environmental factors. Aquat. Toxicol. 2: 47–67.

Chapman, P. M., M. A. Farrell & R. O. Brinkhurst, 1982b. Relative tolerances of selected aquatic oligochaetes to combinations of pollutants and environmental factors. Aquat. Toxicol. 2: 69–78.

Chapman, P. M., M. A. Farrell & R. O. Brinkhurst, 1982c. Effects of species interactions on the survival and respiration of *Limnodrilus hoffmeisteri* and *Tubifex tubifex* (Oligochaeta:Tubificidae) exposed to various pollutants and environmental factors. Wat. Res. 16: 1405–1408.

Chapman, P. M., R. N. Dexter & E. R. Long, 1987. Synoptic measures of sediment contamination, toxicity and infaunal community composition (the Sediment Quality Triad) in San Francisco Bay. Mar. Ecol. Prog. Ser. 37: 75–96.

Chapman, K. K., M. J. Benton, R. O. Brinkhurst & P. R. Scheuerman, 1999. Use of the aquatic oligochaetes *Lumbriculus variegatus* and *Tubifex tubifex* for assessing the toxicity of copper and cadmium in a spiked-artificial-sediment toxicity test. Environ. Toxicol. 14: 271–278.

Chua, K. E. & R. O. Brinkhurst, 1973. Evidence of interspecific interactions in the respiration of tubificid oligochaetes. J. Fish. Res. Bd Can. 30: 617–622.

Ciarelli, S., B. J. Kater & N. M. Van Straalen, 2000. Influence of bioturbation by the amphipod *Corophium volutator* on fluoranthene uptake in the marine polychaete *Nereis virens*. Environ. Toxicol. Chem. 19: 1575–1581.

Coler, R. A., M. S. Coler & P. T. Kostecki, 1988. Tubificid behavior as a stress indicator. Wat. Res. 22: 263–267.

Connell, D. W., M. Bowman & D. W. Hawker, 1988. Bioconcentration of chlorinated hydrocarbons from sediment by oligochaetes. Ecotox. Environ. Safe. 16: 293–302.

Cross, W. H., 1976. A study of predation rates of leeches on tubificid worms under laboratory conditions. Ohio J. Sci. 76: 164–166.

D'Angelo, A. M. & G. Sinorile, 1974. Investigations on mercury localization in a freshwater oligochaete (*Tubifex tubifex*). Igiene. Mod. 66: 286–294.

Dabrowska, H., S. W. Fisher, K. Dabrowski & A. E. Staubus, 1996. Dietary uptake efficiency of HCBP in channel catfish: the effect of fish contaminant body burden. Environ. Toxicol. Chem. 15: 746–749.

Day, K. E., R. S. Kirby & T. B. Reynoldson, 1995a. The effect of manipulations of freshwater sediments on responses of benthic invertebrates in whole-sediment toxicity tests. Environ. Toxicol. Chem. 14: 1333–1343.

Day, K. E., B. J. Dutka, K. K. Kwan, N. Batista, T. B. Reynoldson & J. L. Metcalfe-Smith, 1995b. Correlations between solid-phase microbial screening assays, whole-sediment toxicity tests with macroinvertebrates, and *in situ* benthic community structure. J. Great Lakes Res. 21: 192–206.

Dean, J. M., 1974. The accumulation of ^{65}Zn and other radionuclides by tubificid worms. Hydrobiologia 45: 33–38.

Dermott, R. & M. Munawar, 1992. A simple and sensitive assay for evaluation of sediment toxicity using *Lumbriculus variegatus*. Hydrobiologia 235/236: 407–414.

Diamond, M. L., D. Mackay, R. J. Cornett & L. A. Chant, 1990. A model of the exchange of inorganic chemicals between water and sediments. Environ. Sci. Technol. 24: 713–722.

Diaz, R. J., 1980. Ecology of tidal freshwater and estuarine Tubificidae (Oligochaeta). In Brinkhurst, R. O. & D. G. Cook (eds), Aquatic Oligochaete Biology. Plenum Press, New York: 319–330.

Diaz, R. J., 1989. Pollution and tidal benthic communities of the James River estuary, Virginia. Hydrobiologia 180: 195–211.

Drewes, C. D., 1997. Sublethal effects of environmental toxicants on oligochaete escape reflexes. Am. Zool. 37: 346–353.

Dubilier, N., O. Giere & M. Grieshaber, 1994. Concomitant effects of sulfide and hypoxia on the aerobic metabolism of the marine oligochaete *Tubificoides benedii* to sulfidic sediments. J. exp. Zool. 269: 287–297.

Dubilier, N., O. Giere & M. Grieshaber, 1995. Morphological and ecophysiological adaptations of the marine oligochaete *Tubificoides benedii* to sulfidic sediments. Am. Zool. 35: 163–173.

Dubilier, N., R. Windoffer, M. Grieshaber & O. Giere, 1997. Ultrastructure and anaerobic metabolism of mitochondria in the marine oligochaete *Tubificoides benedii*: effects of hypoxia and sulfide. Mar. Biol. 127: 637–645.

Eaton, A. D., L. C. Clesceri & A. E. Greenberg, 1995. Standard Methods for the Examination of Water and Wastewater. 19th edn. American Public Health Association, Washington, DC: 1346 pp.

ECETOC, 1997. The value of aquatic model ecosystem studies in ecotoxicology. European Center for Ecotoxicology and Toxicology of Chemicals, Technical Report 73, Brussels: 112 pp.

Egeler, Ph., J. Römbke, M. Meller, T. Knacker, C. Franke, G. Studinger & R. Nagel, 1997. Bioaccumulation of lindane and hexachlorobenzene by tubificid sludgeworms (Oligochaeta) under standardised laboratory conditions. Chemosphere 35: 835–852.

Egeler, Ph., M. Meller, J. Roembke, P. Spoerlein, B. Streit & R. Nagel, 2000. Biomagnification of hexachlorobenzene and terbutryn in Stickleback (*Gasterosteus aculeatus*) from spiked sediment and contaminated invertebrates. Third Society of Environmental Toxicology & Chemistry (SETAC) World Congress, Brighton, U.K., May 21–25, 2000. Abstract Book: 156 pp.

Ewell, W. S., J. W. Gorsuch, R. O. Kringle, K. A. Robillard & R. C. Spiegel, 1986. Simultaneous evaluation of the acute effects of chemicals on seven aquatic species. Environ. Toxicol. Chem. 5: 831–840.

Fargasová, A., 1994. Toxicity of metals on *Daphnia magna* and *Tubifex tubifex*. Ecotox. Environ. Safe. 27: 210–213.

Fischer, E. & I. Horvàth, 1977. Cytochemical studies on the cuticle and epidermis of *Tubifex tubifex* Müll. with special regard to the localization of polysaccharides, heavy metals and DAB-reactivity. Histochemistry 54: 259–271.

Fisher, S. W., S. W. Chordas III & P. F. Landrum, 1999. Lethal and sublethal body residues for PCB intoxication in the oligochaete, *Lumbriculus variegatus*. Aquat. Toxicol. 45: 115–126.

Giere, O., J.-H. Preusse & N. Dubilier, 1999. *Tubificoides benedii* (Tubificidae, Oligochaeta) – a pioneer in hypoxic and sulfidic environments. An overview of adaptive pathways. Hydrobiologia 406: 235–241.

Giesy, J. P. & R. A. Hoke, 1989. Freshwater sediment toxicity bioassessment: Rationale for species selection and test design. J. Great Lakes Res. 15: 539–569.

Gunn, A. M., D. T. E. Hunt & D. A. Winnard, 1989. The effect of heavy metal speciation on bioavailability to tubificid worms. Hydrobiologia 188/189: 487–496.

Graefe, U., 1991. Ein Enchytraeentest zur Bestimmung der Säure- und Metalltoxizität im Boden. Mitt. Dt. Bodenkdl. Ges. 66: 487–490.

Hansen, D. J., W. J. Berry, J. D. Mahony, W. S. Boothman, D. M. DiToro, D. L. Robson, G. T. Ankley, D. Ma, Q. Yan & C. E. Pesch, 1996. Predicting the toxicity of metal-contaminated field sediments using interstitial concentration of metals and acid-volatile sulfide normalizations. Environ. Toxicol. Chem. 15: 2080–2094.

Harkey, G. A., P. F. Landrum & S. J. Klaine, 1994. Comparison of whole-sediment, elutriate and pore-water exposures for use in assessing sediment-associated organic contaminants in bioassays. Environ. Toxicol. Chem. 13: 1315–1329.

Hickey, C. W. & M. L. Martin, 1995. Relative sensitivity of five benthic invertebrate species to reference toxicants and resin-acid contaminated sediments. Environ. Toxicol. Chem. 14: 1401–1409.

Hill, R. A., P. M. Chapman, G. L. Mann & G. S. Lawrence, 2000. Level of detail in ecological risk assessments. Mar. Pollut. Bull. 40: 471–477.

Hynes, H. B. N., 1960. The Biology of Polluted Waters. Liverpool University Press, Liverpool U.K.

Ingersoll, C. G., G. T. Ankley, D. A. Benoit, G. A. Burton, F. J. Dwyer, I. F. Greer, T. J. Norberg-King & P. V. Winger, 1995. Toxicity and bioaccumulation of sediment-associated contaminants with freshwater invertebrates: a review of methods and applications. Environ. Toxicol. Chem. 14: 1885–1894.

Ingersoll, C. G., T. Dillon & G. R. Biddinger, 1997. Ecological Risk Assessment of Contaminated Sediments. SETAC Press, Pensacola, Fl.

Jones, J. R. E., 1938. Antagonism between two heavy metals in their toxic action on freshwater animals. Proc. zool. Soc. Lond. A: 481–497.

Kairesalo, T. & I. Kosimis, 1987. Grazing by oligochaetes and snails on epiphytes. Freshwat. Biol. 17: 317–324.

Karickhoff, S. W. & K. R. Morris, 1985. Impact of tubificid oligochaetes on pollutant transport in bottom sediments. Environ. Sci. Technol. 19: 51–56.

Kaster, J. L., 1989. Observations on predator–prey interaction on dispersal of an oligochaete prey, *Limnodrilus hoffmeisteri*. Hydrobiologia 180: 191–193.

Keilty, T. J., D. S. White & P. F. Landrum, 1988a. Short-term lethality and sediment avoidance assays with endrin-contaminated sediment and two oligochaetes from Lake Michigan. Arch. Envir. Contam. Toxicol. 17: 95–101.

Keilty, T. J., D. S. White & P. F. Landrum, 1988b. Sublethal responses to endrin in sediment by *Limnodrilus hoffmeisteri* (Tubificidae), and in mixed-culture with *Stylodrilus heringianus* (Lumbriculidae). Aquat. Toxicol. 13: 227–250.

Keilty, T. J., D. S. White & P. F. Landrum, 1988c. Sublethal responses to endrin in sediment by *Stylodrilus heringianus* (Lumbriculidae) as measured by ^{137}cesium marker layer technique. Aquat. Toxicol. 13: 251–270.

Kennedy, C. R., 1969. Tubificid oligochaetes as food for Dace *Leuciscus leuciscus* (L.). J. Fish. Biol. 1: 11–15.

Khangarot, B. S., 1991. Toxicity of metals to a freshwater tubificid worm, *Tubifex tubifex* (Müller). Bull. envir. Contam. Toxicol. 46: 906–912.

Klerks, P. L., 1999. Acclimation to pollutants by the grass shrimp *Palaemonetes pugio*: individual contaminants vs. mixtures. Ecotoxicology 8: 277–286.

Klerks, P. L. & J. S. Levinton, 1989. Rapid evolution of metal resistance in a benthic oligochaete inhabiting a metal-polluted site. Biol. Bull. 176: 135–141.

Klerks, P. L. & P. R. Bartholomew, 1991. Cadmium accumulation and detoxification in a Cd-resistant population of the oligochaete *Limnodrilus hoffmeisteri*. Aquat. Toxicol. 19: 97–112.

Klok, C. & A. M. de Roos, 1996. Population level consequences of toxicological influences on individual growth and reproduction in *Lumbricus rubellus* (Lumbricidae, Oligochaeta). Ecotox. Environ. Safe. 33: 118–127.

Klump, J. V., J. R. Krezoski, M. E. Smith & J. L. Kaster, 1987. Dual-tracer studies of the assimilation of an organic contaminant from sediments by deposit feeding oligochaetes. Can. J. Fish. aquat. Sci. 44: 1574–1583.

Kosian, P. A., E. A. Makynen, P. D. Monson, D. R. Mount, A. Spacie, O. G. Mekenyan & G. T. Ankley, 1998. Application of toxicity-based fractionation techniques and structure-activity relationship models for the identification of phototoxic polycyclic aromatic hydrocarbons in sediment pore water. Environ. Toxicol. Chem. 17: 1021–1033.

Kukkonen, J. & P. F. Landrum, 1994. Toxicokinetics and toxicity of sediment-associated pyrene to *Lumbriculus variegatus* (Oligochaeta). Environ. Toxicol. Chem. 13: 1457–1468.

Kukkonen, J. & P. F. Landrum, 1995. Effects of sediment-bound polydimethylsiloxane on the bioavailability and distribution of benzo[a]pyrene in lake sediment to *Lumbriculus variegatus*. Environ. Toxicol. Chem. 14: 523–531.

Landis, W. G., A. J. Markiewicz, R. A. Matthews & G. B. Matthews, 2000. A test of the community conditioning hypothesis: persistence of effects in model ecological structures dosed with the jet fuel JP-8. Environ. Toxicol. Chem. 19: 327–336.

Lang, C., 1997. Oligochaetes, organic sedimentation and trophic state: How to assess the biological recovery of sediments in lakes? Aquat. Sci. 59: 26–33.

Lee, B.-G., S. B. Griscom, J.-S. Lee, H. J. Choi, C.-H. Koh, S. N. Luoma & N. S. Fisher, 2000. Influences of dietary uptake and reactive sulfides on metal bioavailability from aquatic sediments. Science 287: 282–284.

Lenat, D. R., 1993. A biotic index for the Southeastern United States: derivation and list of tolerance values, with criteria for assigning water-quality ratings. J. n. am. Benthol. Soc. 12: 279–290.

Leppänen, M. T. & J. V. K. Kukkonen, 1998a. Relative importance of ingested sediment and pore water as bioaccumulation rates for pyrene to an oligochaete (*Lumbriculus variegatus*, Müller). Envir. Sci. Technol. 32: 1503–1508.

Leppänen, M. T. & J. V. K. Kukkonen, 1998b. Relationship between reproduction, sediment type and feeding of *Lumbriculus variegatus* (Müller): implications for sediment toxicity testing. Envir. Toxicol. Chem. 17: 2196–2202.

Leppänen, M. T. & J. V. K. Kukkonen, 1998c. Factors affecting feeding rate, reproduction and growth of an oligochaete *Lumbriculus variegatus* (Müller). Hydrobiologia 377: 183–194.

Leslie, H. A., T. I. Pavluk, A. bij de Vaante & M. H. S. Kraak, 1999. Triad assessment of the impact of chromium contamination on benthic macroinvertebrates in the Chusovaya River (Urals, Russia). Arch. Envir. Contam. Toxicol. 37: 182–189.

Levinton, J. S. & S. Stewart, 1982. Marine succession: The effect of two deposit-feeding gastropod species on the population growth of *Paranais litoralis* Müller 1784 (Oligochaeta). J. exp. mar. Biol. Ecol. 59: 231–241.

Leynen, M., T. Van den Berckt, JU. M. Aerts, B. Castelein, D. Berckmans & F. Ollevier, 1999. The use of Tubificidae in a biological early warning system. Envir. Pollut. 105: 151–154.

Loden, M. S., 1974. Predation by chironomid (Diptera) larvae on oligochaetes. Limnol. Oceanogr. 19: 156–159.

Lotufo, G. R. & J. W. Fleeger, 1996. Toxicity of sediment-associated pyrene and phenanthrene to *Limnodrilus hoffmeisteri* (Oligochaeta: Tubificidae). Environ. Toxicol. Chem. 15: 1508–1516.

Lucan-Bouché, M.-L., S. Biagianti-Risbourg, F. Arsac & G. Vernet, 1999. An original decontamination process developed by the aquatic oligochaete *Tubifex tubifex* exposed to copper and lead. Aquat. Toxicol. 45: 9–17.

Mac, M. J., G. E. Noguchi, R. J. Hesselberg, C. C. Edsall, J. A. Shoesmith & J. D. Bowker, 1990. A bioaccumulation bioassay for freshwater sediments. Environ. Toxicol. Chem. 9: 1405–1414.

MacMichael, G. J., L. R. Brown & C. M. Ladner, 1988. An oligochaete as a potential food source for fish in aquaculture. Prog. Fish. Cult. 50: 31–35.

Maltby, L., 1999. Studying stress: the importance of organism-level responses. Ecol. Appl. 9: 431–440.

Malueg, K. W., G. S. Schuytema, D. F. Krawczyk & J. H. Gakstatter, 1984. Laboratory sediment toxicity tests, sediment chemistry and distribution of benthic macroinvertebrates in sediments from the Keweenaw Waterway, Michigan. Environ. Toxicol. Chem. 3: 233–242.

Marchese, M. R. & R. O. Brinkhurst, 1996. A comparison of two tubificid species as candidates for sublethal bioassay tests relevant to subtropical and tropical regions. Hydrobiologia 334: 163–168.

Marian, M. P. & T. J. Pandian, 1984. Culture and harvesting techniques for *Tubifex tubifex*. Aquaculture 42: 303–315.

Markwell, R. D., D. W. Connell & A. J. Gabric, 1989. Bioaccumulation of lipophilic compounds from sediments by oligochaetes. Wat. Res. 23: 1443–1451.

Martínez-Madrid, M., P. Rodriguez, J. I. Perez-Iglesias & E. Navarro, 1999. Sediment toxicity bioassays for assessment of contaminated sites in the Nervion River (Northern Spain). 2. *Tubifex tubifex* (Müller) reproduction sediment bioassay. Ecotoxicology 8: 111–124.

McLusky, D. S., S. C. Hull & M. Elliott, 1993. Variations in the intertidal and subtidal macrofauna and sediments along a salinity gradient in the upper Forth estuary. Neth. J. aquat. Ecol. 27: 101–110.

McMurtry, M. J., 1984. Avoidance of sublethal doses of copper and zinc by tubificid oligochaetes. J. Great Lakes Res. 10: 267–272.

McPherson, C. A. & P. M. Chapman, 2000. Copper effects on potential sediment test organisms: the importance of appropriate sensitivity. Mar. Pollut. Bull. 40: 656–665.

Meller, M., Ph. Egeler, J. Römbke, I. I. Schallnass, R. Nagel & B. Streit, 1998. Short-term toxicity of lindane, hexachlorobenzene and copper sulfate to tubificid sludgeworms (Oligochaeta) in artificial media. Ecotox. Environ. Safe. 39: 10–20.

Milani, D., T. B. Reynoldson & J. Kolasa, 1998. The relative sensitivity of four benthic invertebrates to selected metals in water-only and in spiked sediment tests. 19th Annual Meeting of the Society of Environmental Toxicology and Chemistry, Charlotte, NC, November 15–19, 1998. Abstract Book: 204.

Milbrink, G., 1987. Biological characterization of sediments by standardized tubificid bioassays. Hydrobiologia 155: 267–275.

Mohanty, S., D. D. Reible, K. T. Valsaraj & L. J. Thiboreaux, 1998. A physical model for the simulation of bioturbation and its comparison to experiments with oligochaetes. Estuaries 21: 255–262.

Motalib, A., M. A. Rida & M. B. Bouche, 1997. Earthworm toxicity: from acute to chronic tests. Soil Biol. Biochem. 29: 699–703.

Mount, D. R., T. D. Dawson & L. P. Burkhard, 1999. Implications of gut purging for tissue residues determined in bioaccumulation testing of sediment with *Lumbriculus variegatus*. Environ. Toxicol. Chem. 18: 1244–1249.

Nebeker, A. V., W. L. Griffis, C. M. Wise, E. Hopkins & J. A. Barbitta, 1989. Survival, reproduction and bioconcentration in invertebrates and fish exposed to hexachlorobenzene. Environ. Toxicol. Chem. 8: 601–611.

Odum, E. P., 1984. The mesocosm. BioScience 34: 558–562.

Oliver, B. G., 1984. Uptake of chlorinated organics from anthropogenically contaminated sediments by oligochaete worms. Can. J. Fish. aquat. Sci. 41: 878–883.

Oliver, B. G., 1987. Biouptake of chlorinated hydrocarbons from laboratory-spiked and field sediments by oligochaete worms. Environ. Sci. Technol. 21: 785–790.

Pascoe, D., A. Wentzel, C. Janssen, A. E. Girling, I. Jüttner, A. Fliedner, S. J. Blockwell, S. J. Maund, E. J. Taylor, M. Diedrich, G. Persoone, P. Verhelst, R. R. Stephenson, N. O. Crossland, G. C. Mitchell, N. Pearson, L. Tattersfield, J.-P. Lay, A. Peither, B. Neumeier, & A.-R. Velletti, 2000. The development of toxicity tests for freshwater pollutants and their validation in stream and pond mesocosms. Wat. Res. 34: 2323–2329.

Patrick, F. M. & M. Loutit, 1976. Passage of metals, in effluents, through bacteria to higher organisms. Wat. Res. 10: 333–335.

Patrick, F. M. & M. Loutit, 1978. Passage of metals to freshwater fish from their food. Wat. Res. 12: 395–398.

Peterson, G. S., G. T. Ankley & E. N. Leonard, 1996. Effect of bioturbation on metal-sulfide oxidation in surficial freshwater sediments. Environ. Toxicol. Chem. 15: 2147–2155.

Phillips, G. R. & D. R. Buhler, 1978. Influences of dieldrin on the growth and body composition of fingerling rainbow trout (*Salmo gairdneri*) fed Oregon Moist Pellets or Tubificid worms (*Tubifex* sp.). J. Fish. Res. Bd Can. 36: 77–80.

Phipps, G. L., G. T. Ankley, D. A. Benoit & V. R. Mattson, 1993. Use of the aquatic oligochaete *Lumbriculus variegatus* for assessing the toxicity and bioaccumulation of sediment-associated contaminants. Environ. Toxicol. Chem. 12: 269–274.

Phipps, G. L., V. R. Mattson & G. T. Ankley, 1995. Relative sensitivity of three freshwater benthic macroinvertebrates to ten contaminants. Arch. envir. Contam. Toxicol. 28: 281–286.

Poddubnaya, T. L., 1962. The consumption of tubificids by fish. Vop Ikhtiol. 2: 560–562 (in Russian).

Popchenko, V. I., 1971. Consumption of Oligochaeta by fishes and invertebrates. J. Ichtyol. 11: 75–80.

Prygiel, J., A. Rosso-Darmet, M. Lafont, C. Lesniak, A. Durbec & B. Ouddane, 2000. Use of oligochaete communities for assessment of ecotoxicological risk in fine sediment of rivers and canals of the Artois-Picardie water basin (France). Hydrobiologia 410: 25–37.

Quammen, M. L., 1984. Predation by shorebirds, fish, and crabs on invertebrates in intertidal mudflats: an experimental test. Ecology 65: 529–537.

Ram, R. N. & J. W. Gillett, 1993. Comparison of alternative models for predicting the uptake of chlorinated hydrocarbons by oligochaetes. Ecotox. Environ. Safe. 26: 166–180.

Reinhold, J. O., A. J. Hendriks, L. K. Slager & M. Ohm. 1999. Transfer of microcontaminants from sediment to chironomids, and the risk for the Pond bat *Myotis dasycneme* (Chiroptera) preying on them. Aquat. Ecol. 33: 363–376.

Reynoldson, T. B., 1987. The role of environmental factors in the ecology of tubificid oligochaetes – an experimental study. Holarct. Ecol. 10: 241–248.

Reynoldson, T. B. & P. Rodriguez, 1999. Field methods and interpretation for sediment bioassessment. In Mudroch, A. M., J. M. Azcue & P. Mudroch (eds), Manual of Bioassessment of Aquatic Sediment Quality. Lewis Publishers, Boca Raton: 135–175.

Reynoldson, T. B., S. P. Thompson & J. L. Bamsey, 1991. A sediment bioassay using the tubificid oligochaete worm *Tubifex tubifex*. Environ. Toxicol. Chem. 10: 1061–1072.

Reynoldson, T. B., K. E. Day, C. Clarke & D. Milani, 1994. Effect of indigenous animals on chronic end points in freshwater sediment toxicity tests. Environ. Toxicol. Chem. 13: 973–977.

Rinderhagen, M. & W. Butte, 1995. Kinetics of accumulation and elimination of isomeric hexachlorocylohexanes by tubificids. SAR and QSAR in Environmental Research 4: 131–138.

Rodriguez, P. & T. B. Reynoldson, 1999. Laboratory methods and criteria for sediment bioassessment. In Mudroch, A. M., J. M. Azcue & P. Mudroch (eds), Manual of Bioassessment of Aquatic Sediment Quality. Lewis Publishers, Boca Raton: 83–133.

Rofritz, D. J., 1977. Oligochaeta as a winter food source for the Old Squaw. J. Wildlife Manage. 41: 590–591.

Rogge, R. W. & C. D. Drewes, 1993. Assessing sublethal neurotoxicity effects in the freshwater oligochaete, *Lumbriculus variegatus*. Aquat. Toxicol. 26: 73–90.

Römbke, J. & Th. Knacker, 1989. Aquatic toxicity test for Enchytraeids. Hydrobiologia 180: 235–242.

Rosso, A., M. Lafont & A. Exinger, 1994. Impact of heavy metals on benthic oligochaete communities in the river Ill and its tributaries. Wat. Sci. Technol. 29: 241–248.

Schaeffer, D. J., W. H. Ettinger, W. J. Tucker & H. W. Kerster, 1985. Evaluation of a community-based index using benthic indicator-organisms for classifying stream quality. J. Wat. Pollut. Cont. Fed. 57: 167–171.

Schubauer-Berigan, M. K., P. D. Monson, C. W. West & G. T. Ankley, 1995. The influence of pH on the toxicity of ammonia to *Chironomus tentans* and *Lumbriculus variegatus*. Environ. Toxicol. Chem. 14: 713–717.

Schuytema, G. S., D. F. Krawczyk, W. L. Griffis, A. V. Nebeker, M. L. Robideaux, B. J. Brownalee & J. C. Westall, 1988. Comparative uptake of hexachlorobenzene by fathead minnows, amphipods and oligochaete worms from water and sediment. Environ. Toxicol. Chem. 7: 1035–1045.

Seys, J., M. Vinex & P. Meire, 1999. Spatial distribution of oligochaetes (Clitellata) in the tidal freshwater and brackish parts of the Schelde estuary (Belgium). Hydrobiologia 406: 119–132.

Smith, D. P., J. H. Kennedy & K. L. Dickson, 1991. An evaluation of a naidid oligochaete as a toxicity test organism. Environ. Toxicol. Chem. 10: 1459–1465.

Standley, L. J., 1997. Effect of sedimentary organic matter composition on the partitioning and bioavailability of dieldrin to the oligochaete *Lumbriculus variegatus*. Environ. Sci. Technol. 31: 2577–2583.

Strangenberg, M., 1956. The growth and the summer food of the economically most important fishes of the Goplo Lake. Zool. Pol. 7: 63–120.

Swift, M. C., T. J. Canfield & T. W. LaPoint, 1996. Sampling benthic communities for sediment toxicity assessments using grab samplers and artificial substrates. J. Great Lakes Res. 22: 557–564.

Thompson, D. J., 1978. The natural prey of larvae of the damselfly, *Ischnura elegans* (Odonata: Zygoptera). Freshwat. Biol. 8: 377–384.

Thompson, K. A., D. A. Brown, P. M. Chapman & R. O. Brinkhurst, 1982. Histopathological effects and cadmium-binding protein synthesis in the marine oligochaete *Monopylephorus cuticulatus* following cadmium exposure. Trans. am. micros. Soc. 101: 10–26.

U.S. EPA, 1992. Framework for ecological risk assessment. U.S. Environmental Protection Agency, Office of Research and Development, Washington, DC. EPA/630/R-92/001.

U.S. EPA, 2000. Methods for measuring the toxicity and bioaccumulation of sediment-associated contaminants with freshwater invertebrates. U.S. Environmental Protection Agency, Office of Research and Development, Washington, DC. EPA/600/R-99/064.

Vecchi, M., T. B. Reynoldson, A. Pasteris & G. Bonomi, 1999. Toxicity of copper-spiked sediments to *Tubifex tubifex* (Oligochaeta, Tubificidae): comparison of the 28-day reproductive bioassay with an early-life-stage bioassay. Environ. Toxicol. Chem. 18: 1173–1179.

Verdonschot, P. F. M. & C. J. T. Ter Braak, 1994. An experimental manipulation of oligochaete communities in mesocosms treated with chlorpyrifos or nutrient additions: multivariate analyses with Monte Carlo permutation tests. Hydrobiologia 278: 251–266.

Voparil, I. M. & L. M. Mayer, 2000. Dissolution of sedimentary polycyclic aromatic hydrocarbons into the lugworm's (*Arenicola marina*) digestive fluids. Environ. Sci. Technol. 34: 1221–1228.

Wallace, W. G., G. R. Lopez & J. S. Levinton, 1998. Cadmium resistance in an oligochaete and its effect on cadmium trophic transfer to an omnivorous shrimp. Mar. Ecol. Prog. Ser. 172: 225–237.

Wang, X. & G. Matisoff, 1997. Solute transport in sediments by a large freshwater oligochaete, *Branchiura sowerbyi*. Environ. Sci. Technol. 31: 1926–1933.

Warren, L. A., A. Tessier & L. Hare, 1998. Modelling cadmium accumulation by benthic invertebrates *in situ*: the relative contributions of sediment and overlying water reservoirs to organism cadmium concentrations. Limnol. Oceanogr. 43: 1442–1454.

Watzin, M. C., A. W. McIntosh, E. A. Brown, R. Lacey, D. C. Lester, K. L. Newbrough & A. R. Williams, 1997. Assessing sediment quality in heterogeneous environments: a case study of a small urban harbor in Lake Champlain, Vermont, U.S.A. Environ. Toxicol. Chem. 16: 2125–2135.

West, C. W. & G. T. Ankley, 1998. A laboratory assay to assess avoidance of contaminated sediments by the freshwater oligochaete *Lumbriculus variegatus*. Arch. envir. Contam. Toxicol. 35: 20–24.

West, C. W., V. R. Mattson, E. N. Leonard, G. L. Phipps & G. T. Ankley, 1993. Comparison of the relative sensitivity of three benthic invertebrates to copper-contaminated sediments from the Keweenaw Waterway. Hydrobiologia 262: 57–63.

West, C. W., G. T. Ankley, J. W. Nichols, G. E. Elohen & D. E. Nessa, 1997. Toxicity and bioaccumulation of 2,3,7,8-tetrachloro-dibenzo-*p*-dioxin in long-term tests with the freshwater benthic invertebrates *Chironomus tentans* and *Lumbriculus variegatus*. Environ. Contam. Toxicol. 16: 1287–1294.

Weston, D. P., D. L. Penry & L. K. Gulmann, 2000. The role of ingestion as a route of contaminant bioaccumulation in a deposit-feeding polychaete. Arch. envir. Contam. Toxicol. 38: 446–454.

White, D. S. & T. J. Keilty, 1988. Burrowing avoidance assays of contaminated Detroit River sediments using the freshwater oligochaete, *Stylodrilus heringianus* (Lumbriculidae). Arch. envir. Contam. Toxicol. 17: 673–681.

Whiteman, F. W., G. T. Ankley, M. D. Kahl, D. M. Rau & M. D. Balcer, 1996. Evaluation of interstitial water as a route of exposure for ammonia in sediment tests with benthic macroinvertebrates. Environ. Toxicol. Chem. 15: 794–801.

Whitley, L. S., 1967. The resistance of tubificid worms to three common pollutants. Hydrobiologia 32: 193–205.

Wiederholm, T. & G. Dave, 1989. Toxicity of metal polluted sediments to *Daphnia magna* and *Tubifex tubifex*. Hydrobiologia 176/177: 411–417.

Wiederholm, T., A. M. Wiederholm & G. Milbrink, 1987. Bulk sediment bioassays with five species of freshwater oligochaetes. Wat. Air Soil Pollut. 36: 131–154.

Wildhaber, M. L. & C. J. Schmitt, 1996. Hazard ranking of contaminated sediments based on chemical analysis, laboratory toxicity tests and benthic community composition: prioritizing sites for remedial action. J. Great Lakes Res. 22: 639–652.

Wildhaber, M. L. & C. J. Schmitt, 1998. Indices of benthic community tolerance in contaminated Great Lakes sediments: relations with sediment contaminant concentrations, sediment toxicity and the Sediment Quality Triad. Environ. Monit. Assess. 49: 23–49.

Wisniewski, R. J., 1978. Effect of predators on Tubificidae groupings and their production in lakes. Ekol. pol. 26: 493–512.

Yaroshenko, M. F., O. I. Val'kovskaya & V. Kh. Chokyrlan, 1980. Fresh-water Oligochaeta and their importance as food for fish. In Aquatic Oligochaete Worms: Taxonomy, Ecology and Faunistic Studies in the U.S.S.R. Amerind Publ. Co., New Delhi: 191–196.

Hydrobiologia **463:** 171–184, 2001.
P. Rodriguez & P.F.M. Verdonschot (eds), Aquatic Oligochaete Biology VIII.

Tubifex tubifex as a link in food chain transfer of hexachlorobenzene from contaminated sediment to fish

Philipp Egeler[1], Michael Meller[1], Joerg Roembke[1], Peter Spoerlein[1], Bruno Streit[2] & Roland Nagel[3]
[1]*ECT Oekotoxikologie GmbH, Floersheim/Main, Germany*
Tel: 49 6145 9564 / 65. Fax: 49 6145 9564 / 99. E-mail: ph-egeler@ect.de
[2]*Johann-Wolfgang-Goethe-University, Frankfurt/Main, Germany*
[3]*Dresden University of Technology, Dresden, Germany*

Key words: tubificids, sediment, bioaccumulation, food chain, hexachlorobenzene

Abstract

Sediments contaminated with poorly water-soluble organic chemicals pose a risk to aquatic food chains. Sediment-associated chemicals can be accumulated by endobenthic, sediment-ingesting invertebrates. Some tubificid species – or other benthic annelids – serve as food for benthivorous fish, which thereby ingest the sediment-borne chemicals and may accumulate contaminant concentrations far higher than from water exposure only, and transfer them to organisms of higher trophic levels. For measurement of biomagnification, a sediment based food chain was developed and established in the laboratory. The two-step food chain included the sediment-dwelling freshwater oligochaete *Tubifex tubifex* (Müller) as a representative species of benthic infauna. The three-spined stickleback (*Gasterosteus aculeatus*, Linné), a small teleost fish which often feeds primarily on benthic invertebrates, served as a model predator. Spiked artificial sediment and reconstituted water as the overlying medium were used. Experiments were performed using ^{14}C-labelled hexachlorobenzene, a hydrophobic pollutant as a model compound. To examine the influence of benthic prey on the bioaccumulation of the test substance in the predator, fish were exposed to spiked water, spiked sediment, pre-contaminated prey organisms, or combinations of these exposure routes. The results of these experiments indicate that for hexachlorobenzene, the presence of contaminated *Tubifex tubifex* as a food source in combined exposure leads to significantly higher accumulation in fish than exposure to single pathways.

Introduction

Hydrophobic organic chemicals can adsorb to suspended particles and sediments. Endobenthic organisms are exposed to sediment-associated substances, and may subsequently bioaccumulate such compounds by ingestion of, and contact with contaminated sediment and porewater (e.g. Ingersoll et al., 1995). This bioaccumulation can lead to direct toxic effects, or to trophic transfer of the accumulated chemicals to predators (benthivorous fish) as reported on the basis of various laboratory studies, e.g. by Patrick & Loutit (1976, 1978), Clements et al. (1994), DiPinto & Coull (1997) and Sijm & Hermens (2000), as well as field investigations, (e.g. Russel et al. 1995): biomagnification and/or secondary poisoning can occur (e.g. Nendza et al., 1997).

This poses a risk for the aquatic environment and for piscivorous top consumers including humans (e.g. EC, 1996; Muir et al., 1999). Possible steps within the risk assessment of biomagnification before approval of such substances by government authorities may be based on physical-chemical properties (e.g. solubility in water and lipids, adsorptive properties, biodegradability, etc.) and subsequent modelling, and/or based on data from laboratory tests and field measurements.

Standardised laboratory test methods covering biomagnification of sediment-bound substances are currently not available. Therefore, the main task of this study was the development of such a test method for investigating accumulation of sediment-associated organic chemicals in sediment dwellers, including the subsequent transfer to organisms of higher trophic levels.

Among the multitude of models applied to estimate bioaccumulation and biomagnification, one approach utilises the fugacity of a chemical (for details see e.g. Mackay, 1991; Gobas et al., 1993, 1999; Campfens & Mackay, 1997). According to this approach, the knowledge of the degree of thermodynamic equilibrium between chemical concentrations in predator and prey may assist in an interpretation and assessment of experimentally derived biomagnification data.

Target substances which should be evaluated for biomagnification potential include persistent, bioaccumulative, organic chemicals, which associate with particles and sediments. These compounds often show a medium to high solubility in lipids or n-octanol (e.g. Biddinger & Gloss, 1984; Romijn et al., 1993; Russel et al., 1995; Fisk et al., 1998; Berglund et al., 2000), represented by the n-octanol-water-partition coefficient, K_{OW}. For the substances in question, the log K_{OW} ranges between between 4.0 and 6.0 (highly lipophilic). However, superlipophilic substances (log K_{OW} > 6.0), which are virtually insoluble in water, and substances from chemical classes of known bioaccumulation potential, e.g. surface-active or highly adsorptive substances, may also be candidates to be tested for biomagnification in the laboratory. Chemicals which are known to be mineralised rapidly, or degraded to compounds of low bioaccumulation potential, are assumed to be of lower tendency to bioaccumulate in biota, and may therefore be excluded from such a test procedure (Beek et al., 2000).

Representing freshwater benthic infauna worldwide, *Tubifex tubifex* (Müller) was chosen for the experiments as a trophic link between contaminated sediment and benthivorous fish. Due to their endobenthic lifestyle, these oligochaetes are in intimate contact with the solid phase and the pore water of the sediment. They ingest the substrate and utilise adhering micro-organisms and detrital material (e.g. Kosiorek, 1974; Patrick & Loutit, 1976). However, the main reason for using these organisms in this study is that *T. tubifex* and other aquatic oligochaetes – besides their conveyor-belt mode of pollutant transport in sediments (Karickhoff & Morris, 1985; Matisoff et al., 1999) – are well known to be an important prey item for benthic and pelagic predators and are, therefore, prone to transfer sediment-associated contaminants to higher trophic levels. *T. tubifex* has been utilised in both sediment toxicity and bioaccumulation studies (e.g. Oliver, 1984, 1987; Connell et al., 1988; Reynoldson et al., 1991; Rinderhagen & Butte, 1995; Egeler et al., 1997; Meller et al., 1998; Martinez-Madrid et al., 1999; Rodriguez & Reynoldson, 1999; Bouché et al., 2000).

In contrast to other possible sediment test organisms which require sophisticated culture conditions (e.g. *Chironomus riparius*, *Hyalella azteca*, as described in U.S. EPA, 2000), *T. tubifex* can be cultured and handled easily in the laboratory (Kosiorek, 1974; Reynoldson et al., 1991; Martinez-Madrid et al., 1999).

The three-spined stickleback (*Gasterosteus aculeatus* Linné) was selected as a model predator because of its feeding habits, its convenient size, and mainly because it occurs in freshwater as well as in estuarine and marine habitats of the northern hemisphere (Wootton, 1984; Sterba, 1987). This small fish feeds on small macroinvertebrates, with benthic organisms often comprising the main portion of its prey items (Wootton, 1984; Arnoult, 1986; Thiel et al., 1996). Sticklebacks, in turn, are preyed upon by larger fish, and by piscivorous top-predators like the European otter (Taastrøm & Jacobson, 1999). Furthermore, several ecotoxicological studies have been performed with this species (e.g. Ekelund et al., 1990; Holm, 1994; Allner, 1997), and it is recommended as a test species in standardised test guidelines (e.g. ASTM, 1988; Van den Dikkenberg et al., 1989; Environment Canada, 1990; OECD, 1992, 1996). Recent references (Froese & Pauly, 2000) report *Gasterosteus aculeatus aculeatus* as the currently valid species name according to Masuda et al. (1984). In order to avoid further confusion, this present paper will use the traditional species name, *Gasterosteus aculeatus*.

Hexachlorobenzene (HCB) is found in water bodies, mainly adsorbed to particulate organic matter and to sediments (e.g. BUA, 1994), and it can be detected in significant concentrations in piscivorous mammals of top trophic levels, e.g. dolphins and minks (e.g. Kuehl et al., 1991; Moore et al., 1997). This contaminant still poses a significant risk for the environment (e.g. CEPA, 1993; Tchounwou et al., 1998). Its lipophilicity (log K_{OW}: 5.73; De Bruijn et al., 1989) and its resistance to be metabolised in biota were the main reasons for using HCB as a model substance in this present study.

Table 1. Holding and breeding conditions for *Gasterosteus aculeatus*

Temperature	18±2 °C
Containers	glass aquaria 20 l, 40 l, 60 l, 250 l
Feeding	TetraMin®; *Artemia salina* (deep-frozen or alive)
Light regime	16 h light/8 h dark
Water type	reconstituted freshwater
Water renewal	once per week

Table 2. Nominal ^{14}C-HCB concentrations in spiked sediment and in spiked water

	dpm g^{-1} (ww)	dpm g^{-1} (dw)	$\mu g\ g^{-1}$ (ww)	$\mu g\ g^{-1}$ (dw)
Sediment	10 000	14 600	0.26	0.38
Water	≤195	–	≤0.005	–

Materials and methods

Sediment

Artificial sediment based on the artificial soil used for earthworm toxicity testing according to the OECD guideline 207 (OECD, 1984) as described by Egeler et al. (1997) and Meller et al. (1998) was used for the culture of *T. tubifex* and in the experiments. The main components are quartz sand (76% of sediment dry weight (dw)), kaolinite clay (22% of sediment dw) and peat powder (2% of sediment dw). The pH of the sediment was adjusted to 6 ± 0.5 with $CaCO_3$, and the water content was 46% of sediment dw.

Overlying water

Reconstituted freshwater for tests and culture was prepared and used according to the OECD guideline 203 (OECD, 1992). The freshly prepared water was aerated vigorously for at least 48 h before use in culture or tests. The reconstituted freshwater used for permanent culture of the sticklebacks was amended with 1% reconstituted seawater (resulting in a salinity of 0.77 g l^{-1}) using a commercially available sea salt blend (Tropic Marin, Dr Biener GmbH Aquarientechnik, Wartenberg, Germany) to compensate for micro-elements missing in the reconstituted freshwater alone.

Test organisms

T. tubifex was cultured in the laboratory as described by Egeler et al. (1997) using artificial sediment and reconstituted freshwater. For the experiments, adult *T. tubifex* (age 8–12 weeks) were used.

The original batch of sticklebacks (*Gasterosteus aculeatus*) was purchased from Aquarien Glaser, Rodgau-Weiskirchen, Germany. After an eight-week quarantine and acclimation period under laboratory conditions (see Table 1), breeding was started with mature animals.

For the experiments, juvenile sticklebacks (3–8 months old, approx.: 300–500 mg wet weight (ww) per fish) were separated from the culture tanks and weighed. According to OECD (1996), the test fish were chosen so that the smallest had at least 2/3 of the weight of the biggest animals. These fish were then randomly distributed to test containers and acclimated to test conditions for at least 2 weeks prior to the start of the exposure.

Chemicals, sample processing & radioanalysis

Hexachlorobenzene ([U-^{14}C-]HCB) was purchased from International Isotopes Munich (IIM, Unterschleißheim, Germany). Its specific radioactivity was 185 MBq $mmol^{-1}$ or 0.65 MBq mg^{-1}, respectively; the radiochemical purity was ≥97% as re-determined by high performance liquid chromatography (HPLC). ^{14}C-HCB was dissolved in cyclohexane (HPLC grade, Merck, Darmstadt, Germany) before use in the tests.

For bioconcentration testing in water only, ^{14}C-HCB was dissolved in acetone (p.a. grade, Roth, Karlsruhe, Germany), and thoroughly mixed with the water body of the test containers (<0.1 ml acetone l^{-1} test water). The sediment was spiked by mixing ^{14}C-HCB dissolved in cyclohexane with the quartz sand fraction of the sediment, giving a nominal HCB concentration of 0.38 $\mu g\ g^{-1}$ sediment dw (see Table 2). After complete evaporation of the solvent, the coated quartz sand was mixed with the other sediment components until the sediment was homogeneous. Then the total radioactive concentration was measured by 4 replicate samples to ensure homogeneous distribution of the test substance.

Total radioactive residues were measured by liquid scintillation counting (LSC; Tri Carb 2500 TR, Canberra Packard, Dreieich, Germany). Water samples were directly measured by LSC. Sediment samples were mixed with LSC cocktail (Ultima Gold, Packard, Dreieich, Germany), ultrasonicated, left to settle overnight, and measured by LSC. Biological samples were solubilised (Soluene 350, Canberra Packard, Dreieich, Germany) prior to LSC. All worm and fish samples were processed without gut purging, since predators are exposed to whole organisms including their gut content (e.g. ASTM, 2000). Samples containing potential non-extractable or bound residues were combusted in a sample oxidiser (TriCarb 307, Canberra Packard, Dreieich, Germany). The released $^{14}CO_2$ was trapped in an absorbent (Carbosorb, Canberra Packard, Dreieich, Germany), mixed with scintillation cocktail (PermaFluor, Canberra Packard, Dreieich, Germany) and measured by LSC. Radioactive concentrations were calculated by subtracting background radioactivity from the measured values and dividing the resulting disintegrations per minute (dpm) by the sample wet weight. All concentrations are reported on a wet weight basis unless otherwise indicated.

Samples of sediment, worms, fish and water were extracted with n-hexane followed by methanol-extraction, and analysed for the percentage of radioactivity associated with parent compound (PRP) by HPLC (Merck Hitachi, Darmstadt, Germany: normal phase silica column with n-hexane, UV-detection) and subsequent LSC analysis of fractioned and sampled eluates. In all extracts, PRP was higher than 98%, therefore the ^{14}C-based bioaccumulation factors (BAF) were not corrected for parent compound.

Dry weight, lipid content and organic carbon content (OC) were determined in subsamples of control sediment, worms and fish sampled at the end of the uptake phase. For measuring dry weight, the samples were heated at 65 °C in a drying cabinet until weight constancy was reached. The lipid content of worm and fish tissue was measured gravimetrically according to De Boer et al. (1999). A mixture of 20 ml cyclohexane and 16 ml 2-propanol was used to extract lipids from wet, homogenised tissue. Deionised water was added to the homogenate; the water volume to be added was calculated as follows:

$$x = 22 - \frac{\text{specimen ww [g]} \times \text{water content [\%]}}{100} \text{ [ml]}$$

After phase separation, the cyclohexanic phase was removed and volumetrically quantified. A second extraction of the water-2-propanol homogenate was performed with 20 ml of cyclohexane containing 13% 2-propanol (w/w). The resulting cyclohexanic phase was combined with the one resulting from the first extraction, the solvent was evaporated using a gentle nitrogen stream, and the residue was dried at 105±2 °C and weighed. The lipid content (F_{lipid}) was calculated according to:

$$\frac{\text{(lipid weight [g])} \times 100}{\text{specimen ww [g]}} = F_{\text{lipid}} \text{ in \% ww}$$

Worm lipid content (% of dry weight) was calculated by correcting the wet weight of the specimens used for lipid determination for the dw:ww ratio as outlined above.

The organic carbon content (OC) of the sediment was assessed according to DIN (1985). The dried sediment specimens were transferred to pre-heated, pre-weighed porcelain crucibles, and weighed. HCl (4 mol l^{-1}) was mixed with the specimens to remove carbonate-bound carbon. After 2–4 h, the specimens were dried overnight at 60±2 °C in a drying cabinet. The specimens were weighed after cooling in a desiccator for 1–2 h and ashed in a muffle furnace at 550±10 °C for 1–2 h. After cooling in a desiccator for 1–2 h, the specimens were weighed again. The weight loss (loss on ignition, LOI) in percent of dry weight of the sediment was calculated by subtracting the ash weight of the specimen from the dry weight of the specimen. The LOI was divided by the dry weight of the specimen, and multiplied by 100 to result in the LOI in percent of dry weight of the specimen. The LOI was divided by a factor of 1.72 to correct for volatile sediment components other than CO_2 to result in the OC in percent of dry weight of sediment.

Exposure

In order to examine the influence of different uptake pathways on the overall bioaccumulation in sticklebacks, uptake kinetics were examined resulting from exposure to different single and combined uptake pathways (Table 3). Acclimated fish were added to the test containers, and samples of four fish per sampling date were taken on at least 6 sampling dates. Uptake from water only (bioconcentration) was studied for 28 days using spiked water (semi static or recirculating system), while the fish were fed uncontaminated TetraMin®. A second scenario included spiked sediment, where the fish were fed uncontaminated TetraMin® to simulate sediment-influenced bioconcentration during a 63-day period. For a separate experiment (dietary exposure, biomagnification), *T. tubifex* were pre-exposed in spiked sediment. Worms

Table 3. Experimental exposure situations for bioaccumulation in fish

Exposure scenario	Spiked compartment(s)	Uptake pathway(s)	Duration
Bioconcentration	Water	Water	28 d
Bioconcentration	Sediment	Sediment, water incl. particulate matter	63 d
Biomagnification	Worms	Prey	60 d
Biomagnification/ Bioaccumulation (combined exposure)	Sediment, worms	Prey, sediment, water (incl. particulate matter)	63 d

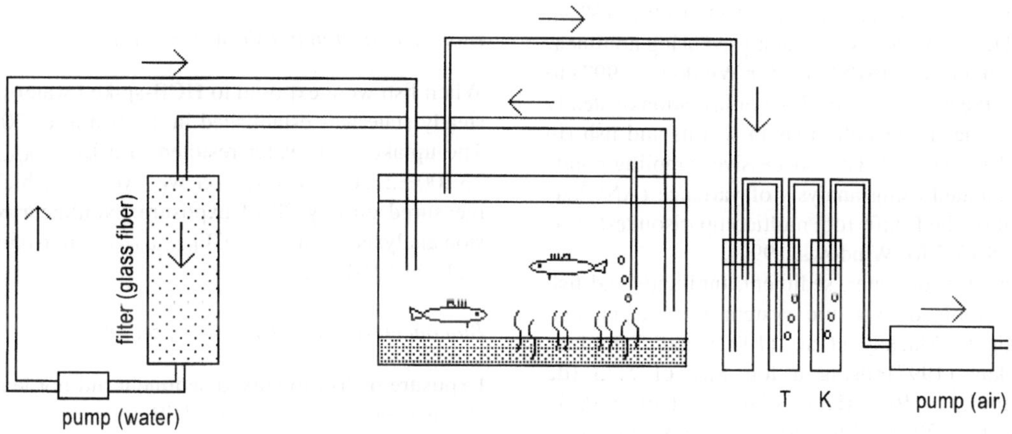

Figure 1. Combined exposure (bioaccumulation/biomagnification): test setup; T: triethylene-glycol, K: KOH.

were added to the test containers after a settling period of 24 h. Subsamples of the worms were analysed for total ^{14}C-activity until an equilibrium of uptake and elimination processes in the worms (steady state) had been established, i.e. within a period of 6–12 days. The fish were kept in uncontaminated water for 60 days and were fed pre-exposed, contaminated live *T. tubifex* twice a day. During this long-term exposure, the worms were fed once per week. Finally, the sticklebacks were kept in a system containing each of the potential uptake pathways, i.e. spiked sediment, contaminated water, and pre-exposed worms. In this combined exposure (see Fig. 1), bioaccumulation including bioconcentration and biomagnification processes was measured over a 63-day period. In the experiments containing spiked sediment, the overlying water was recirculated over glass wool filters and renewed once per week. All tests were conducted at 18 °C, under a light regime of 16 h light:8 h dark. In all experiments with sediment, the ^{14}C-concentration was determined in the sediment before the animals were added to the system in order to prevent excess resuspension of sediment as well as disturbing of the test organisms caused by sampling.

The freshly spiked sediment was distributed to the test containers, and carefully covered with the overlying water, avoiding excess resuspension of fine-grained sediment particles. After a 24-h settling period, weighed worms were introduced and pre-exposed until a steady state was established. Table 2 presents an overview of the nominal ^{14}C-HCB concentrations in the spiked compartments of the different experiments.

A system of trapping agents (triethylene glycol and potassium hydroxide solution, 10%) was used to retain volatilised ^{14}C-activity contained in the exhaust air of the test systems (Egeler et al., 1997).

Data analysis

For evaluation of the uptake kinetics, the wet weight based ^{14}C-concentrations in biota (C_{fish} or C_{worm}) were divided by the corresponding ^{14}C-concentration in the respective compartment (C_{water}, $C_{sediment}$, or C_{worm}) at each sampling date. The resulting ratio, termed the accumulation factor (AF) at each specific sampling date, was plotted against time. The term accumulation factor or AF is used for any time point of uptake when the steady state has not been reached, while the bioaccumulation factor (BAF) is defined for steady state conditions only (OECD, 1996; ASTM, 2000). These AF values were analysed using nonlinear regression analysis (SPSS 7.5 for Windows, 1997) to calculate the uptake curves. For confirmation of steady state, the mean concentrations in worms and fish (in dpm g^{-1} ww) of at least 3 successive sampling points were evaluated using analysis of variance (ANOVA) followed by the Bonferroni multicomparison test ($p <$ 0.05; SPSS 7.5 for Windows, 1997).

Fugacities for water, sediment, tubificids and fish in the combined exposure system were calculated according to Mackay et al. (1992) and Campfens & Mackay (1997) using a log K_{ow} of 5.73 (de Bruijn et al., 1989), a Henry's law constant of 10.32 Pa $m^3 mol^{-1}$ (20 °C; Montgomery, 1993), the HCB concentrations in the different media, the lipid content of organism tissue, and the organic carbon content of the sediment, respectively. Fugacities in the sediment were calculated assuming K_{oc} equals 0.41 K_{ow} (Karickhoff, 1981), and a density of sediment particles of 1.5 kg l^{-1}.

Results

Bioaccumulation in T. tubifex

Figure 2 shows the uptake kinetics of ^{14}C-HCB in *T. tubifex* exposed to 0.26 μg g^{-1} (nominal) sediment, assessed from data from a test container which was run in parallel, but separately from the combined exposure with fish. A steady state was reached within 10 d. The maximum $AF_{worm/sediment}$ was 10.8±1.2 (standard error, se) on day 19, and the $BAF_{worm/sediment}$ calculated from the kinetic parameters by nonlinear regression analysis was 7.8±0.4 (se).

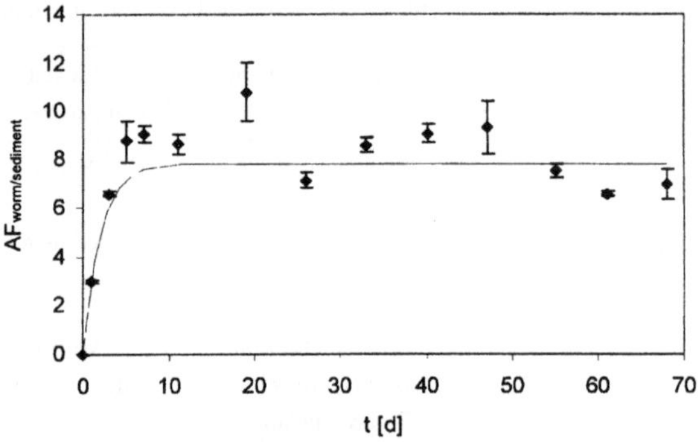

Figure 2. Bioaccumulation kinetics in *T. tubifex* in spiked sediment: accumulation factor (AF as ratio of C_{worm} and $C_{sediment}$; ww/ww; mean±se, n = 4–6); time measured in days (d).

Bioconcentration in fish (water only)

When fish were exposed to HCB-spiked water only, a steady state was established in the fish after 20 days. The uptake from water resulted in a $BCF_{fish/water}$ of 23 100±2100 (se) based on wet weight, which was measured on day 28 of exposure. Nonlinear regression analysis revealed a $BCF_{fish/water}$ of approximately 22 100±1250 (se).

Bioconcentration in fish (water, sediment)

Exposure of fish to spiked sediment did not lead to a steady state by day 63. The HCB concentration in the fish at the end of this experiment approached 3.1±0.2 (se) μg g^{-1} (ww). Using this value, an $AF_{fish/sediment}$ of 13±0.9 (se), and an $AF_{fish/water}$ of 36 800±2570 (se) was calculated.

Dietary exposure

During dietary exposure, separately pre-exposed *T. tubifex* were fed to the sticklebacks over a period of 60 d. The resulting linear increase of radioactive concentration in the fish is shown in Figure 3a, while the accumulation factors $AF_{fish/worm}$ are plotted against time in Figure 3b.

A steady state was not reached during the 60 d dietary uptake period. The ^{14}C-HCB concentration in the fish increased until the experiment was stopped. Based on radioactive residues in prey (ww), an $AF_{fish/worm}$ of 1.7±0.12 (se) was calculated for the end of the uptake phase by linear regression analysis. The $AF_{fish/worm}$ based on the day 60 samples was only 2.1±0.1 (se). Based on the mean ^{14}C-concentration in the prey, an $AF_{fish/worm}$ of 1.3±0.1 (se) can be determined.

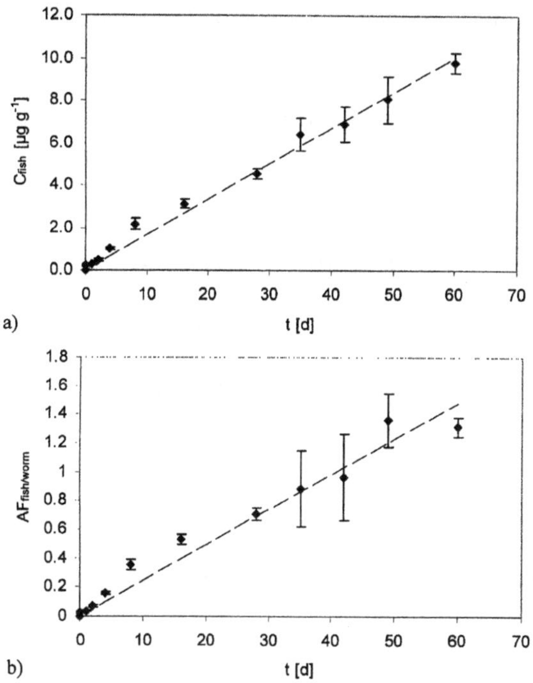

Figure 3. Dietary exposure: biomagnification kinetics of ^{14}C-HCB in *G. aculeatus* fed pre exposed *T. tubifex*; (a) HCB-concentration in fish (C_{fish}); (b) accumulation factors (AF as ratio of C_{fish} and C_{worm}; ww/ww; mean±se, n = 4–6); time measured in days (d).

Table 4. Observed HCB fugacities in the bioaccumulation scenario (combined exposure)

	Water	Sediment	Worm	Fish
Fugacity [μPa]	7.8	3.5	8.7	11.3

Combined exposure

Figure 4 shows the time course of uptake in stickleback under combined exposure. A steady state was established after 30 days; after that, the concentration in fish remained constant. Nonlinear regression analysis of the accumulation factors $AF_{fish/worm}$ over time indicated that 96% of the steady state had been reached by day 63 of exposure. Regression analysis revealed a $BAF_{fish/worm}$ of 3.5±0.3 (se) based on the ^{14}C-concentration in worms, which was reached after approximately 60 d. The ratio of C_{fish} and C_{worm} on day 63 resulted in a $BAF_{fish/worm}$ of approximately 3±0.3 (se) on a wet weight basis. The HCB fugacities observed in the bioaccumulation scenario (combined exposure) are listed in Table 4. In fish, the highest

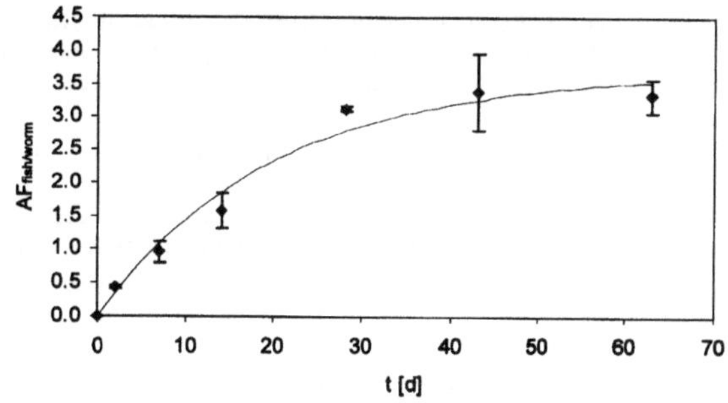

Figure 4. Combined exposure: Bioaccumulation kinetics of ^{14}C-HCB in *G. aculeatus*; accumulation factors (AF as ratio of C_{fish} and C_{worm}; ww/ww; mean±se, n = 4–6); time measured in days (d).

fugacity was found (11.3 μPa), while the fugacity in the sediment (3.5 μPa) showed the lowest value of all compartments.

In Table 5, the measured concentrations of ^{14}C-HCB in water, sediment, tubificids and fish in the different exposure scenarios are listed on a wet weight basis, and normalised for OC and lipid content as shown in Table 6. The accumulation factors in the sticklebacks determined under these exposure conditions are illustrated in Table 7.

Discussion

Several studies (e.g. Jepsen et al., 1995; Chen, 2000) reported that true equilibrium of HCB between sediment, porewater, and overlying water may take several days to months. The 24 h settling period before pre-exposure of the tubificids was not intended to generate equilibrium. However, a true equilibrium at this time of the uptake phase is not necessary for this specific purpose of the exposure (pre-exposure of worms until equilibrium). After the worms had been added to the system (at non-equilibrium conditions), samples of worms and sediment were taken until the concentration in the worms remained constant (i.e. after 6 – 12 days); by then, an equilibrium was assumed to have been established. During this long-term exposure (75 d including the 12 d pre-exposure) the worms were fed once weekly. Therefore, the variation in accumulation factors observed throughout the uptake phase may be due to changes of the lipid content of the worms during uptake. Nevertheless, the values of HCB uptake are in good accordance with previous findings, where a mean BAF of 6.5±1.7 (se) had been measured in tubificids under comparable exposure conditions (Egeler et al., 1997).

Table 5. Accumulation of ^{14}C-HCB in fish in different exposure scenarios; measured ^{14}C-concentrations (C) based on wet weight, and normalised for dry weight, lipid content and OC; mean values, se in brackets

		scenario:			
		Bioconcentration	Bioconcentration incl. sediment	Biomagnification	Bioaccumulation
C_{water}	[μg L^{-1}]	**1.796** (0.056)	**0.09** (0.01)	**0.054** (0.009)	**0.215** (0.009)
$C_{sediment}$	[μg g^{-1} ww]	–	**0.238** (0.003)	**0.266** (0.002)	**0.238** (0.003)
	[μg g^{-1} dw]	–	**0.348** (0.004)	**0.388** (0.002)	**0.348** (0.004)
	[μg g^{-1} OC]	–	**20.720** (0.261)	**23.158** (0.174)	**20.720** (0.261)
C_{worm}	[μg g^{-1} ww]	–	–	**7.48** (0.18)	**3.51** (0.12)
	[μg g^{-1} dw]	–	–	**53.42** (1.29)	**25.07** (0.86)
	[μg g^{-1} lipid]	–	–	**275.14** (6.62)	**129.11** (4.41)
C_{fish}	[μg g^{-1} ww]	**41.55** (3.9)	**3.08** (0.21)	**9.81** (0.49)	**11.14** (0.95)
	[μg g^{-1} dw]	**118.86** (11.16)	**8.81** (0.60)	**28.06** (1.40)	**31.87** (2.72)
	[μg g^{-1} lipid]	**628.92** (59.03)	**46.62** (3.18)	**148.49** (7.42)	**168.62** (14.38)

The $AF_{fish/water}$ in the presence of spiked sediment (36 800±2570) was considerably higher than the steady state BCF measured in water-only exposure (23 100±2100), even though in the study presented here a steady state was not reached during 63 d exposure to spiked sediment. This fact is interpreted as a consequence of the particulate organic matter present in the overlying water. In several studies, a negative correlation between accumulation of organic chemicals and the organic content in the test water was found (e.g. Neff, 1984; Landrum et al., 1987; Weinstein & Oris, 1999). However, in the studies presented here, accumulation in fish increased when particulate organic matter from spiked sediment was present in the overlying water. Similar findings were reported by Schuytema et al. (1988). These authors exposed fathead minnows to HCB-spiked sediment and measured significantly higher $BCF_{fish/water}$ values for HCB than in treatments where spiked water was the only available uptake route. Lores et al. (1993) found that the HCB body concentration in fish was not reduced by the presence of humic acids in water. Clements et al. (1994) investigated short-time bioaccumulation of benzo[a]pyrene and fluoranthene in chironomid larvae and fish and reported elevated fish tissue levels when fish were exposed directly to spiked sediment. Possible reasons of elevated accumulation in the presence

Table 6. Fractions of dry weight (dw), organic content (F_{oc}), and lipid content (F_{lipid}) in control sediment, fish and worms at the end of exposure; mean values, standard error (se) in brackets

	Sediment	Worm	Fish
dw/ww	0.685 (0.002)	0.140 (0.01)	0.350 (0.04)
F_{oc} of ww	0.011 (0.0001)	–	–
F_{oc} of dw	0.017 (0.0002)	–	–
F_{lipid} of ww	–	0.027 (0.008)	0.066 (0.026)
F_{lipid} of dw	–	0.197 (0.064)	0.192 (0.076)

of contaminated sediment are an intensive contact of the gills with contaminated particles, the enhanced desorption of HCB from suspended material due to a high exchange surface area, as well as ingestion of contaminated sediment by the fish (e.g. Kolok et al., 1996).

With field bioaccumulation data, the concentration in the biota is often simply divided by the concentration in water, sediment or prey (as performed in Table 7) to derive BCF, BAF or BMF, respectively, regardless of the actual source of the bioaccumulated contaminant. This makes sense, since in the 'real world', there is never only one source of accumulated chemicals. If comparable data from bioconcentration studies and from dietary accumulation studies are available, the relative contribution of the different uptake pathways to the overall bioaccumulation in fish exposed to multiple pathways may be estimated. In the following, this estimation will be calculated for HCB using the data from the different exposure scenarios.

The $BCF_{fish/water}$ from water-only exposure is multiplied by the HCB concentration in the overlying water of the sediment water (sw) exposure, $C_{water(sw)}$. The product is then assumed to be the portion of the concentration in the fish which results from uptake via contaminated water, ($C_{fish(water)(sw)}$):

$$C_{fish(water)(sw)} = BCF_{fish/water} \times C_{water(sw)}.$$

Here, the resulting $C_{fish(water)(sw)}$ is subtracted from the measured concentration in the fish to give the portion of the concentration in the fish which results from uptake via contaminated sediment, $C_{fish(sed)(sw)}$:

$$C_{fish(sed)(sw)} = C_{fish(measured)(sw)} - C_{fish(water)(sw)}.$$

The same procedure can be applied to assess the influence of contaminated water on overall fish body concentration in a combined exposure, $C_{water(ce)}$ exposure:

$$C_{fish(water)(ce)} = BCF_{fish/water} \times C_{water(ce)}.$$

The portion of the concentration in fish originating from dietary accumulation or biomagnification can be estimated in two ways. The $AF_{fish/worm(de)}$ from the pure dietary exposure scenario is multiplied by the concentration in the worms of the combined exposure to result in $C_{fish(worm)(ce)}$:

$$C_{fish(worm)(ce)} = AF_{fish/worm(de)} \times C_{worm(ce)}$$

Now each of these portions of the measured concentration in the fish can be divided by the concentration in the corresponding compartment of the combined exposure (water, sediment, worms) to give more realistic, source-based BAF values:

$$BAF_{fish/sed} = C_{fish(sed)(sw)} / C_{sed(sw)}$$

The source-based BAF values for the combined exposure are given in Table 8. When $C_{fish(water)(ce)}$ and $C_{fish(sed)(ce)}$ are known, $C_{fish(worm)(ce)}$ can also be calculated by subtracting ($C_{fish(water)(ce)} + C_{fish(sed)(ce)}$) from the measured concentration in the fish. The calculations outlined above lead to relative contributions of the uptake pathways (expressed in percent) to the overall body concentration in fish, which are also shown in Table 8.

These data show that for HCB in the combined exposure scenario uptake from ambient water (45%) and from diet (41%) contributed approximately equal portions to the overall body concentration measured in the fish. In comparison to these values, the presence of spiked sediment had a relatively weak influence on the stickleback's body concentration. However, when the dietary uptake route was not involved – as in the water–sediment scenario – the presence of spiked sediment considerably increased bioaccumulation in the fish compared to uptake from spiked water only. According to the calculations outlined above, the source-based accumulation factors were estimated to be 23 000 ($AF_{fish/water(ce)}$), 6.6 ($AF_{fish/sed(ce)}$) and 1.3 ($AF_{fish/worm(ce)}$) on a wet weight basis. Since this last accumulation factor is higher than 1, i.e. the concentration in the predator exceeds the concentration in the prey, it indicates biomagnification of the test compound via the dietary pathway.

When using the lipid-normalised concentrations in fish and worms (as recommended e.g. by Landrum & Fisher, 1998; Gobas et al., 1999) for the calculation of source-based AF, the lipid based $AF_{fish/worm(de)}$ of 0.53, which results is smaller than 1. Therefore, biomagnification is not evident in the pure dietary exposure scenario. Nevertheless, the accumulation factor derived from lipid normalised concentrations measured in fish under combined exposure, $AF_{fish/worm(ce)}$, has a value of 1.3 (see Table 7), which indicates biomagnification.

The biomagnification factors ($BMF_{fish/zooplankton}$) of PCB-congeners found by Berglund et al. (2000) in pelagic freshwater food chains were strongly correlated with the K_{ow}. PCBs with a log $K_{ow} > 6$ showed a log BMF > 0, while less lipophilic compounds were not biomagnified (dry weight- and lipid-normalised). The results found in our study are in accordance with Berglund et al. (2000), since the log K_{ow} of HCB approaches a value of 6 (5.73). Nevertheless, when comparing the strictly pelagic food chain investigated by Berglund et al. (2000) with the sediment-based food chain in the present paper, the additional exposure of fish to sediment borne chemical has to be considered. This additional pathway leads to higher fish body concentrations than dietary exposure (or dietary + water borne exposure) alone, resulting in a lipid normalised $BAF_{fish/worm(ce)}$ of 1.3.

In addition to the concentration-based calculations above, the results of the experiments were evaluated using the fugacities in the different compartments. The fugacity approach suggests that equal fugacities of a

Table 7. Accumulation of ^{14}C-HCB in fish in different exposure scenarios; measured accumulation factors (AF) based on wet weight, and normalised for dry weight, lipid content and OC; mean values, se in brackets

		Scenario:			
		Bioconcentration	Bioconcentration incl. sediment	Biomagnification	Bioaccumulation
AF$_{fish/water}$	(ww/ww)	**23 100** (2100)	**36 800** (2570)	–	**52 500** (4390)
	(dw/ww)	**66 080** (6 007)	**105 270** (7352)	–	**150 190** (12 559)
	(lipid/ww)	**349 650** (31 787)	**557 020** (38 901)	–	**794 660** (66 449)
AF$_{fish/sediment}$	(ww/ww)	–	**13** (0.9)	**37**[a] (1.8)	**47** (4.0)
	(dw/dw)	–	**25** (1.8)	**72**[a] (3.9)	**92** (7.8)
	(lipid/OC)	–	**2** (0.2)	**6**[a] (0.3)	**8** (0.7)
AF$_{fish/worm}$	(ww/ww)	–	–	**1.3** (0.1)	**3.2** (0.27)
	(dw/dw)	–	–	**0.5** (0.04)	**1.3** (0.11)
	(lipid/lipid)	–	–	**0.5** (0.04)	**1.3** (0.11)

[a] Calculated indirectly, sediment and fish in separate chambers.

Table 8. Accumulation of ^{14}C-HCB in fish (combined exposure) based on relative contribution of different uptake pathways (sources); concentrations based on wet weight; accumulation factors (AF) as C_{fish}/C_{source}; se in brackets

		C_{water}	$C_{sediment}$	C_{worm}
	C_{source} in $\mu g\ kg^{-1}$	0.215 (0.009)	238 (3)	3,586 (120)
	C_{fish}[a]	$C_{fish(water)}$	$C_{fish(sediment)}$	$C_{fish(worm)}$
In $\mu g\ g^{-1}$	11.1 (0.95)	5.0	1.6	4.6
In% of total C_{fish}	**100**	**45**	**14**	**41**
AF$_{fish/source(ce)}$		**23 100**	**6.6**	**1.3**

[a] Total ^{14}C-HCB per g fish (measured).

chemical in different media reflect a thermodynamic equilibrium (Morrison et al., 1996). A fugacities ratio of 1 indicates an equilibrium between two different media such as fish and tubificids, whereas a ratio greater than 1 implies that the concentration of HCB in fish is higher than the equilibrium concentration with the corresponding medium; a ratio less than 1 suggests that the concentration of chemical in fish has not reached its equilibrium concentration.

Inspection of the fugacities of HCB measured in the bioaccumulation-scenario (Table 4) allows an estimation of whether or not the HCB concentration in fish is in equilibrium with any of the single uptake routes. In the bioconcentration scenario including sediment (without worms), the ratio of HCB fugacities, $f_{sediment}/f_{water}$, was calculated to be 1.1, showing that a thermodynamic equilibrium between sediment and water had been established. The ratio ($f_{sediment}/f_{water}$) calculated from observed data in sediment and water of the combined exposure system including *T. tubifex*, was 0.4. Obviously, the concentration in the overlying water exceeded the equilibrium with the concentration in the sediment. This may be the result of the sediment-reworking activity of the worms (Matisoff et al., 1999), which induces a flux of HCB from sediment to the overlying water (Karickhoff & Morris, 1985).

In the combined exposure scenario, the ratio of HCB fugacities calculated for fish and water (f_{fish}/f_{water}) was 1.5, and 1.3 between fish and worms (f_{fish}/f_{worm}), respectively. This indicates that the HCB concentration in fish is near equilibrium with the concentration in water and worms. However, it is still slightly higher than a tissue concentration resulting exclusively from thermodynamic equilibrium with water or with food. The fugacities ratio of fish and sediment ($f_{fish}/f_{sediment}$ = 3.3) demonstrates that the HCB concentration in fish tissue is far from an equilibrium with sediment. These fugacity-based considerations are in close agreement with the percentages of relative contribution of the single contaminant sources as calculated on the basis of observed concentrations for the combined exposure system (see Table 8).

The findings derived from the experimental data, based on concentrations and fugacities suggest that the major uptake routes of HCB in fish are the overlying water and the food, whereas the contribution of the presence of sediment to the overall accumulation in fish is relatively small. Consequently, it is not possible to explain the bioaccumulation of HCB in stickleback with one single, significant exposure route. Therefore, sediment-influenced bioaccumulation, in particular chemical uptake from food, can not be easily predicted using the Equilibrium Partitioning (EqP) Model, which was evaluated by DiToro et al. (1991). In order to predict bioaccumulation and biomagnification in aquatic food chains, it is necessary to use models which differentiate between the various uptake pathways, such as the models described by Gobas (1993) and Campfens & Mackay (1997).

The results derived from this study were compared with predicted HCB accumulation in a benthic freshwater food chain using the 'Food Web Bioaccumulation Model' according to Gobas (1993), including the submodule for benthic bioaccumulation (Morrison et al., 1996), and the 'Fugacity-Based-Model' by Campfens & Mackay (1997). The predicted values were in good agreement with the accumulation factors found in the laboratory tests. The model by Campfens & Mackay (1997) generated stickleback accumulation factors lower than the laboratory derived values by a factor of approximately 1.4. The accumulation factors estimated by the Gobas (1993) model were lower than the data from our study by a factor of approximately 1.7, while the results derived from the combination of Gobas (1993) and Morrison et al. (1996) were in closer agreement (i.e. lower by a factor of approximately 1.1) with the experimental data presented in this paper. Further details regarding a comparison of observed and predicted HCB bioaccumulation in the described laboratory food chain are currently being investigated.

Conclusions

HCB was accumulated from contaminated tubificids in the sticklebacks by a factor of 1.3 and 0.5 (wet weight and lipid weight basis, respectively). Combined exposure of fish to HCB spiked sediment and contaminated *T. tubifex* revealed considerably higher bioaccumulation than exposure to single or two uptake routes alone, resulting in $BAF_{fish/worm(ce)}$ of 3 and 1.3 (wet weight and lipid weight basis, respectively). This indicates that HCB was biomagnified in the tested laboratory food chain. It could be demonstrated that contaminated water and prey contributed notably to the overall bioaccumulation of HCB in benthivorous fish.

The combined exposure approaches field conditions more closely than bioconcentration or dietary accumulation studies. Therefore, data obtained under combined exposure, including all possible uptake routes, may be interpreted as conservative with respect to an environmental risk assessment. If information on specific uptake routes (food, sediment, water) is required, single or two-way exposures should be performed additionally. However, the applicability of this test system needs to be confirmed for different substances, species, and/or sediments.

Application of the presented test systems may be considered when an Environmental Risk Assessment reveals the potential of a chemical to bioaccumulate by trophic transfer. The methods presented in this paper could then be used as an experimental tool to confirm or refute an estimated risk for biomagnification.

Acknowledgements

Financial support was kindly granted by the German Federal Environmental Agency (R+D-No. 296 28 308). The authors wish to thank the following persons: Carola Kussatz, Dieter Schudoma and Christian Franke (Federal Environmental Agency, Berlin, Germany) for constructive supervision of the project; Eric Bruns (ECT Oekotoxikologie GmbH) for proofreading the manuscript; two anonymous reviewers for inspiring comments on an earlier version of the manuscript; John P. Giesy and Kevin Henry (Michigan State University, East Lansing, Michigan, U.S.A.), Lawrence P. Burkhard, Dave R. Mount and David L. DeFoe (U.S. EPA, MED Duluth, Minnesota, USA) for stimulating discussions; Mandy Tauscher (ECT Oekotoxikologie GmbH) for taking care of the sludgeworms; Bernhard Allner (HLUG, Wiesbaden, Germany) for useful advice on breeding the sticklebacks; Jan Brodsky (CAU, Dreieich-Sprendlingen, Germany) and Veit Hultsch (Technische Universität Dresden, Germany) for technical advice with HPLC-techniques.

References

Allner, B., 1997. Toxikokinetik von 3,4-Dichloranilin beim dreistachligen Stichling (*Gasterosteus aculeatus*) unter besonderer Berücksichtigung der Fortpflanzungsphysiologie. Ph. D. Thesis, Johannes Gutenberg-University, Mainz, Germany: 137 pp.

Arnoult, J., 1986. Gasterosteidae. In Daget, J., J.-P. Gosse & D. F. E. Thys van den Audenaerde (eds), Check-list of the Freshwater Fishes of Africa (CLOFFA). ISNB, Brussels; MRAC, Tervuren; and ORSTOM, Paris. Vol. 2.: 280 pp.

ASTM, 1988. Standard guide for conducting acute toxicity tests with fishes, macroinvertebrates, and amphibians. American Society for Testing and Materials, E 729 - 88^{a}: 20 pp.

ASTM, 2000. Standard guide for determination of the bioaccumulation of sediment-associated contaminants by benthic invertebrates. American Society for Testing and Materials, E 1688 - 00a: 54 pp.

Beek, B., S. Boehling, U. Bruckmann, C. Franke, U. Joehncke & G. Studinger, 2000. The assessment of bioaccumulation. In Hutzinger, O. (ed.), The Handbook of Environmental Chemistry, Vol. 2 Part J (Vol. ed.: B. Beek): Bioaccumulation – New Aspects and Developments. Springer-Verlag, Berlin: 235–276.

Berglund, O., P. Larsson, G. Ewald & L. Okla, 2000. Bioaccumulation and differential partitioning of polychlorinated biphenyls in freshwater, planctonic foodwebs. Can. J. Fish. aquat. Sci. 57: 1160–1168.

Biddinger, G. R. & S. P. Gloss, 1984. The importance of trophic transfer in the bioaccumulation of chemical contaminants in aquatic ecosystems. Residue Reviews 91: 103–145.

Bouché, M.-L., F. Habets, S. Biagianti-Risbourg & G. Vernet, 2000. Toxic effects and bioaccumulation of cadmium in the aquatic oligochaete *Tubifex tubifex*. Ecotox. Environ. Safe. 46. 246–251.

BUA, 1994. Hexachlorbenzol. BUA-Stoffbericht 119. Hrsg.: Beratergremium für umweltrelevante Altstoffe (BUA) der Gesellschaft Deutscher Chemiker. Hirzel, Wissenschaftliche Verlagsgesellschaft, Stuttgart: 294 pp.

Campfens, J. & D. Mackay, 1997. Fugacity-based model of PCB bioaccumulation in complex aquatic foodwebs. Environ. Sci. Technol. 31: 577–583

CEPA, 1993. Canadian Environmental Protection Act. Priority substances list assessment report: Hexachlorobenzene. Canada Communication Group - Publishing, Ottawa, Canada, 52 pp.

Chen, W., A. T. Kan, G. Fu & M. B. Tomson, 2000. Factors affecting the release of hydrophobic organic contaminants from natural sediments. Environ. Toxicol. Chem. 19: 2401–2408.

Clements, W. H., J. T. Oris & T. E. Wissing, 1994. Accumulation and food chain transfer of fluoranthene and benzo[a]pyrene in *Chironomus riparius* and *Lepomis macrochirus*. Arch. Environ. Contam. Toxicol. 26: 261–266.

Connell, D. W., M. Bowman & D. W. Hawker, 1988. Bioconcentration of chlorinated hydrocarbons from sediment by oligochaetes. Ecotox. Environ. Saf. 16: 293–302.

De Boer, J., F. Smedes, D. Wells & A. Allan, 1999. Report on the QUASH interlaboratory study on the determination of total-lipid in fish and shellfish. Round 1 SBT-2. Exercise 1000. EU, Standards, Measurement and Testing Programme: 20 pp.

De Bruijn, J., F. Busser, W. Seinen & J. Hermens, 1989. Determination of octanol/water partition coefficients for hydrophobic organic chemicals with the 'slow-stirring' method. Environ. Toxicol. Chem. 8: 499–512.

DIN, 1985. DIN No. 38 414. Schlamm und Sedimente (Gruppe S), Teil 2: Bestimmung des Wassergehaltes und des Trockenrückstandes bzw. der Trockensubstanz (S2). Teil 3: Bestimmung des Glührückstandes und des Glühverlustes der Trockenmasse eines Schlammes (S3). In Fachgruppe Wasserchemie in der Gesellschaft Deutscher Chemiker (GDCh) (eds), Deutsche Einheitsverfahren (DEV) zur Wasser-, Abwasser- und Schlammuntersuchung. VCH, Weinheim, New York: 12 pp.

DiPinto, L. M. & B. C. Coull, 1997. Trophic transfer of sediment-associated polychlorinated biphenyls from meiobenthos to bottom-feeding fish. Environ. Toxicol. Chem. 16: 2568–2575.

DiToro, D. M., C. S. Zarba, D. J. Hansen, W. J. Berry, R. C. Swartz, C. E. Cowan, S. P. Pavlou, S. E. Allen, N. A. Thomas & P. R. Paquin, 1991. Technical basis for establishing sediment quality criteria for nonionic organic chemicals using equilibrium partitioning. Environ. Toxicol. Chem. 10: 1541–1583.

EC, 1996. Technical Guidance Documents in Support of Commission Directive 93/67/EEC on Risk Assessment for New Notified Substances and Commission Regulation (EC) No. 1488/94 on Risk Assessment for Existing Substances. Office for Official Publications of the EC (European Commission), Luxembourg; Part I – IV, 739 pp.

Egeler, Ph., J. Roembke, M. Meller, Th. Knacker, C. Franke, G. Studinger & R. Nagel, 1997. Bioaccumulation of lindane and hexachlorobenzene by tubificid sludgeworms (Oligochaeta) under standardised laboratory conditions. Chemosphere 35: 835–852.

Ekelund, R., Å. Bergman, Å. Granmo & M. Berggren, 1990. Bioaccumulation of 4-nonylphenol in marine animals – a re-evaluation. Environ. Pollut. 64: 107–120.

Environment Canada, 1990. Biological Test Method: Acute Lethality Test using the Threespine Stickleback (*Gasterosteus aculeatus*). Environmental Protection Series. Report EPS 1/RM/10: 45 pp.

Fisk, A. T., R. J. Norstrom, C. D. Cymbalisty & D. C. G. Muir, 1998. Dietary accumulation and depuration of hydrophobic organochlorines: bioaccumulation parameters and their relationship with the octanol/water partition coefficient. Environ. Toxicol. Chem. 17: 951–961.

Froese, R. & D. Pauly (eds), 2000. FishBase 2000. World Wide Web electronic publication. www.fishbase.org, as modified Sept. 15, 2000.

Gobas, F. A. P. C., 1993. A model for predicting the bioaccumulation of hydrophobic organic chemicals in aquatic food webs: application to Lake Ontario. Ecol. Model. 69: 1–17.

Gobas, F. A. P. C., J. R. McCorquodale & G. D. Haffner, 1993. Intestinal absorption and biomagnification of organochlorines. Environ. Toxicol. Chem. 12: 567–576.

Gobas, F. A. P. C., J. B. Wilcockson, R. W. Russel & D. D. Haffner, 1999. Mechanism of biomagnification in fishes under laboratory and field conditions. Environ. Sci. Technol. 33: 133–141.

Holm, G., 1994. The three-spined stickleback in ecotoxicological test systems. Doctoral Dissertation, Stockholm University: 141 pp.

Ingersoll, C. G., G. T. Ankley, D. A. Benoit, E. L. Brunson, G. A. Burton, F. J. Dwyer, R. A. Hoke, P. F. Landrum, T. J. Norberg-King & P. V. Winger, 1995. Toxicity and bioaccumulation of sediment-associated contaminants using freshwater invertebrates: a review of methods and applications. Environ. Toxicol. Chem. 14: 1885–1894.

Jepsen, R., S. Borglin, W. Lick & D. L. Swackhamer, 1995. Parameters affecting the adsorption of hexachlorobenzene to natural sediments. Environ. Toxicol. Chem. 14: 1487–1497.

Karickhoff, S. W., 1981. Semi-empirical estimation of sorption of hydrophobic pollutants on natural sediments and soils. Chemosphere 10: 833–846.

Karickhoff, S. W. & K. R. Morris, 1985. Impact of tubificid oligochaetes on pollutant transport in bottom sediments. Environ. Sci. Technol. 19: 51–56.

Kolok, A. S, K. J. Groetsch & J. T. Oris, 1996. The role of water ventilation and sediment ingestion on the uptake of hexachlorobenzene by Gizzard Shad (*Dorosoma cepedianum*). Environ. Toxicol. Chem. 15: 1760–1762.

Kosiorek, D., 1974. Development cycle of *Tubifex tubifex* Müll. in experimental culture. Pol. Arch. Hydrobiol. 21: 411–422.

Kuehl, D. W., R. Haebler & C. Potter, 1991. Chemical residues in dolphins from the U.S. Atlantic coast including Atlantic Bottlenose obtained during the 1987/88 mass mortality. Chemosphere 22: 1071–1084.

Landrum, P. F., S. R. Nihart, B. J. Eadie & L. R. Herche, 1987. Reduction in bioavailability of organic contaminants to the amphipod *Pontoporeia hoyi* by dissolved organic matter of sediment interstitial waters. Environ. Toxicol. Chem. 6: 11–20.

Landrum, P. F. & S. W. Fisher, 1998. Influence of lipids on the bioaccumulation and trophic transfer of organic contaminants in aquatic organisms. In Arts, M. & B. Wainman (eds), Lipids in Freshwater Ecosystems. Springer-Verlag, New York: 203–234.

Lores, E. M., J. M. Patrick & J. K. Summers, 1993. Humic acid effects on uptake of hexachlorobenzene and hexachlorobiphenyl by sheepshead minnows in static sediment/water systems. Environ. Toxicol. Chem. 12: 541–550.

Mackay, D., 1991. Multimedia environmental models – The fugacity approach. Lewis Publishers, Inc., Chelsea, Michigan: 257 pp.

Martinez–Madrid, M., P. Rodriguez, J. I. Perez–Iglesias & E. Navarro, 1999. Sediment toxicity bioassays for assessment of contaminated sites in the Nervion river (Northern Spain). 2. *Tubifex tubifex* (Müller) reproduction sediment bioassay. Ecotoxicology 8: 111–124.

Masuda, H., K. Amaoka, C. Araga, T. Uyeno & T. Yoshino, 1984. The Fishes of the Japanese Archipelago. Vol. 1. Tokai University Press, Tokyo: 437 pp.

Matisoff, G., X. Wang, & P. L. McCall, 1999. Biological redistribution of lake sediments by tubificid oligochaetes: *Branchiura sowerbyi* and *Limnodrilus hoffmeisteri/Tubifex tubifex*. J. Great Lakes Res. 25: 205–209.

Meller, M., Ph. Egeler, J. Roembke, H. Schallnass, R. Nagel & B. Streit, 1998. Short-term toxicity of lindane, hexachlorobenzene and copper sulfate to tubificid sludgeworms (Oligochaeta) in artificial media. Ecotox. Environ. Safe. 39: 10–20.

Montgomery, J. H., 1993. Agrochemical Desk Reference: Environmental Data. Lewis Publishers, Boca Raton: 625 pp.

Moore, D. R. J., R. L. Breton & K. Lloyd, 1997. The effects of hexachlorobenzene on mink in the Canadian environment: an ecological risk assessment. Environ. Toxicol. Chem. 16: 1042–1050.

Morrison, H. A., F. A. P. C. Gobas, R. Lazar & G. D. Haffner, 1996. Development and verification of a bioaccumulation model for organic contaminants in benthic invertebrates. Environ. Sci. Technol. 30: 3377–3384.

Muir, D., B. Braune, B. DeMarch, R. Norstrom, R. Wagemann, L. Lockhart, B. Hargrave, D. Bright, R. Addison, J. Payne, K. & Reimer, 1999. Spatial and temporal trends of contaminants in the Canadian Arctic marine ecosystem: a review. Sci. Total Environ. 230: 83–144.

Neff, J. M., 1984. Bioaccumulation of organic micropollutants from sediments and suspended particles by aquatic animals. Fresenius Z. Anal. Chem. 319: 132–136.

Nendza, M., T. Herbst, C. Kussatz & A. Gies, 1997. Potential for secondary poisoning and biomagnification in marine organisms. Chemosphere 35: 1875–1885.

OECD, 1984. Guidelines for Testing of Chemicals No. 207. Earthworm, Acute Toxicity Test. Organisation for Economic Cooperation and Development (OECD), Paris: 9 pp.

OECD, 1992. Guidelines for Testing of Chemicals No. 203. Fish, Acute Toxicity Test. OECD, Paris: 9 pp.

OECD, 1996. Guidelines for Testing of Chemicals No. 305. Bioconcentration: Flow-through Fish Test. OECD, Paris: 23 pp.

Oliver, B. G., 1984. Uptake of chlorinated organics from anthropogenically contaminated sediments by oligochaete worms. Can. J. Fish. aquat. Sci. 41: 878–883.

Oliver, B. G., 1987. Biouptake of chlorinated hydrocarbons from laboratory-spiked and field sediments by oligochaete worms. Environ. Sci. Technol. 21: 785–790.

Patrick, F. M. & M. Loutit, 1976. Passage of metals in effluents, through bacteria to higher organisms. Wat. Res. 10: 333–335.

Patrick, F. M. & M. Loutit, 1978. Passage of metals to freshwater fish from their food. Wat. Res. 12: 395–398.

Reynoldson, T. B., S. P. Thompson & J. L. Bamsey, 1991. A sediment bioassay using the tubificid oligochaete worm *Tubifex tubifex*. Environ. Toxicol. Chem. 10: 1061–1072.

Rinderhagen, M. & W. Butte, 1995. Kinetics of accumulation and elimination of isomeric hexachlorocyclohexanes by tubificids. SAS and QSAR in Environmental Research 4: 131–138.

Rodriguez, P. & T. B. Reynoldson, 1999. Laboratory methods and criteria for sediment bioassessment. In Mudroch, A., J. M. Azcue & P. Mudroch (eds), Manual of Bioassessment of Aquatic Sediment Quality. Lewis Publishers/CRC Press LLC, Boca Raton: 83–133.

Romijn, C. A. F. M., R. Luttik, D. Van de Meent, W. Slooff & J. H. Canton, 1993. Presentation of a General Algorithm to Include Effect Assessment on Secondary Poisoning in the Derivation of Environmental Quality Criteria. Part 1: Aquatic food chains. Ecotox. Environ. Safe. 26: 61-85.

Russell, R. W., R. Lazar & D. G. Haffner, 1995. Biomagnification of organochlorines in Lake Erie white bass. Environ. Toxicol. Chem. 14: 719–724.

Schuytema, G. S., D. F. Krawczyk, W. L. Griffis, A. V. Nebeker & M. L. Robideaux, 1988. Comparative uptake of hexachlorobenzene by fathead minnows, amphipods and oligochaete worms from water and sediment. Environ. Toxicol. Chem. 7: 1035–1045.

Sijm, D. T. H. M. & J. L. M. Hermens, 2000. Internal effect concentration: link between bioaccumulation and ecotoxicity of organic chemicals. In Hutzinger, O. (ed.), The Handbook of Environmental Chemistry, Vol. 2 Part J (Vol. ed.: B. Beek), Bioaccumulation – New Aspects and Developments. Springer-Verlag Berlin Heidelberg: 167–200.

Sterba, G., 1987. Süßwasserfische der Welt. Urania-Verlag Leipzig: 915 pp.

Streit, B., 1998. Bioaccumulation of contaminants in fish. In Braunbeck, T., D. E. Hinton & B. Streit (eds), Fish Ecotoxicology. Birkhäuser Verlag Basel, Switzerland: 353–387.

Taastrøm, H.-M. & L. Jacobsen, 1999. The diet of otters (*Lutra lutra* L.) in Danish freshwater habitats: comparisons of prey fish populations. J. Zool. Lond. 248: 1–13.

Tchounwou, P. B., A. A. Abdelghani, Y. V. Pramar & L. R. Heyer, 1998. Health risk assessment of hexachlorobenzene and hexachlorobutadiene residues in fish collected from a hazardous waste contaminated wetland in Louisiana, U.S.A. In Little, E. E., A. J. DeLonay & B. M. Greenberg (eds), Environmental Toxicology and Risk Assessment 7. ASTM STP 1333: 368–382.

Thiel, R., T. Mehner, B. Koepcke & R. Kafemann, 1996. Diet niche relationship among early life stages of fish in German estuaries. Mar. Freshwat. Res. 47: 123–136.

U.S. EPA, 1999. AQUIRE (Aquatic Toxicity Information Retrieval). In U.S. EPA, ECOTOX (Ecotoxicology Database System). Last updated May 1999. Website: http://www.epa.gov/eco-tox.

U.S. EPA, 2000. Methods for measuring the toxicity and bioaccumulation of sediment-associated contaminants with freshwater invertebrates. Second Edition. EPA 600/R-99/064, U.S. Environmental Protection Agency, Duluth, March 2000: 192 pp.

Van den Dikkenberg R. P., H. H. Canton, L. A. M. Mathijssen–Spiekman & C. J. Roghair, 1989. The usefulness of *Gasterosteus aculeatus* – the three-spined stickleback – as a test organism in routine toxicity testing. Report No. 718625003, National Institute of Public Health and Environmental Protection, Bilthoven, The Netherlands: 22 pp.

Weinstein, J. E. & J. T. Oris, 1999. Humic acids reduce the bioaccumulation and phototoxicity of fluoroanthene to fish. Environ. Toxicol. Chem. 18: 2087–2094.

Wootton, R. J., 1984. A Functional Biology of Sticklebacks. Functional Biology Series. Croom Helm, London & Sydney: 265 pp.

Hydrobiologia **463:** 185–196, 2001.
P. Rodriguez & P.F.M. Verdonschot (eds), Aquatic Oligochaete Biology VIII.

Bioaccumulation of lindane and hexachlorobenzene by the oligochaetes *Enchytraeus luxuriosus* and *Enchytraeus albidus* (Enchytraeidae, Oligochaeta, Annelida)

Eric Bruns, Philipp Egeler*, Joerg Roembke, Adam Scheffczyk & Peter Spoerlein
ECT Oekotoxikologie GmbH, Böttgerstr. 2-14, D-65439 Flörsheim, Germany
Tel: 49 6145 9564 / 65. Fax: 49 6145 9564/ 99. E-mail: ph-egeler@ect.de
*(*Author for correspondence)*

Key words: Enchytraeidae, bioaccumulation, lindane, hexachlorobenzene, method standardisation

Abstract

The uptake of chemicals in soil organisms, especially earthworms, has been studied many times. However, in Europe no internationally accepted standardised test guideline for the assessment of bioaccumulation in the soil ecosystem exists. Therefore, the German Federal Environmental Agency recently funded a project in which a standardisable test method for measuring bioaccumulation of chemicals using earthworms and enchytraeids is being developed. In this contribution, initial results with the new method are presented, using two model chemicals (the insecticide lindane and the fungicide hexachlorobenzene). Two enchytraeid species (*Enchytraeus luxuriosus* and *Enchytraeus albidus*) were selected as test organisms due to their easy handling and their important ecological role in the soil compartment. Artificial soil and a natural standard soil were used as test substrates. Test concentrations were based on previous results of acute and reproduction toxicity tests performed with the same species. Uptake as well as the elimination of the test substances were examined under standardised conditions in a closed test system. The first results show that both chemicals were accumulated considerably by both enchytraeid species. The bioaccumulation factors (BAFs) of lindane and hexachlorobenzene found for enchytraeids are significantly higher compared to those for lumbricid earthworms. Evaluation of the preliminary data suggests that the smaller species *E. luxuriosus* accumulated the two chemicals to a greater extent than *E. albidus*. In most cases, both chemicals were eliminated completely. The use of this new test method appears suitable for the ecotoxicological risk assessment of bioaccumulative chemicals.

Introduction

Enchytraeid oligochaetes were selected as test organisms due to their easy handling and their important ecological role in the soil compartment (Didden, 1993). Enchytraeids have been found to contribute significantly to soil respiration, to constitute the second-highest biomass in many soils, and to affect cycling of nutrients and community metabolism considerably (e.g. Bengtsson & Rundgren, 1982; Römbke, 1992; Didden, 1993). A major change in the abundance of soil invertebrates such as enchytraeids can have serious, adverse ecological effects on the entire terrestrial system, impairing a proper function of trophic energy transfer and nutrient cycling. These worms can accumulate and may be affected by a variety of organic and inorganic compounds (Pirhonen & Huhta, 1984; Römbke & Federschmidt, 1994; Amorim, 2000; Didden & Römbke, 2001). In addition, this group of worms is an important component in terrestrial food webs. A wide variety of organisms, including birds, mammals, reptiles, amphibians, fish, insects, nematodes and centipedes prey upon enchytraeids. Taking into consideration this central role within terrestrial food webs, bioaccumulation data are important tools for assessment of exposure (internal concentrations) and food chain transfer (Romijn et al., 1994; Kramarz & Laskowski, 1999) within the risk assessment of chemicals.

Several national guidelines address the issue of bioaccumulation in organisms other than fish, e.g. ASTM (2000a) and U.S.EPA (2000). More importantly, a guideline covering bioaccumulation in earthworms (*Eisenia fetida* Savigny) is currently being developed (ASTM, 2000b). Since within the EU, adopted standardised guidelines exist only for fish bioaccumulation tests up to now (OECD, 1996), only fish data are considered for this purpose. However, an extrapolation of fish bioaccumulation data to terrestrial organisms like earthworms is difficult, if possible at all. Therefore, for the assessment of bioaccumulation of chemicals in the soil ecosystem, model calculations are currently used (e.g. Connell & Markwell, 1990). The lipophilicity of chemicals is known as one of the most important factors influencing bioaccumulation (e.g. Nendza, 1991). Usually, it is presented as the logarithm of the octanol-water-partitioning coefficient, the log K_{OW}. The development and standardisation of a test method for the experimental determination of bioaccumulation in soil organisms can, therefore, improve considerably risk assessment for chemicals in terrestrial ecosystems.

The main objectives of the presented study are the standardisation of an easily applicable test method for the determination of the bioaccumulation of organic pollutants in enchytraeids from spiked soil. Although the experiments presented in this study were also performed with two earthworm species (*Eisenia fetida, Lumbricus rubellus*), only the results of the work with the two enchytraeid species will be reported in detail here. Eight bioaccumulation tests with different enchytraeid species and two pesticides in two different soils were performed in order to evaluate the influence of species and soil type, and the reproducibility and suitability of the new test method. Bioaccumulation data generated with the method presented in this paper may improve substantially the assessment of bioaccumulation of substances which show a certain persistence in soil systems.

Materials and methods

In accordance with ASTM (2000a), OECD (1996) and U.S. EPA (1996, 2000), the term bioaccumulation is defined here as the increase in concentration of the test substance in or on an organism relative to the concentration of the test substance in the surrounding medium, including uptake of a chemical from both medium and food via all possible pathways. All results presented here are based on the bioaccumulation of chemicals, regardless of the accumulation via the surrounding medium or the food, a differentiation of which is nearly impossible when testing terrestrial oligochaetes.

The bioaccumulation factor (BAF) is calculated at steady state as the quotient of the concentration of the test chemical in the animal divided by the concentration in the soil (both based on wet weight or dry weight). At steady state, the concentration of the test substance in the organism is constant, provided that the concentration in the surrounding medium is not changing significantly. If the steady state was not reached within the test period, the same ratio is defined as the accumulation factor (AF). The loss of a chemical from the organism is termed elimination.

Test substrates

Since one main aspect of this study was the standardisation of the test method employed, two standardised soils, having well-defined properties and being already used regularly in terrestrial ecotoxicology, were used in the experiments: an artificial soil as developed for use in earthworm toxicity tests by Edwards (1984), and the natural standard soil 'Lufa 2.2', which is one of several reference soils recommended for testing in Germany (BBA, 1986). Both soils have been used extensively in ecotoxicological tests with earthworms and enchytraeids (Ebing et al., 1984; Högger & Ammon, 1994; Beyer, 1996; Fitzgerald et al., 1996; Løkke & van Gestel, 1998; Augustsson & Rundgren, 1998; Römbke & Moser, 1999). The OECD soil was prepared according to OECD (1984) from 10% air-dried sphagnum peat (as close to pH 5.5–6.0 as possible, no visible plant remains, finely ground), 20% kaolinite clay (kaolinite content preferably above 30%), and 70% quartz sand (more than 50% of the particles between 50 and 200 μm in diameter). The pH of the soil was adjusted to 6.0±0.5 by addition of calcium carbonate. Its final organic carbon content (OC) was 5.6% of dry weight as calculated from the organic matter content of the peat as given by the supplier (95%), which was then divided by a factor of 1.72 to correct for volatile soil components other than carbon to result in the OC in percent of dry weight of soil. Lufa 2.2 is a sandy loam with a pH value of about 5.7 and an organic carbon content (OC) of 2.2% of dry weight, and was obtained from Landwirtschaftliche Untersuchungs- und Forschungsanstalt Speyer, Germany. Both soils were moistened with reconstituted water prepared according to OECD (1992) to 50% of dw (OECD soil) and to 38.8% of dw (Lufa soil 2.2). The dry weight of the soils was determined on subsamples of control substrate, which were heated at 105±5 °C until weight constancy was obtained.

Test organisms

Two enchytraeid oligochaete species, *Enchytraeus albidus* Henle, 1837 and *Enchytraeus luxuriosus* Schmelz & Collado, 1999, were chosen for the development of the new test method. *E. albidus*, mainly living in soils rich in organic matter, has been cultured for decades, whereas *E. luxuriosus* was just recently described as a new species from a meadow site in Northern Germany (Schmelz & Collado, 1999). Both species are recommended as test organisms in reproduction tests according to ISO (2001) and OECD (2000) draft guidelines. Their small size supports a small scale laboratory test set-up, use of small amounts of test substances, and exposure to both solid phase and porewater of the soil. The enchytraeids are kept in a permanent in-house culture, where they are bred in single species cultures in commercial garden soil without additives at 20±2 °C and 16h:8h light/dark cycle. They are fed with finely ground rolled oats once weekly. Details of culturing of the enchytraeids are described by Römbke & Moser (1999). For the tests, mature worms (with clitellum) were taken from the cultures and were transferred to uncontaminated soil for a 14 d-acclimation period.

Dry weight and lipid content were determined in control organisms of *E. albidus*. For measurement of dry weight, the samples were heated at 105±2 °C in a drying cabinet until weight constancy was reached. The lipid content of worm tissue was measured gravimetrically based on the method according to De Boer et al. (1999). A mixture of 20 ml cyclohexane and 16 ml 2-propanol was used to extract lipids from wet, homogenised tissue. Deionised water was added to the homogenate; the water volume to be added was calculated as follows:

$$x = 22 - \frac{\text{specimen ww [g]} \times \text{water content [\%]}}{100} \text{[ml]}$$

After phase separation, the cyclohexanic phase was removed and volumetrically quantified. A second extraction of the water-2-propanol homogenate was done with 20 ml of cyclohexane containing 13% 2-propanol (w/w). The resulting cyclohexanic phase was combined with the one resulting from the first extraction, the solvent was evaporated by N_2, and the residue was dried at 105±2 °C and weighed. The lipid content (F_{lipid}) was calculated according to:

$$\frac{\text{(lipid weight [g]} \times 100}{\text{specimen ww [g]}} = F_{\text{lipid}} \text{ in \% ww}$$

Worm lipid content in % of dry weight (dw) was calculated by correcting the wet weight (ww) of the specimens used for lipid determination for the dw:ww ratio as outlined above.

Chemicals

The insecticide lindane, and the fungicide hexachlorobenzene (HCB) were chosen due to their wide distribution in the environment, and their different lipophilicity. Lindane represents chemicals with a medium lipophilicity (i.e. a log K_{ow} of about 3.6; Rippen, 1999). With a log K_{ow} of about 5.7 (De Bruijn et al., 1989), hexachlorobenzene belongs to a class of highly lipophilic chemicals. All bioaccumulation tests were performed with ^{14}C-labelled test substances in order be able to detect very low amounts of test substances in small samples. Hexachlorobenzene ([U-^{14}C-]HCB; specific radioactivity: 185 MBq mmol^{-1} or 0.65 MBq mg^{-1}; radiochemical purity: ≥97%) and lindane ([U-^{14}C-] HCH; 1.11 GBq mmol^{-1} or 3.82 MBq mg^{-1}; radiochemical purity: ≥95%) were purchased from International Isotopes Munich (IIM, Unterschleißheim, Germany). ^{14}C-HCB was dissolved in cyclohexane, while n-hexane served as solvent for ^{14}C-HCH before spiking to the soil. The solvents were HPLC grade, purchased from Merck, Darmstadt, Germany. When using OECD soil, each of the dissolved chemicals was applied by coating a part of the quartz sand fraction corresponding to 30% of the whole dry weight of the soil. Since for the Lufa 2.2 soil the sand fraction could not easily be separated from the soil, 30% of the complete soil (air dried) was soaked with a solution of the labelled test substance. After complete evaporation of the solvent, the coated fractions were mixed with the remaining 70% of the soil until the soil was visually homogeneous. Then the total radioactive concentration was measured by 3 replicate soil samples to ensure homogeneous distribution of the test substance in the substrate. The nominal test substance concentrations were approximately 333 Bq g^{-1} soil ww. The lindane concentrations in the uptake experiments were within a range of 31–126 μg kg^{-1} soil (dw) and for HCB, concentrations between 770 and 1100 μg kg^{-1} soil (dw) were used. The lower lindane concentrations were achieved with a separate charge of lindane with a higher specific radioactivity ([U-

^{14}C-]HCH; 2.33 GBq mmol^{-1} or 8.01 MBq mg^{-1}; radiochemical purity: ≥95%), which was used in one experiment (*E. albidus* in spiked OECD soil). The test substance concentrations were at least 100 fold below the No Observed Effect Concentrations (NOEC) previously determined in acute and reproduction studies with the same enchytraeid species in the same soils in our laboratory (unpublished results).

Total radioactive residues were measured by liquid scintillation counting (LSC; Tri Carb 2500 TR, Canberra Packard, Dreieich, Germany). Soil samples (0.5–1 g ww) were weighed, mixed with 3 ml of the solvent used for spiking, ultrasonicated, thoroughly mixed with LSC cocktail (Rotiszint EcoPlus, Roth, Karlsruhe, Germany), left to settle for 2–3 d (7–8 d for OECD soil due to higher OC) at ambient temperature, and measured by LSC. For sampling of the worms, the soil of each beaker was spread on a steel plate, and the animals were picked from the substrate using watchmaker forceps. After rinsing with reconstituted water to remove soil particles, the worms were counted, gently blotted dry on a cellulose tissue, and transferred to pre-weighed LSC-vials (one vial per test beaker). The worm samples were weighed, killed at ≤−18 °C, and solubilised (Soluene 350, Canberra Packard, Dreieich, Germany) prior to LSC, as described previously by Egeler et al. (1997). Samples containing potential non-extractable or bound residues were combusted in a sample oxidiser (TriCarb 307, Canberra Packard, Dreieich, Germany); the released $^{14}CO_2$ was trapped in an absorbent (CarboSorb, Canberra Packard, Dreieich, Germany), mixed with scintillation cocktail (PermaFluor, Canberra Packard, Dreieich, Germany) and measured by LSC. Control samples were processed in an identical manner in order to provide background values of the different matrices. Radioactive concentrations were corrected by subtracting background values from the measured values and dividing the resulting disintegrations per minute (dpm) by the sample wet weight. All concentrations are reported on a wet weight basis unless otherwise indicated. All enchytraeid samples were processed without gut purging, in order to avoid losses of accumulated radioactivity during the purging period, and since possible predators are exposed to whole organisms including their gut content (as outlined in ASTM, 2000a).

Uptake and elimination experiments

Both ^{14}C-lindane and ^{14}C-HCB were tested for uptake and elimination with each species using spiked OECD soil and Lufa 2.2, respectively. Glass beakers (50 ml) with a height of 7 cm and a diameter of 3.8 cm covered by watch glass dishes were used as test vessels. The vessels were filled with 10 or 20 g dw freshly spiked test substrate, which was adjusted for the desired moisture as described above, for experiments with *E. luxuriosus* and *E. albidus*, respectively. Twenty acclimated adult enchytraeids (with clitellum) of similar size were added to each vessel. Mean biomass per replicate was 127±33 (SD) mg ww for *E. albidus*, and 12±6 (SD) mg ww for *E. luxuriosus*. Table 1 summarises the amounts of substrate, biomass and food applied to the test vessels. The beakers were placed in a stainless steel box which was covered by a glass lid (Fig. 1). Air supply was provided by a vacuum pump delivering ambient air through the steel box. The exhaust air was lead through a series of washing bottles containing triethylene glycol and potassium hydroxide solution (10%) for trapping of volatile organic compounds and radioactive CO_2, respectively. All washing bottles were assembled in an ice bath throughout the individual test periods. This test setup minimised leaking of radioactivity and allowed the determination of a mass balance of the radioactivity at the end of the tests. The tests were carried out at 20±2 °C (16 h:8 h light/dark cycle). The worms were fed rolled oats once per week. Water lost during the test – as determined by weekly weighing of the test vessels – was compensated for by adding the evaporated volume of demineralised water once per week.

The worms were exposed to the spiked soil for an uptake period of 21–28 d (for details see Table 2). The duration of the uptake period was not fixed, since reliable information on the time needed to reach steady state was not available before the start of the tests. During the uptake period, four replicates were analysed for total radioactivity in worms and soil on each of at least six sampling dates (e.g. on days 1, 2, 4, 7, 14 and 21). The exposure period did not exceed 28 days in order to avoid the presence of newly-hatched juvenile worms in the test vessels. Nevertheless, juvenile enchytraeids appeared after 21 d in two experiments. It was, therefore, necessary to restrict the exposure period to approximately 21 days to ensure that all individuals of a test were exposed for the same time period. At the end of the exposure period, the worms of the remaining replicates

Table 1. Average biomass, amounts of substrate, and food rations applied per test vessel

Species	Average biomass per vessel [mg ww]	OECD-soil [g]	Lufa 2.2 [g]	Weekly food ration [mg]
E. albidus	127 ± 33 (SD)	20 (dw)	18 (dw)	50
		30 (ww)	25 (ww)	
E. luxuriosus	12 ± 6 (SD)	10 (dw)	9 (dw)	20
		15 (ww)	12.5 (ww)	

Table 2. Uptake parameters of ^{14}C-lindane and ^{14}C-HCB in *Enchytraeus albidus* and *Enchytraeus luxuriosus*. All values refer to steady state. BAF, bioaccumulation factor for ^{14}C-activity; k_s , soil uptake rate [g soil g^{-1} worm d^{-1}] as BAF × k_e of uptake phase; SD: standard deviation; SE: standard error; BSAF: biota-soil accumulation factor

	Soil type	Duration [d]	C_{soil} [μg kg^{-1} ww] (*SD*)	C_{soil} [μg kg^{-1} dw] (*SD*)	C_{worm} [μg g^{-1} worm ww] (*SD*)	BAF (*SE*)	BSAF [g OC g^{-1} lipid]	k_s [g soil g^{-1} worm d^{-1}]
Lindane								
E. albidus	OECD	28	21.1 (2.1)	31.7 (3.2)	0.25 (0.07)	12 ± 0.7	9.5	8.6
E. albidus	Lufa 2.2	21	32.1 (2.0)	44.6 (2.8)	0.71 (0.08)	22 ± 0.4	7.7	9.7
E. luxuriosus	OECD	21	34.8 (3.3)	52.2 (4.9)	0.42 (0.05)	12 ± 1.1	9.7	36.9
E. luxuriosus	Lufa 2.2	21	30.6 (1.2)	42.5 (1.7)	0.37 (0.02)	36 ± 1.1	12.5	32.4
HCB								
E. albidus	OECD	21	598.7 (139.2)	898.1 (208.7)	8.20 (2.82)	14 ± 4.4	11.0	1.8
E. albidus	Lufa 2.2	21	545.6 (32.1)	757.3 (44.5)	15.28 (0.88)	28 ± 0.8	9.7	7.6
E. luxuriosus	OECD	28	678.0 (65.8)	1017.0 (98.7)	18.31 (2.01)	27 ± 0.9	21.7	9.3
E. luxuriosus	Lufa 2.2	21	421.6 (24.8)	585.3 (34.4)	14.76 (1.89)	35 ± 0.7	12.2	6.7

Table 3. Elimination parameters of ^{14}C-Lindane and ^{14}C-HCB in *Enchytraeus albidus* and *Enchytraeus luxuriosus*. A, 'fast' compartment [% of steady state concentration]; k_a, elimination rate constant for compartment A; B, 'slow' compartment [% of steady state concentration]; k_b, elimination rate constant for compartment B; residue, concentration in worms at end of elimination phase [% of steady state concentration]; r^2, coefficient of determination; *SE*, standard error

	Soil type	Duration [d]	Residue [%]	A ± SE [%]	k_a ± SE [d^{-1}]	B ± SE [%]	k_b ± SE [d^{-1}]	r^2
Lindane								
E. albidus	OECD	32	50.1	63 ± 14	1.4 ± 0.9	37 ± 8	–0.01 ± 0.01	0.883
E. albidus	Lufa 2.2	21	8.6	43 ± 4	2.8 ± 1.2	58 ± 3	0.09 ± 0.01	0.997
E. luxuriosus	OECD	21	2.6	91 ± 5	0.8 ± 0.1	9 ± 5	0.06 ± 0.04	0.998
E. luxuriosus	Lufa 2.2	21	4.1	60 ± 7	0.9 ± 0.2	40 ± 7	0.11 ± 0.02	0.998
HCB								
E. albidus	OECD	21	14.3	29 ± 28	0.6 ± 0.7	71 ± 28	0.08 ± 0.03	0.978
E. albidus	Lufa 2.2	25	10.0	17 ± 16	1.0 ± 1.8	80 ± 15	0.08 ± 0.02	0.970
E. luxuriosus	OECD	28	0.02	48 ± 21	24.0 ± 11.3	52 ± 20	0.28 ± 0.13	0.984
E. luxuriosus	Lufa 2.2	21	8.4	103 ± 4	0.1 ± 0.01	–	–	0.983

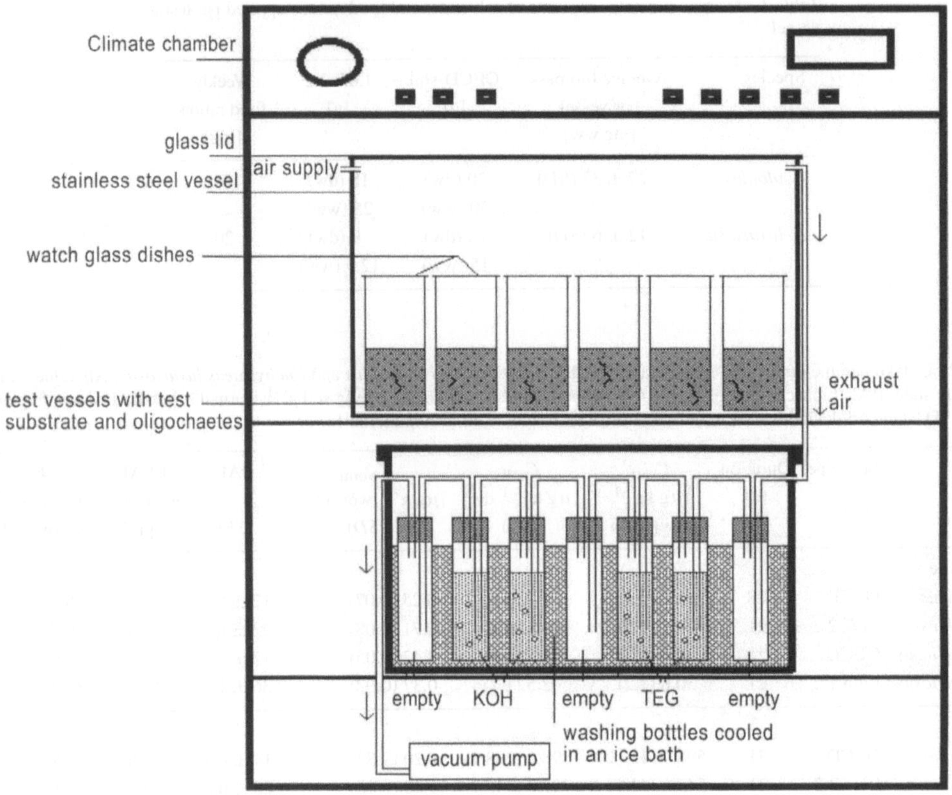

Figure 1. Schematic set-up of the test apparatus for bioaccumulation studies with oligochaetes. Abbreviations: KOH – 10% solution of potassium hydroxide; TEG – triethylene-glycol.

were transferred from the spiked soil to test vessels containing untreated substrate in order to determine elimination kinetics.

During elimination, the sampling procedures were performed according to the methods used during the uptake phase. The total radioactivity in the worms and the test substrate was measured at six sampling dates (e.g. days 1, 2, 4, 7, 14 and 21 of the elimination phase) in four replicates each.

Data evaluation

For each replicate, the radioactivity measured in the worms was divided by the wet weight of these animals. The obtained radioactive concentration in the worms (dpm g^{-1} ww) of each replicate was divided by the radioactive concentration in the corresponding soil (dpm g^{-1} ww). This resulted in 4 replicate biota-soil-concentration ratios (accumulation factors, AF) for each sampling point. The ratio of radioactivity per g worm and radioactivity per g test substrate was calculated for each replicate. All toxicokinetics, accumulation and bioaccumulation factors were based on total radioactive residues on a wet weight basis. Uptake and elimination curves were fitted to the mean values of these 4 ratios plotted against time with nonlinear regression analysis using commercially available software (SPSS 7.5 for Windows, 1997). To describe the time course of uptake with a one compartment model, Equation 1 was used:

$$\frac{dC_a}{dt} = k_s C_s - k_e C_a$$

or

$$C_a = \frac{k_s}{k_e} C_s (1 - e^{k_e t}), \qquad (1)$$

where C_a is the concentration of substance in animals [g substance g^{-1} tissue]; C_s is the concentration of substance in soil [g substance g^{-1} soil]; k_s is the soil uptake rate coefficient [g soil g^{-1} tissue d^{-1}]; k_e is the elimination rate constant [d^{-1}].

When steady state is reached during the uptake phase, i.e. when an equilibrium is established between the uptake and the simultaneously occurring elimination (i.e. $t = \infty$ and C_a remains constant), Equation (1) may be reduced to:

$$C_a = \frac{k_s}{K_e} C_s$$

or

$$\frac{C_a}{C_s} = \frac{k_s}{k_e} = BAF. \qquad (2)$$

Then $k_s/k_e{}^*C_s$ is an approach to the concentration of the test substance in the worms at steady state ($C_{a,ss}$). If steady state is not reached during the uptake phase, the accumulation factor (AF) can be calculated for any given sampling period (t) according to:

$$\mathrm{AF} = \frac{C_a(t)}{C_s} = \frac{k_s}{k_e}(1 - e^{-k_e t}). \qquad (3)$$

When the test animals are transferred from contaminated soil to a vessel free of the test substance, the accumulated substance can be eliminated from the organism without the influence of uptake processes. The test animals can be abstracted to compartments which retain the test substance to different degrees. If the data points plotted against time indicate a constant exponential decline of the test substance concentration in the animals, a one-compartment model (Equation (4)) can be used to describe the time course of elimination.

$$C_o(t) = C_{a,ss} e^{-k_e t}. \qquad (4)$$

If the test substance is obviously eliminated with different elimination rate constants, a two-compartment model is required (Equation (5), Spacie & Hamelink, 1982):

$$C_o = Ae^{-k_a t} + Be^{-k_b t}. \qquad (5)$$

A and B represent the size of the compartments (in percent), where A is the compartment with rapid, and B the compartment with slow loss of test substance. The sum of A and B equals 100% of the whole-animal compartment volume at steady state. The corresponding elimination rate constants are represented by k_a and k_b [d^{-1}]. If the two-compartment model is fitted to the measured elimination data, Equation (1) may be replaced by Equation (6) to determine the uptake rate coefficient k_s (Zok et al., 1991):

$$k_s = \frac{(Ak_a + Bk_b)\,BAF}{A + B} \qquad (6)$$

The coefficient of determination (r^2) was used as a measure of the goodness of fit of the model equations to the experimental data. It also served as a decisive tool to judge whether the elimination followed the one compartment or the two-compartment model. In cases where the application of either of the two models produced a good fit to the data, the model with the higher value of r^2 was assumed to suit better.

For interspecies comparison of bioaccumulation data, the normalisation of BAFs on the test organism lipid content and on the organic carbon fraction of the soil may be useful. Belfroid et al. (1994), Füll (1996) and Tracey & Hansen (1996) used the biota-sediment (or -soil) accumulation factor (BSAF) as the lipid-normalised concentration of test substance in the test organism divided by the organic carbon-normalised concentration of the substance in the soil at steady state. C_a is then expressed as mg g^{-1} lipid content of the organism, and C_s as mg g^{-1} organic content (OC) of the soil. The data on lipid content of *E. albidus* measured during this study were used for calculation of BSAF of both enchytraeid species.

Results

The results of the different experiments are described in the following (for details see Tables 2 and Table 3). *E. albidus* accumulated ^{14}C-lindane from OECD soil by a factor of 11.8 (ww-based). Steady state was reached after 6–10 days of exposure. The soil uptake rate constant was 8.6 g soil g^{-1} worm d^{-1}. The uptake curve could be fitted to the experimental data with a determination coefficient (r^2) of 0.92. After transfer to uncontaminated OECD soil, 63% of the steady state concentration was eliminated from the worms after approximately 4 days, while until the end of the elimination phase, which in this case was run for 32 days, no further decrease of ^{14}C-lindane concentration could be measured in the worms. This means that after 32 d, 50% of the steady state concentration still emained in the worms. A two-compartment model was fitted to the data with r^2 of 0.88. The elimination rate constants, k_a and k_b were 1.4 d^{-1} and –0.009 d^{-1}, respectively.

The uptake and elimination of lindane from spiked Lufa 2.2 soil by *E. albidus* are shown in Figure 2a,b. The steady state was reached about 7 days after test initiation, leading to an accumulation of 1.4 μg lindane g^{-1} wet weight, which corresponds to a wet weight based BAF of 21.9 (r^2= 0.99). About

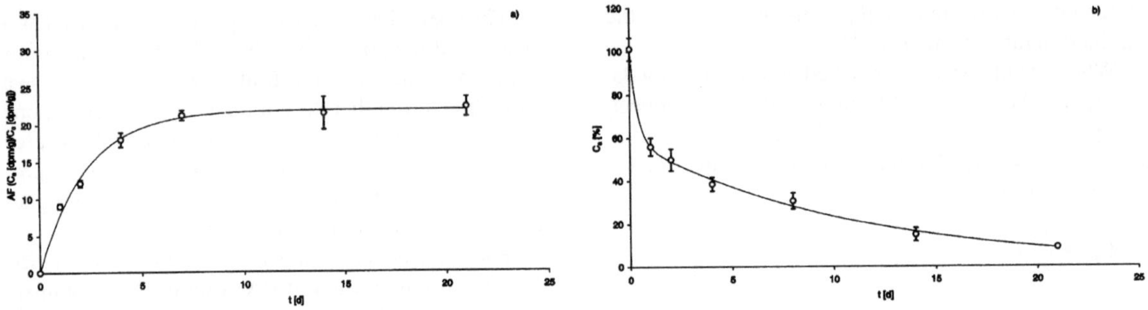

Figure 2. ^{14}C-lindane kinetics in *Enchytraeus albidus* using Lufa 2.2 natural standard soil during the uptake (a) and elimination (b) phases. Abbreviations: AF – accumulation factor; C_a – ^{14}C-concentration in% of steady state concentration in worms (ww); $n = 4$; error bars = SE.

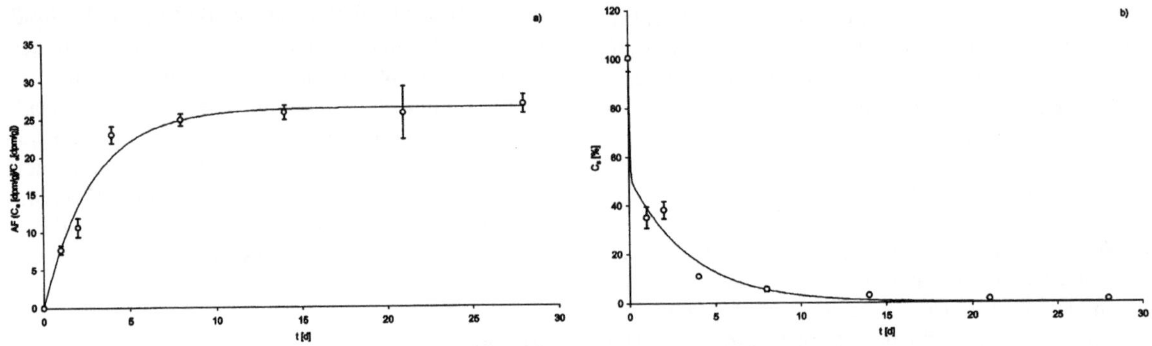

Figure 3. ^{14}C-hexachlorobenzene (HCB) kinetics in *Enchytraeus luxuriosus* in OECD soil during the uptake (a) and elimination (b) phase. Abbreviations: AF – accumulation factor; C_a – ^{14}C-concentration in% of steady state concentration in worms (ww); $n = 4$; error bars = SE.

0.6 μg g^{-1} body weight were eliminated within 24 h. Thereafter, the elimination velocity was significantly lower, so that after 21 days, about 0.1 μg lindane g^{-1} wet weight (i.e. 8.6% of steady state concentration) were still measured in the worms. This elimination pattern could be described best with a two compartment model (r^2= 0.997).

In *E. luxuriosus*, lindane uptake from spiked OECD soil reached a BAF of 12. Initial uptake was very fast (k_s of uptake phase = 36.9 g g^{-1} d^{-1}), so that uptake and elimination processes were at steady state after approximately 3 days. The coefficient of determination of the uptake curve was 0.80. Elimination was described best with a two compartment model (r^2 = 0.998). After 5 d, approximately 90% of steady state body concentration was lost from the worms (k_a= 0.82 d^{-1}). At the end of the elimination phase, i.e. after 21 d, 2.6% of steady state body concentration was measured in the worm samples (k_b= 0.06 d^{-1}).

When *E. luxuriosus* was exposed to lindane-spiked Lufa 2.2, steady state was reached after 7 d, resulting in a BAF of 36. The uptake rate was similar to the experiment with *E. luxuriosus* in OECD soil (k_s= 32.4 g g^{-1} d^{-1}; r^2= 0.98). According to the applied two compartment model (r^2= 0.98), 60% was eliminated after 2–3 days (k_a= 0.9 d^{-1}; k_b= 0.1; r^2= 0.998), leaving 4.1% of steady state body concentration in the worms after 21 d of elimination.

In *E. albidus*, the concentration in the worms remained constant after 7 d (k_s= 1.8 g g^{-1} d^{-1}) when exposed to HCB spiked OECD soil. HCB was accumulated in the worms by a factor of 22 based on wet weight. The regression curve fitted the uptake data very well (r^2= 0.99). During the elimination phase, HCB was lost from the worms according to a two-compartment model (r^2= 0.978). The fast compartment comprised 29% of steady state activity in the worms and the corresponding rate constant, k_a, was 0.6 d^{-1}. The concentration in the worms measured at the end of the elimination phase amounted to 14.3% of the value at steady state.

In spiked Lufa 2.2, *E. albidus* accumulated HCB by a factor of 28 as based on worm and soil wet weight. Steady state was reached after 10–15 d, with a k_s of 7.6 g g^{-1} d^{-1} (r^2= 0.99). Elimination had been extended to 25 days, and 10% of steady state body concentration was detected in the worms. A two-compartment model fitted the data best (r^2= 0.97), showing a small fast compartment (17%), leaving 10% of steady state activity in the worms at the end of the elimination phase.

Hexachlorobenzene spiked to OECD soil (28 d uptake phase) was accumulated and eliminated by *E. luxuriosus* as shown in Figure 3a,b. The uptake of hexachlorobenzene by *E. luxuriosus* in OECD soil reached a steady state after about 7 days at a concentration of 17 μg g^{-1} body wet weight. This value corresponds to a bioaccumulation factor of 26.5 at an uptake rate coefficient of 9.3 g g^{-1} d^{-1}. Based on a two compartment elimination model (r^2= 0.98), a nearly complete elimination of HCB was observed after 28 days.

In HCB spiked Lufa 2.2, *E. luxuriosus* accumulated the test substance by a factor of 35. The steady state was reached after 14 d of exposure. The uptake rate coefficient was 6.7 g g^{-1} d^{-1} (r^2= 0.99). HCB was eliminated from the worms according to a one compartment kinetic model: an elimination rate constant k_e of 0.12 d^{-1} was calculated (r^2= 0.98). 8.4% of steady state activity could be measured in the worm samples taken at the end of the elimination phase.

The determination of lipid content in *E. albidus* resulted in a value of 4.7% based on wet weight, and 23.0% based on dry weight. The biota-soil factors calculated from the *E. albidus* lipid percentage ranged from 7.7 in *E. albidus* exposed to lindane in Lufa 2.2 to 21.7 in E. luxuriosus exposed to HCB in OECD soil, and are shown in Table 2.

Discussion

In general, the uptake of lindane by *E. albidus* in Lufa 2.2 (Fig. 2) is quite similar to the HCB kinetics in *E. luxuriosus* exposed to OECD soil. However, the elimination of lindane by *E. albidus* is quite different compared to the HCB elimination by *E. luxuriosus*.

The complete results for the uptake of the two chemicals by the two enchytraeid species are presented in Table 2, where bioaccumulation factors as well as body concentrations in μg test substance per g worm (ww) are shown. When these data are examined, it appears that the BAFs for Lindane and HCB in OECD soil are in the same range for both species. The BAF for lindane determined in OECD soil (12.2) and Lufa 2.2 (21.9) in *E. albidus* differ by a factor of approximately 2. This result may be due to the lower OC and the subsequent higher bioavailability in Lufa 2.2 (Amorim, 2000).

A comparison between the data of the two enchytraeid species shows that the BAFs determined for *E. luxuriosus* are higher than those for *E. albidus*. There are several possible reasons. These small enchytraeids may selectively feed on small particles, which may carry higher amounts of test substance than coarser material. Furthermore, the smaller size of *E. luxuriosus* leads to a large body surface area relative to the weight which supports an enhanced diffusion of test substance via the body wall.

Referring to the different Log K_{ow} values for lindane and HCB, the differences between the BAF values were expected to be higher. For example, the BAF for lindane and HCB measured in *E. albidus* in OECD soil were quite similar (12.2 and 13.7, respectively), but differed by a factor of two for HCB in *E. luxuriosus* determined in the same soil. This indicates that HCB is associated to the organic fraction in OECD soil much more strongly than lindane.

The results of the elimination experiments as summarised in Table 3 show that the elimination of lindane in all experiments as well as elimination of HCB in OECD soil by *E. albidus* were described best with a two compartment model. In these cases, the coefficient of determination of nonlinear regression varied between 0.956 and 0.999. Only in the case of HCB elimination by *E. luxuriosus* in Lufa 2.2, a one compartment model fitted the data best. It is obvious that in the experiments with *E. luxuriosus*, the test substances were eliminated within 21 days to a higher extent than in *E. albidus*. In two cases (HCB in OECD soil and Lindane in Lufa 2.2), the fast compartment consisted of about 40–50% and the slow compartment of about 50–60% of the total radioactivity in the worms. About 8 – 9% of the steady state concentration was still measured in the organisms after an elimination period of 21 days. In the experiment with lindane in OECD soil, 38% of the steady state concentration was still incorporated by the worms after 25 days.

In a recent study, Amorim (2000) investigated uptake and elimination of ^{14}C-lindane in *E. albidus* exposed to OECD soil, and to a natural soil. The uptake and elimination phase in Amorim's (2000) experiments lasted 10 d each and the nominal lindane concentration was 25 mg kg^{-1} soil dw. After the uptake phase, BAFs of 6.4 and 9.6 were reported in OECD and natural soil, respectively. These values are by a factor of appr. 2 lower than the lindane BAFs in *E. albidus* found in the study presented here. In contrast to the results described in Table 3, Amorim (2000) found that in both soil types, more than 95% of the accumulated lindane was eliminated after 10 d. Considering the different test designs, the results of both studies are in relatively good agreement.

Metabolisation of both test compounds was considered neglectable taking into account previous results on bioaccumulation of these chemicals in aquatic oligochaetes (Egeler et al., 1997). HCB has been shown previously to be transformed very slowly by earthworms (Khalil, 1990). However, some studies (e.g. Visvanathan et al., 1988; Füll, 1996) revealed substantial metabolisation of lindane in earthworms. Therefore, a final confirmation of compound stability will be necessary to definitely assign the measured BAFs to the parent compounds, especially to lindane.

A comparison of the four enchytraeid BAFs found in the present study with data for tubificids, obtained by Egeler et al. (1999), who determined a mean BAF of 4.5 for ^{14}C-lindane, and of 6.3 for ^{14}C-HCB, shows that the enchytraeid values are considerably higher than tubificid BAFs, but remain within the same order of magnitude. This is confirmed by the data published on bioaccumulation in other terrestrial oligochaetes up to now. Viswanathan et al. (1988) reported a wet weight based BAF of 2.1 for lindane in *Eisenia fetida* using a humus soil enriched with 10% cattle manure. Füll (1996) determined lindane accumulation factors of 1.5 and 2.1 for *Lumbricus rubellus* after exposure periods of 22 and 21 days, respectively. In both cases, she determined a further increase of the lindane uptake and a steady state was not reached within the test period. For HCB, Ebing et al. (1984) determined BAFs of 1.3 and 2 in spiked sand, and of 0.5 in Lufa 2.2 in *Lumbricus terrestris*. BAFs of 1.8 and 3.0 for HCB were reported by Khalil (1990). Belfroid et al. (1994) determined a BAF of 215 based on the lipid content, in experiments carried out with *Eisenia andrei* in OECD soil. If a lipid content of 2% is assumed for *E. andrei*, a BAF of about 4 can be calculated for the fresh weight.

All BAFs determined in enchytraeids up to now were higher than tubificid BAFs (Oliver, 1987; Egeler et al., 1997), and considerably higher when compared to earthworms (up to a factor of 10, see above). Further studies have to be performed to determine if this difference is high enough in order to be considered for the risk assessment of chemicals in soil.

The present results show that the developed test system is a suitable tool for the investigation of bioaccumulation in soil. In all experiments except *E. albidus* uptake of HCB from OECD soil, a steady state was established and BAFs could be determined. The description of the elimination processes by different nonlinear regression models was possible in all cases, and the reported coefficients of determination show that the models employed describe the elimination of both lindane and HCB in the two different soils and the two different species very well. Both enchytraeid species appeared to be suitable bioaccumulation indicators. Uptake and elimination data did not differ considerably between the tested species; although there was a tendency that the smaller species (*E. luxuriosus*) showed higher BAFs than the large *E. albidus*, this indicates that bioaccumulation data gained with *E. albidus* may be extrapolated to *E. luxuriosus*. This is especially important for the environmental risk assessment of pesticides, since the latter species is more common in meadow soils than *E. albidus* (Schmelz & Collado, 1999). *E. albidus* also showed some advantages over *E. luxuriosus* with respect to easier handling and counting during the tests, as well as higher biomass, which facilitates analysis of accumulated test substances. The results reported by Amorim (2000) support the suitability especially of *E. albidus* as a test organism.

The presented method, being specific for the soil ecosystem, overcomes the extrapolation of fish BCF to other compartments like soil and sediment as it is still performed for the environmental risk assessment of chemicals.

Conclusions

A test system for standardised measuring of bioaccumulation of chemicals in terrestrial oligochaetes was established. The test system is meant as an instrument in the risk assessment of chemicals for the soil compartment. The results produced so far show that the bioaccumulation factors determined in Enchytraeids are higher than those known for earthworms. The comparably low bioaccumulation factors determined for soil-inhabiting oligochaetes may still lead to high body burdens, so that a risk for the environment can not be excluded. Further research is needed to verify the reproducibility and the applicability of the test system to other chemicals. The data set for the two chemicals, the two soil types and the two model chemicals is currently being improved by analogous studies using the two earthworm species *Lumbriculus rubellus* and *Eisenia fetida*. The complete data set will be used to optimise the presented test system and will finally allow the preparation of a draft test guideline for the measurement of bioaccumulation in soil-inhabiting oligochaetes.

Acknowledgements

Financial support was kindly granted by the German Federal Environmental Agency (R+D-No. 298 64 416). In particular, we appreciate the kind cooperation with Dr G. Studinger and Dr C. Franke.

References

Amorim, M., 2000. Chronic and toxicokinetic behavior of lindane (γ-HCH) in the enchytraeid *Enchytraeus albidus*. For the Degree of Master of Science in Ecology. Universidade de Coimbra, Faculdade de Ciencias e Tecnologia, Departamento de Zoologia. Coimbra, Portugal, 2000: 82 pp.

ASTM, 2000a. Standard Guide for Determination of the Bioaccumulation of Sediment-Associated Contaminants by Benthic Invertebrates. American Society for Testing and Materials, E 1688-00a: 54 pp.

ASTM, 2000b. Standard Guide for conducting laboratory soil toxicity or bioaccumulation tests with the lumbricid earthworm *Eisenia fetida* and the enchytraeid potworm *Enchytraeus albidus*. E 1676–97, draft version, July 17, 2000: 37 pp.

Augustsson, A. & S. Rundgren, 1998. The enchytraeid *Cognettia sphagnetorum* in risk assessment: advantages and disadvantages. Ambio 27: 62–69.

BBA, 1986. Richtlinien für die amtliche Prüfung von Pflanzenschutzmitteln. Teil IV, 4-1: Verbleib von Pflanzenschutzmitteln im Boden-Abbau, Umwandlung und Metabolismus. ACO Druck, Braunschweig: 35 pp.

Beyer, W. N., 1996. Accumulation of Chlorinated Benzenes in Earthworms. Bull. Environ. Contam. Toxicol. 57: 729–736.

Belfroid, A., M. Sikkenk, W. Seinen, C. Van Gestel & J. Hermens, 1994. The toxicokinetic behaviour of chlorobenzenes in earthworms (*Eisenia andrei*): experiments in soil. Environ. Toxicol. Chem. 13: 93–99.

Bengtsson, G. & S. Rundgren, 1982. Population density and species number of enchytraeids in coniferous forest soils polluted by a brass mill. Pedobiologia 24: 211–218.

Connell, D. W. & R. D. Markwell, 1990. Bioaccumulation in the soil to earthworm system. Chemosphere 20: 91–100.

De Boer, J., F. Smedes, D. Wells & A. Allan, 1999. Report on the QUASH interlaboratory study on the determination of total-lipid in fish and shellfish. Round 1 SBT-2. Exercise 1000. EU, Standards, Measurement and Testing Programme: 20 pp.

De Bruijn, J., F. Busser, W. Seinen & J. Hermens, 1989. Determination of octanol/water partition coefficients for hydrophobic organic chemicals with the 'slow-stirring' method. Environ. Toxicol. Chem. 8: 499–512.

Didden, W. A. M., 1993. Ecology of terrestrial enchytraeidae. Pedobiologia 37: 2–29.

Didden, W. A. M. & J. Römbke, 2001. Enchytraeids as test organisms for chemical stress in terrestrial ecosystems. Ecotox. Environ. Safety 50: 25–43.

Ebing, W., J. Pflugmacher & A. Haque, 1984. Der Regenwurm als Schlüsselorganismus zur Messung der Bodenbelastung mit organischen Fremdchemikalien. Berichte über Landwirtschaft 62: 222–255.

Edwards, C. A., 1984. Report of the second stage in development of a standardized laboratory method for assessing the toxicity of chemical substances to earthworms. In Commission of the European Communities (ed.), Environment and the Quality of Life. Report EUR 9360 EN. ECSC-EEC-EAEC, Brussels, Luxembourg: 99 pp.

Egeler, Ph., J. Römbke, M. Meller, Th. Knacker, C. Franke, G. Studinger & R. Nagel, 1997. Bioaccumulation of lindane and hexachlorobenzene by tubificid sludgeworms (Oligochaeta) under standardised laboratory conditions. Chemosphere 35: 835–852.

Egeler, Ph., J. Römbke, M. Meller, Th. Knacker & R. Nagel, 1999. Bioaccumulation test with tubificid sludgeworms in artificial media-development of a standardisable method. Hydrobiologia 406: 271–280.

Fitzgerald, D. G., K. A. Warner, R. P. Lanno & D. G. Dixon, 1996. Assessing the effects of modifying factors of pentachlorophenol toxicity to earthworms: Application of body residues. Environ. Toxicol. Chem. 15: 2299–2304.

Franke, C., G. Studinger, G. Berger, S. Böhling, U. Bruckmann, D. Cohors-Fresenborg & U. Jöhnke, 1994. The assessment of bioaccumulation. Chemosphere 29: 1501–1514.

Füll, C., 1996. Bioakkumulation und Metabolismus von γ–1,2,3,4,5,6-Hexachlorcyclohexan (Lindan) und 2,4-(2,4-Dichlorphenoxy)-Propionsäure (Dichlorprop) beim Regenwurm *Lumbricus rubellus* (Oligochaeta, Lumbricidae). PhD-Thesis, Johannes Gutenberg-University, Mainz: 156 pp.

Högger, Ch. & H. U. Ammon, 1994. Testing the toxicity of pesticides to earthworms in laboratory and field tests. IOBC WPRS Bull. 17: 157–178.

ISO (International Organisation for Standardization), 2000. Soil Quality – Effects of Pollutants on Enchytraeidae (*Enchytraeus* sp.) Determinations of Effects on Reproduction (Enchytraeid reproduction test (ERT). Draft ISO / WD 16387: 31 pp.

Khalil, A. M., 1990. Aufnahme und Metabolismus von ^{14}C-Hexachlorbenzol und ^{14}C-Pentachlornitrobenzol in Regenwürmern. Ph.D.-Thesis, University Munich: 137 pp.

Kramarz, P. & R. Laskowski, 1999. Toxicity and possible food-chain effects of copper, dimethoate and a detergent (LAS) on a centipede (*Lithobius mutabilis*) and its prey (*Musca domestica*). Appl. Soil Ecol. 13: 177–185.

Løkke, H. & C. A. M. Van Gestel, 1998. Handbook of soil invertebrate toxicity tests. Ecological & Environmental Toxicology Series. Wiley & Sons, New York: 281 pp.

Nendza, M., 1991. QSARs of bioaccumulation: Validity assessment of log P_{OW}/log BCF correlations. In Nagel, R. & R. Loskill (eds), Bioaccumulation in aquatic systems. Contributions to the assessment. Proceedings of an international workshop, Berlin 1990. VCH, Weinheim: 43–66.

OECD, 1984. Guidelines for Testing of Chemicals No. 207. Earthworm, Acute Toxicity Test. Organisation for Economic Cooperation and Development (OECD), Paris: 9 pp.

OECD, 1992. Guidelines for Testing of Chemicals No. 203. Fish, Acute Toxicity Test. OECD, Paris: 9 pp.

OECD, 1996. Guidelines for Testing of Chemicals No. 305. Bioconcentration: Flow-through Fish Test. OECD, Paris: 9 pp.

OECD, 2000. Guidelines for the Testing of Chemicals 220 (Proposal for a new Guideline) Enchytraeidae Reproduction Test. OECD, Paris: 24 pp.

Oliver, B. G., 1987. Biouptake of chlorinated hydrocarbons from laboratory-spiked and field sediments by oligochaete worms. Environ. Sci. Technol. 21: 785–790.

Pirhonen, R. & V. Huhta, 1984. Petroleum fractions in soil: Effects on populations of Nematoda, Enchytraeidae and Microarthropoda. Soil Biol. Biochem. 16: 347–350.

Rippen, G., 1999. Handbuch der Umweltchemikalien: Physikalisch-chemische und ökotoxikologische Daten ausgewählter chemischer Stoffe. Loose-leaf book, updated periodically. Ecomed Verlagsgesellschaft, Landsberg.

Römbke, J., 1992. Estimates of the Enchytraeidae (Oligochaeta, Annelida) contribution to energy flow in the soil system of an acid beech wood forest. Biol. Fert. Soil 11: 255–260.

Römbke, J. & A. Federschmidt, 1994. Effects of the fungicide Carbendazim on Enchytraeidae in laboratory and field tests. Newsletter on Enchytraeidae 4: 79–96.

Römbke, J. & T. Moser, 1999. Organisation and Performance of an International Ringtest for the Validation of the Enchytraeid Reproduction Test. UBA-Text 4/99: 150 pp. (+ annex: 249 pp.).

Romijn, C. F. A. M., R. Luttik & J. R. Canton, 1994. Presentation of a general algorithm to include effect assessment of secondary poisoning in the derivation of environmental quality criteria. Part 2. Terrestrial food chains. Ecotox. Environ. Safety 27: 107–127.

Schmelz, R. & R. Collado, 1999. *Enchytraeus luxuriosus* sp. nov., a new terrestrial oligochaete species (Enchytraeidae, Clitellata, Annelida). Carolinea 57: 93–100.

Spacie, A. & J. L. Hamelink, 1982. Alternative models for describing the bioconcentration of organics in fish. Environ. Ecotox. Chem. 1: 309–320.

Tracey, G. A. & D. J. Hansen, 1996. Use of biota-sediment accumulation factors to assess similarity of nonionic organic chemical exposure to benthically-coupled organisms of differing trophic mode. Arch. Environ. Contam. Toxicol. 30: 467–475.

U.S. EPA, 1996. Ecological Test Guidelines OPPTS 850.1730 Fish BCF (Public Draft): 23 pp. U.S. EPA, 2000. Methods for measuring the toxicity and bioaccumulation of sediment-associated contaminants with freshwater invertebrates. 2nd edn. EPA 600/R-99/064, U.S. Environmental Protection Agency, Duluth, MN. March 2000: 192 pp.

Viswanathan, R., S. Ray, I. Scheunert & F. Korte, 1988. Investigations on accumulation and biotransformation by earthworms of lindane occurring as soil contaminant. In Abbou, R. (ed.), Hazardous Waste: Detection, Control, Treatment. Elsevier Science Publishers, Amsterdam: 759–765.

Zok, S., G. Görge, W. Kalsch & R. Nagel, 1991. Bioconcentration, metabolism and toxicity of substituted anilines in the zebrafish (*Brachydanio rerio*). Sci. Total. Environ. 109/110: 411–421.

Hydrobiologia **463:** 197–205, 2001.
P. Rodriguez & P.F.M. Verdonschot (eds), Aquatic Oligochaete Biology VIII.

Spatial and temporal variations of oligochaetes of the Ivinhema River and Patos Lake in the Upper Paraná River Basin, Brazil

Maria Célia Montanholi-Martins & Alice M. Takeda
Department of Biology, State University of Maringá/Nupelia, Av. Colombo 5790, Maringá, PR, CEP 87020-900, Brazil
E-mail: mcmontanholi@terra.com.br   alice@nupelia.uem.br

Key words: oligochaete, spatial and temporal variations, Ivinhema River, floodplain lake, Paraná River, Brazil

Abstract

The composition and spatial and temporal variation of oligochaete community were analyzed in Patos Lake and Ivinhema River, a tributary of the Paraná River Brazil, during the period March 1993 to February 1994. Lentic and lotic environments were studied, with the objective of analyzing species association and temporal changes in the function of the abiotic variables at nine sites studied. Seventeen species of oligochaetes were identified. Oligochaete association was analyzed by Detrended Correspondence Analysis (DCA) and abiotic variables, by Principal Components Analysis (PCA). This method showed that *Aulodrilus pigueti* predominated in Patos Lake, probably because it has adapted to mud with high organic matter in low currents. *Brinkhurstia americanus* predominated at the right bank of the Ivinhema River while *Paranadrilus descolei* predominated at the left bank. Temporal variation of organism density was influenced by hydrological cycle, and spatial variation was determined by grain size and quantities of organic matter in the sediments.

Introduction

Floodplains are highly dynamic topographic surfaces caused by fluvial systems and the adjustments that they develop on the solid and liquid sections of the basin (Drago, 1976). This dynamism favors the formation of various types of water bodies and influences the physical and chemical characteristics of water and sediments, and consequently the oligochaete community.

Little research has been done on the oligochaete community of floodplains in Brazil (Alves & Strixino, 1997 on the Mogi-Guaçu River floodplain; Montanholi-Martins & Takeda, 1999 and Takeda, 1999 on the alluvial plain of the Upper Paraná River).

Due to variations of hydrological levels, interactions between lentic and lotic environments occur, such as in Patos Lake and the Ivinhema River which have some special abiotic characteristics. The objective of this paper is analyzing the species association and temporal changes in relation to the abiotic variables at nine sites.

Study area

The Paraná River has an extensive alluvial plain with secondary channels, tributaries, like the Ivinhema River, and floodplain lakes, like Patos Lake. Two sampling stations were chosen in the Ivinhema River (22° 49′ 01″ S and 53° 33′ 42″ W): one on the right bank (I_1) without aquatic macrophytes; the other on the left bank (I_2) characterized by macrophytes as *Polygonum* spp. and *Eichhornia azurea* (Fig. 1).

Patos Lake (22° 49′ 19″ S and 53° 31′ 33″ W), with an area of 785 553. 5 m^2 is connected permanently to the Ivinhema River. It lies on the left bank and is influenced by its hydrological regime. Four sites on the bank (P_1, P_5, P_6 and P_7) and three in the central region (P_2, P_3 and P_4) were sampled.

Patos Lake has water macrophytes margins at some sites, with *E. azurea* and *Salvinia* sp. (P_1). At site P_6 *E. azurea* and *Polygonum* spp. grow and at site P_7 only *E. azurea*.

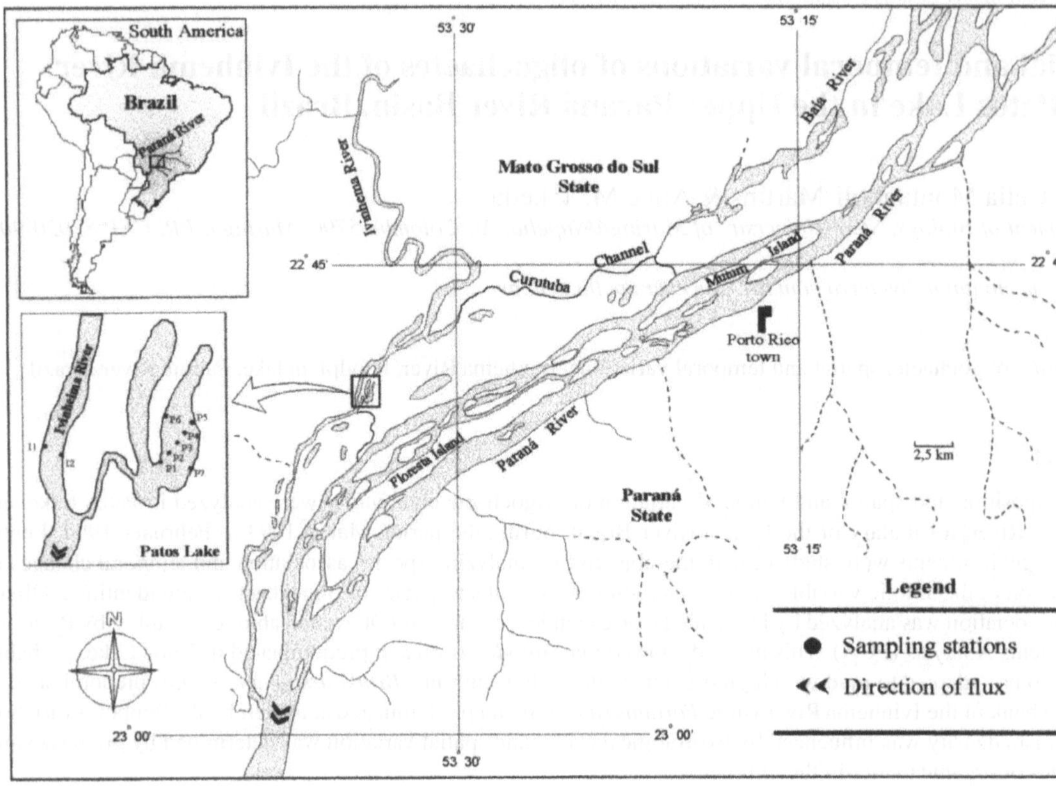

Figure 1. Study area Ivinhema River and Patos Lake, and the 9 sampling sites.

Materials and methods

Collections were carried out monthly with a Petersen grab (0.0345 m^2) from March 1993 up to and including February 1994. At each station, four samples were taken, three for biological studies and one for sediment granulometry and organic matter content analyses.

Water transparency (Secchi disc), depth and temperature were measured. Water samples were collected with a Van Dorn bottle to determine pH, conductivity, and dissolved oxygen concentration (by the Winkler method, as modified by Golterman et al., 1978). Daily values of the Ivinhema River water level were obtained from the National Department of Waters and Electrical Energy.

The samples used for composition granulometry and organic matter content were stored in a freezer and later oven-dried at 80 °C. The granulometry was determined according to the Wentworth Scale (Wentworth, 1922), and organic matter content by incinerating 10 g of sediment at 560 °C for 4 h in a muffle furnace.

The three samples for biological analysis were washed through a series of sieves with the smallest mesh size 0.2 mm. The material retained was fixed in a 10% formalin solution buffered with calcium carbonate. In the laboratory, the samples were sorted under stereoscopic microscope. Identification of oligochaetes followed Brinkhurst & Marchese (1992).

The abiotic data (53 observations × 16 variables) were standardized and subjected to Principal Components Analysis (PCA) using the program Statistics for Windows version 5.0. The biotic data were normalized by logarithmic transformation log $(x+1)$.

The oligochaete mean density of the species which were found in more than three samples were analyzed with Detrended Correspondence Analysis (DCA), (53 observations × 9 variables). Statistical programs 5.0 and PC-ORD 2.0 were used to calculate PCA and DCA. According to Gauch (1982), the PCA and DCA

Table 1. Mean ($\bar{x}$), standard deviation (s), maximum and minimum of the abiotic variables measured in the Ivinhema River and Patos Lake during March 1993 to February 1994

Samples sites	Banks						Middle of Lake		
	River		Lake						
	I1	I2	P1	P5	P6	P7	P2	P3	P4
	$\bar{x} = 23.60$	$\bar{x} = 23.57$	$\bar{x} = 23.56$	$\bar{x} = 23.16$	$\bar{x} = 23.64$	$\bar{x} = 23.52$	$\bar{x} = 22.58$	$\bar{x} = 22.19$	$\bar{x} = 22.58$
Water temperature	$S = 3.70$	$S = 3.44$	$S = 3.62$	$S = 3.67$	$S = 3.47$	$S = 3.54$	$S = 4.02$	$S = 4.02$	$S = 3.88$
(°C)	17.10 (June)	18.10 (June)	17.90 (June)	8.10 (June)	19.30 (June)	18.50 (June)	17.00 (June)	16.70 (June)	16.10 (June)
	30.00 (March)	29.30 (March)	29.60 (March)	29.90 (March)	29.90 (March)	29.80 (March)	29.70 (March)	29.60 (March)	29.40 (March)
	$\bar{x} = 1.33$	$\bar{x} = 2.68$	$\bar{x} = 1.62$	$\bar{x} = 2.26$	$\bar{x} = 1.70$	$\bar{x} = 1.85$	$\bar{x} = 3.88$	$\bar{x} = 4.53$	$\bar{x} = 3.94$
Depth	$S = 0.83$	$S = 1.22$	$S = 0.72$	$S = 0.93$	$S = 0.75$	$S = 0.84$	$S = 0.87$	$S = 0.80$	$S = 0.93$
(m)	0.60 (Sep., Nov)	0.80 (July)	0.80 (Nov.)	1.30 (Nov.)	1.00 (Sep. , Dec.)	1.00 (November)	2.80 (November)	3.50 (November)	3.00 (Aug. , Nov.)
	3.20 (March)	5.80 (March)	3.20 (March)	4.60 (March)	3.60 (March)	4.20(March)	5.80 (March)	6.20 (March)	6.20 (March)
	$\bar{x} = 0.82$	$\bar{x} = 0.85$	$\bar{x} = 0.97$	$\bar{x} = 1.09$	$\bar{x} = 1.01$	$\bar{x} = 1.11$	$\bar{x} = 1.16$	$\bar{x} = 1.16$	$\bar{x} = 1.17$
Water transparency	$S = 0.58$	$S = 0.66$	$S = 0.70$	$S = 0.68$	$S = 0.65$	$S = 0.71$	$S = 0.70$	$S = 0.73$	$S = 0.73$
(m)	0.40 (February)	0.45 (December)	0.30 (Nov.)	0.40 Nov., Dec)	0.37 (November)	0.40 (Nov., Dec.)	0.45 (Nov., Dec.)	0.45 (December)	0.45 (Nov., Dec.)
	2.50 (March)	2.85 (March)	2.90 (March)	2.90 (March)	2.90 (March)	3.00 (March)	2.70 (March)	2.85 (March)	2.90 (March)
pH	$\bar{x} = 6.88$	$\bar{x} = 6.74$	$\bar{x} = 6.53$	$\bar{x} = 6.41$	$\bar{x} = 6.50$	$\bar{x} = 6.48$	$\bar{x} = 6.40$	$\bar{x} = 6.48$	$\bar{x} = 6.38$
	$S = 0.17$	$S = 0.25$	$S = 0.29$	$S = 0.43$	$S = 0.50$	$S = 0.38$	$S = 0.40$	$S = 0.41$	$S = 0.43$
	6.60 (February)	6.30 (February)	6.00 (February)	5.70 (June)	5.80 (June, Feb.)	5.80 (June)	5.70 (June)	5.60 (June)	5.60 (June)
	7.10 (Apr., Aug., Nov.)	7.10 (August)	7.00 (August)	7.10 (August)	7.50 (September)	7.20 (August)	7.10 (August)	6.90 (Aug.., Nov.)	7.10 (August)
	$\bar{x} = 42.33$	$\bar{x} = 42.10$	$\bar{x} = 34.25$	$\bar{x} = 34.08$	$\bar{x} = 33.67$	$\bar{x} = 34.33$	$\bar{x} = 35.08$	$\bar{x} = 35.08$	$\bar{x} = 34.83$
Conductivity	$S = 3.14$	$S = 3.12$	$S = 8.35$	$S = 8.62$	$S = 8.17$	$S = 8.47$	$S = 9.98$	$S = 10.23$	$S = 9.27$
(μS cm^{-1})	36 (June)	35 (June)	23 (August)	23 (July, Aug.)	23 (July, Aug.)	24 (July, Aug.)	23 (August)	23 (August)	23 (August)
	47 (September)	47 (September)	49 (February)	48 (February)	49 (February)	49 (February)	52 (Apr., Feb.)	52 (February)	51 (February)
	$\bar{x} = 7.26$	$\bar{x} = 7.03$	$\bar{x} = 5.12$	$\bar{x} = 5.09$	$\bar{x} = 5.32$	$\bar{x} = 5.36$	$\bar{x} = 5.10$	$\bar{x} = 4.61$	$\bar{x} = 4.79$
Dissolved oxygen	$S = 1.18$	$S = 1.28$	$S = 2.93$	$S = 3.18$	$S = 3.25$	$S = 3.24$	$S = 2.98$	$S = 3.05$	$S = 3.08$
(mg l^{-1})	4.73 (March)	4.54 (March)	0.81 (April)	0.63 (April)	0.84 (February)	0.65 (April)	0.91 (February)	0.36 (April)	0.53 (April)
	8.64 (August)	8.62 (August)	9.23 (August)	9.16 (August)	9.40 (September)	8.99 (August)	9.18 (August)	9.24 (August)	8.97 (August)
	$\bar{x} = 3.30$	$\bar{x} = 3.51$	$\bar{x} = 17.11$	$\bar{x} = 11.13$	$\bar{x} = 13.19$	$\bar{x} = 4.48$	$\bar{x} = 14.60$	$\bar{x} = 18.41$	$\bar{x} = 12.26$
Organic matter	$S = 1.42$	$S = 4.08$	$S = 2.64$	$S = 1.49$	$S = 2.26$	$S = 1.77$	$S = 6.30$	$S = 2.11$	$S = 6.16$
(%)	1.60 (January)	0.80 (Oct., Feb.)	13.50 (Feb.)	8.70 (Nov.)	8.60 (April)	1.95 (January)	3.20 (September)	13.10 (December)	1.70 (October)
	6.10 (October)	15.50 (March)	23.40 (April)	14.60 (July)	17.30 (February)	7.70 (December)	25.90 (December)	22.50 (April)	19.60 (January)

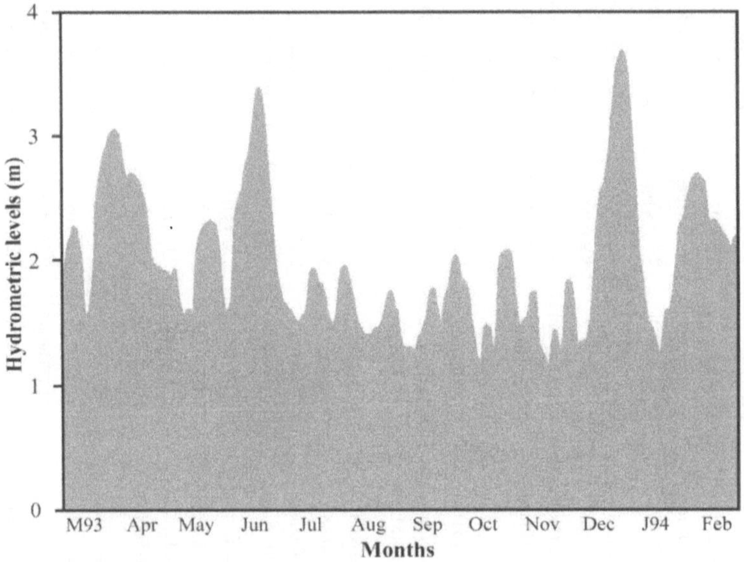

Figure 2. Hydrometric levels (m) of the Paraná River measured between March 1993 and February 1994.

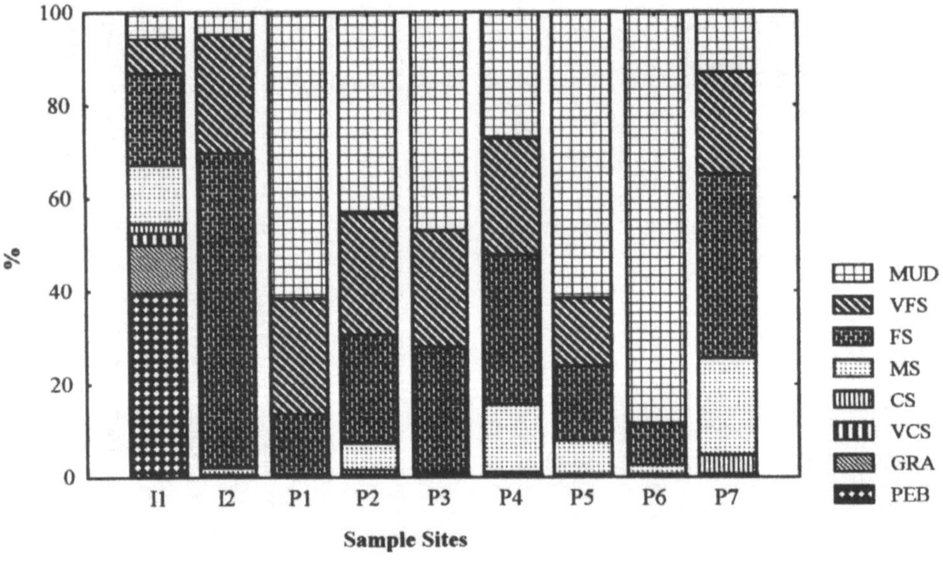

Figure 3. Annual percentage of grain sizes (%) at the sampling points. PEB – pebbles; GRA – gravel; VCS – very coarse sand; CS – coarse sand; MS – medium sand; FS – fine sand; VFS – very fine sand; M – mud.

have similar concepts and their ordination techniques of the multivariables are complementary. The relation between abiotic and biotic variables was calculated with Pearson's correlation with scores of factors 1 and 2 for PCA and DCA, respectively.

Kruskal–Wallis analysis was used to show the difference between oligochaete density in Patos Lake and the Ivinhema River. The Shannon–Wiener diversity index (Pielou, 1975), equitability (Pielou, 1966), and the

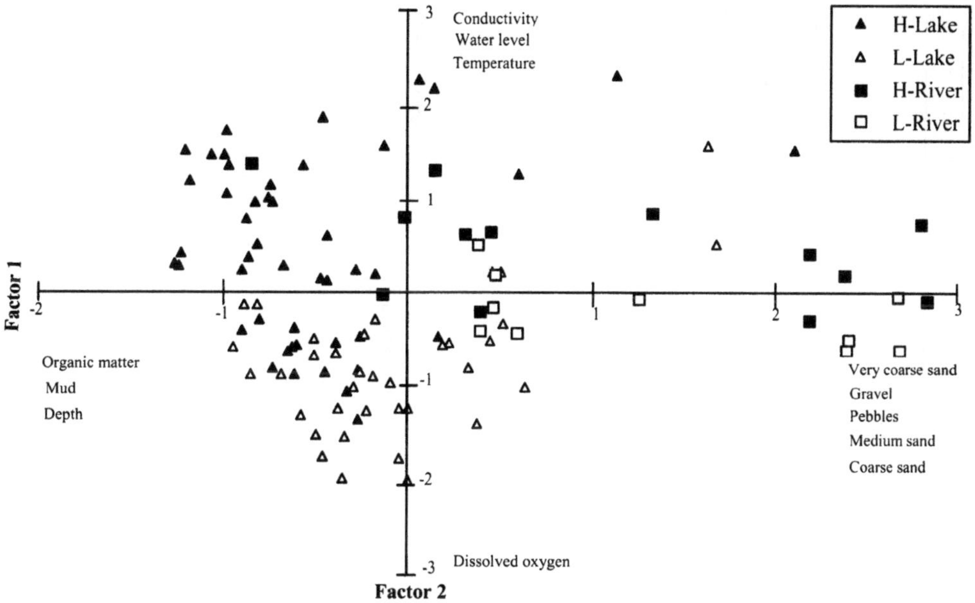

Figure 4. PCA scores of the sampling station and month, during the high water (H) and low water periods (L).

dominance index of Kownacki (1971) were calculated using the oligochaete densities at each site.

Results

According to the daily hydrological level of the Ivinhema River, two phases were established: high water phase from March to June 1993 and December 1993 to February 1994 and low water phase for the remaining months (Fig. 2).

The main differences between abiotic factors were depth, dissolved oxygen, organic matter and granulometric texture. The water temperatures varied from 30 °C (March at site I_1) to 8.10 °C (June at site P_5). The greatest depths were observed in the central region of the lake with a maximum of 6.20 m (March at sites $P_3$and P_4) and the lowest depth with a minimum of 0.60 m occurred in September, November at site I_1. Water transparency varied from 3.00 (March at site P_7) to 0.30 (November at site P_1); pH varied from 7.50 (September at site P_6) to 5.60 (June at sites P_3, P_4). Conductivity varied from 52 μS cm^{-1} (April, February at site P_2; February at site P_3) to 23 μS cm^{-1} (August at sites P_1, P_2, P_3, P_4) and (July, August at sites P_5, P_6); The average concentration of dissolved oxygen was higher in the Ivinhema River than, varied from 9.40 mg l^{-1} (September at site P_6) to 0.36 mg l^{-1} (April at site P_3). The organic matter content found in the sediment was higher in the lake and varied between 25.90% (December at site P_2) to 0.80% (October and February at site I_2) Table 1. The granulometric texture of Ivinhema River (I_1) was very different from the other sites because pebbles were predominant (39.82%). In the Patos Lake, mud predominated with a maximum of 88.52% at site P_6, and a minimum of 13.17% at site P_7 (Fig. 3).

The first two PCA factors explained 28% and 17% of total variability of the data. Significant positive correlation (>0.50) was observed in factor 1 for very coarse sand, gravel, pebbles, medium sand and coarse sand; and negative correlation for organic matter, mud and depth. In factor 2, significant positive correlation for conductivity, water level and temperature; and significant negative correlation for dissolved oxygen were observed (Fig. 4).

We recorded 17 species of aquatic oligochaetes belonging to 7 families (Table 2). The highest monthly mean densities of oligochaetes were observed in the Ivinhema River during the high water period, represented mainly by Enchytraeidae and *Paranadrilus descolei;* in Patos Lake, *Aulodrilus pigueti* was found mainly during the low water period. Moreover, *A. pigueti* dominated at all lake sites. *Opistocysta serrata, Haemonais waldvogeli, Dero (D.) pectinata, D. (D.) righii, D. (A.) furcatus, Nais schubarti,*

Table 2. Values of mean density (ind.m^{-2}) of oligochaete ($\bar{x}$), standard deviation (s) and species dominance index (d) at the collecting points from March 1993 to February 1994

Sample sites	I1			I2			P1			P2			P3			P4			P5			P6			P7		
	$\bar{x}$	s	d	$\bar{x}$	s	d	$\bar{x}$	s	d	$\bar{x}$	s	d	$\bar{x}$	s	d	$\bar{x}$	s	d	$\bar{x}$	s	d	$\bar{x}$	s	d	$\bar{x}$	s	d
Haplotaxidae																											
Haplotaxis aedeochaeta (Brinkhurst &Marchese, 1987)	6.0	11.0	2.8																								
Enchytraeidae	13.0	29.0	6.4																						1.0	3.0	0.2
Opistocystidae																											
Opistocysta serrata (Harman, 1969)																			1.0	3.0	0.3				1.0	3.0	0.2
Naididae																											
Pristina americana (Cernosvitov, 1937)	4.0	8.0	2.0																2.0	8.0	1.0				1.0	3.0	0.2
Haemonais waldvogeli (Bretscher, 1900)																			2.0	6.0	0.7						
Dero (D.) pectinata (Aiyer, 1929)																			4.0	14.0	1.7				2.0	6.0	0.4
Dero (D.) righii (Varela, 1990)																									3.0	11.0	0.8
Dero (Aulophorus) furcatus (Müller, 1773)							1.0	2.8	0.7																		
Nais schubarti (Marcus, 1944)	4.0	14.0	0.7																								
Stephensoniana trivandrana (Aiyer, 1926)																									4.0	10.0	24.6
Bratislavia unidentata (Harman, 1973)																			3.0	4.0	3.2	1.0	2.8	0.4			
Pristinella menoni (Aiyer, 1929)	1.0	3.0	0.7																								
Pristinella longidentata (Harman, 1965)				2.0	6.0	1.5							1.0	3.0	0.9												
Tubificidae																											
Tubificidae imaturo										1.0	3.0	0.3										2.0	6.0	0.7			
Aulodrilus limnobius (Bretscher, 1899)	6.0	11.0	2.8																								
Aulodrilus pigueti (Kowalewski, 1914)	3.0	6.0	1.6	2.0	4.0	3.0	10.0	23.0	23.0	22.0	43.0	67.0	8.0	17.0	30.2	7.0	11.0	42.0	8.0	12.0	17.3	18.0	41.0	40.1	21.0	30.0	42.2
Paranadrilus descolei (Gavrilov, 1955)				6.0	14.0	16.0																					
Narapidae																											
Narapa bonettoi (Righi &Varela,1983)	3.0	5.0	2.2																								
Alluroididae																											
Brinkhurstia americanus (Brinkhurst, 1964)	11.0	12.0	12.3																						1.0	2.8	0.2

Dominance index, according to Kownacki (1971): Dominants (10–100); Subdominants (1.00–9.99); Non-dominants A (0.10–0.99); Non-dominants B (0–0.099).

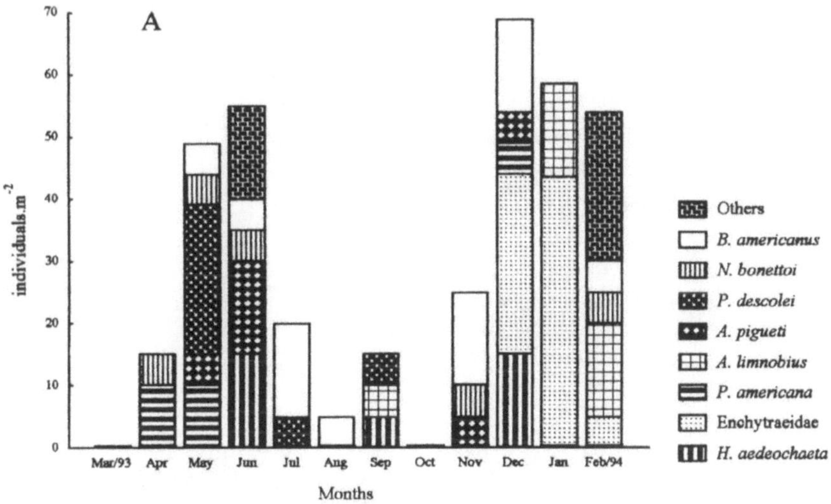

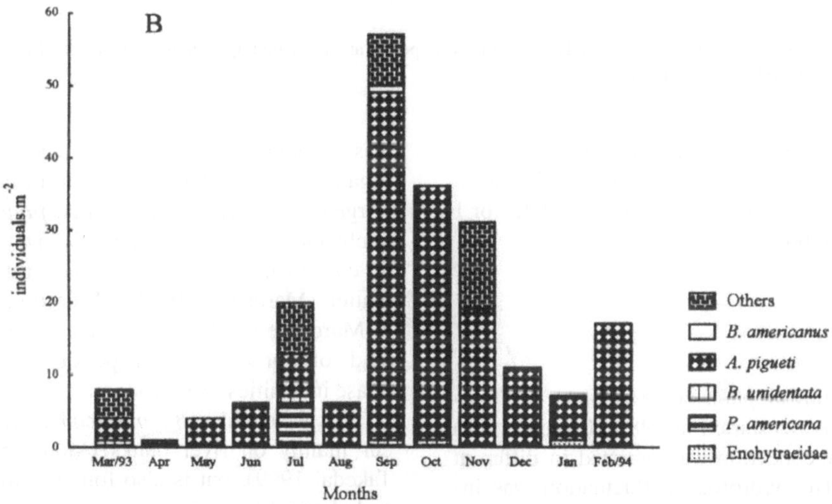

Figure 5. Mean density of individuals at all river sites (A) and all lake sites of the (B).

Stephensoniana trivandrana, *Bratislavia unidentada*, *Pristinella menoni* and *P. longidentata* were not observed in more than three collections and were therefore grouped in the category of other species (Fig. 5 A,B).

Value of diversity index ($H' = 1.34$) and equitability ($E = 0.37$)were lower in Patos Lake than Ivinhema River ($H' = 2.99$ and $E = 0.90$).

The Kruskal–Wallis ANOVA test showed a significant difference between densities in Patos Lake and the Ivinhema River ($P = 0.024$ and $H' = 5.08$).

The Detrended Correspondence Analysis with eigenvalues 0.83 (factor 1) and 0.32 (factor 2), showed two groups caused as a result of differences in hydrological levels (Fig. 6). The first group formed by *Narapa bonettoi*, *Haplotaxis aedeochaeta*, *A. limnobius* and Enchytraeidae characterized the banks of the Ivinhema River and was found mainly in the high water phase. Another group was formed in Patos Lake by an abundance of *A. pigueti* in the low water phase and *B. unidentata* in the high water phase.

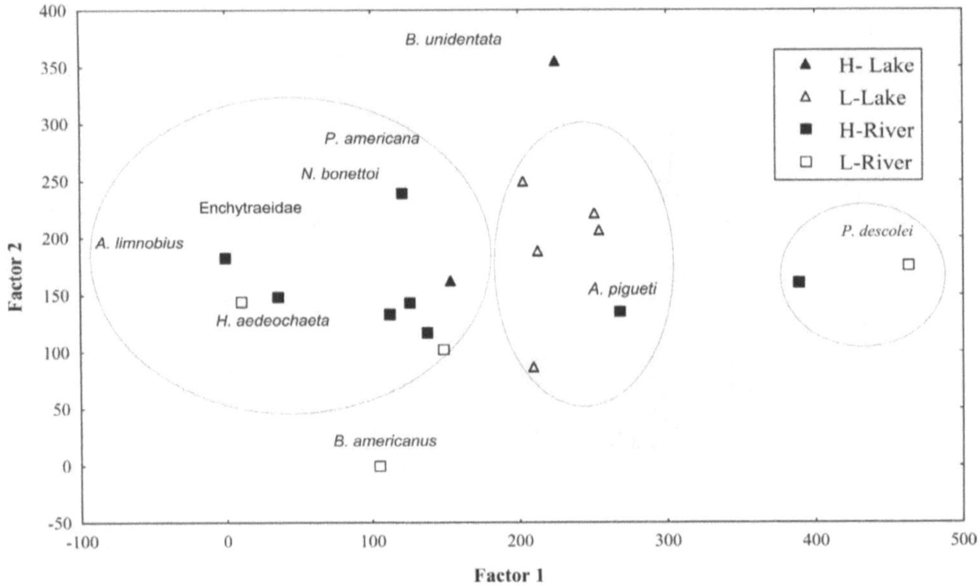

Figure 6. DCA scores for density of the most abundant oligochaetes species at each sampling station in the lake and in the river, during the periods of high water (H) and low water phases (L).

Pearson's correlation was shown to be significant (5%) between biotic and abiotic variables, however, only factor 1 of the DCA was significant with factor 1 of the PCA ($r = 0.66$).

Discussion

The study of alluvial plain zoobenthos of the Upper Paraná River was started in 1987 (Takeda, 1999) when a monthly collection was initiated in different environments. The hydrological fluctuation was intense, constantly changing the environment, especially the bottom of the lakes and rivers. *B. unidentata*, *D. (D.) pectinata*, *D. (D.) righii* and *O. serrata* are new records for Brazilian waters.

Oligochaete density of the region studied was low when compared Alves & Strixino (1997) on the Mogi-Guaçu River floodplain (São Paulo, Brazil) and Marchese (1987) in the tributaries of the Paraná River (Argentina).

The average monthly densities of the Ivinhema River and those of Patos Lake were significantly different, exhibiting pronounced variation in the distinct phases of the hydrological cycle. The higher density of *A. pigueti* in the lake during the low water phase and its low density in lotic environments suggest that it is adapted to muddy lentic environment, with high organic matter and a low current. In the Paraná River (Argentina), high densities of *A. pigueti* were found in channels of high conductivity, low to moderate water current and places with abundant mud and organic matter (Marchese, 1987). According to Brinkhurst & Marchese (1992), this species also has an asexual made of reproduction. It is possible that the low water phase intensifies its reproduction, becoming abundant.

A species like *B. americanus* is known to occur mainly on river banks (Montanholi–Martins & Takeda, 1999), but is also found in margins of lakes and channels with semi-lotic characteristics (Takeda, 1999). Predominance of *P. descolei* at site I_2 was also observed in 1987–1988 by Takeda (1999).

The substitution of *P. descolei* in 1987–1988 (Takeda, 1999) by *A. pigueti* (1993–1994) in this region, mainly Patos Lake, was probably caused by a catastrophic flood that occurred in the Paraná River in 1990 with 21 959 $m^3\ s^{-1}$. Normally, the mean discharge of the Paraná River is 9014 $m^3\ s^{-1}$ (mean of 1964 – 1997).

Species predominance varied between the right and left banks. The right bank (I_1) represents the erosional zone of the river with high flow velocity, and coarse sediment, while the left bank (I_2) is a depositional zone with fine sediment. These conditions produce dominance of *B. americanus* on the right bank and of *P. descolei* on the left bank. The analysis showed the preference of *A. limnobius*, Enchytraeidae, *N. bonettoi*, *P. americana*, *H. aedeochaeta* and *B. americanus* for coarse sediment. High densities of *N. bonettoi* and *H. aedeochaeta* were found in a secondary channel of the Upper Paraná River (Montanholi–Martins & Takeda, 1999). The relationship between abundance of *N. bonettoi* and middle-sized sand was emphasized by Bonetto et al. (1985–86) and Marchese (1994) in the Paraná River (Argentina) and by Takeda et al. (1997) in the Upper Paraná River (Brazil).

The temporal variation of organism density was influenced mainly by the hydrological cycle, and spatial variation was affected mainly by differences in grain size and organic matter quantities in the sediments.

Acknowledgements

We would like to thank Mercedes Rosa Marchese for identification of Oligochaeta, the researchers of our NUPELIA laboratory of physics and chemistry for water analyses, M.Sc. Rosângela G. Santana for help in statistical analysis and Dr Luiz C. Gomes for suggestions of the manuscript. Support by CI-AMB/PADCT/CNPq is gratefully acknowledged.

References

Alves, R. G. & G. Strixino, 1997. Chironomidae e Oligochaeta de uma lagoa da planície de inundação do rio Mogi-Guaçu: Lagoa do Diogo (Estação Ecológica de Jataí, Luiz Antonio, SP.). Anais do VIII Seminário Regional de Ecologia 8: 141–149.

Bonetto, A., M. E. Varela & J. A. Bechara, 1985/86. El bentos del Paraná medio en el tramo Corrientes-Esquina. ECOSUR 12/13: 37–57.

Brinkhurst, R. O. & M. R. Marchese, 1992. Guía para la identificación de oligoquetos acuáticos continentales de Sud y Centro America. Asociación de Ciencias Naturales del Litoral, Santo Tomé: 207 pp.

Drago, E. C., 1976. Origen y clasificación de ambientes leníticos en llanuras aluviales. Rev. Asoc. Cienc. Nat. Litoral 7: 123–137.

Gauch, H. G. Jr., 1982. Multivariate Analysis in Community Ecology. Cambridge University Press, Cambridge, England.

Golterman, L., R. S. Clyno & M. A. M. Ohmstad, 1978. Methods for Physical and Chemical Analysis of Fresh Waters. 2nd edn. Blackwell Scientific Publications, Oxford: 215 pp.

Kownacki, A., 1971. Taxocens of Chironomidae in streams of the Polish High Tatra Mts. (Str.). Acta. Hydrobiol. 13: 439–464.

Marchese, M. R., 1987. The ecology of some benthic Oligochaeta from the Paraná River. Hydrobiologia 155: 209–214.

Marchese, M., 1994. Population dynamics of *Narapa bonettoi* Righi and Varela, 1983 (Oligochaeta: Narapidae) from the main channel of the middle Paraná River, Argentina. Hydrobiologia 278: 303–308.

Montanholi–Martins, M. C. & A. M. Takeda, 1999. Communities of benthic oligochaetes in relation to sediment structure in the Upper Paraná river, Brazil. Stud Neotrop. Fauna Environ. 34: 52–58.

Pielou, E. C., 1966. The measurements of diversity in different types of biological collections. J. Theoret. Biol. 13: 131–144.

Pielou, E. C., 1975. Ecological diversity. Wiley, New York: 165 pp.

Takeda, A. M., G. Shimizu & J. Higuti, 1997. Zoobentos da planície aluvial do alto rio Paraná. In Agostinho, A. A., A. E. M. Vazzoler & N. S. Hahn (eds), A Planície de Inundação do Alto Rio Paraná: Aspectos Físicos, Químicos, Biológicos e Sócio-econômicos. Editora da Universidade Estadual de Maringá, Brasil: 460 pp.

Takeda, A. M., 1999. Oligochaete community of alluvial upper Paraná River, Brazil: spatial and temporal distribution (1987–1988). Hydrobiologia: 35–42.

Wentworth, C. K., 1922. A scale of grade and class terms for clastic sediments. J. Geol. 30: 377–392.

Hydrobiologia **463:** 207–216, 2001.
P. Rodriguez & P.F.M. Verdonschot (eds), Aquatic Oligochaete Biology VIII.

The distribution and ecology of littoral Oligochaeta and Aphanoneura (Annelida) of the Natural and Historical Reserve of Isla Martín García, Río de la Plata River, Argentina

Laura C. Armendáriz & Inés I. César
Departamento Científico Zoología Invertebrados, Facultad de Ciencias Naturales y Museo, Paseo del Bosque s/n, La Plata (1900), Argentina
Fax: 54 0221 425 7527. E-mail: lauraarm@netverk.com.ar

Key words: Oligochaeta, Aphanoneura, distribution, ecology, Neotropical region, Argentina

Abstract

The objective of this paper is to provide data on the distribution and ecology of oligochaetes and aphanoneurans occurring in the Río de la Plata, Argentina, including details regarding, their densities, faunal characteristics, species richness, diversity and equitability. Between 1995 and 1997, a total of five stations on the Río de la Plata were sampled on eight different occasions. The taxocenosis of Oligochaeta and Aphanoneura included 30 species belonging in six families: Naididae, Tubificidae, Narapidae, Enchytraeidae, Lumbricidae and Aeolosomatidae. The density varied between 44 495 ind m^{-2} and 57 ind m^{-2}. Species richness varied between 1 and 17, the diversity ranged from 0.07 to 2.55, and equitability varied between 0.08 and 0.64. According to dominance and frequency, the dominant, constant and widely distributed species were: *Nais variabilis, Stephensoniana trivandrana, Amphichaeta leydigi, Chaetogaster diastrophus, Limnodrilus hoffmeisteri, Aulodrilus pigueti, Narapa bonettoi* and the species of Enchytraeidae and Aeolosomatidae.

Introduction

The island of Martín García lies at the confluence of the Uruguay and Paraná rivers (Upper Río de la Plata River), 34° 11′ S, 58° 15′ W. This island is an outcrop of the crystalline basement, unconformably overlain by Pleistocene and Holocene sediments (Ravizza, 1984). Knowledge of the island's wildlife is generally limited to trees and bushes (Lahitte & Hurrell, 1994) and vertebrates (Juárez, 1995; Lahitte et al., 1995; Lahitte & Hurrell, 1998). Aquatic invertebrates mentioned include freshwater mollusks (Rumi et al., 1996), temnocephalous platyhelminths (Damborenea et al., 1997) and aquatic and semi-aquatic insects (Fernández & López Ruf, 1999) among others. The aim of this work is to provide data on the distribution and ecology of oligochaetes and aeolosomatids, including their density, faunal characteristics, species richness, diversity and equitability.

Description of sites studied

The coastline around the island of Martín García is asymmetrical. The western and northern shores constantly receive silty material which is deposited on a rocky bottom and is subsequently consolidated by vegetation. The eastern and southern shorelines are rocky, and because of strong southeasterly wind, they are only occasionally covered by silt-sand deposits. The functional shoreline asymmetry includes a differential distribution of vegetation. Thus, along the northern shore (Site 4, Puerto Viejo Bay, Fig. 1), areas of reedy vegetation are prevalent. These are composed of approximately 16 different species of hydrophytes (e.g. *Typha latifolia* L., *Echinodorus grandiflorus* (Cham & Schtdl.) Micheli, *Panicum pernambucense* (Spreng.) Mez ex Pilg., *Cyperus virens* Michx., *Pontederia rotundifolia* L.f., *Ludwigia elegans* (Cambess.) Hara, *Alternanthera philoxeroides* (Mart.) Griseb, *Ranunculus flageliformis* Sm.). Submerged vegetation such as *Potamogeton gayii* A. Benn, *Egeria densa* Planch, and *Myriophyllum aquaticum* (Vell.) Verdc.

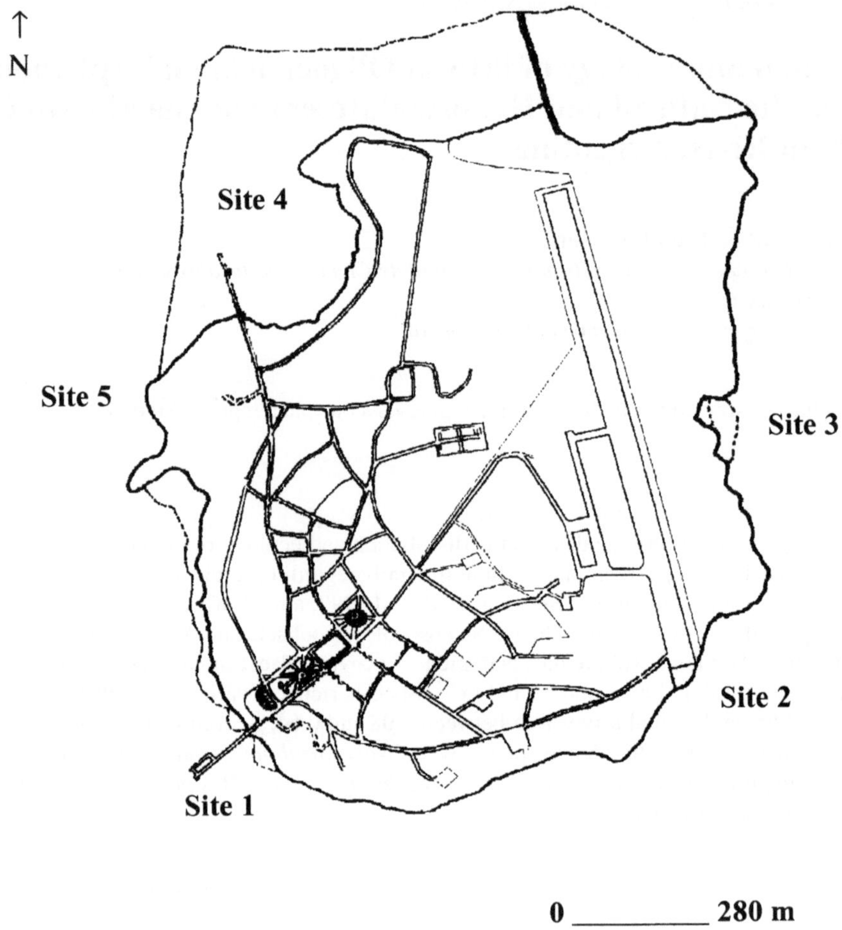

Figure 1. Map showing the sampling sites on the Island of Martín García.

grow in the permanently flooded areas of the reedy sectors (Lahitte & Hurrell, 1996). Site 5 is an area with sand-silt sediments nearby an anthropogenic garbage disposal operation. Towards the southern coast of the island is Site 1 (Puerto Nuevo jetty), which is sandy and supports a few reedy areas with *Schoenoplectus californicus* (C. A. Mey.) Soják. The southeastern end is a small beach of clean sand (Site 2); an extensive area of reeds occurs farther north along the coast (Site 3).

Materials and methods

Sampling was conducted on eight occasions between 1995 and 1997. Five sites were established along the island coast, each assigned by substrate characteristics (fine sands, silty sands, reeds and silted areas with great development of hydrophytes). The samples (three replicates per site) were taken using an Eckman hand dredge (175 cm^{-2}) and were fixed using 10% formaldehyde. The physical habitat and water quality parameters at each site were assessed and measured during each visit (Table 1).

Processing of samples in the laboratory included washing and sieving, using either a 125 μm mesh screen or the flotation techniques of Anderson (1959); the specific methodology employed depended on the dominant substrate of each sample. Floated material was sieved and then dyed with erythrosin. Oligochaetes were analyzed quantitatively and qualitatively using conventional microscope techniques. Taxonomic identifications follow Brinkhurst & Marchese (1992).

Table 1. Physical and chemical variables measured at sites 1–5 on Isla Martin Garcia

Site	Sampling date	T (°C)	Conductivity (μS cm^{-1})	DO (mg l^{-1})	pH	NH_4^+ (μg l^{-1})	NO_3^- (μg l^{-1})	SRP (μg l^{-1})	Ca^{2+} (mg l^{-1})	Mg^{2+} (mg l^{-1})	Na^+ (mg l^{-1})	K^+ (mg l^{-1})	HCO_3^- (mg l^{-1})	TP (mg l^{-1})	TOC (mg l^{-1})	SO_4^{2-} (mg l^{-1})	SS (mg l^{-1})
1	Mar-95	21.5	85.0	-	7.87	14.0	256.0	43.0	7.4	2.5	7.0	2.3	34.0	222.0	3.7	-	-
	May-95	19.5	117.4	7.9	5.8	10.0	183.0	40.0	8.7	3.0	15.4	2.4	49.0	140.0	6.4	4.9	67.2
	Aug-95	13.5	91.5	9.8	6.8	30.0	306.0	39.0	6.9	2.9	9.0	2.2	34.0	102.0		4.9	19.0
	Nov-96	29.5	123.5	-	7.17	–	–	–	–	–	–	–	–	–	–	–	62.4
	Mar-97	28.05	112.95	5.97	9.05	–	–	–	–	–	–	–	–	–	–	–	56.65
	Jun-97	15.5	116.0	–	–	–	–	–	–	–	–	–	–	–	–	–	58.3
2	Mar-95	29.0	92.3	8.2	8.22	–	185.0	40.0	6.6	3.5	8.1	2.7	35.0	162.0	3.5	–	–
	Nov-96	22.0	378.0	–	7.25	–	–	–	–	–	–	–	–	–	–	–	195.0
	Mar-97	26.93	108.2	6.87	8.26	–	–	–	–	–	–	–	–	–	–	–	54.4
	Jun-97	15.1	108.1	–	–	–	–	–	–	–	–	–	–	–	–	–	54.4
3	Mar-95	31.2	109.6	8.2	9.2	31.0	68.0	57.0	7.6	3.8	9.2	2.9	42.0	–	–	–	–
	May-95	20.8	139.2	8.9	5.7	12.0	179.0	44.0	9.4	3.9	16.9	2.8	50.7	291.0	7.7	4.6	–
	Aug-95	12.5	98.4	8.7	7.2	21.0	278.0	48.0	7.6	3.3	13.0	2.3	41.0	137.0	5.7	5.3	–
	Nov-96	24.4	118.3	–	8.09	–	–	–	–	–	–	–	–	–	–	–	59.4
	Mar-97	–	–	–	–	–	–	–	–	–	–	–	–	–	–	–	
	Jun-97	14.9	120.6	–	–	–	–	–	–	–	–	–	–	–	–	–	60.7
4	May-95	23.0	143.2	10.8	5.9	13.2	78.0	212.0	8.7	3.8	16.7	2.7	46.9	122.0	5.1	4.6	55.2
	Aug-95	14.0	118.2	10.7	7.1	19.0	100.0	48.0	7.3	3.4	14.6	2.9	40.0	110.0	5.7	5.0	14.0
	Nov-96	22.3	108.3	–	7.22	–	–	–	–	–	–	–	–	–	–	–	54.5
	Mar-97	25.9	101.5	3.93	6.11	–	–	–	–	–	–	–	–	–	–	–	50.9
	Jun-97	15.34	330.0	–	–	–	–	–	–	–	–	–	–	–	–	–	166.0
5	Mar-95	31.0	99.8	11.0	9.24	3.0	58.0	71.0	7.4	8.0	10.3	2.0	41.0	222.0	4.0	–	–
	Nov-96	28.1	120.8	–	7.59	–	–	–	–	–	–	–	–	–	–	–	60.2
	Mar-97	26.17	99.6	2.5	5.84	–	–	–	–	–	–	–	–	–	–	–	50.1
	Jun-97	15.2	144.6	–	–	–	–	–	–	–	–	–	–	–	–	–	74.4

Specific diversity (H') was calculated according to Shannon–Weaver, equitability (J) as $J = (e^{H'-1})(S-1)^{-1}$, where S is the number of species (Begon et al., 1988).

The calculated parameters (Cornet, 1986) included: mean density ($D = [\sum(n_i\ a)^{-1}]\ M^{-1}$, with n_i, number of specimens per sample; a, area sampled; M, total number of samples); dominance ($Dm = 100\ (n\ N^{-1})$, with n, total number of specimens for a given species; N, total number of specimens of all species) and frequency ($F = 100\ (m\ M^{-1})$, with m, number of samples containing a given species; M, total number of samples). Considering dominance and frequency (Rodriguez et al., 1980), species were classified as follows:

(a) Dominant: Dm $\geq$ 1%
(b) Constant: $F \geq 50\%$
(c) Accessory: $25\% < F < 50\%$
(d) Accidental: $F < 25\%$
(e) Expansive: $F > 15\%$ and Dm > 25%
(f) Diffuse: $F > 15\%$ and Dm < 25%

Cluster analysis of sites was performed using UPGMA and the Manhattan Distance coefficient (Crisci & López Armengol, 1983) pairing the values of numerical density of each species in the different seasons for each sampling locality.

Results

Thirty species of Oligochaeta and Aphanoneura representing six families (Naididae, Tubificidae, Narapidae, Enchytraeidae, Lumbricidae and Aeolosomatidae) were collected during this study.

The lowest total density of specimens (57 ind m^{-2}) was recorded at Site 3 in November 1995, whereas the highest total density of specimens (44 495 ind m^{-2}) was recorded at Site 4 in August of the same year. Diversity (H') ranged between 0.07 in Site 2 (March 1995) and 2.25 in Site 4 (June 1997) (Table 2); species richness varied from 1 (Sites 2 and 3) to 17 (Site 4). Equitability ranged from a low of 0.08 at Site 2 (November 1995) to a high of 0.64 at Site 1 (March 1996).

Table 2. Diversity, species richness and equitability of Annelida at the sampled sites

Site	Sampling date	Diversity	Species richness	Equitability
1	Aug/95	1.29	5	0.34
	Nov/95	1.04	3	0.52
	Mar/96	0.56	2	0.64
	Nov/96	0.37	5	0.13
	Mar/97	1.8	9	0.28
	Jun/97	1.56	5	0.44
2	Mar/95	0.07	2	0.39
	May/95		1	
	Aug/95	0.19	2	0.45
	Nov/95	0.08	6	0.08
	Mar/96		1	
	Nov/96	0.68	3	0.36
	Mar/97		1	
	Jun/97	1.5	5	0.41
3	Mar/95	1.7	7	0.33
	May/95	0.85	4	0.29
	Aug/95	0.97	10	0.1
	Nov/95		1	
	Mar/96	0.2	2	0.45
	Nov/96	1.14	9	0.14
	Jun/97	1.1	7	0.18
4	May/95	1.92	12	0.23
	Aug/95	1.08	10	0.12
	Mar/96	1.1	11	0.11
	Nov/96	1.5	11	0.16
	Jun/97	2.25	17	0.21
5	May/95	1.85	14	0.18
	Aug/95	1.00	7	0.17
	Nov/95	1.33	6	0.28
	Mar/96	1.23	5	0.32
	Nov/96	1.00	9	0.13
	Jun/97	1.05	9	0.13

Site 1

Fourteen taxa of oligochaetes and aeolosomatids were recorded from Site 1 during this study (Table 3). The density and diversity of species varied from one sampling period to the next. The total density of specimens ranged from 76 ind m^{-2} (March 1996) to 6457 ind m^{-2} (November 1996). For the highest of these values, 92% (5942 ind m^{-2}) correspond to the naidid *Amphichaeta leydigi* (Fig. 2). The lowest diversity occurred in November 1996, whereas the highest was observed in March 1997 (Table 2). Species richness ranged from 2 to 9 species in two consecutive years (March 1996 and 1997).

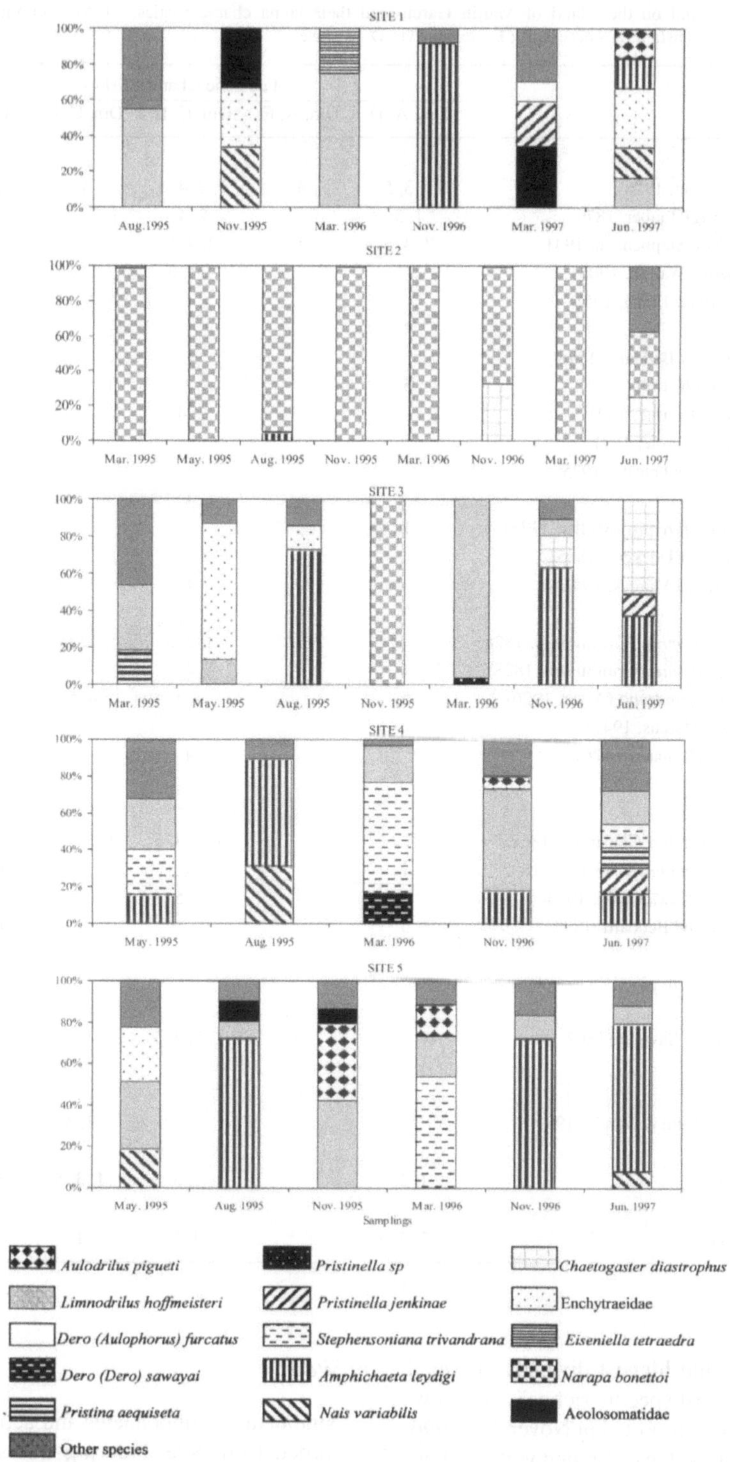

Figure 2. Temporal changes in abundance of the most relevant species recorded at the sampling sites during the study.

Table 3. Species found on the island of Martin Garcia and their fauna characteristics. (1–5), sampling sites 1–5; Dm – dominant; C – constant; A – accessory; E – expansive; D – diffuse

Species	Faunistic Characteristics					
	Dm, A, D	Dm, A, E	Dm, C, D	Dm, C, E	A, D	C, D
Naididae						
Nais variabilis Piguet, 1906	1, 3, 5	4	1, 3, 4, 5	1	3, 4	1, 3
Amphichaeta leydigi Tauber, 1879	1, 5		2, 4	1, 3, 4, 5		2
Pristinella jenkinae Stephenson, 1931	3, 4, 5	1	3, 4, 5		3, 4	
Pristinella osborni (Walton, 1906)	2, 5		3, 4, 5			
Pristinella acuminata Liang, 1958					5	
Pristinella sp.	3					
Pristina proboscidea Beddard, 1896			1			
Pristina leidyi Smith, 1896	5					2, 4
Pristina aequiseta Bourne, 1891			3, 4		3	5
Pristina breviseta Bourne, 1891	5				4	
Pristina synclites Stephenson, 1925					4	
Pristina sp.			4		4	
Dero (Aulophorus) furcatus (Müller, 1773)	1				4	
Dero Dero obtusa (d'Udekem, 1885)						4
Dero (Dero) sawayai Marcus, 1943			4		4	
Dero sp.	5				3	2
Chaetogaster diastrophus (Gruithuisen, 1828)			1, 2, 3	2, 3	5	2, 5
Chaetogaster diaphanus (Gruithuisen, 1828)	1		4			4
Stephensoniana trivandrana (Aiyer, 1926)	5		4, 5	4, 5		
Slavina evelinae (Marcus, 1942)			4		5	
Stylaria lacustris (Linnaeus, 1767)			4		5	4
Tubificidae						
Limnodrilus hoffmeisteri Claparède, 1862	1, 2		1, 3, 4, 5	1, 3, 4, 5		2
Limnodrilus udekemianus Claparède, 1862	1		4		4, 5	4
Aulodrilus pigueti Kowalewski, 1914	1		4, 5		3, 4	
Branchiura sowerbyi Beddard, 1892	5				4, 5	4
Inmature Tubificidae					4	
Lumbricidae						
Eiseniella tetraedra Savigny, 1867		1	1, 3, 4, 5		3, 4	3
Narapidae						
Narapa bonettoi Righi & Varela, 1983				2, 3	3	1
Enchytraeidae	1, 2, 3		1, 3	1, 3, 5	3, 4	4
Aeolosomatidae	2, 3, 5		4, 5	1	2, 5	5

During this study, the highest diversity at Site 1 coincided with the highest specific richness. The lowest equitability value was reported in November 1996. The highest equitability value coincided with the lowest species richness value. Site 1 was the site most affected by anthropogenic influences.

Site 2

Nine taxa of oligochaetes and aeolosomatids were recorded from Site 2 during this study (Table 3). Of these, *Narapa bonettoi* was the only species present in all samples. The lowest density of *N. bonettoi* was recorded in June 1997 (57 ind m^{-2}, 37%); the highest density of this species was recorded in November 1995 (25 971 ind m^{-2}, 99%) (Fig. 2). The lowest diversity and species richness values during this study were recorded from Site 2 (Table 2). Although the highest values of species richness were recorded at this site in November 1995, the lowest equitability value was recorded during this sampling period. The only species recorded from samples collected in May 1995, March 1996 and March 1997 was *N. bonettoi*.

Site 3

Fifteen taxa of oligochaetes and aeolosomatids were recorded from Site 3 during this study (Table 3). Total density ranged from 57 ind m^{-2} (November 1995) to 9333 ind m^{-2} (November 1996). *Narapa bonettoi* was the only species recorded in November 1995. In November 1996, the highest abundance was recorded for *A. leydigi* (5942 ind m^{-2}, 64%) (Fig. 2). The lowest diversity value was recorded in March 1996, coincident with the highest equitability value (Table 2). The highest species richness value was recorded in August 1995, when equitability was lowest. Total specimen density recorded in November 1995 was particularly low, given the high values of density exhibited in this station (Fig. 2).

Site 4

Twenty-three taxa of oligochaetes and aeolosomatids were recorded from Site 4 during this study (Table 3). Mature specimens of *Limnodrilus hoffmeisteri* appeared in March 1996 and June 1997, representing 14 and 2.5% of the population, respectively. This is a station with high total densities, where the lowest value was recorded in May 1995 (4200 ind m^{-2}). The highest density occurred in August of the same year (44 495 ind m^{-2}) and the most abundant species then were *A. leydigi* (58%) and *Nais variabilis* (31.5%) (Fig. 2). This was the station with the highest species richness presenting a maximum value in June 1997, when the sampling yielded the highest diversity too (Table 2).

Site 5

Twenty taxa of oligochaetes and aeolosomatids were recorded from Site 5 during this study (Table 3). Mature specimens of *L. hoffmeisteri* were recorded in May 1995 and March 1996 (7.4 and 25% of the population, respectively). The lowest total abundance was recorded in March 1996 (991 ind m^{-2}). The highest numerical density was recorded in June 1997 (9600 ind m^{-2}), of which *A. leydigi* comprised 71% (Fig. 2). The highest diversity is coincident with the highest species richness in May 1995 and the highest equitability was observed in March 1996, coincident with the lowest recorded species richness (Table 2).

Following Cornet (1986), the dominant, constant and widely distributed species were *N. variabilis*, *Stephensoniana trivandrana*, *A. leydigi*, *Chaetogaster diastrophus*, *L. hoffmeisteri*, *Aulodrilus pigueti*, *N. bonettoi* and the species of Enchytraeidae and Aeolosomatidae. *Pristina synclites* showed an accessory and diffuse presence (Site 4, March 1996), while *Dero (D.) obtusa* was constant and diffuse turning up only once in Site 4 (November 1996) (Table 3).

For similarity analysis between seasons and sampling sites, 29 of the 30 annelid species were considered. *L. hoffmeisteri* was excluded because it was present in all sampling stations. Two groups could be distinguished (Fig. 3). Group A includes Site 2 in all sampling periods and Site 1 in autumn and winter. Group B includes Site 3 (autumn, winter and spring), Site 1 (spring) and Site 5 in winter. All other stations do not form homogeneous groups.

Discussion

Two sectors can be distinguished in the Río de la Plata River. In the inner sector, where the subfluvial delta is developed, the predominant sediments are fine sand, silt and clay (Boschi, 1988). In this sector, the marine effects are perceived only by the tides, which reach further upstream from the island of Martín García to the town of San Pedro (province of Buenos Aires). However, the organisms living in this area are typically freshwater organisms. The amount of suspended particulate matter is fairly large, most of it coming from the Paraná River (Boschi, 1988). This has the effect of increasing water turbidity and thus limiting primary production, despite the availability of nutrients and oxygen. The values of particulate matter and, among others physical and chemical variables measured, the conductivity and pH showed similar values (Table 1) to those registered for the Middle Paraná and Río de la Plata rivers (Marchese & de Drago, 1992; Arraga et al., 1997).

Although Marchese & de Drago (1992) recorded only slight fluctuations in the temperature of the Paraná River, the Río de la Plata River system is

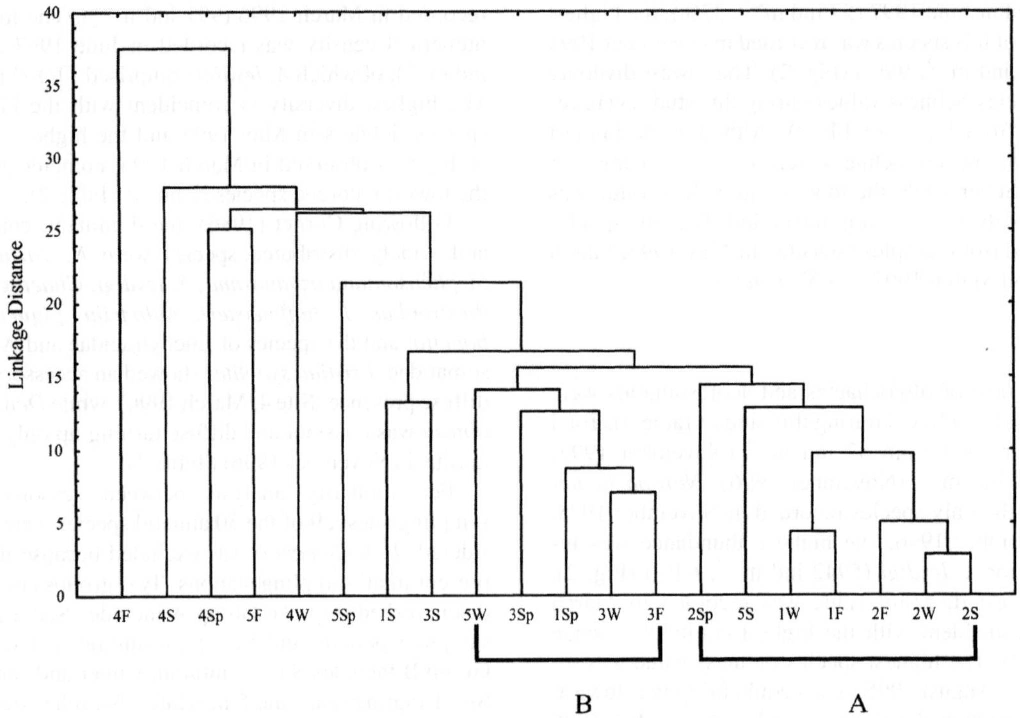

Figure 3. Dendrogram of cluster analysis of sites using UPGMA and the Manhattan Distance coefficient in the different seasons for each sampling locality. 1, 2, 3, 4 and 5, sampling sites; F – fall; S – summer; Sp – spring; W – winter.

strongly influenced by seasonal variations and the dynamics of the ecosystem (Arraga et al., 1997). With an annual temperature fluctuation of ±22.6 °C, the thermal characteristics of the Río de la Plata River are highly variable. Around the coast of Martín García, we have recorded a temperature difference of 18.7 °C between summer and winter (Site 3).

The coast around the island supports a much more diverse oligochaete fauna than is supported by the Middle Paraná. Among the species common to both areas are *Limnodrilus hoffmeisteri, Aulodrilus pigueti, Pristinella jenkinae, Dero (Aulophorus) furcatus, Narapa bonettoi* and *Eiseniella tetraedra* (Marchese & de Drago, 1983; Marchese, 1986).

Most of the species found around the island are cosmopolitan and have been associated with diverse aquatic environments at different latitudes (Collado et al., 1999; Erséus et al., 1999).

Narapa bonettoi, considered to be endemic, being dominant and occurring in high numbers in the Paraná River (Righi & Varela, 1983; Marchese & de Drago, 1983, 1992; Marchese, 1987; Paggi et al., 1998), has also been found in the Río de la Plata River. It is typically rheophilous and psammophilous, living in clear bottoms where the current velocity is high. During the present study, we found *N. bonettoi* most commonly at Site 2, in sandy sediments. There, it was the most commonly collected and widely distributed species during each collecting period and frequently the only species found in this station. It was occasionally recorded from Sites 1 and 3, where the sediment was a mixture of sand and silt. It is also found near the deltaic area of the river, always associated with the same kind of sediment (Rodrigues Capítulo et al., 1997).

Another of the species recorded from Martín García was *Limnodrilus hoffmeisteri*, an ubiquitous and cosmopolitan species (Probst, 1987; Sang, 1987; Milbrink, 1994; Rieradevall & Real, 1994). It was found also in the Middle Paraná River (Marchese & de Drago, 1983, 1992; Marchese, 1986, 1987). In the Río de la Plata River, primarily in areas receiving discharge from sewage systems, it is one of the most abundant tubificids (Rodrigues Capítulo et al., 1997). It was reported in association with the byssus of *Limnoperna fortunei* (Mytilidae) by Darrigran et al. (1998). Around the island, *L. hoffmeisteri* was recorded from all the sampling stations and in all seasons, with one exception (Site 2, November 1995), and it was the dominant species during this study. The highest density of specimens was always recorded from Site 4 (silty sediment colonized by hydrophytes). Mature specimens were recorded in summer and autumn, as also recorded in Bagliardi Beach associated with *Limnoperna fortunei* (pers. obs.).

Aulodrilus pigueti, also an ubiquitous species (Fomenko, 1972; Rieradeval & Real, 1994), has been found in high densities in the Middle Paraná River (Marchese, 1986, 1987; Marchese & de Drago, 1992). During this study, *A. pigueti* was recorded in all seasons and at all sampling localities except for Site 2. Its faunal characteristics were variable (Table 3). The highest densities of specimens were recorded in Site 4 in autumn and spring.

Other tubificids recorded from the island, although less numerous, were *Branchiura sowerbyi* and *Limnodrilus udekemianus*, both found in stations with silty sediments, rich in organic matter and vegetation. Sang (1987) and Probst (1987) also found these species in eutrophic and partially polluted environments. *B. sowerbyi*, recorded only during the temperate and warm seasons, may be an indicator of more temperate and perhaps thermally influenced waters (Sang, 1987).

Eiseniella tetraedra was recorded in the Middle Paraná River and in the Río de la Plata River (Marchese & de Drago, 1992; Darrigran et al., 1998). During this study, it was recorded in stations with silt or silt-sand sediments tolerating eutrophic conditions similar to those discussed by Milbrink (1994), although its faunal characteristics were highly variable (Table 3).

Among the naidids collected during this study, several are considered to be cosmopolitan, adapted to a range of environmental conditions (Fomenko, 1972; Verdonschot, 1987; Sang, 1987; McCall & Soster, 1990; Snimschikova & Akinshina, 1994). The dominant, constant and most widely distributed species of naidids around the island were *Nais variabilis*, *Stephensoniana trivandrana*, *Amphichaeta leydigi* and *Chaetogaster diastrophus*. *Amphichaeta leydigi* was found in all the sampling localities, and it had been previously reported by Fomenko (1972) in sand and sand-silt sediments, in both temperate and cold seasons. Other naidids appeared only occasionally and showed variable faunal features (Table 3).

Although high values of species richness were recorded during this study, diversity was not variable. This could be explained by the high relative abundance exhibited by some naidid species and the tubificid *Limnodrilus hoffmeisteri* on several occasions during this study. Statzner (1981) and Sang (1987) considered that the diversity index may be reduced by the high relative abundance values of some species or by the effects of pollution, which result in lower densities of sensitive species and an increase in densities of tolerant species.

From the cluster analysis performed, group A is the only one clearly distinguishable from the other groups, probably because Site 2 always exhibited the lowest species richness values related to the other stations. Site 1 is incorporated into the group during autumn and winter, with average values of species richness and density of specimens similar to those of Site 2. Site 5 is similar to Sites 1 and 2 during the summer, when the highest value for species richness was determined. Group B included stations of similar, intermediate species richness and average abundances. Although the rest of the stations did not comprise homogeneous groups, they presented the highest species richness, diversity and abundance values.

Acknowledgements

We wish to thank the Ranger Corps of Martín García and the Dirección de Recursos Naturales del Ministerio de Agricultura, Ganadería y Alimentación, Provincia de Buenos Aires. We would also like to thank M. Sc. Mercedes Marchese for her support and Dr Alejandra Rumi for her useful suggestions on a first draft of the manuscript. Financial support was provided by the Facultad de Ciencias Naturales y Museo, the UNLP and the Comisión de Investigaciones Científicas de la Provincia de Buenos Aires.

References

Anderson, R. O., 1959. A modified flotation technique for sorting bottom fauna samples. Limnol. Oceanogr.: 223–225.

Arraga, E., J. Bazán, M. Cavallo, J. Degli Alberti, A. Domenech, A. Gando, E. Germinaro & J. López, 1997. El oxígeno disuelto y sus demandas. Calidad de las aguas de la Franja Costera Sur del Río de la Plata (San Fernando-Magdalena). Consejo Permanente para el Monitoreo de la Calidad de las Aguas de la Franja Costera Sur del Río de la Plata (ed.): 39–45.

Begon, M., J. L. Harper & C. R. Townsend, 1988. Ecología. Individuos, poblaciones y comunidades. Omega, S. A., Barcelona: 886 pp.

Boschi, E. E., 1988. El ecosistema estuarial del Río de la Plata (Argentina y Uruguay). An. Inst. Cienc. Mar Limnol. 15: 159–182.

Brinkhurst, R. O. & M. R. Marchese, 1992. Guia para la identificación de oligoquetos acuáticos continentales de Sud y Centroamérica. Col. Climax 6, Santo Tomé: 207 pp.

Collado, R., P. Kasprzak & R. M. Schmelz, 1999. Oligochaeta and Aphanoneura in two Northern German hardwater lakes of different trophic state. Hydrobiologia 406: 143–148.

Cornet, M., 1986. Evolution des parametres de structure des populations de mollusques bivalves des fonds à Abra Alba au large de Bassin d' Arcachon. Vie Milieu 36: 15–25.

Crisci, J. V. & M. F. López Armengol, 1983. Introducción a la teoría y práctica de la taxonomía numérica. Monografía No 26, Serie Biología, OEA, Washington D. C.: 132 pp.

Damborenea, M. C., I. I. César & L. C. Armendáriz, 1997. Especies de *Temnocephala* (Platyhelminthes Temnocephalidae) de la Isla Martín García, Buenos Aires, Argentina. Neotropica 43: 123–124.

Darrigran, G., S. M. Martín, B. Gullo & L. Armendáriz, 1998. Macroinvertebrates associated with *Limnoperna fortunei* (Dunker, 1857) (Bivalvia, Mytilidae) in Río de la Plata, Argentina. Hydrobiologia 367: 223–230.

Erséus, C., R. Grimm, B. Healey, S. Lundberg, E. Rota & T. Timm, 1999. Clitellate diversity in Nationalstadsparken, an urban national park in Stockholm, Sweden. Hydrobiologia 406: 101–110.

Fernández, L. A. & M. López Ruf, 1999. Coleoptera y Heteroptera acuáticos y semiacuáticos de la Isla Martín García (Provincia de Buenos Aires). Physis (Buenos Aires), Secc. B, 57: 1–4.

Fomenko, N. V., 1972. Grupos ecológicos de Oligochaeta del Río Dniepper. Oligochaeta Acuáticos (Sistemática, ecología, investigaciones de la fauna de URSS). Acad. Cienc. URSS, Asoc. Hidrobiol. Sov. Trab.17. 'Ciencia', Moscú (in Spanish, translation from Russian).

Juárez, M. C., 1995. Estatus de residencia y categorización trófica de las aves en la Reserva Natural isla Martín García, Río de la Plata Superior. Neotropica 41: 83–88.

Lahitte, H. B. & J. A. Hurrell, 1994. Los Arboles de la isla Martín García. Arboles y arbustos arborescentes (nativos y naturalizados) de la Reserva Natural y Cultural Isla Martín García. Comisión de Investigaciones Científicas (CIC), La Plata: 135 pp.

Lahitte, H. B., J. A. Hurrell, G. Canda & C. Trucco Alemán, 1995. Poblaciones de *Tupinambis teguixin* (Sauria-Teidae) de la Isla Martín García (Buenos Aires, Argentina) I. Etnobiología y encuadre metodológico-cognitivo. Pinaco 2: 45–62.

Lahitte, H. B. & J. A. Hurrel, 1996. Plantas Hidrófilas de la Isla Martín García (Buenos Aires, República Argentina). Comisión de Investigaciones Científicas (CIC), La Plata: 236 pp.

Lahitte, H. B. & J. A. Hurrell, 1998. Catálogo de las Aves de la Isla Martín García (Buenos Aires, Argentina). Comisión de Investigaciones Científicas Ser. Inf. 53, La Plata: 69 pp.

Marchese, M., 1986. Nuevos aportes al conocimiento de los Oligoquetos del Río Paraná Medio y algunos tributarios. Studies Neotrop. Fauna Envir. 21: 231–249.

Marchese, M., 1987. The ecology of some benthic Oligochaeta from the Paraná River, Argentina. Hydrobiologia 155: 209–214.

Marchese, M. & I. E. de Drago, 1983. Zoobentos de los principales tributarios del Río Paraná Medio en el tramo Goya-Diamante. Su relación con el cauce principal y cauces secundarios. Rev. Asoc. Cs. Nat. Lit. 14: 95–109.

Marchese, M. & I. E. de Drago, 1992. Benthos of the lotic environments in the middle Paraná River system: transverse zonation. Hydrobiologia 237: 1–13.

McCall, P. L. & F. M. Soster, 1990. Benthos response to disturbance in Western Lake Erie: regional faunal surveys. Can. J. Fish. aquat. Sci. 47: 1996–2009.

Milbrink, G., 1994. Oligochaetes and water pollution in two deep Norwegian lakes. Hydrobiologia 278: 213–222.

Paggi, A. C., I. I. César & A. Rodrigues Capítulo, 1998. Benthic studies in the zone of islands of Yacyretá prior to impoundment of the Upper Paraná River (Argentina). Verh. int Ver. Limnol. 26: 1089–1094.

Probst, L., 1987. Sublitoral and profundal Oligochaeta fauna of the Lake Constance (Bodensee-Obersee). Hydrobiologia 155: 277–282.

Ravizza, G. B., 1984. Principales aspectos geológicos del cuaternario en la isla Martín García, Río de la Plata Superior. Asoc. Geol. Arg. Rev. 39: 125–130.

Rieradevall, M. & M. Real, 1994. On the distribution patterns and population dynamics of sublittoral and profundal oligochaeta fauna from Lake Banyoles (Catalonia, NE Spain). Hydrobiologia 278: 139–149.

Righi, G. & M. E. Varela, 1983. *Narapa bonettoi* gen. nov. sp. nov. (Oligochaeta, Narapidae, fam. nov.) de agua doce da Argentina. Rev. Asoc. Cs. Nat. Litoral 14: 7–15.

Rodrigues Capítulo, A., I. César, M. P. Tassara, A. C. Paggi & M. Remes Lenicov, 1997. Zoobentos. Calidad de las aguas de la Franja Costera Sur del Río de la Plata (San Fernando-Magdalena). Consejo Permanente para el Monitoreo de la Calidad de las Aguas de la Franja Costera Sur del Río de la Plata (ed.), Buenos Aires: 157 pp.

Rodriguez, V., M. Ibáñez & J. Rodríguez, 1980. Ecologie des Annélides Polychètes de quelques plages de la Baie D' Algeciras (Espagne). Vie Milieu 30: 131–138.

Rumi, A., S. M. Martin, M. P. Tassara & G. Darrigran, 1996. Moluscos de agua dulce de la Reserva Natural e Histórica Isla Martín García, Río de la Plata, Argentina. Com. Soc. Malac. Urug. 7: 7–12.

Sang, Q., 1987. Some ecological aspects of aquatic oligochaetes in the Lower Pearl River (People's Republic of China). Hydrobiologia 155: 199–208.

Snimschikova, L. N. & T. W. Akinshina, 1994. Oligochaete fauna of Lake Baikal. Hydrobiologia 278: 27–34.

Statzner, B., 1981. Shannon–Weaver diversity of the macrobenthos in the Schierenseebrooks (North Germany) and problems of its use for the interpretation of the community structure. Verh. int. Ver. Limnol. 21: 782–786.

Verdonschot, P. F. M., 1987. Aquatic oligochaetes in ditches. Hydrobiologia 155: 283–292.

Hydrobiologia **463:** 217–222, 2001.
P. Rodriguez & P.F.M. Verdonschot (eds), Aquatic Oligochaete Biology VIII.

Effects of *Nais elinguis* on the performance of an activated sludge plant

Christa H. Ratsak
Faculty of Biology, Free University, De Boelelaan 1087, 1081 HV Amsterdam, The Netherlands
Present address: Sub-department of Environmental Technology, Wageningen University, P.O. Box 8129, 6700 EV Wageningen, The Netherlands
E-mail: christa.ratsak@algemeen.mt.wag-ur.nl

Key words: oligochaete worms, *Nais elinguis*, activated sludge, sludge reduction, sludge volume-index, energy consumption

Abstract

The oligochaete worm *Nais elinguis* was counted during a year and a half in a full-scale, completely mixed, municipal activated sludge plant consisting of four aeration tanks connected in parallel. Simultaneously the operating variables, i.e. effluent quality, energy costs in kWh for oxygen supply in the aeration tanks, and sludge-disposal were measured. The number of worms varied both seasonally and among the aeration tanks. A major worm bloom resulted in a low sludge volume-index, lower energy consumption for oxygen supply expressed in kWh and, depending on the temperature, less sludge-disposal. The worms had no influence on the effluent quality.

Introduction

Aerobic wastewater treatment is not only a technical or (bio)chemical process but also a biological process. Although activated sludge contains many organisms, the only organisms extensively studied are bacteria. However, intra- and inter-specific interactions between bacteria and predator organisms are important. Nowadays, the relationships between the different organisms in municipal wastewater such as bacteria, protozoa and metazoa are receiving more and more attention. First, because of the possible negative impact of grazers due to decreasing the amount of bacteria (Lee & Welander, 1993) and, second, because predator organisms have a positive influence on several processes such as removal of suspended bacteria, bacterial growth, and floc formation (Curds, 1975; Learner, 1979; Eikelboom, 1988; Verdonschot, 1989; Ratsak et al., 1993; Ratsak et al., 1994).

During aerobic wastewater treatment, a lot of waste-sludge is produced. Because of the high cost of surplus sludge disposal, it is important to search for ways to reduce sludge production. One option to reduce sludge production is to introduce organisms in the sewage treatment process which feed on the (sludge) bacteria. Sludge is converted into energy, new biomass, water and carbon dioxide through predation. Energy loss occurs because the biomass conversion from prey- to predator-biomass is not 100%. The total loss of energy will be maximal (mineralization) and biomass production minimal under specific, imposed conditions.

Oligochaete worms occur in many municipal wastewater treatment systems in The Netherlands. However, their presence is unpredictable and uncontrollable. The primary aim of this study was to evaluate the presence of oligochaete worms in an activated sludge plant. It is important to learn if these organisms indeed are able to persist in the plant and thus, can be used to regulate biomass production. The influence of these organisms on sludge-quality, sludge-disposal and other important parameters was also studied. This study may be regarded as a feasibility study pointing to useful roles of predator organisms in activated sludge plants.

Materials and methods

The activated sludge plant

The activated sludge plant studied is a municipal

Table 1. Specifications of Deventer wastewater treatment plant. MLSS – mixed liquor suspended solids; p.e. – population equivalents; COD – chemical oxygen demand; BOD – biological oxygen demand; N-Kj – Kjeldahl-nitrogen

Parameter	Range	Unit
Total volume aeration tanks	6000	m^3
Design capacity	100 000	p.e.
Mean sludge age	15	days
Setpoint biomass concentration	5	g MLSS l^{-1}
Setpoint oxygen concentration	2	mg l^{-1}
Sludge loading rate	0.173	kg COD kg^{-1} MLSS.day
(annual average 1992)	0.062	kg COD kg^{-1} MLSS.day
	0.033	kg BOD kg^{-1} MLSS.day
Mean yearly sludge production	200 000	m^3
	10 000	ton MLSS

wastewater treatment plant located in Deventer (The Netherlands). It is a completely mixed system, consisting of four aeration tanks (ATs) connected in parallel, each provided with two aerators. The sludge load of the four lines is always the same, and varies depending on the diurnal fluctuations in both influent flow rate and strength. The return activated sludge flows are completely separated. The sludge wastage rate is 80 m^3 day^{-1}. The specifications of the treatment plant are shown in Table 1.

The organisms

In order to follow the behaviour of the worm populations, 1 litre of mixed liquor suspended solids (MLSS) was collected weekly from each aeration tank over one and a half years. Subsequently, five aliquots of 200 μl MLSS were investigated under the binocular microscope. The worms in the MLSS were counted and an average density was calculated.

The analysis

During the period of oligochaete counting, the following data were collected from the database in Deventer: dry weight (DW) of MLSS, amount of sludge-disposal, temperature, sludge-age, sludge volume-index, oxygen concentration and energy consumption (kWh) to maintain an oxygen concentration of 2 mg l^{-1}. At the end of the experiment, the effluent composition of two aeration tanks with and without worms was analysed three times a week over 1 month. The following variables were analysed: BOD_5, COD, Kjeldahl-N, NH_4-N, NO_4-N and PO_{4tot}-P.

Results

Number of oligochaetes

In October 1992, sludge wastage was nil from AT2 and AT3 of the activated sludge plant in Deventer, over a period of 3 weeks (weeks 40–42). Conspicuously, in the same period very large numbers of small worms were observed in these tanks (personal communication). In the other two ATs, the sludge-disposal continued, and in these tanks many fewer worms were present, or they were even absent. The MLSS concentration of the ATs with worms did not differ from the ATs without worms. In week 43, the number of worms in the aeration tanks was counted for the first time. Unfortunately, the number of worms declined quickly during the next few weeks.

The most common worms in the activated sludge tanks at Deventer were oligochaetes. Three different annelid species were found: *Nais elinguis*, *Pristina* sp. and *Aeolosoma hemprichi*. Nematodes were observed only occasionally. Since *N. elinguis* dominated in the ATs, the number of this oligochaete species was counted weekly.

The number of *N. elinguis* differed per AT (Fig. 1). In September/October 1992 (weeks 38–40), a major worm bloom was observed in AT1 and AT2. At the

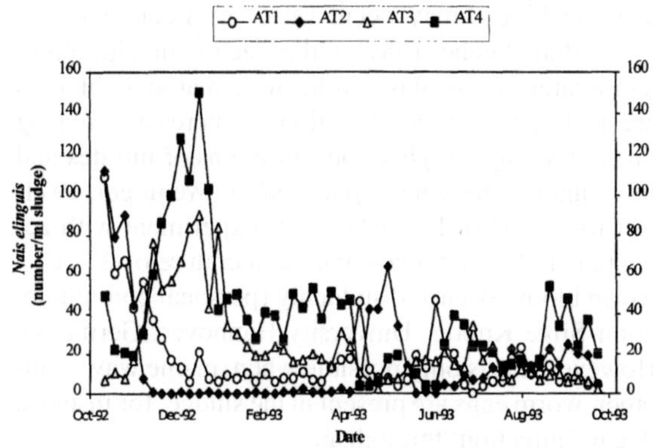

Figure 1. The number of worms in all four ATs during the measurement period of one and a half years.

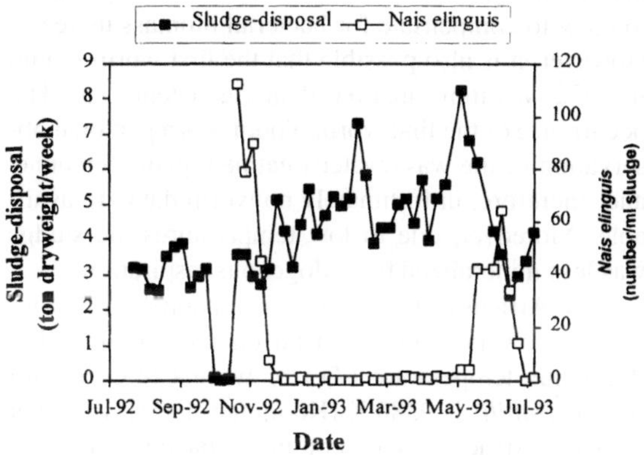

Figure 2. The course of the worm numbers and the sludge-disposal of AT2.

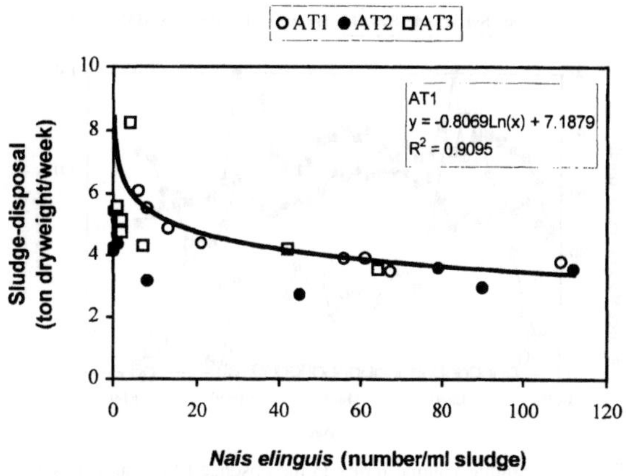

Figure 3. Sludge-disposal expressed against the number of worms during blooms in 3 ATs. The Pearson correlation coefficient, R^2, was valid only for the data set originating from AT1.

end of 1992, the decline of worms in AT1 and AT2 was followed by a worm bloom in AT3 and AT4. In 1993, a small worm bloom occurred in May. Extreme rainfall probably prohibited worm bloom in August and September. Only a minor worm bloom was observed in AT4. Besides these incidental worm blooms, the numbers of worms in the different ATs showed an oscillating pattern over time with minor worm blooms of up to 50 worms ml^{-1} sludge.

Sludge-disposal

The amount of sludge-disposal was inversely proportional to the number of worms (Fig. 2). There was no sludge-disposal at all in AT2 and AT3 in October 1992 (as described earlier). A large number of worms was present in the tanks during the same period. The increase of worms coincided with a decrease in the sludge-disposal in May 1993. In June, the number of worms declined and sludge-disposal increased. The amount of sludge-disposal is expressed against the number of worms in the ATs in Figure 3. Because there were problems with the sludge recycle pump of one aeration tank in this period, the data of this aeration tank are not included. The sludge-disposal of an AT with worms is about twice as low as that of an AT without worms. All three ATs show a negative relationship between the sludge-disposal and the number of worms. When less worms are present in the AT, the waste-sludge production increases and vice versa.

Temperature

A worm bloom in winter had no marked influence on sludge-disposal, but a major worm bloom resulted in a significant or even complete surplus sludge reduction especially in the summer, see September/October 1992 (Fig. 2). The temperature of the sludge during this first worm bloom varied between 18 and 19 °C. The subsequent worm bloom in AT4 did not have the same effect. During this bloom, the temperature varied between 8 and 11 °C. The third, small bloom in May/June 1993 which was also accompanied by a considerable surplus sludge reduction, occurred when the temperature was 17–21 °C (Fig. 2).

Sludge volume-index

The sludge volume-index (SVI) was considerably lower when worms were present in the ATs (Fig. 4). As an example, the SVI of AT4 and AT2 are given.

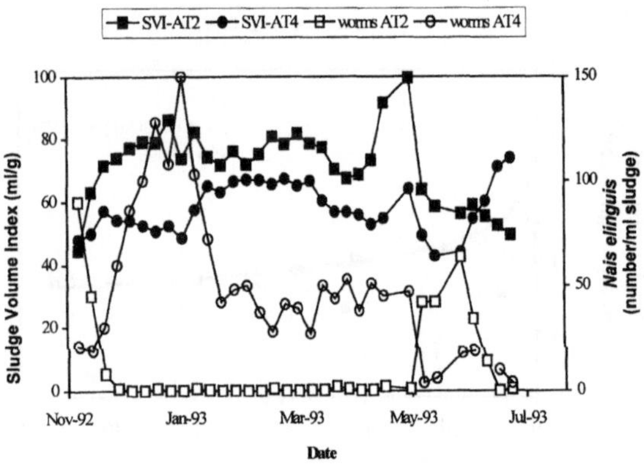

Figure 4. Sludge volume-index in the absence (AT2) and presence (AT4) of worms.

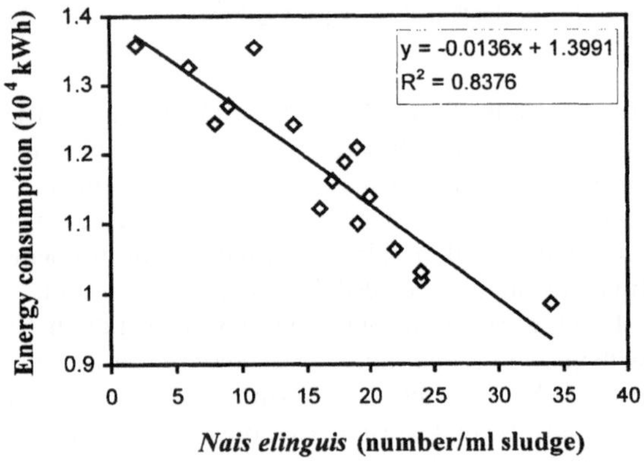

Figure 5. Energy consumption in kWh for the oxygen supply expressed against the number of worms in AT3.

Energy consumption

Figure 5 illustrates that the energy consumption for oxygen supply (kWh) decreased with increasing numbers of worms in AT3. When worms were present, energy consumption was 20–30% lower.

Effluent quality

There were no large differences in effluent composition.

Discussion

The number of worms in an aeration tank varied over time (within an aeration tank) and space (between different aeration tanks). Under full-scale conditions even in winter, when the temperature is relatively low, a worm bloom may occur (also see Learner et al., 1978; Pfannkuche, 1981). Other factors in addition to temperature must play a role. In nematodes, blooms are fostered by a combination of increasing dosing rates of sewage, higher concentrations of nitrates and more intense growth of *Sphaerotilus* (Weninger, 1971, in Curds, 1975). In a laboratory experiment with activated sludge and a constant temperature of 27 °C, a worm bloom occurred suddenly (personal communication Silke Knoop, University Hannover, Germany). However, this is only possible when, in one way or another, worm eggs are present in the sludge, for instance due to 'infection' through air.

A worm bloom in winter did not have the same effect as a worm bloom in summer. The rate of grazing and the growth of worms in winter were probably too low to compensate for bacterial biomass increase. However, it is also possible that the first worm bloom in 1992 was more marked than the second one. The occurrence of the first worm bloom was a personal observation of the wastewater treatment plant personnel and, therefore, unfortunately no worm data are available. Moreover, due to low temperatures the sludge was less mineralized by endogenous respiration.

The Arrhenius temperature is a parameter that indicates how processes depend on temperature. The Arrhenius temperature (T_A) of worms (e.g. T_A for *Nais variabilis* is 9380 K [Kooijman, 1993]) is higher than the Arrhenius temperature of bacteria (T_A varies between 6000 and 7000 K) (Kooijman, 1993). When this parameter is large the rate of physiological processes is more strongly influenced by temperature. Thus, at low temperatures, the metabolic rate of worms is slackened compared to the metabolic rate of the bacteria. However, on the other hand, when the temperature is rising, the metabolic rate of worms considerably accelerates compared to that of bacteria. This could explain the excessive sludge reduction in a period with prolonged higher temperatures. Other experiments with predator and prey cultured in a chemostat at a constant temperature and with a constant food supply showed oscillations in prey and predator populations. The cyclic behaviour of predator and prey organisms is extensively documented in population dynamics literature (Kooijman, 1993).

There was evidence that sludge production decreased when worms were present in the ATs. Rensink & Rulkens (1997) and Janssen et al. (1998) found sludge reductions in the range of 10–75% in pilot plant experiments with *Tubifex*. Despite the fact that *Nais* and *Tubifex* belong to families with specific ecological niches, comparable results concerning sludge reduction are feasible.

The observations described in the present work show that the relationship between worm bloom and temperature is still difficult to interpret. Certainly, as mentioned above, other still overlooked effects also play a part. It is, however, obvious that a major worm bloom only led to substantial surplus sludge reduction when the temperature was favourable.

The sludge volume-index was considerably lower when worms were present. In activated sludge plants in the Netherlands, including 'Kampen', 'Horstermeer' and 'Nieuwe Waterweg', worm blooms also yielded low sludge indices (personal communication from the managers). Several authors have mentioned this (positive) effect of ciliates and other predator organisms on the sludge volume-index. This is due to excretion products (faeces) of the organisms and the forming of larger flocs by the bacterial population in order to withstand grazing pressure, which have a positive influence on floc forming processes and thus yield lower sludge indices (Ratsak et al., 1996).

The energy consumption for oxygen supply in the aeration tanks decreased with increasing numbers of worms. This observation is somewhat unexpected since mineralization of the sludge increases when worms are present and thus the oxygen consumption also increases. There are two possible explanations for this phenomenon. First, in terms of the sludge composition changes due to the nature of excretion products. This could result in a lower density of the sludge and thus a lower resistance. This means that the oxygenation capacity increases at the same kWh setpoint. Second, in terms of the carbon/oxygen (C/O) ratio of bacteria and worm biomass. The C/O ratio of bacteria is 0.5 and that of polychaete worms is 3.12 (resembling oligochaete worms) (Percy & Fife, 1981). When biomass conversion is 100%, 1 mol C of bacteria is converted into 1 mol C of worms. This results in a release of oxygen when worms graze bacteria. However, these hypotheses, are only theoretical and have to be confirmed experimentally.

Conclusions

By monitoring the presence of worms (predators) in an activated-sludge plant, it appeared that:

1. The number of worms in an aeration tank was inversely related to sludge-disposal.
2. An overall biomass reduction of 25–50% was possible.
3. No sludge-disposal was necessary when major worm-blooms occurred in the summer.
4. The oxygen demand of sludge was lower when predators were present.
5. The threshold value of worms influencing the sludge-disposal was different in summer and winter.
6. Prolonged rainfall resulted in a wash-out of the worms not because of the low residence time but because of diluted sludge (the sludge structure changes).
7. Grazing of the sludge flocs is limited by the mechanistic limits of worm ingestion properties (size of the mouth and pharynx). When flocs are too small or too large an oligochaete worm is not able to take up food.

Grazing by predators can diminish sludge production to a considerable extent but further research on worm-blooms is needed:

1. To assess precisely the environmental conditions which stimulate a worm bloom.
2. To maintain worm-blooms under controlled full-scale conditions.

Further research on the feasibility of regulating the sludge-production in pilot-plants by oligochaetes grazing activity is now in progress in co-operation with DHV Water consultants, The Netherlands.

Acknowledgements

This work was financially supported by the 'Institute for Inland Water Management and Waste Water Treatment (RIZA)' and the 'Foundation for Water Research (STOWA)' within the framework of the research programme 'Future Treatment Techniques for Municipal Waste Water (RWZI 2000)' in The Netherlands. The author is grateful to the co-operators and the manager of the Deventer wastewater treatment plant, a division of the Water Authority 'West-Overijssel', for their hospitality and co-operative support during the weekly sampling of the four aeration tanks. I also want to thank the crew of the analytical laboratory at Zwolle for analysing the effluent samples. Ir. D. Eikelboom of the TNO Institute of Environmental and Energy research, the Netherlands and Dr P. Verdonschot of

Alterra Green World research, Team Freshwater Ecology, The Netherlands, are also gratefully acknowledged for carefully reading this manuscript and for their helpful discussions.

References

Curds, C. R., 1975. Protozoa. In: Curds, C. R. & H. A. Hawkes (eds), Ecological Aspects of Used-water Treatment, Vol. 1, The Organisms and their Ecology. London Academic Press Ltd.: 203–268.

Eikelboom, D. H., 1988. Extra toepassingsmogelijkheden voor Protozoa en Metazoa bij de zuivering van afvalwater. In Dutch. TNO-rapport nr.: R 88/286. Projektnummer: 18614.

Janssen, P. M. J., W. H. Rulkens, J. H. Rensink & H. F. van der Roest, 1998. The potential significance of Metazoa in biological waste water treatment. Sludge Management Sept/Oct.: 25–27.

Kooijman, S. A. L. M., 1993. Dynamic Energy Budgets in Biological Systems. Cambridge University Press.

Learner, M. A., G. Lochhead & B. D. Hughes, 1978. A review of the Biology of British Naididae (Oligochaeta) with emphasis on the lotic environment. Freshwat. Biol. 8: 357–375.

Learner, M. A., 1979. The distribution and ecology of the Naididae (Oligochaeta) which inhabit the filter-beds of sewage-works in Britain. Wat. Res. 13: 1291–1299.

Lee, N. M. & T. Welander, 1993. Influence of predators on nitrification in aerobic biofilm processes. In: First International Specialized Conference on Microorganisms in Activated Sludge and Biofilm Processes, September 27–28, 1993, Paris: 337–345.

Percy, J. A. & F. J. Fife, 1981. The biochemical composition and energy content of Arctic Marine Macrozooplankton. Arctic 34: 307–313.

Pfannkuche, O., 1981. Distribution, abundance and life cycles of aquatic Oligochaeta (Annelida) in a freshwater tidel flat of the Elbe Estuary. Arch. Hydrobiol. Suppl. 43: 605–524.

Ratsak, C. H., S. A. L. M. Kooijman & B. W. Kooi, 1993. Modelling the growth of an oligochaete on activated sludge. Wat. Res. 27: 739–747.

Ratsak, C. H., B. W. Kooi & H. W. Van Verseveld, 1994. Biomass reduction and mineralization increase due to the ciliate *Tetrahymena pyriformis* grazing on the bacterium *Pseudomonas fluorescens*. In First International Specialized Conference on Micro-organisms in Activated Sludge and Biofilm Processes, September 27–28, 1993, Paris. Wat. Sci. Technol. 29: 119–128.

Ratsak, C. H., K. Maarsen & S. A. L. M. Kooijman, 1996. The effect of Protozoa on carbon mineralization in activated sludge. Wat. Res. 30: 1–12.

Rensink, J. H. & W. H. Rulkens, 1997. Using Metazoa to reduce sludge production. Wat. Sci. Technol. 36: 171–179.

Verdonschot, P. F. M., 1989. The role of oligochaetes in the management of waters. Hydrobiologia 180: 213–227.

Hydrobiologia **463:** 223–234, 2001.
P. Rodriguez & P.F.M. Verdonschot (eds), Aquatic Oligochaete Biology VIII.

Half a century of oligochaete research in Estonian running waters

Tarmo Timm[1], Ado Seire[2] & Peeter Pall[2]
[1]*Institute of Zoology and Botany, Võrtsjärv Limnological Station, EE61101 Rannu, Tartumaa, Estonia*
E-mail: ttimm@zbi.ee
[2]*Institute of Zoology and Botany, Department of Hydrobiology, Riia 181, EE51014 Tartu, Estonia*
E-mail: peeter@zbi.ee

Key words: Oligochaeta, stream fauna, distribution, long-term changes, Estonia

Abstract

About 51 500 specimens from 1542 samples, collected over the years 1954–1975 and 1986–1999 in different running water bodies throughout Estonia, were identified. Tubificidae prevailed in the material, with *Limnodrilus hoffmeisteri* forming about 40%. This species was followed by the tubificids *Tubifex tubifex, Potamothrix hammoniensis, Psammoryctides barbatus, L. udekemianus* and *Spirosperma ferox*, the naidid *Stylaria lacustris*, and the lumbriculid *Stylodrilus heringianus*. Two main ecological assemblages were distinguished: the pelophilous assemblage, dominated by *L. hoffmeisteri*, and the psammophilous one, where usually *P. barbatus* was dominant. The relationships between different species and the chemical parameters of water were usually weak but in contrast, correlated well with sediment preferences. In organically enriched reaches, *L. hoffmeisteri* usually dominated. The fauna of the streams of the islands was poorer in species due to their small size rather than geographical isolation. Some recent antropochorous Ponto-Caspian invaders have only reached the lowermost reaches of the two largest rivers. Some brackish water species were found in the mouth of the Pärnu River. No essential differences were found between the comparable sets of oligochaete samples collected in 1954–1975 and 1987–1997 in the Estonian running waters.

Introduction

Oligochaetes are regarded as common animals in fresh water, their abundance and species content indicating the quality of water and bottom sediments (e.g. Liebmann, 1962; Brinkhurst, 1966; Johnson et al., 1993). In Estonia, oligochaetes of running waters have been studied by the first author, mostly during the period 1956–1962 (Timm, 1962, 1963, 1967a, 1967b, 1970, 1987; Tõlp, 1976).

Beginning in 1987, the River Biology Group of the Institute of Zoology and Botany, under the supervision of Dr A. Järvekülg, including the second and the third authors, has surveyed the biota and hydrochemical background of all important Estonian river systems (e.g. Järvekülg, 1994). Oligochaetes collected from among benthic animals were identified by the first author. This enabled us to evaluate the oligochaete fauna in relation to the contemporary state of the Estonian rivers and streams, as well as to examine possible changes during the roughly thirty years that had passed from the earlier sampling period.

Materials and methods

Four sets of oligochaete material from 1954–1999, with a total of 51 526 ethanol-preserved specimens in 1542 samples, were studied (Table 1).

I. Samples collected by the River Biology Group in 1987–1997. These were mostly large qualitative samples taken usually with a hand net but in the first three years also with a small bottom grab. They covered almost all important streams and rivers in Estonia except for the largest ones (the Narva, the Emajõgi, and the lowest course of the Pärnu River; Fig. 1).

II. Other recent samples collected from different streams and rivers, mainly by Dr Henn Timm (Võrtsjärv Limnological Station) and his students in 1986–

Table 1. Overview of the material included in the survey

Series	Designation	Years	Samples with worms	No of specimens per sample			No of taxa		
				Total	Average	Maximum	Total	Average	Maximum
I	Samples collected by the River Biol. Group	1987–1997	415	12 069	29.1	>330	60	3.2	11
II	Other recent samples	1986–1999	302	3548	11.7	711	48	1.9	15
III	Earlier comparable samples	1954–1975	594	17 850	30.0	582	67	3.2	13
IV	Earlier repeated samples from mud	1958–1962	231	18 079	78.3	875	41	3.8	12
	All samples together	**1954–1999**	**1542**	**51 546**	**33.4**	**875**	**85**	**3**	**15**

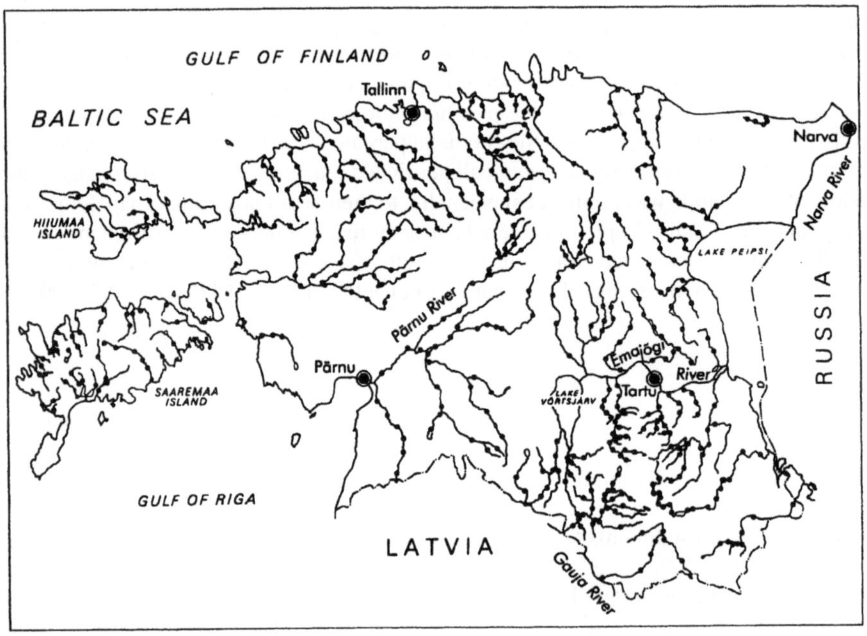

Figure 1. Major river systems of Estonia with sampling locations used by the River Biology Group, 1987–1997.

1999. These were qualitative, mostly kick samples taken with a hand net; usually smaller than those of the previous series.

III. Earlier comparable samples collected by the first author as well as by the other researchers of the Institute of Zoology and Botany in 1954–1975. These were taken in many running waters, mainly with a hand net, but in the three above mentioned large rivers, with bottom grabs. They can be used for comparison with more recent material.

IV. Earlier, repeatedly or seasonally collected quantitative material taken with bottom grabs from the muddy bottom of four rivers in 1958–1962. These should be excluded from comparison with numbers I–II to avoid any selective predomination of the pelophilous fauna.

The worms were picked macroscopically from the sieve residue, either alive or after preservation in ethanol or formalin, and were identified mostly in glycerine mounts under magnification. Relative frequency and abundance (in percentages) were calculated for each species in a particular series. Co-existence (percentage of the findings of some species together with some other species from the same series) was calculated pairwise for the most abundant species. Absolute abundance per m^2 was seldom available, except for series IV. In series I, the possible relationship between a particular species and water quality was tested by comparing the values of selected parameters (temperature, dissolved oxygen, pH, chemical and biological oxygen demand, content of N_{tot}, P_{tot}, NO_3–N, NO_2–N, NH_4–N, PO_4–P, N_{org} and P_{org}, and number of saprophytic bacteria) for reaches where the species were found with the respective values for reaches where they were absent. When results from the *F*-test and the Kruskal–Wallis test were statistically significant ($P < 0.05$), the parameter was considered to exert an effect on the distribution of the species.

Results

A total of 78 species of Oligochaeta, as well as 7 taxa unidentified to the species level, were found in Estonian flowing waters (Table 2).

Series I:

Material collected by the River Biology Group, 1987–1997. Altogether 60 taxa, including 55 identified species, were found. The most frequent species were:

Limnodrilus hoffmeisteri – in 206 samples (49.6%);
Psammoryctides barbatus – in 180 samples (43.4%);
Tubifex tubifex – in 137 samples (33.0%);
Stylodrilus heringianus – in 99 samples (23.9%);
Spirosperma ferox – in 93 samples (22.4%);
Potamothrix hammoniensis – in 92 samples (22.2%).

The same six species were also the most abundant:

Limnodrilus hoffmeisteri – 4390 individuals (36.4%);
Tubifex tubifex – 2063 individuals (17.1%);
Psammoryctides barbatus – 1676 individuals (13.9%);
Potamothrix hammoniensis – 1200 individuals (9.9%);
Stylodrilus heringianus – 440 individuals (3.6%);
Spirosperma ferox – 304 individuals (2.5%).

Limnodrilus hoffmeisteri was dominant in both frequency and abundance, followed by *Psammoryctides barbatus* in frequency, but by *Tubifex tubifex* in abundance. The most common taxon, *L. hoffmeisteri,* served as the main background species, often in more than 50% of cases, for many other species: *L. udekemianus, T. tubifex, Potamothrix hammoniensis,* etc. (Table 3). The next most abundant species, *Psammoryctides barbatus*, did not accompany it so often as was expected. *P. barbatus* itself was the preferred background species for *Rhyacodrilus coccineus, Rhynchelmis tetratheca, Nais bretscheri, Spirosperma ferox* and *Stylodrilus heringianus.* Among other frequent taxa, only *T. tubifex* formed a common background (>50% of cases) for *T. ignotus*, and *S. ferox* for *R. tetratheca.* Thus, two major ecological groupings are evident: the pelophilous group preferring soft sediments with slow current (strongly dominated by *L. hoffmeisteri*), and the psammophilous group inhabiting harder sediments with faster current (*P. barbatus* and others).

Among the studied environmental parameters (Table 4), water temperature appeared to be an important factor for many species. *Stylaria lacustris, Limnodrilus hoffmeisteri* and *Spirosperma ferox* clearly preferred warmer habitats. So did *Cognettia glandulosa*, while its sister species *C. sphagnetorum* was bound to cooler (smaller) streams, like the groundwater-bound *Trichodrilus* sp. In most cases, weak preferences for the habitats with lower content of dissolved oxygen were observed. Oxygen-rich water was preferred by *C. sphagnetorum* and *Stylodrilus heringianus. Psammoryctides barbatus* was more often found in stream reaches with higher pH values, while *Tubifex tubifex* was bound to reaches with lower pH. A relation with the chemical oxygen demand (COD-Cr) occurred only in a few species. *Spirosperma ferox* preferred higher values of COD-Cr. Numerous species showed weak preferences for biological oxygen demand (BOD_5): the pelophilous *L. hoffmeisteri, T. tubifex* and *T. ignotus,* and the phytophilous *Nais communis* preferred higher values. The psammophilous *P. barbatus, Rhyacodrilus coccineus* and *S. heringianus*, but also the pelophilous *P. hammoniensis* and phytophilous *Rhynchelmis limosella* were more often found at lower BOD_5 values. The phytophilous *Nais communis* and *N. barbata* were the

Table 2. List of species, with their absolute frequency and abundance in the four sample series

Series	I		II		III		IV		Total	
Taxa	Sampl.	Indiv.	Sampl.	Indiv.	Sampl.	Indiv.	Sampl.	Indiv.	Sampl.	Indiv
Naididae										
Stylaria lacustris (L., 1767)	27	183	30	316	99	989	17	44	173	1532
Arcteonais lomondi (Martin, 1907)					2	2			2	2
Ripistes parasita (Schmidt, 1847)					2	2			2	2
Vejdovskyella comata (Vejd., 1883)					1	1			1	1
Slavina appendiculata (Udek., 1855)	2	7	2	10	5	6			9	23
Dero dorsalis Ferronière, 1899					3	8	5	5	8	13
Dero digitata (Müller, 1773)			1	1			1	1	2	2
Dero obtusa Udekem, 1855	2	7					5	7	7	14
Nais alpina Sperber, 1948	6	21	4	37	7	21			17	79
Nais barbata Müller, 1773	12	63	7	687	13	42	8	93	40	885
Nais behningi Michaelsen, 1923	1	1	1	2					2	3
Nais bretscheri Michaelsen, 1899	29	162	2	2	4	14	2	2	37	180
Nais christinae Kasprzak, 1973	2	4							2	4
Nais communis Piguet, 1906	7	21	4	31	9	15	1	24	21	91
Nais elinguis Müller, 1773	5	153	14	205	2	58			21	416
Nais pardalis Piguet, 1906	15	62	2	7	6	84			23	153
Nais pseudobtusa Piguet, 1906	2	2	1	2	16	73	3	6	22	83
Nais simplex Piguet, 1906	2	2			3	3			5	5
Nais variabilis Piguet, 1906	1	1			22	89	7	80	30	170
Specaria josinae (Vejdovský, 1883)	2	2	1	1	1	1			4	4
Piguetiella blanci (Piguet, 1906)			1	2					1	2
Ophidonais serpentina (Müller, 1773)	13	54	8	25	19	42	10	24	50	145
Uncinais uncinata (Oersted, 1842)	5	7	5	12	23	45	1	3	34	67
Paranais litoralis (Müller, 1784)					1	3			1	3
Paranais frici Hrabě, 1941					2	3	4	4	6	7
Chaetogaster diaphanus (Gr., 1828)			3	13	10	23	3	29	16	65
Chaetogaster diastrophus (Gr., 1828)			2	2	9	38			11	40
Chaetogaster langi Bretscher, 1896	1	1			5	6			6	7
Pristina longiseta Ehrenberg, 1828					2	2			2	2
Pristina aequiseta Bourne, 1891	1	1			1	3			2	4
Pristinella rosea (Piguet, 1906)	1	2			1	3			2	5
Naididae indet.					13	14			13	14
Tubificidae										
Rhyacodrilus coccineus (Vejd., 1875)	39	164	16	26	36	144	6	15	97	349
Rhyacodrilus falciformis Bret., 1901	4	4			1	1			5	5
Bothrione. vejdovskyanum St., 1888	2	10	4	94	3	45	18	274	27	423
Limnodrilus hoffmeisteri Clap., 1862	206	4390	77	839	241	4447	199	10805	723	20481
Limnodrilus profundicol.(Verr., 1871)	1	3	1	3	6	83	8	211	16	300
Limnodrilus claparedean. Ratz., 1868	3	9			20	96	8	81	31	186
Limnodrilus udekemianus Cl., 1862	55	248	12	63	113	758	114	1512	294	2581
Potamothrix hammoniensis (M., 1901)	92	1200	66	228	163	1715	97	2135	418	5278
Potamothrix heuscheri (Bret., 1900)					6	14	15	34	21	48
Potamothrix bedoti (Piguet, 1913)	1	1	1	1	4	17			6	19
Potamothrix bavaricus (Os., 1913)			1	1	10	93	17	144	28	238
P. moldaviensis (Vejd.et Mr., 1903)			1	3	8	116	5	34	14	153
Psammoryctides barbatus (G., 1861)	179	1673	65	169	189	1864	32	68	465	3774
Psammoryctides albicola (M., 1901)	20	61	7	16	46	393	19	35	92	505

Continued on p. 227

Table 2. Continued

Series Taxa	I Sampl.	 Indiv.	II Sampl.	 Indiv.	III Sampl.	 Indiv.	IV Sampl.	 Indiv.	Total Sampl.	 Indiv
Psammoryctides moravicus (Hr.,1934)			1	2	3	7			4	9
Spirosperma ferox Eisen, 1879	93	304	14	18	164	935	54	293	325	1550
Tubifex tubifex (Müller, 1774)	137	2063	77	345	110	2940	81	1657	405	7005
Tubifex ignotus (Štolc, 1886)	42	135	5	17	29	132	27	136	103	420
Tubifex smirnowi Lastočkin, 1927					1	12			1	12
Tubifex newaensis (Michaelsen, 1902)					18	359	20	39	38	398
Isochaetides michaelseni (Last., 1936)					1	50			1	50
Ilyodrilus templetoni (South., 1909)	3	3	5	15	3	3	5	13	16	34
Tubificoides heterochaetus (M., 1926)							7	40	7	40
Aulodrilus pluriseta (Piguet, 1906)	12	157	5	21	6	18	8	11	31	207
Aulodrilus limnobius Bretsch., 1899	7	15	2	5	6	7	1	3	16	30
Aulodrilus pigueti Kowalewski, 1914							1	1	1	1
Tubificidae indet.					80	735	14	50	94	785
Propappidae										
Propappus volki Michaelsen, 1916	3	8			2	29			5	37
Enchytraeidae										
Mesenchytraeus armatus(Lev., 1884)	2	2							2	2
Cernosvitoviella sp.	1	1							1	1
Cognettia glandulosa (Mich., 1888)	17	34							17	34
Cognettia sphagnetorum (Vej., 1878)	7	7							7	7
Henlea ventriculosa (Udekem, 1854)	3	4	1	1					4	5
Henlea perpusilla Friend, 1911	9	9							9	9
Fridericia spp.	6	8	8	11	4	6			18	25
Timmodrilus oligoseta Doz.-F., 1997	1	1							1	1
Lumbricillus lineatus (Müller, 1774)	2	12	8	67					10	79
Marionina riparia (Bretscher, 1899)	33	52							33	52
Enchytraeidae indet.	7	9	1	1	43	100	5	9	56	119
Lumbriculidae										
Lumbriculus variegatus (Müll., 1774)	71	164	40	60	108	355	21	42	240	621
Trichodrilus sp.	3	3	1	1	2	2			6	6
Stylodrilus heringianus Clap., 1862	99	440	39	132	90	558	25	89	253	1219
Rhynchelmis limosella Hoff., 1843	4	8	2	3	13	33	10	23	29	67
Rhynchelmis tetratheca Mich., 1920	27	57	8	11	21	52	1	1	57	121
Branchiobdellidae										
Branchiobdella parasita Hen., 1835					1	1			1	1
Lumbricidae										
Eiseniella tetraedra (Savigny, 1826)	29	35	22	30	25	69	2	2	78	136
Aporrectodea caliginosa (Sav., 1826)	2	3	1	2	1	1			4	6
Allolobophora chlorotica (Sav., 1826)	2	12	1	1	1	1			4	14
Dendrobaena octaedra (Sav., 1826)	2	4							4	4
Dendrodrilus rubidus (Sav., 1826)					1	2			1	2
Lumbricus terrestris L., 1758	1	1							1	1
Lumbricus rubellus Hoff., 1843					1	2			1	2
Lumbricidae indet.	6	6	6	7	19	38			31	51
Total	**415**	**12 076**	**302**	**3548**	**594**	**17 823**	**231**	**18 079**	**1542**	**51 526**

Table 3. Co-existence of 15 species with 10 background species in 415 samples of series I (material of the River Biology Group) in percentages. Numbers appear in bold when %-occurrence >50

	Background species									
Co-existence of...	*Lim.hof*	*Ps.bar*	*T.tub*	*Styl.her*	*Sp.fer*	*Pot.ham*	*Lum.var*	*Lim.ud*	*T.ign*	*Rh.coc*
Limnodrilus hoffmeisteri	**100**	43.2	42.7	16.5	28.6	27.1	16	18.4	15.5	7.8
Psammoryctides barbatus	49.7	**100**	25.1	31.2	33	24	15.6	10.6	11.7	19
Tubifex tubifex	**64.2**	32.8	**100**	16.8	19	23.3	21.9	19.7	16.8	4.4
Stylodrilus heringianus	34.3	**56.6**	23.2	**100**	26.3	20.2	21.2	12.1	12.1	20
Spirosperma ferox	**63.4**	**63.4**	28	28	**100**	31.2	24.7	15	14	17.2
Potamothrix hammoniensis	**60.9**	46.7	34.8	21.7	31.5	**100**	15.2	13	13	13
Lumbriculus variegatus	46.5	39.4	42.3	29.6	32.4	19.7	**100**	19.7	8.5	9.9
Limnodrilus udekemianus	**69.1**	34.5	49.1	21.8	25.4	21.8	25.5	**100**	14.5	5.5
Tubifex ignotus	**76.2**	50	**54.8**	28.6	30.1	28.6	14.3	19	**100**	11.9
Rhyacodrilus coccineus	41	**87.2**	15.4	46.2	38.1	30.8	17.9	7.7	12.8	**100**
Marionina riparia	**69.7**	42.4	48.5	33.3	24.2	18.2	18.2	15.1	21.2	12.1
Nais bretscheri	**54.5**	**69**	31	41.4	27.6	39.4	13.8	10.3	13.8	17.2
Eiseniella tetraedra	27.6	44.8	27.6	31	17.2	13.8	10.3	10.3	0	17.2
Stylaria lacustris	**51.9**	44.4	25.9	18.5	29.6	34.5	29.6	3.7	3.7	3.7
Rhynchelmis tetratheca	**66.7**	**74**	14.8	29.6	**51.2**	29.6	22.2	22.2	11.1	22.2

Table 4. Results of the Kruskal–Wallis' test illustrating the relationship between the distribution of some species and environmental parameters for series I samples, as statistic *H*. The statistic is given when $P < 0.05$, and in bold when $P < 0.001$. (+) means that higher values, and (–) that lower values of the corresponding parameter are preferred

	T	O_2	pH	COD-Cr	BOD_5	Sapro. bacter.	Ntot	NO_3	NO_2	NH_4	Norg	Ptot	PO_4	Porg
Stylaria lacustris	**19.9 (+)**													
Nais barbata						7 (+)								
Nais communis	6 (+)				7 (+)	8.3 (+)			4 (+)			3.9 (+)		7.8 (+)
Rhyacodrilus coccineus					5.7 (–)									
Limnodrilus hoffmeisteri	**28.1(+)**	6.5 (–)			9 (+)			**11.8 (–)**	5 (+)	**13.1 (+)**	5 (+)	**36.4 (+)**	**15.9 (+)**	**26.7 (+)**
Limnodrilus udekemianus														3.8 (+)
Potamothrix hammoniensis		4.6 (–)			4.3 (–)				4.1 (+)					
Psammoryctides barbatus	10.4 (+)		**25 (+)**		8.9 (–)		8.8 (+)	**13.2 (+)**	5.2 (+)					
Psammoryctides albicola									6.8 (+)			**12.8 (+)**	9.7 (+)	5.9 (+)
Spirosperma ferox	**25.7 (+)**	**5.7 (–)**	3.9 (+)	6.3 (+)										
Tubifex tubifex		3.9 (–)	**26.3 (–)**		5 (+)				5.1 (+)	**43.6 (+)**		**27.1 (+)**	**12.4 (+)**	**23.9 (+)**
Tubifex ignotus					6.1 (+)				5.6 (+)		3.9 (+)	4.4 (+)		7.3 (+)
Cognettia glandulosa	**12.5 (+)**						4.9 (–)	4.2 (–)						
Cognettia sphagnetorum	10.2 (–)	5.3 (+)												
Marionina riparia												5.7 (+)	5.6 (+)	
Lumbriculus variegatus						7.8 (–)								
?*Trichodrilus* sp.	4.7 (–)													
Stylodrilus heringianus		4.3 (+)		3.9 (–)	8.6 (–)			5.4 (+)				4.3 (–)		4.2 (–)
Rhynchelmis limosella					5.8 (–)					4.1 (–)				
Rhynchelmis tetratheca	8.2 (+)		3.9 (+)										4.1 (+)	

only species preferring waters with higher numbers of saprophytic bacteria, whereas *Lumbriculus variegatus* preferred lower numbers of these bacteria; all sediment-dwelling species were indifferent in this respect. The distribution of a few species was weakly related to the content of total nitrogen, while different forms of nitrogen had different effects. Thus, *Limnodrilus hoffmeisteri* was more often found in streams with lower nitrate content, and *Psammoryctides barbatus* in streams with higher nitrate content. *L. hoffmeisteri* and *Tubifex tubifex* clearly preferred river reaches with a higher ammonia content. *L. hoffmeisteri* and *T. tubifex*, but to a lesser extent also several other pelophilous tubificids and other species, had a strong preference for a higher content of both organic and inorganic phosphorus. The only species that preferred lower phosphorus content was *Stylodrilus heringianus*.

The most common species can be divided into two major groups according to their relation to water quality: pelophilous species (*Limnodrilus hoffmeisteri*, et cetera) preferred more eutrophic and warmer habitats, while the psammophilous group (*Psammoryctides barbatus*, etc.) was typical of less eutrophied and cooler habitats, usually with a sandy or stony bottom and considerable current velocity. *P. barbatus* was particularly common and abundant in the limestone karst areas of North Estonia where groundwater is contaminated with nitrogen fertilizers. This explains its preference for the habitats with higher pH and nitrate content. The strong preference of the opportunistic *Tubifex tubifex* for higher ammonium content reflects the abundance of this species in rapidly decaying organic sediments. The reason for the preference for warmer habitats of *Spirosperma ferox* (otherwise known as a common species in the cool waters of the North Palearctic) remains obscure.

The average abundance of oligochaetes in 51 quantitative samples taken from different smaller rivers in 1987–1989 within series I was 1,191$\pm$254 (*SE*) ind. m^{-2}.

Series II:

Other recent material, 1986–1999. Altogether 48 taxa, including 44 identified species, were found. The most frequent species were:

Limnodrilus hoffmeisteri – in 77 samples (25.5%);
Tubifex tubifex – in 77 samples (25.5%);
Potamothrix hammoniensis – in 66 samples (21.9%);
Psammoryctides barbatus – in 65 samples (21.5%);
Lumbriculus variegatus – in 40 samples (13.2%);
Stylodrilus heringianus – in 39 samples (12.9%);
Stylaria lacustris – in 30 samples (9.9%);
Eiseniella tetraedra – in 22 samples (7.3%).

The most abundant species in this series were:

Limnodrilus hoffmeisteri – 839 individuals (23.6%);
Nais barbata – 687 individuals (19.4%) but 600 of them in one single sample!;
Tubifex tubifex – 345 individuals (9.7%);
Stylaria lacustris – 316 individuals (8.9%);
Potamothrix hammoniensis – 228 individuals (6.4%);
Nais elinguis – 205 individuals (5.8%);
Psammoryctides barbatus – 169 individuals (4.8%);
Stylodrilus heringianus – 132 individuals (3.7%).

Samples of this series were usually small and were often taken from among the vegetation. Hence, the number of taxa per sample and the degree of coexistence in samples were also lower. *Limnodrilus hoffmeisteri* was the most common species, but the share of tubificids was lower and that of phytophilous taxa (*Stylaria lacustris, Nais* spp., *Lumbriculus variegatus*) higher than in series I. The degree of coexistence was 50% or higher in two cases only: for *S. lacustris* with *L. hoffmeisteri*, and for *Rhyacodrilus coccineus* with *Psammoryctides barbatus*.

Series III:

Earlier comparable samples, 1954–1975. Altogether 67 taxa, including 61 species, were documented in this series. All Enchytraeidae, as well as many juvenile Tubificidae and Naididae remained unidentified. Nevertheless, the species number was the highest due to the diverse fauna of the lower courses of the large rivers of Pärnu and Narva. The most frequent taxa were:

Limnodrilus hoffmeisteri – in 241 samples (40.5%);
Psammoryctides barbatus – in 189 samples (31.8%);
Spirosperma ferox – in 164 samples (27.6%);
Potamothrix hammoniensis – in 163 samples (27.4%);
Limnodrilus udekemianus – in 113 samples (19.0%);
Tubifex tubifex – in 110 samples (18.5%);

Lumbriculus variegatus – in 108 samples (18.2%);
Stylaria lacustris – in 99 samples (16.7%);
Stylodrilus heringianus – in 90 samples (15.1%).

The most abundant taxa were:

Limnodrilus hoffmeisteri – 4447 individuals (24.9%);
Tubifex tubifex – 2940 individuals (16.5%);
Psammoryctides barbatus – 1864 individuals (10.4%);
Potamothrix hammoniensis – 1715 individuals (9.6%);
Stylaria lacustris – 989 individuals (5.5%);
Spirosperma ferox – 935 individuals (5.2%);
Limnodrilus udekemianus – 758 individuals (4.2%);
Tubificidae gen. sp. indet. – 735 individuals (4.1%);
Stylodrilus heringianus – 558 individuals (3.1%).

Limnodrilus hoffmeisteri was dominant as in both previous series. The list of the other common taxa was similar also. The relative frequency and abundance of *Spirosperma ferox* were somewhat higher and those of *Stylodrilus heringianus* lower in this earlier period.

Co-existence of common species revealed a pattern similar to the one observed in more recent samples collected by the River Biology Group: *Limnodrilus hoffmeisteri* served as the background species (>50% of cases) for *L. udekemianus, Potamothrix hammoniensis* and *T. tubifex*, while *Psammoryctides barbatus* was the common dominant in samples with *Rhyacodrilus coccineus, Stylodrilus heringianus, Spirosperma ferox* and *Eiseniella tetraedra*. Two species, *Psammoryctides albicola* and *Tubifex ignotus*, accompanied both dominants at an equal frequency.

In quantitative samples taken from the three largest Estonian rivers, the average abundance of oligochaetes differed significantly. For the upper course of the Emajõgi River, beginning from Lake Võrtsjärv and lined mostly with fine sand and clay, it was only 686±214 (*SE*) ind. m^{-2} (or only 545 if empty samples were taken into account). For the Narva River, with more diverse sediments, the respective parameter was 1303±275 (*SE*) ind. m^{-2}, and for the lower course of the Pärnu River in 1959, polluted in places (oligosaprobic to polysaprobic according to Tõlp, 1976), 3017±794 (*SE*) ind. m^{-2}.

Series IV:

Earlier repeated samples from muddy bottom, 1958–1962. This set includes repeated samples collected from the lower course of the Pärnu River as well as monthly samples from some muddy stations in three smaller rivers. Forty-one taxa, among them 39 species, were distinguished. The most frequent species were:

Limnodrilus hoffmeisteri – in 199 samples (86.1%);
Limnodrilus udekemianus – in 114 samples (49.4%);
Potamothrix hammoniensis – in 97 samples (42.0%);
Tubifex tubifex – in 81 samples (35.1%);
Spirosperma ferox – in 54 samples (23.4%);
Psammoryctides barbatus – in 32 samples (13.9%).

The most abundant species were:

Limnodrilus hoffmeisteri – 10 805 individuals (59.8%);
Potamothrix hammoniensis – 2135 individuals (11.8%);
Tubifex tubifex – 1657 individuals (9.2%);
Limnodrilus udekemianus – 1512 individuals (8.4%);
Spirosperma ferox – 293 individuals (1.6%);
Bothrioneurum vejdovskyanum – 274 individuals (1.5%).

The series was strongly dominated by pelophilous species, with *Limnodrilus hoffmeisteri* occurring almost everywhere and accounting for more than half of the specimens. However, the fauna was quite diverse in the lower course of the Pärnu River. Several brackish-water species (*Tubificoides heterochaetus, Paranais litoralis, P. frici, Potamothrix bavaricus*) and rheophilous Ponto-Caspian invaders (*Tubifex newaensis, Potamothrix heuscheri*), not yet found in inland rivers, occurred here.

In the polluted mouth of the Pärnu River within the town of Pärnu, oligochaetes formed 37–99% of the bottom macrofauna. *Limnodrilus hoffmeisteri* and *L. udekemianus* were the most abundant species everywhere, except in a few polysaprobic stations dominated by *Tubifex tubifex*. Upstream of the town, the bottom fauna had mostly an oligotrophic appearance, with the share of oligochaetes constituting 8–69%; *L. hoffmeisteri* occupied the first place even here (Tõlp, 1976).

Repeated quantitative samples from the Pärnu River revealed an average abundance very similar to that recorded in 1959: 3236±623 ind. m^{-2}, with an absolute maximum of 38 889 ind. m^{-2} at a polluted station. The average abundance at yearly stations in three smaller rivers was 1157±220, 4065±855, and 4347±696 ind. m^{-2}, respectively.

Distribution of oligochaetes in Estonian running waters

The Estonian rivers can be divided into three major hydrological regions: Lake Peipsi basin (with the outflowing Narva River), continental rivers falling directly in the Baltic Sea, and small streams on the islands (Fig. 1).

The Peipsi watershed in southeastern Estonia lies on sandstone covered with moraine hills. The water of these rivers is relatively acid and iron-rich, and the stream beds are usually sandy. Rivers flowing directly into the sea lie mostly on limestone lowlands of northern and western Estonia. Their water is mostly alkaline and their bottom is often stony. The streams of Saaremaa Island are similar to the former ones, while in the small brooks of Hiiumaa Island sandy bottom prevails, like in southeastern Estonia.

In samples collected by the River Biology Group (series I), the pelophilous *Limnodrilus hoffmeisteri, Tubifex tubifex* and *Aulodrilus pluriseta* (but not *Potamothrix hammoniensis*), as well as the ubiquitous *Lumbriculus variegatus* occurred most often in the Lake Peipsi watershed where sediments are generally finer. On the contrary, the psammophilous *Psammoryctides barbatus* and *Rhyacodrilus coccineus* were twice as frequent and abundant in the lowland group of rivers with numerous riffles. *Spirosperma ferox* was also more frequent and abundant here. However, the psammophilous *Stylodrilus heringianus* was evenly distributed in both regions, and the lithophilous *Nais bretscheri* was even much more common in the Lake Peipsi watershed.

In series III (earlier comparable material), the number of species was considerably larger in rivers flowing into the sea. This faunistic diversity was mostly accounted for by samples from the lower courses of the Pärnu and Narva Rivers, which are not included in the former series. Pelophilous species as well as *S. heringianus* were represented equally in both regions. *L. variegatus* was far more common in the Lake Peipsi watershed, while *P. barbatus, S. ferox* and *R. coccineus* were more frequent in the rivers flowing into the sea, as in the former series.

In recent times, *R. coccineus* appears to be spreading in the flowing waters of the Lake Peipsi watershed, as far as most of its findings here belong to the more recent period (Fig. 2).

Propappus volki has a limited distribution range in Estonia: it occurs in only a few rivers flowing into the sea as well as in Lake Peipsi but not in its inflows (Fig. 2).

Several rheophilous species of Ponto-Caspian origin have evidently been introduced with ships in the mouths of the two largest rivers only in recent centuries. *Isochaetides michaelseni* and *Psammoryctides moravicus* are found only in the Narva River, *Potamothrix heuscheri* in the Pärnu River, and *Tubifex newaensis* as well as *Potamothrix moldaviensis* in both rivers. *P. moldaviensis* has been the most successful, this species has reached also the mouth of a smaller river (Sõtke) and the lakes of Peipsi and Võrtsjärv (like *Rhyacodrilus coccineus*) but not yet the Emajõgi River connecting these lakes (Fig. 2). The brackish-water species *Potamothrix bavaricus,* found in the mouth of the Pärnu River, is probably of Ponto-Caspian origin as well.

The streams of the Estonian islands are relatively small and poor in species. Their fauna on Saaremaa Island is similar to the fauna of continental rivers flowing into the sea; species number is limited but is evidently accidental, including also some rare freshwater taxa (*Rhyacodrilus falciformis*). In some cases, an elsewhere very common species appears to be replaced by its rarer sister taxon (e.g. *Limnodrilus hoffmeisteri* replaced by *L. claparedeanus* in the Leisi River). This paucity of fauna seems to be caused by the short age and isolation of separate streams rather than by the isolation of the island from continental fresh waters. In sandy brown-water brooks of Hiiumaa Island, oligochaetes are often represented only by *L. hoffmeisteri* and/or *T. tubifex.*

Comparison of the fauna of two periods (1954–1975 and 1987–1997)

Differences in the qualitative content of comparable samples taken in two periods, series I and III (Table 2), are evidently caused by differences in the methods of sampling and identification. Large species-rich lower courses of the Pärnu and Narva Rivers were sampled only in the earlier period. Conventional 'taxa' of undetermined Naididae, Tubificidae and Enchytraeidae were common in earlier samples, while the species of Enchytraeidae (among them *Marionina riparia* occurring regularly in the water of the studied rivers) were identified only in the later period.

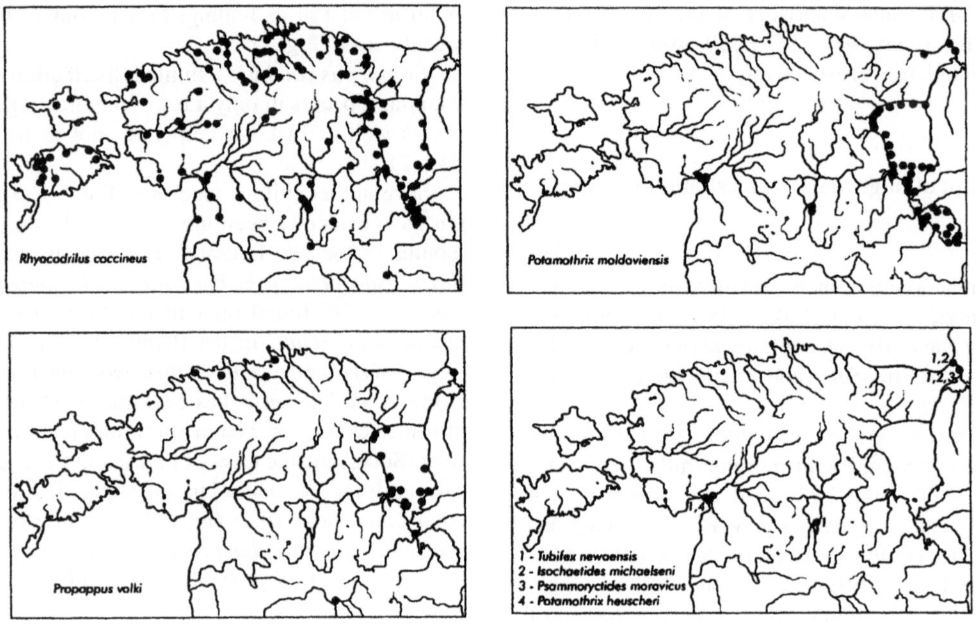

Figure 2. Findings of some oligochaete species in Estonian rivers and lakes.

The relative abundance of the most common species has remained similar, with a considerable predominance of *Limnodrilus hoffmeisteri*. The six next most frequent taxa were also similar in both periods (*Psammoryctides barbatus – Spirosperma ferox – Potamothrix hammoniensis – Limnodrilus udekemianus – Tubifex tubifex – Lumbriculus variegatus* in 1954–1975; *P. barbatus – T. tubifex – S. heringianus – S. ferox – P. hammoniensis – L. variegatus* in 1987–1997). It is remarkable that the oxygen demanding, psammophilous *S. heringianus* was the fourth most frequent species in the last period (30 years ago: the ninth), while the pelophilous *L. udekemianus* fell from the fifth to the eight place (this species was particularly abundant in the Pärnu River not sampled in the second period).

The phytophilous naidid *Stylaria lacustris* and the psammophilous *Uncinais uncinata* were much less frequent in the second period, while the lithophilous *Nais bretscheri* became much more common. These differences may be caused by some subjective factors related to sampling (macrovegetation was not particularly sampled in the second period).

Human impact on the oligochaete fauna

Data related to this issue are fragmentary in Estonia. Tõlp (1956) has reported a significant predominance of Tubificidae in the bottom fauna of the Emajõgi River downstream of the town of Tartu, with a maximum abundance of 50 000 ind. m^{-2}. In 1962, an average of only 686 ind. m^{-2} was found in the unpolluted upper course of this river (see above), and in 1980 a number of 489 ind. m^{-2} was registered in the river mouth 40 km downstream of the town. No bottom grab samples are available from the 1990's when purification of waste water was started in Tartu. However, the semi-qualitative hand net samples of 1998–1999 (from series II) revealed a species-rich but non-abundant fauna (100–300 m^{-2}) in the macrovegetation zone both upstream and downstream of the town, with a prevalence of *Limnodrilus hoffmeisteri* and *Potamothrix hammoniensis*.

Quantitative samples from the Pärnu River collected in 1959-1962 revealed an average abundance of over 3000 ind. m^{-2} in all years, with an absolute maximum of 38 889 ind. m^{-2} at one of the polluted stations (Tõlp, 1976). No data were collected in the second period.

Nais elinguis, characteristic of cool spring brooks in the limestone karst area of North Estonia, was accompanied by the enchytraeid *Lumbricillus lineatus* in conditions of heavy organic pollution in the second period. *Tubifex tubifex* (not *L. hoffmeisteri*) was the main background species in these brooks.

Tubifex tubifex and *Limnodrilus hoffmeisteri* have been reported as dominants in polluted rivers of England, the former indicating particularly heavy pollution. *Potamothrix hammoniensis, Psammoryctides barbatus* and *Limnodrilus udekemianus* were also abundant but somewhat less pollution tolerant, while *Rhyacodrilus coccineus* and *Stylodrilus heringianus* appeared considerably more sensitive to pollution. The naidid *Nais elinguis* can also be very abundant in polluted streams (Brinkhurst, 1962). *T. tubifex* had a very broad ecological range in the Upper Danube, but was particularly abundant in extremely polluted places. *L. hoffmeisteri* revealed a more limited ecological range, preferring somewhat higher oxygen content and lower concentrations of organic matter as well as phosphorus and ammonium ions in water (Korn, 1963). Although Estonian streams and rivers were very seldom heavily enough polluted for *T. tubifex* to be abundant and dominating there, its preference for the ammonium ion (related to readily decaying organic sediment) was evident. *T. tubifex* was a single or prevailing species as the first colonizer in the smallest brooks and ditches.

Finogenova (1976) claims that both *T. tubifex* and *L. hoffmeisteri* serve as indicators of heavy pollution (in Russia) only when highly abundant. The other species of *Limnodrilus* are somewhat less tolerant to pollution but can still accompany the former two. *Potamothrix hammoniensis* and *Stylaria lacustris* indicate a moderate pollution level; *Stylodrilus heringianus* and *Propappus volki* are characteristic of clean water. Many common species, e.g. the eurybiontic *Nais elinguis*, have no indication value according to Finogenova (1976). In our experience, *Nais elinguis* is sensitive to competition (like *Tubifex tubifex*, see also Milbrink, 1978) and, therefore, abundant in habitats devoid of related species.

Riverine oligochaetes and their relationship to water quality have been thoroughly investigated in the neighbouring country of Latvia by Kačalova (1963), Kačalova & Parele (1964), Parele (1972, 1973, 1981, 1986), Parele & Astapenok (1975). The oligosaprobic *Stylodrilus heringianus* and *Nais* spp. prevailed on riffles in the large Daugava River. At slow current, the abundant *Limnodrilus hoffmeisteri* was accompanied by *Tubifex newaensis, Potamothrix hammoniensis, P. moldaviensis* and many other tubificids. This diverse assemblage inhabited even the lowest river stretch, polluted by the town of Riga, where additionally several brackish water species appeared (Kačalova, 1963; Kačalova & Parele, 1964; Parele, 1973). The lowest stretch of the Pärnu River in Estonia is well comparable to the Daugava River with respect to oligochaete fauna. The second largest river of Latvia, Lielupe, is almost lentic in its lower course, the most polluted places being almost devoid of animals (Parele, 1972). There are no such rivers in Estonia.

Several habitat types were distinguished in the smaller rivers of North Latvia by Parele (1986). In fast-flowing rivers, the role of oligochaetes in the fauna was insignificant; several naidids (*Stylaria lacustris, Nais* spp. etc.) but also the tubificids *Rhyacodrilus coccineus, Limnodrilus hoffmeisteri, Spirosperma ferox*, and the lumbriculids*Lumbriculus variegatus* and *Stylodrilus heringianus* were the most common. It is remarkable that the calciphilous *Psammoryctides barbatus,* very characteristic of North Estonia but less abundant in the Lake Peipsi watershed and in other South Estonian rivers, is absent from this list. In rivers with a slower current and finer sediment, the share of tubificids (*L. hoffmeisteri, Potamothrix hammoniensis, P. barbatus, Tubifex ignotus* etc.) was larger. The presence of some recent invaders (*Potamothrix moldaviensis* and *Tubifex newaensis*) in small rivers was more common in Latvia than in Estonia. In polluted stretches of Latvian small rivers, the pelophilous tubificids *L. hoffmeisteri, P. hammoniensis, Tubifex tubifex, T. ignotus,* et cetera, were very abundant (up to 30 000–40 000 ind. m^{-2}) in such places, as they were in Estonia.

Parele & Astapenok (1975) have calculated several pollution indices based on oligochaetes, but they assess only correctly for large Latvian rivers. The ratio of tubificid abundance to the abundance of all oligochaetes, recommended by these authors, would be 0.86 for our samples of series I, 0.84 for samples of series III, and as high as 0.97 for samples of series IV. All these figures seem to indicate heavy pollution of the Estonian rivers, which is evidently not the case. For series II samples the ratio is only 0.52, which corresponds to a low pollution level. Obviously, in this case, the aim of the collector was to record as large a number of taxa as possible rather than to collect a large number of specimens. As a matter of fact, the mean value of BOD_5, 3.0 ± 1.4 mg O_2 l^{-1}, for the studied Estonian streams corresponds to the β-mesosaprobic level after Sládeček (1973), while polysaprobic stations were lacking altogether.

Conclusions

A total of 78 species and 7 unidentified forms were documented in the material collected from Estonian rivers and streams during two sampling periods, 1954–

1975 and 1986–1999. Specimens of the family Tubificidae prevailed with *Limnodrilus hoffmeisteri* being the most frequent and abundant taxon. Two ecological groups were distinguished among common species: the pelophilous group dominated by *L. hoffmeisteri* and the psammophilous group dominated by *Psammoryctides barbatus*. In the case of some species, slow spreading from river mouths towards the inland direction was evident. Small streams on the islands are often poor in species due to their isolation. The relative frequency and abundance of common species were similar in both investigation periods, except for a slight increase in the clean-water species *Stylodrilus heringianus* in the second period. Thus, the ecological state of the Estonian running waters has remained the same, or has even slightly improved during the last thirty years. Regrettably, no comparable data on particular polluted stretches were available for both periods. The riverine oligochaete fauna of Estonia and Latvia appeared similar.

Acknowledgements

We are greatly indebted to the leader of the River Biology Group, Dr Arvi Järvekülg, for organizing the research of Estonian rivers, as well as for his great contribution to the collecting of the material used in this paper. We thank the staff members of the group for participating in the collecting and sorting of animals in different years. We are obliged to Dr Henn Timm for collecting a large number of additional samples. We are grateful to Mrs Ester Jaigma who kindly checked our English text. Suggestions of two reviewers have considerably improved the quality of manuscript. The project was supported by Grant No. 3690 of the Estonian Science Foundation.

References

Brinkhurst, R. O., 1962. The biology of the Tubificidae with special reference to pollution. In Biological Problems in Water Pollution, Third Seminar, Cincinnati: 57–65.

Brinkhurst, R. O., 1966. Detection and assessment of water pollution using oligochaete worms. 2. Water and Sewage Works 113: 438–441.

Brinkhurst, R. O. & B. G. M. Jamieson, 1971. Aquatic Oligochaeta of the World. Oliver & Boyd, Edinburgh: 860 pp.

Finogenova, N. P., 1976. Značenie oligohet kak indikatorov zagrjaznennyh vod. In Gidrobiologičeskie osnovy samoočiščenija vod, Leningrad: 51–59.

Johnson, R. K., T. Wiederholm & D. M. Rosenberg, 1993. Freshwater biomonitoring using individual organisms, populations, and species assemblages of benthic macroinvertebrates. In Rosenberg, D. M. & V. H. Resh (eds), Freshwater Biomonitoring and Benthic Macroinvertebrates, Chapman & Hall, New York & London: 40–158.

Järvekülg, A., 1994. Eesti jõgede vee troofsusaste ja algproduktsiooni tase suvel. In Järvekülg, A. (ed.), Eesti jõgede ja järvede seisund ning kaitse. TA Kirjastus, Tallinn: 148–165.

Kačalova, O.,1963. Materialy po faune presnovodnyh oligohet Latvijskoj SSR. Izvestija Akademii Nauk Latvijskoj SSR 5(190): 85–90.

Kačalova, O. & È. A. Parele, 1964. Daugavas mazsaru tarpi (Oligochaeta) un to nozíme zoobentosā. Latvijas PSR Zinātnu Akadēmijas Véstis 9(206): 64–70.

Korn, H., 1963. Studien zur Ökologie der Oligochaeten in der oberen Donau unter Berücksichtigung der Abwassereinwirkungen. Arch. Hydrobiol. Suppl. 27: 131–182.

Liebmann, H., 1962. Handbuch der Frischwasser- und Abwasser-Biologie I. Jena: 588 pp.

Milbrink, G., 1978. Indicator communities of oligochaetes in Scandinavian lakes. Verh. int. Ver. Limnol. 20: 2406–2411.

Parele, È. A., 1972. Malosščetinkovye červi r. Lielupe. In Vodnye malosščetinkovye červi, Materialy Vtorogo vsesojuznogo simpoziuma, Borok, 27-30 ijunja 1972 g. Jaroslavl': 60–74.

Parele, È. A.,1973. Malosščetinkovye červi (Oligochaeta) ust'evoj oblasti reki Daugava. In Limnologija Severo-Zapada SSSR (Materialy XVII naučnoj konferencii po izučeniju vnutrennih vodoemov Pribaltiki) III. Tallinn: 7–10.

Parele, È. A., 1981. Oligohetofauna kak pokazatel' saprobnosti malyh rek. In Gidrobiologičeskij režim malyh rek v uslovijah antropogennogo vozdejstvija. Zinatne, Riga: 127–135.

Parele, È. A., 1986. Èkologičeskie gruppirovki donnyh biontov malyh rek Latvii. In V S"ezd VGBO (Tol'jatti, 15-19 sentjabrja 1986 g.), tezisy dokladov II. Kujbyšev: 277–278.

Parele, È. A. & E. B. Astapenok, 1975. Tubificidy (Oligochaeta: Tubificidae) – indikatory kačestva vodoema. Izvestija Akademii Nauk Latvijskoj SSR 9: 44–46.

Sládeček, V., 1973. System of water quality from the biological point of view. Arch. Hydrobiol., Beiheft Ergebnisse der Limnologie 7: 1–218.

Timm, T., 1962. Eesti NSV magevee-väheharjasusside faunast, ökoloogiast ja levikust. Tartu Riikliku Ülikooli Toimetised 120, Zooloogia-alaseid Töid II: 63–107.

Timm, T., 1963. Eesti vooluveekogude väheharjasussidest. Üliõpilaste zooloogia-alaste tööde kogumik I. Tartu Riiklik Ülikool, Tartu: 24–30.

Timm, T., 1960. Andmeid Eesti NSV magevete oligoheetide (Oligochaeta limicola) faunast. Tartu Riikliku Ülikooli Matemaatikaloodusteaduskonna üliõpilastööde kogumik 1. Tartu: 55–65.

Timm, T., 1967a. O rasprostranenii malosščetinkovyh červej v tekučih vodoemah Èstonii. Eesti NSV Teaduste Akadeemia juures asuva Loodusuurijate Seltsi Aastaraamat 58: 164–174.

Timm, T., 1970. On the fauna of the Estonian Oligochaeta. Pedobiologia 10: 52–78.

Timm, T., 1987. Malosščetinkovye červi (Oligochaeta) vodoemov Severo-Zapada SSSR. Valgus, Tallinn: 299 pp.

Timm, V., 1967b. Märkmeid Narva jõe bentosest. Eesti NSV Teaduste Akadeemia juures asuva Loodusuurijate Seltsi Aastaraamat 58: 154–163.

Tõlp, Õ., 1956. Emajõe bentosest. Eesti NSV Teaduste Akadeemia juures asuva Loodusuurijate Seltsi Aastaraamat 49: 143–160.

Tõlp, Õ., 1976. Põhjaloomastik ja selle jaotumus Pärnu jõe alamjooksul. Eesti NSV Teaduste Akadeemia juures asuva Loodusuurijate Seltsi Aastaraamat 64: 198–233.

Hydrobiologia **463:** 235–240, 2001.
P. Rodriguez & P.F.M. Verdonschot (eds), Aquatic Oligochaete Biology VIII.

Oligochaeta of the Morava and Odra River basins (Czech Republic): species distribution and community composition

Jana Schenková, Ondřej Komárek & Světlana Zahrádková
Masaryk University, Faculty of Science, Department of Zoology and Ecology, Kotlářská 2, 611 37 Brno, Czech Republic
Fax: +420-5-41211214. E-mail: schenk@sci.muni.cz

Key words: aquatic Oligochaeta, distribution, ordination, Czech Republic

Abstract

The oligochaete fauna of 57 small streams of the Morava and Odra River basins, Czech Republic, was studied during 1997 and 1998. TWINSPAN and CCA (CANOCO) were used to characterize the relationship between oligochaete species, habitat diversity and environmental variables. Chemical oxygen demand (COD), water hardness, NO_3^-–N, Ca^{2+}, Mg^{2+}, altitude, and substrate type were found to be the primary factors influencing the distribution of aquatic Oligochaeta.

Introduction

The ecological requirements of aquatic Oligochaeta (Clitellata) and their relationship to abiotic factors in streams of the Czech Republic (central Europe) have received little attention in previous studies. While extensive taxonomic reports on oligochaetes have been published by Hrabě (1979) and Lišková (1976), and on the oligochaete family Enchytraeidae, by Chalupský (1988), few ecological studies on this diverse and often dominant group of lotic benthic organisms have been conducted (Brinkhurst, 1978; Uzunov et al., 1988; Hörner et al., 1995).

The objectives of this study were to determine the distribution of oligochaete species in small, relatively unpolluted streams in the Czech Republic, and to evaluate the impact of environmental factors on the composition of oligochaete taxocenoses using multicriterial analyses.

* The paper is dedicated to the memory of Prof. Dr Sergej Hrabě, on the 100th anniversary of his birth. Dr Hrabě, a major contributor to aquatic oligochaete taxonomy during his extensive career in the Department of Zoology and Anthropology at Masaryk University, actively published from 1927 up to 1984.

Description of sites studied

Study sites were located on small streams in the Morava River basin (*sensu lato*, which means the Morava and Dyje River basins together) and in the Odra River basin (Czech Republic) (Fig. 1). The studied area represents an interesting region since all biogeographical subprovinces found in the Czech Republic, are present: Hercynicum, Pannonicum, Carpathicum and Polonicum (Culek, 1996) (Fig. 1). All sites were either unpolluted or moderately influenced by treated and untreated domestic wastewater effluent. Studied localities presented the following characteristics: they were situated from 175 to 750 m a.s.l.; the distance from the source was 0.7 to 16.8 km; the mean width of the streams ranged from 0.93 to 4.25 m; and the mean depth of the streams ranged from 0.06 to 1.1 m.

Materials and methods

Samples of macrozoobenthos were collected from 27 localities during spring, summer, and autumn of 1997, and from 30 localities during the same seasons in 1998. Sampling at each station was completed using

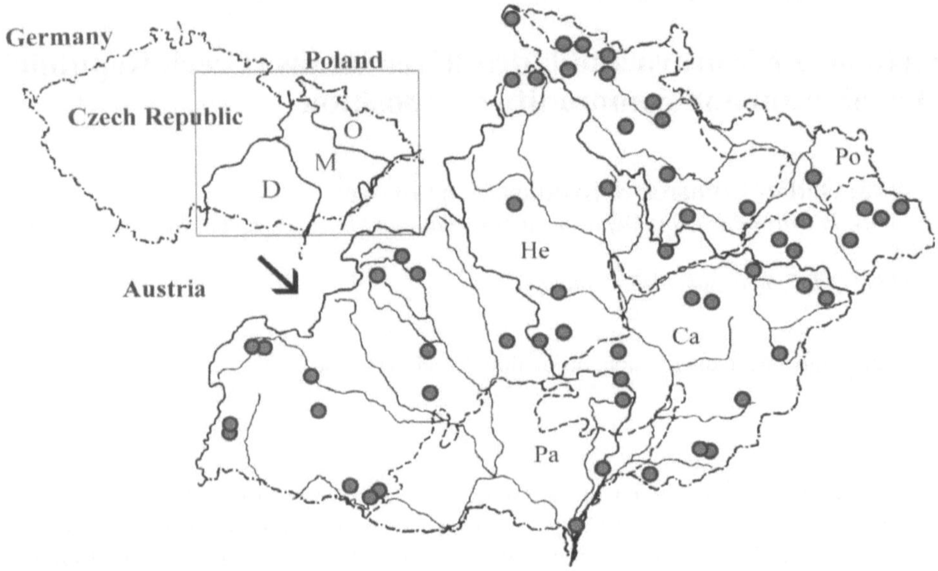

Figure 1. Map of the study area showing site localities and river basins. D – Dyje; M – Morava; O – Odra River basin. Biogeographical subprovinces named according to Culek (1996). He – Hercynicum; Po – Polonicum; Ca – Carpathicum; Pa – Pannonicum.

a standard method of kick sampling for a period of 3 min (Kershaw & Frost, 1968).

During each sampling period, water temperature, width and depth of the stream, hydrogen ion concentration (as pH), conductivity and the substrate composition were measured. The evaluation of the substrate composition at each site followed Wright et al. (1984) by expressing the weighted average of percent presence of each particle size-classes (boulders, cobbles, pebbles, granules, sand, mud and clay). The mean value of *phi*, which is the negative logarithm of the particle size in millimetres, characterised each particle size-class. Laboratory analyses for important ion components present at our sampling sites were performed on samples collected during each sampling period. Alkalinity, water hardness, oxygen content, dissolved oxygen (DO), biochemical oxygen demand (BOD_5), chemical oxygen demand (COD), total organic carbon (TOC), dissolved organic carbon (DOC), NH_4^+–N, NO_2^-–N, NO_3^-–N, total nitrogen (N TOT), o-PO_4^{3-}–P, total phosphorus (P TOT), Cl^-, SO_4^{2-}, Ca^{2+} and Mg^{2+} were measured. Current velocity was not included in the analyses; single measurements obtained during sampling had no predicable value because of the frequently changing water level at the studied sites.

Samples of macrozoobenthos were fixed on site with 4% formaldehyde. After the samples had been sorted, oligochaetes were prepared for identification by either mounting permanently on slides using Canada balsam, or prepared as temporary mounts using a glycerin-water (1:5) solution.

With the exception of taxa representing five genera in the family Enchytraeidae, which were determined using the keys of Nielsen & Christensen (1959, 1961, 1963), all oligochaetes were identified to the species level using the keys of Hrabě (1954, 1979). Juvenile forms or damaged individuals were determined only to family level.

The resulting data were evaluated and interpreted using TWINSPAN (Hill, 1979). All options were set to default; maximum group size for division was five and maximum level of divisions was six. Relationships between oligochaete species and environmental variables were evaluated by Canonical Correspondence Analysis (CCA) using the CANOCO program for Windows (Ter Braak & Šmilauer, 1998). The scaling type was biplot and species data were not transformed. With the exception of representatives of Enchytraeidae, taxa unidentified to the species level were omitted from the analyses. Statistical significance was then determined by a Monte Carlo permutation test (Ter Braak & Šmilauer, 1998). 27 environmental vari-

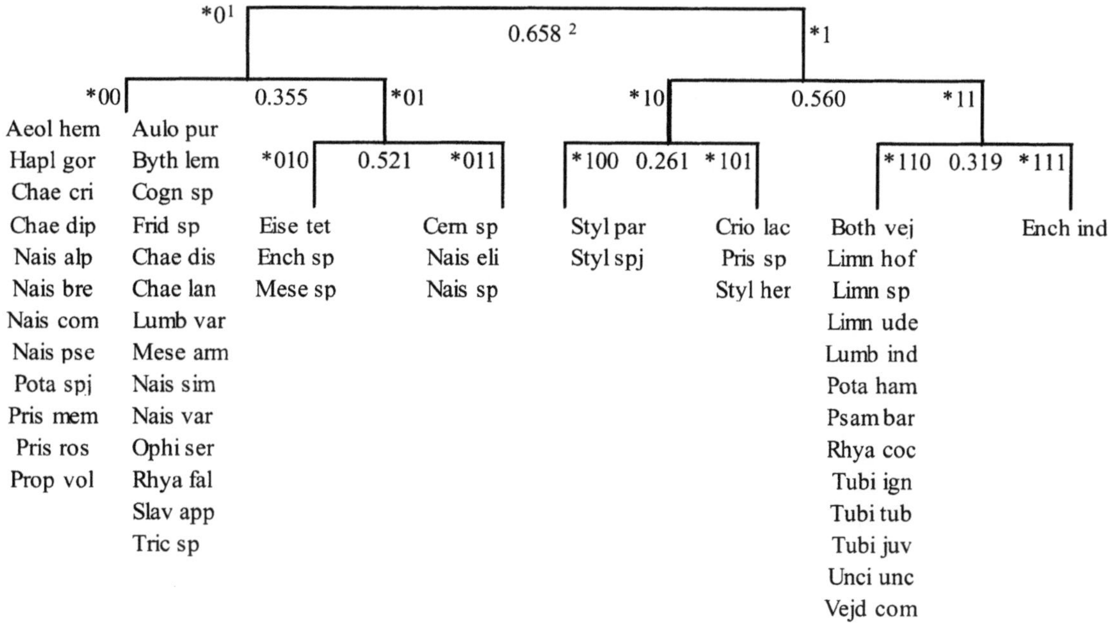

Figure 2. Divisive hierarchic classification (TWINSPAN) of species (for abbreviations see Table 1). [1]TWINSPAN group, [2]eigenvalue.

ables were tested in 999 permutations using automatic forward selection.

Results

During the 1997 and 1998 sampling periods, 22 211 individuals of the class Oligochaeta representing eight families, 27 genera and 36 species were collected (Table 1). One species of Aeolosomatidae (Annelida, Aphanoneura) was also collected. The abbreviation for each identified taxon used in analyses is provided in Table 1.

Hierarchical divisive classification of species

Hierarchical divisive classification of 51 taxa with eigenvalues of divisions is apparent from Figure 2. The eigenvalues express the level of distinction of separated groups at each division. Oligochaeta were separated into seven different groups. Group *00, further division of which expressed eigenvalue 0.196 and hence was not significant, consists of representatives of the family Naididae and other species from sites characterized by a sandy bottom with plant detritus. Group *010 includes the taxa *Enchytraeus*, *Mesenchytraeus* and *Eiseniella tetraedra* (Savigny, 1826). Group *011 consists of the cosmopolitan species *Nais elinguis* Müller, 1773 and *Cernosvitoviella* (Enchytraeidae). Group *100 and group *101, with relatively low eigenvalue of division, include species of the genus *Stylodrilus* – inhabiting unpolluted brooks and rivers with sandy bottom, the species *Criodrilus lacuum* Hoffmeister, 1845 – common in pools, and the genus *Pristinella* – which prefers pristine waters. Group *110 includes species in the family Tubificidae, and *Uncinais uncinata* (Ørsted, 1842) (Naididae). Group *111 contains only unidentified Enchytraeidae species.

Canonical correspondence analysis (CCA)

Canonical correspondence analysis (CCA) concerned 27 environmental variables and 46 species or genera, respectively; five genera that might have included more than one species were omitted (Fig. 3). The eigenvalue for the first axis was 0.535, the second 0.335 and the third 0.255. The percentage variance of species environment relation was 35.8 for the first axis, 22.5 for the second and 17.1 for the third axis.

The statistical significance of the data was evaluated using the Monte Carlo permutation test. The following environmental factors were found to be statistically significant: COD, hardness, NO_3^-–N,

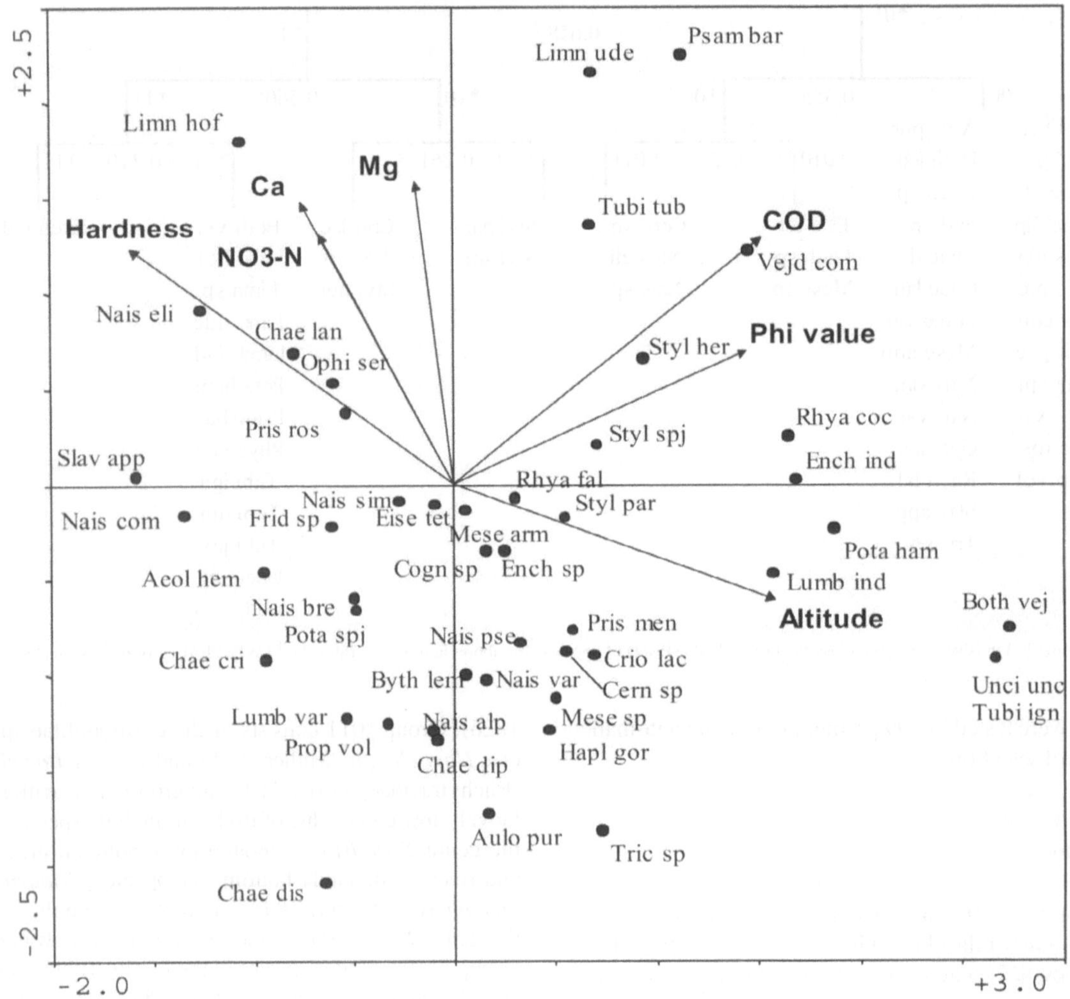

Figure 3. Canonical correspondence analysis (CCA), ordination plot of species (for abbreviations see Table 1) and main environmental variables on the first two CCA axes.

Ca^{2+}, Mg^{2+}, altitude, and substrate type ($P < 0.01$). CCA analysis showed correlations (>0.5): positive for COD, altitude and negative for hardness and the first axis. The distribution of species along the first axis reflects their requirements regarding organic matter supply, altitude and hardness. A positive correlation for Mg^{2+} was shown by the second axis, while positive correlations for Ca^{2+} and Mg^{2+} were shown by the third axis. The distribution of species along the second and third axis can be explained by the water hardness.

Discussion

From the analyses performed, we determined that COD and water hardness together with concentrations of NO_3^-–N, Ca^{2+}, Mg^{2+}, altitude, and substrate type are the primary factors influencing the distribution of Oligochaeta in small streams.

The distribution of the species of the family Tubificidae can be explained by their preference for substrates with higher organic content and fine particle size (Fig. 3). The species are widespread in distribution, from lower to higher altitudes. All species collected are common and abundant, especially in

Table 1. List of oligochaete species from the Morava and Odra River Basins in the Czech Republic and the abbreviation of each species

Species	Abbreviation
Aeolosoma hemprichi Ehrenberg, 1831	Aeol hem
Aulodrilus pluriseta (Piguet, 1906)	Aulo pur
Bothrioneurum vejdovskyanum Štolc, 1886	Both vej
Bythonomus lemani Grube, 1880	Byth lem
Cernosvitoviella sp.	Cern sp
Chaetogaster cristallinus Vejdovský, 1883	Chae cri
Chaetogaster diaphanus (Gruithuisen, 1828)	Chae dip
Chaetogaster diastrophus (Gruithuisen, 1828)	Chae dis
Chaetogaster langi Bretscher, 1896	Chae lan
Cognettia sp.	Cogn sp
Criodrilus lacuum Hoffmeister, 1845	Crio lac
Eiseniella tetraedra (Savigny, 1826)	Eise tet
Enchytraeidae	Ench ind
Enchytraeus sp.	Ench sp
Fridericia sp.	Frid sp
Haplotaxis gordioides (Hartmann, 1821)	Hapl gor
Limnodrilus hoffmeisteri Claparède, 1862	Limn hof
Limnodrilus udekemianus Claparède, 1862	Limn ude
Limnodrilus sp. juv.	Limn spj
Lumbricidae	Lumb ind
Lumbriculus variegatus Grube, 1844	Lumb var
Mesenchytraeus armatus Levinsen, 1884	Mese arm
Mesenchytraeus sp.	Mese sp
Nais alpina Sperber, 1948	Nais alp
Nais bretscheri Michaelsen, 1899	Nais bre
Nais communis Piguet, 1906	Nais com
Nais elinguis O. F. Müller, 1773	Nais eli
Nais pseudoobtusa Piguet, 1906	Nais pse
Nais simplex Piguet, 1906	Nais sim
Nais variabilis Piguet, 1906	Nais var
Nais sp.	Nais sp
Ophidonais serpentina (O. F. Müller, 1773)	Ophi ser
Potamothrix hammoniensis (Michaelsen, 1901)	Pota ham
Potamothrix sp. juv.	Pota spj
Pristinella menoni (Aiyer, 1930)	Pris men
Pristinella rosea (Piguet, 1906)	Pris ros
Pristinella sp.	Pris sp
Propappus volki Michaelsen, 1905	Prop vol
Psammoryctides barbatus (Grube, 1861)	Psam bar
Rhyacodrilus coccineus (Vejdovský, 1875)	Rhya coc
Rhyacodrilus falciformis Bretscher, 1901	Rhya fal
Slavina appendiculata (Udekem, 1855)	Slav app
Stylodrilus heringianus Claparède, 1862	Styl her
Stylodrilus parvus (Hrabě & Černosvitov, 1927)	Styl par
Stylodrilus sp. juv.	Styl spj
Trichodrilus sp.	Tric sp
Tubifex ignotus (Štolc, 1886)	Tubi ign
Tubifex tubifex (O. F. Müller, 1774)	Tubi tub
Tubificidae juv.	Tubi juv
Uncinais uncinata (Ørsted, 1842)	Unci unc
Vejdovskyella comata (Vejdovský, 1883)	Vejd com

polluted waters (Brinkhurst, 1971). COD and the sediment structure, as the most significant factors for oligochaete occurrence, have been mentioned by other authors (Brinkhurst, 1967; Uzunov, 1982).

The distribution of *Limnodrilus hoffmeisteri* Claparède, 1862, *Nais elinguis* Müller, 1773, *Chaetogaster langi* Bretscher, 1896, *Ophidonais serpentina* (Müller, 1773) and *Pristinella rosea* (Piguet, 1906) is positively correlated with water hardness, Ca^{2+}, Mg^{2+} and NO_3^--N and negatively correlated with altitude (Fig. 3). All are known to occur also in stagnant waters. While the values for water hardness, and Ca^{2+} and Mg^{2+} ion concentration are readily influenced by easily eroded bedrock, the presence of these ions also depends also on the concentration of dissolved CO_2, which is related to water eutrophication.

Major changes in the structure of oligochaete assemblages related to water mineralization, substrate type, and current velocity were documented by Martínez-Ansemil & Collado (1996), who studied Oligochaeta inhabiting largely unpolluted watercourses in Galicia (Spain) and Northern Portugal. Dumnicka (1978), in her studies of Oligochaeta in the River Nida, Poland, concluded that bottom type and current speed as well as hardness or Ca^{2+} and Mg^{2+} content determine the composition of oligochaete communities.

With the exception of the species noted above, naidids are distributed in streams with the lower content of organic matter and coarse substratum (Fig. 3), which is typical for unpolluted waters. This group is distributed both at low and high altitudes.

While oligochaete distribution in small streams is determined by COD, water hardness (Martínez-Ansemil & Collado, 1996) or current velocity (Dumnicka, 1978), substrate type seems to be the most important factor governing the presence of Oligochaeta in large rivers (Šporka, 1998; Montanholi-Martins & Takeda, 1999; Rempel et al., 2000). This could also be influenced by the presence of a sufficient food supply. The results of this study also suggest that oligochaetes, a group which demonstrate a broad range of ecological tolerance, are nonetheless more restricted at the family level to certain types of environments.

Acknowledgements

This project was supported by grants from Masaryk University No. 206/99/1522 and from Ministry of Education No. 143100010. We wish to thank our teacher Prof. F. Kubíček for assistance with macroinvertebrate identifications, advice and direction throughout this study. We also thank first author's father Prof. J. Gaisler for critical review of the manuscript and M. J. Wetzel (Illinois Natural History Survey, Champaign, U.S.A.) for language help and suggesting many helpful improvements.

References

Brinkhurst, R. O., 1967. The distribution of aquatic oligochaetes in Saginaw bay, Lake Huron. Limnol. Oceanogr. 12: 137–143.

Brinkhurst, R. O., 1971. A Guide for the Identification of British Aquatic Oligochaeta. Titus Wilson & Son, Kendal: 55 pp.

Brinkhurst, R. O., 1978. Oligochaeta. In J. Illies (ed.), Limnofauna Europea. G. Fischer, Stuttgart, New York: 139–144.

Chalupský, J., 1988. Czechoslovak enchytraeids (Oligochaeta, Enchytraeidae) II. Catalogue of species. Věst. čs. Společ. zool. 52: 81–95. (in Czech).

Culek, M. (ed.), 1996. Biogeografické členení České Republiky. (biogeographical classification of the Czech Republic). Enigma, Praha: 347 pp (in Czech).

Dumnicka, E. & K. Pasternak, 1978. The influence of physicochemical properties of water and bottom sediments in the River Nida on the distribution and numbers of Oligochaeta. Acta Hydrobiol. 20: 215–232.

Hill, M. O., 1979. TWINSPAN – A FORTRAN program for arranging multivariate in an ordered two-way table by classification of individuals and attributes. Cornell University, Ithaca, New York: 99 pp.

Hörner, K. M., O. Moog & F. Šporka, 1995. Oligochaeta. In Moog, O. (ed.), Fauna Aquatica Austriaca. Katalog zur autökologischen Einstufung Aquatischer Organismen Österreichs. Bundesministerium für Land- und Forstwirtschaft, Vienna: 277 pp.

Hrabě, S. (ed.), 1954. Klíč k určování zvířeny ČSR. (key to the Czechoslovak fauna). Vol. 1. ČSAV, Praha: 539 pp (in Czech).

Hrabě, S., 1979. Vodní máloštětinatci (Oligochaeta) Československa. (aquatic Oligochaeta of Czechoslovakia). Acta Univ. Carol., Biol. 1–2: 1–167 (in Czech, with English summ).

Kershaw, W. & S. Frost, 1968. The selective control of *Simulium* larvae by particulate insecticides and its significance in river management. Trans. r. Soc. trop. Med. Hyg. 63: 35–40.

Lisková, E., 1976. Příspěvek k výskytu některých druhů limikolních máloštětinatců (Oligochaeta) na území Čech a Moravy (Československo) (contribution to the existence of some species of Oligochaeta limicola on the Bohemian and Moravian territory (Czechoslovakia)). Acta sci. nat. Mus. Bohem. merid. České Budějovice 16: 17–30 (in Czech, with English summ).

Martínez-Ansemil, E. & R. Collado, 1996. Distribution patterns of aquatic oligochaetes inhabiting watercourses in the Northwestern Iberian Peninsula. Hydrobiologia 334: 73–83.

Montanholi-Martins, M. C. & A. M. Takeda, 1999. Communities of benthic oligochaetes in relation to sediment structure in the upper Paraná River, Brazil. Stud. Neotrop. Fauna Envir. 34: 52–58.

Nielsen, C. O. & B. Christensen, 1959. The Enchytraeidae – critical revision and taxonomy of European species (studies on Enchytraeidae VII). Natura Jutlandica 8–9: 1–160.

Nielsen, C. O. & B. Christensen, 1961. The Enchytraeidae – critical revision and taxonomy of European species. Natura Jutlandica, Suppl. 1: 1–23.

Nielsen, C. O. & B. Christensen, 1963. The Enchytraeidae – critical revision and taxonomy of European species. Natura Jutlandica, Suppl. 2: 1–19.

Rempel, L. L., J. S. Richardson & M. C. Healey, 2000. Macroinvertebrate community structure along gradients of hydraulic and sedimentary conditions in a large gravel-bed river. Freshwat. Biol. 45: 57–73.

Šporka, F., 1998. The typology of floodplain water bodies of the Middle Danube (Slovakia) on the basis of superficial polychaete and oligochaete fauna. Hydrobiologia 386: 55–62.

Ter Braak, C. J. F. & P. Šmilauer, 1998. Canoco reference manual and user's guide to Canoco for Windows. Centre for Biometry, Wageningen: 351 pp.

Uzunov, J., V. Košel & V. Sládeček, 1988. Indicator value of freshwater Oligochaeta. Acta Hydrochim. Hydrobiol. 16: 173–186.

Uzunov, J., 1982. Statistical assessment of the significance of both bottom substrata and saprobity for the distribution of aquatic Oligochaeta in rivers. Limnologia 14: 353–361.

Wright, J. F., D. Moss, P. D. Armitage & M. Furse, 1984. A preliminary classification of running-water sites in Great Britain based on macro-invertebrate species and the prediction of community type using environmental data. Freshwat. Biol. 14: 221–256.

Hydrobiologia **463:** 241–248, 2001.
P. Rodriguez & P.F.M. Verdonschot (eds), Aquatic Oligochaete Biology VIII.

Effect of hydraulics, bed load grain size and water factors on habitat and abundance of *Narapa bonettoi* Righi & Varela, 1983 of the Upper Paraná River, Brazil

Alice M. Takeda[1], José C. Stevaux[2] & Daniele S. Fujita[1]
[1]Department of Biology; [2]Department of Geography; State University of Maringá/Nupelia, Av. Colombo 5790, Maringá, PR, CEP 87020-900, Brazil
Fax: +55-44-2631424. E-mail: alice@nupelia.uem.br

Key words: Narapa, oligochaete, sedimentary dynamic, Paraná River, Brazil

Abstract

A cross-section of the upper Paraná was studied in order to evaluate which hydraulic, sedimentary and water variables influenced on the spatial – temporal distribution and abundance of the *Narapa bonettoi* Righi & Varela, 1983 population. From June 1993 to February 1995, data on discharge, channel morphology, flow velocity, sediment suspended concentration and depth, water variables and benthic community were obtained. Data were analyzed by principal component analysis – PCA. The highest density of *N. bonettoi* occurred at the site with coarser sediments and stable bottom morphology. The lowest density was recorded at the site that presented less stable channel morphology. Temporal variation is controlled by hydrological regime and *N. bonettoi* cycle life whereas spatial distribution is tightly associated with channel stability.

Introduction

Narapa bonettoi Righi & Varela, 1983 is a small oligochaete species, without setae,that inhabits the interstices between sand grains in large fluvial systems, where intense erosional and depositional dynamics is observed. Apparently, this species is highly specialized to environments otherwise hostile to other oligochaetes. *Narapa bonettoi* has been recorded in the Argentinian stretch of the Paraná River, where it was studied mainly by Marchese (1984, 1987) and Marchese & Ezcurra de Drago (1983, 1992). Marchese (1994) studied its life cycle in detail and mentioned that *N. bonettoi* has both sexual and asexual reproduction as a survival strategy. In the Brazilian stretch, *N. bonnetoi* was recorded near the bank of the Paraná and in its secondary channels by Takeda et al. (1997) and Montanholi & Takeda (1999).

The present study aims to answer the following question: what are the factors that influence temporal and spatial variation in the density of *Narapa bonettoi*? For this, the influence of the hydraulic geometry (section morphology, depth, flow velocity, suspended load concentration and bed load grain-size) and water factors (pH, electrical conductivity, temperature, dissolved oxygen and transparency) on the density patterns of *N. bonettoi* were analyzed.

Study area

The area studied lies near the town of Porto Rico (22° 43′ S; 53° 10′ W), State of Paraná, Brazil, 10 km downstream from the mouth of the Paranapanema River (Fig. 1). At the studied section, Paraná River's drainage basin area is about 670 000 km^2. In this area, the river forms a large anastomosed channel, 3.4–4.0 km wide, with an extensive floodplain on its right bank.

The study section is 1200 m wide and constitutes a nodal point between two anastomosed stretches. The channel morphology is asymmetrical with main channel to the left bank. Four sites were chosen to characterize the channel dynamics: site S_1 lies near

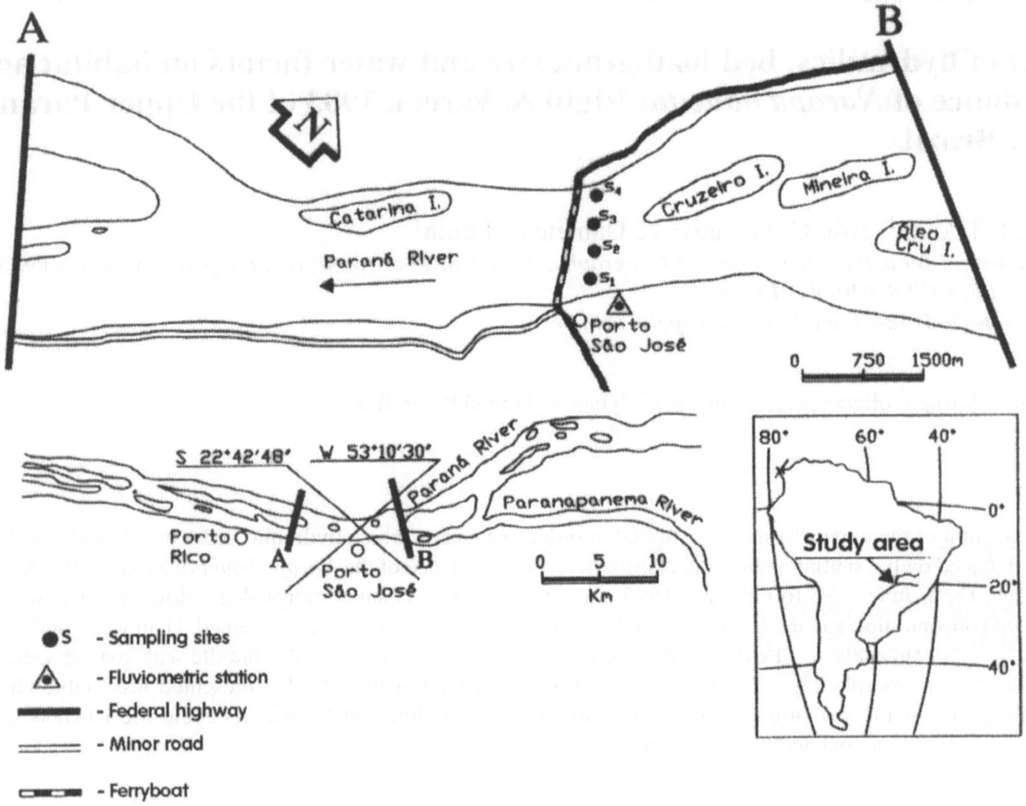

Figure 1. Sampling area within the Upper Paraná River.z

the left bank with a depth of 5–6 m, site S_2 lies in the talweg with depths from 10 to 14 m, site S_3 is situated in the mid-channel, with a depth about 9 m; and site S_4 lies on a flat surface near the right bank with depths of 5–6 m. The channel morphology in the section changed during the study period (Fig. 2). Sites S_2 and S_3 show greater change than sites S_1 and S_4. At S_2, dunes and mega-ripples are the main bedforms and they migrate downstream at about 67 m per month. The exact position of each site is defined every month by echo-sounder surveys.

Materials and methods

Sampling

Sampling was performed monthly, from July 1993 through February 1995 (except for January 1995). Using a modified Petersen grab (0.0345 m^2), four benthic samples were taken at each site, three for the biological analysis and one for grain size analysis.

The three samples collected for biological analysis were washed through a series of sieves (mesh size from 2.0 to 0.2 mm). All organisms retained on sieves 2.0 and 1.0 mm, were immediately picked out. The material retained on 0.5 and 0.2 mm sieves was fixed in 10% neutral formalin and sorted under stereoscopic microscope.

Along with the biological samples, water samples were taken with a Van Dorn bottle, transferred to polyethylene flasks and taken on ice in an insulated box to the laboratory. Electrical conductivity, pH and dissolved oxygen (Winkler method, modified by Golterman et al., 1978) were measured. To estimate suspended sediment load at each site, 2000 ml of water were taken at the middle depth of the water column with a Van Dorn bottle. Each sample was divided into four 1 l sub-samples, filtered through Whatman GF/C micropore filters and processed according to the meth-

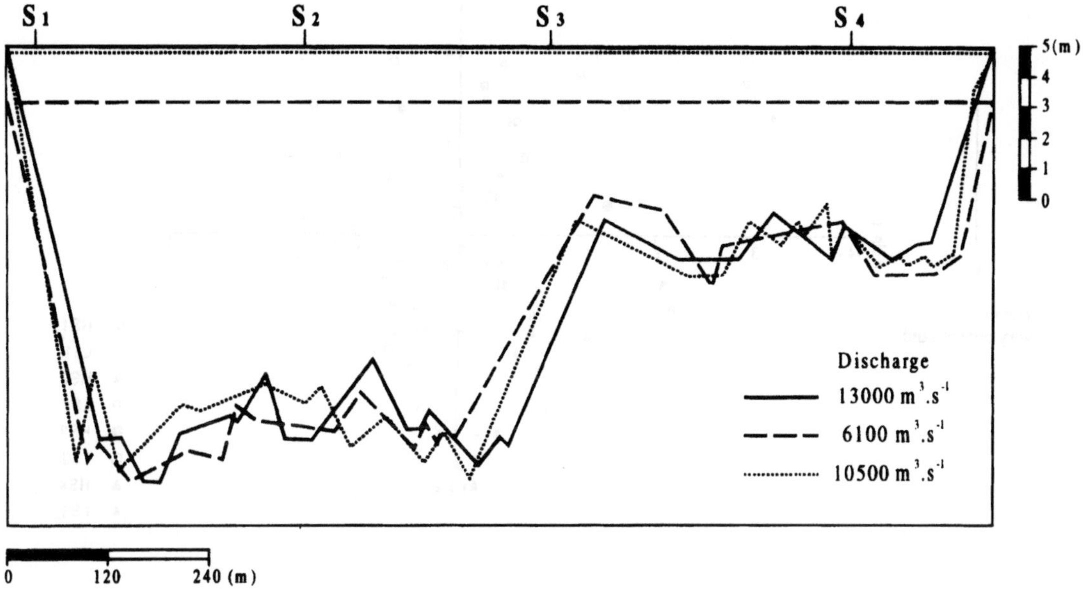

Figure 2. Variation in channel morphology of Paraná River cross-section, according to the river's maximum, medium and minimum discharge. S_1, S_2, S_3 and S_4 are sampling sites.

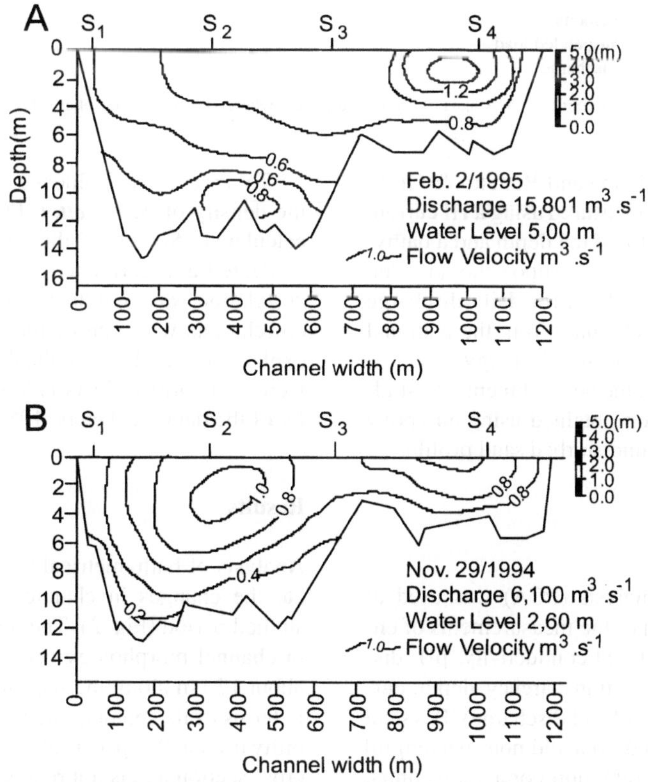

Figure 3. Flow velocity distribution at Porto São José section: (A) High water level corresponding to a discharge of 15 801 $m^3.s^{-1}$, (B) Low water level, corresponding to a discharge of 6100 $m^3.s^{-1}$.

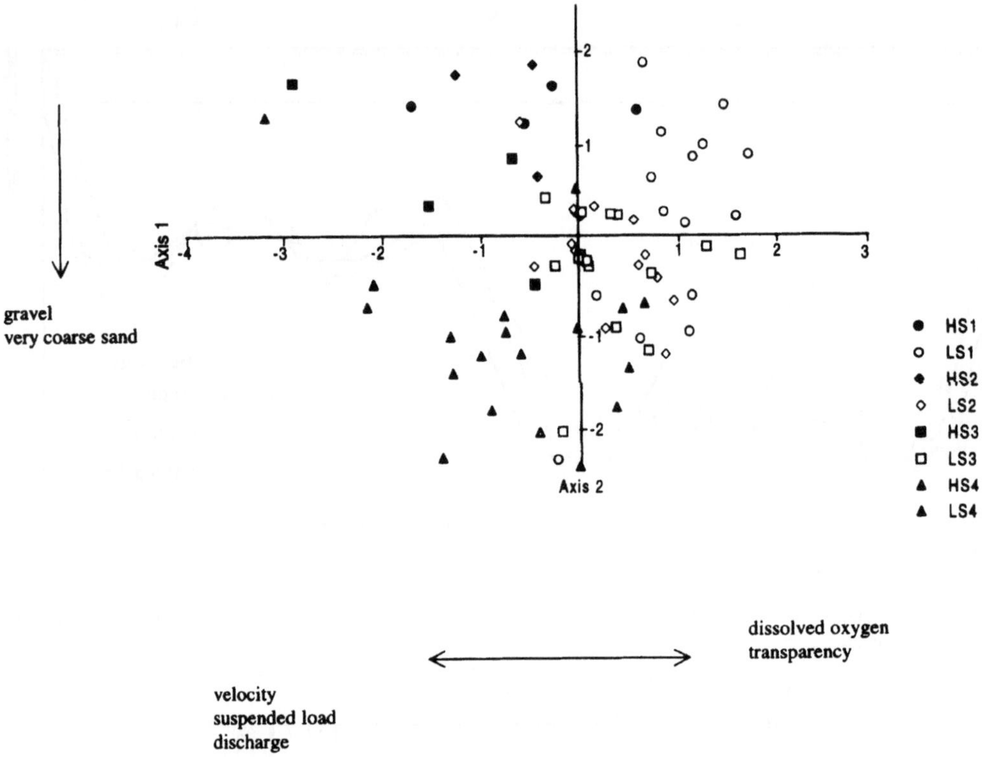

Figure 4. Ordination scores of the first two PCA axes of abiotic variables, where H = high water phase; L = low water phase; S = site.

ods described in APHA (1975) and Bonetto & Orfeo (1984). Flow velocity was measured using a HI current meter at 80%, 50% and 30% of the depth and a bathymetric profile was taken using a Furuno echo-sounder, FE – 881II, 50 kHz model. Data for daily discharge of the Paraná River were obtained from the National Department of Waters and Electrical Energy.

The interstitial space of the bed sediment was studied by taking a thin section obtained using an epoxy resin impregnation of the undisturbed sand profile.

Data processing

Principal Components Analysis (PCA) was used to simplify the interpretation of the measurements of environment variables (electrical conductivity, pH, dissolved oxygen, temperature, transparency, depth, current velocity, suspended load and discharge). Physical variables were standardized; to avoid non-meaningful weighting caused by different, non-comparable units, to make attributes dimensionless, and to reduced them to a scale of comparable ranges.

To assess the influence of the abiotic data on the density of *N. bonettoi*, Pearson's correlation was calculated. Scores of the PCA axes 1 and 2 were generated and correlated with density data. When a correlation was significant, the variables with higher correlations with one of the axis were the ones that explain the variability in the density data. Density data were transformed (ln x+1) to reduce the dimensionality of the data set (Peters, 1986).

Results

Analysis of bathymetrical transects allowed to estimate the changes in channel morphology during the studied period (Fig. 2). The most expressive alteration in channel morphology occurred at S_3 that migrated about 120 m along the section while S_4 stayed stable in terms of its geometry during the year. The term stability used in this paper refers to the morphology of the cross section and is not related to bedform movement.

Flow velocity sections obtained in low and high discharge periods (Fig. 3) showed that flow is divided

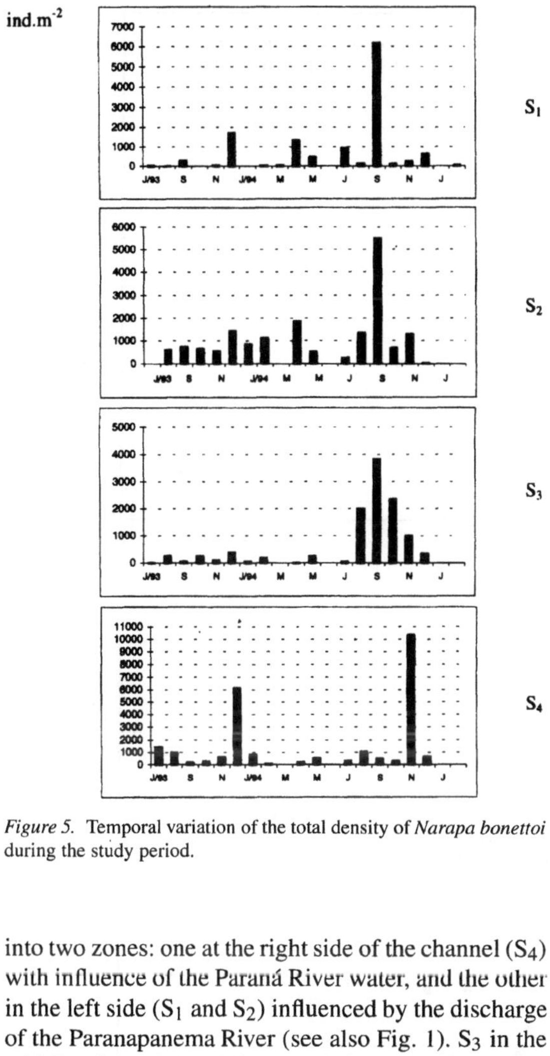

Figure 5. Temporal variation of the total density of *Narapa bonettoi* during the study period.

into two zones: one at the right side of the channel (S_4) with influence of the Paraná River water, and the other in the left side (S_1 and S_2) influenced by the discharge of the Paranapanema River (see also Fig. 1). S_3 in the middle of the channel, is under influence of both flow zones.

The first two axes of the PCA (Fig. 4) explained 46% of the total variability of the physical and chemical data. Considering only correlations higher than 0.60, along axis 1 (25%) there were significant positive correlations with dissolved oxygen and transparency and negative correlations with velocity, suspended load and river discharge. Along axis 2 (21%), there was a significant positive correlation with gravel and very coarse sand.

The ordination diagram of physical and chemical variables showed a sharp difference between high and low water phases. The high water phase was positively correlated with river discharge and suspended load; and the low water phase with high transparency. The sites were separated according to dominant grain size. Fine sand was predominant at the site S_1, while coarse sands predominated at S_4. Sites S_2 and S_3 were intermediate between sites S_1 and S_4. The PCA ordination diagram shows a difference between the high and low water phases. The dots of site S_4 concentrated at lower left corner are characterized with coarser sand, high flow velocity, suspended load and discharge. In the upper right corner, the dots of site S_1 occur and are related to fine to coarse sand, lower velocity, suspended load, and discharge and higher dissolved oxygen and transparency. The dots of sites S_2 and S_3 were not described.

Higher densities of *N. bonettoi* were recorded from July to December. Maximum density of *N. bonettoi* was found in November 1994 at Site S_4 (10 367 ind m^{-2}) (Fig. 5). Density data were negatively correlated with both axis 1 ($r = -0.30$; $P = 0.009$) and axis 2 ($r = -0.23$; $P = 0.04$). Although axis 2 was significantly negative correlated with density, the scatter plot showed several outliers, which do not support this assumption. Therefore, only the correlation with axis 1 (Fig. 6) was analysed. Samples and months with negative scores were similar due to the effect of velocity, suspended load and discharge on density.

S_1, S_2 and S_3 had the same grain size with the same interstitial space varying from 0.2 to 0.8 mm. S_4, constituted by very coarse sand to granules, show to have pores up to 2.0 mm (Fig. 7).

Discussion

The Paraná River discharge varied from 6100 to 15 801 m^3 s^{-1} during the studied period. During intermediate discharge (10 500 m^3 s^{-1}) of February 1994, *Narapa bonettoi* specimens were recorded at all sites, with an average density of 350 ind. m^{-2} (sd = 511.7), on the other hand, on February 1995 during a high discharge period, *N. bonettoi* was almost absent (average density of 12 ind m^{-2}, *sd* = 24.0). This observation suggests that the density of *N. bonettoi* is affected not only by its life cycle but also by the river hydrology.

Flow velocity variation near the bottom of the channel during the high and low discharge periods is larger at the left side (0.2–0.8 m s^{-1}) in comparison to the right one (0.6–0.8 m s^{-1}). According to the classical flow speed versus particle diameter diagram of Hjulström (Friedman & Sanders, 1978, p. 110), a flow velocity of 0.6 m s^{-1} is the erosion threshold for very coarse sand. S_2 and S_3 were submitted to a

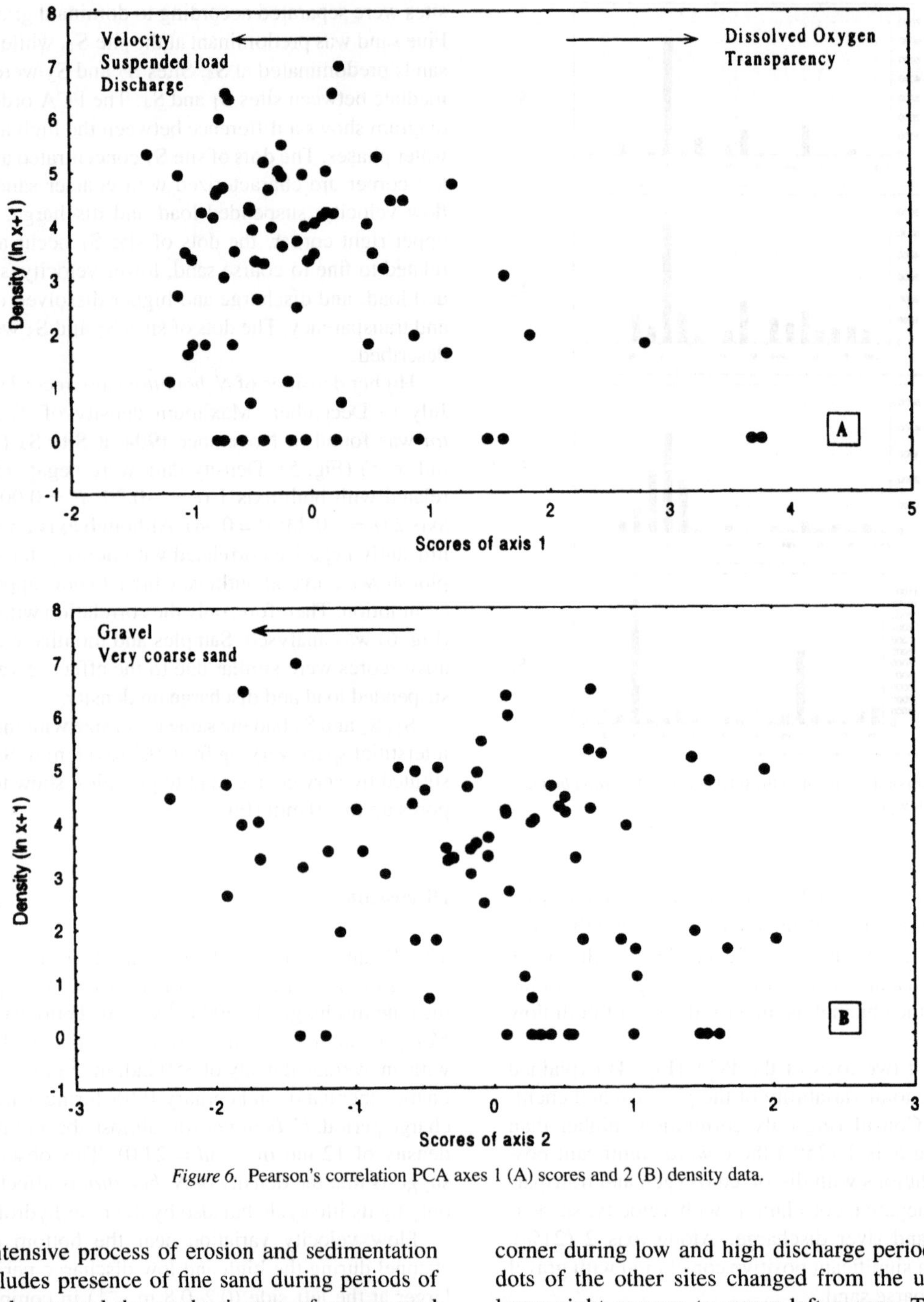

Figure 6. Pearson's correlation PCA axes 1 (A) scores and 2 (B) density data.

more intensive process of erosion and sedimentation that includes presence of fine sand during periods of low discharge and the predominance of coarse sand during floods. Particle size was more stable during the year at S_4, due to the low variation in flow velocity.

This is underlined by the results of the PCA ordination, where the dots of S_4 stayed on the lower left corner during low and high discharge periods, while dots of the other sites changed from the upper and lower right corners to upper left corner. The upper left corner is characterized by the highest values of velocity, suspended load and discharge during the high discharge phase.

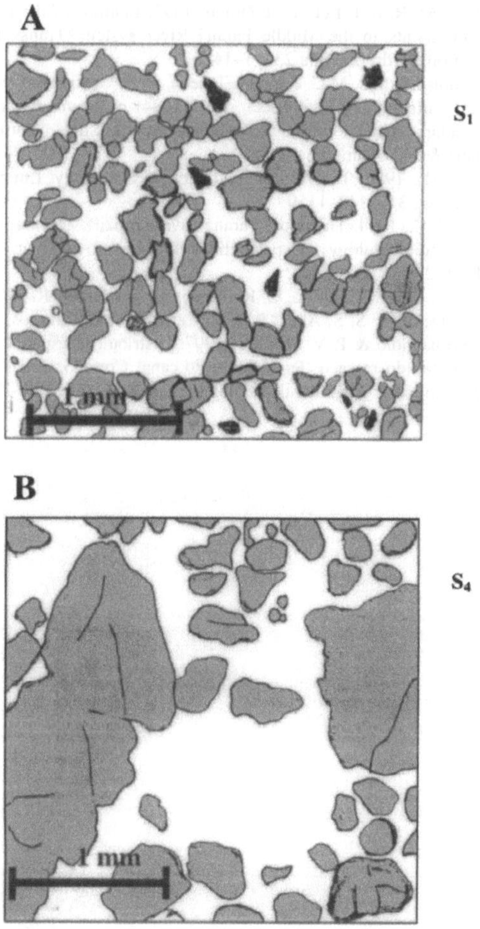

Figure 7. Thin sections made from bed load deposits in sites S_1 (A) and S_4(B). Porosity is oversized in S4 due to coarse sand and granules.

The highest density was recorded at site S_4 ($\bar{x}$ = 106 657 ind m^{-2}; sd = 2581.3), and lowest density occurred at the site S_3 ($\bar{x}$ = 589 ind m^{-2}; sd = 1034.8). Sites S_1 and S_2 ($\bar{x}$ = 654 ind m^{-2}; sd = 1431.2; $\bar{x}$ = 967 ind m^{-2}; sd = 1257.4, respectively) showed low densities, although both had the same grain size as those of S3. At those sites, the density of *N. bonettoi* was controlled by the stability of channel morphology.

Coarse grain sizes of S_4, which were sorted out by high flow velocities, produced the most optimal sediment structure and sediment stability to interstitial organisms such as *N. bonettoi*.

The largest pore size observed at S_4, when compared to the other sites, may have contributed to the high density of organisms. Major bedforms in the Paraná River can migrate about 67 m a month, in this movement, interstitial space can be continuously destroyed and formed causing stress to organisms. Thus, sediment with larger pores, like S_4, can promote better conditions to organism survival.

The highest densities of *N. bonettoi* occurred between July and December coinciding with the low water phase of the Paraná River. During that period, less stress due to the slow movement of the river bedforms allowed greater success of reproduction and survival. Temporal variations of *N. bonettoi* density similar to those observed in the upper Paraná River were recorded by Marchese (1994) in the middle Paraná.

Sandy areas support few species, but often in high densities. Typically, these are small, psammic organisms like *N. bonettoi*, adapted to the conditions imposed by high current velocities and to a correspondingly changeable substratum (Di Persia, 1986). In the upper Paraná River system, the hydrological pulse of the river mainly affected, among other variables, the sediment structure and sediment stability, which both influence the density of *N. bonettoi*.

Acknowledgements

We thank Rosângela G. Santana for helping the statistics analysis, Drs Paul Potter and Luiz Gomes for suggestions. We gratefully acknowledge Dr Mario Sommerhäuser, Dr Rüdger M. Schmelz for their contributions and Dr Janet Reid for assistance with the English text. This project was supported by PADCT/ CIAMB/ CNPq.

References

APHA, 1975. Standard Methods for the Examination of Water and Wastewater. 14th edn. American Public Health Association, Washington: 1194 pp.

Bonetto, A. A. & O. Orfeo, 1984. Caracteres sedimentologicos de la carga en suspension del río Paraná entre Confluencia y Esquina (Prov. de Corrientes, R.A.). Rev. Assoc. Argent. mineral. Petrol. Sediment. 15: 51–61.

Di Persia, H, 1986. Zoobenthos of the Paraná system. In Davies, B. R. & K. F. Walker (eds), The Ecology of River Systems. Dr W. Junk Publishers, Dordrecht: 589–597.

Friedman, G.M. & Sanders, J.E., 1978. Principles of Sedimentology. John Wiley & Sons, New York: 792 pp.

Golterman, H. L., R. S. Clymo & M. A. M. Ohmstad, 1978. Methods for Physical and Chemical Analysis of Fresh Waters. Blackwell Scientific, Oxford: 214 pp.

Junk, W. J., P. B. Bayley, & R. E. Sparks, 1989. The flood pulse concept in river-floodplain systems. In Dodge, D. P. (ed.) Proceedings of the International Large River Symposium (LARS). Can. Spec. Publ. Fish. aquat. Sci. 106: 110–127.

Junk, W.J & K. Furch, 1993. A general review of tropical South American floodplains. Wetlands Ecology and Management 2: 231–238.

Marchese, M., 1994. Population dynamics of *Narapa bonettoi* Righi and Varela, 1983 (Oligochaeta: Narapidae) from the main channel of the middle Paraná River, Argentina. Hydrobiologia 278: 303–308.

Marchese, M.R., 1984. Estudios limnológicos en una seccíon transversal del tramo medio del rio Paraná. XI: Zoobentos. Ver. Asoc. Cienc. Nat. Litoral 15: 157–174.

Marchese, M.R., 1987. The ecology of some benthic Oligochaeta from the Paraná River. Hydrobiologia 155: 209–214.

Marchese, M.R. & I. Ecurra de Drago, 1983. Zoobentos de los principales tributarios del rio Paraná médio en el tramo Goya-diamante. Su relacion com el cauce principal y cauces secundarios. Ver. Asoc. Cienc. Nat. Litoral 14: 95–109.

Marchese M. R. & I. Ecurra de Drago, 1992. Benthos of the lotic environments in the Middle Paraná River system. Transverse zonation. Hydrobiologia 237: 1–14

Montanholi-Martins, M.C. & A. M. Takeda, 1999. Spatial and temporal variation of Oligochaeta (1993–1994) in the main and a secondary channel of the Rio Paraná, Brazil. Stud. Neotrop. Fauna & Environm. 34: 52–58.

Peters, R.H., 1986. The role of prediction in limnology. Limnol. Oceanogr. 31: 1143–1159.

Stevaux, J. C., 1994. The upper Paraná River (Brazil): geomorphology, sedimentology and paleoclimatology. Quat. Internat. 21: 143–162.

Takeda, A. M., J. Higuti, L. C. Rodrigues, A. Bialetzki, R. Pilati, R. L. Delariva, S. S. Abes, J. R. C. Campos, J. B. Campos, M. B. Romagnolo & P. V. Sanches, 1997. Distribuição espacial de zoobentos do ressaco 'Pau Véio'e do canal Cortado. Anais do VIII Seminário Regional de Ecologia. São Carlos, SP: 127–139.

Hydrobiologia **463:** 249–262, 2001.
P. Rodriguez & P.F.M. Verdonschot (eds), Aquatic Oligochaete Biology VIII.

Hydrology and substrates: determinants of oligochaete distribution in lowland streams (The Netherlands)

Piet F.M. Verdonschot
Alterra, Green World Research, Department of Freshwater Ecosystems, P.O. Box 47, 6700 AA Wageningen, The Netherlands
Fax: 317 424988. E-mail: p.f.m.verdonschot@alterra.wag-ur.nl

Key words: Oligochaeta, lowland streams, discharge, substrates, multivariate analysis, dynamics

Abstract

In most soft-bottomed, lowland streams in the Netherlands discharge regimes largely follow the precipitation pattern. Winter discharges are higher and much more dynamic then summer discharges, although rain storms throughout the year cause unexpected peak flows. Minimal precipitation, reduced stream flow and droughts can occur during the summer months. Lowland stream habitat, particularly in The Netherlands, is hydrologically dynamic, with substrates frequently moved or disturbed. Differences in discharge patterns in Dutch soft-bottomed lowland streams are expected to affect distribution patterns of macroinvertebrates and thus oligochaetes. Ten small to medium sized lowland streams, differing from one another in hydrological regime, were studied. Five major habitats in each stream were assessed on three occasions over a 15-month period. Each habitat sampled with a micro-macrofauna shovel; during each sampling period, several environmental parameters, especially hydrological and substrate parameters, were measured. Ordination (CANOCO) (Ter Braak, 1989) and statistical tests (chi-squared test) (Lindgren & McElrath, 1970) were used to determine the major oligochaete distribution patterns between streams as well as between habitats within each stream. Each stream was characterized by its oligochaete assemblage; general distribution patterns and individual oligochaete–substratum relationships were documented. Hydro-morphological parameters together explained the differences in major distribution patterns. Preferences of oligochaetes for specific structural habitats are discussed.

Introduction

Climate change, agricultural land-use and urbanization are all examples of large-scale human activities which affect stream hydrology. With the growing interest in stream restoration, water managers ask questions on the predictability of stream communities under changed hydrological conditions (e.g., Petersen et al. 1992; Boon, 1998), and similar questions are also presented by planners and policy makers. What effect does change in climate have on the hydrology of streams? We face the problem of coupling processes acting at the catchment scale, to those acting at the in-stream habitat. The impact of most human activities acts at the catchment scale, whilst the effects become visible at the scale of the habitat of stream oligochaetes. This habitat is limited to some square centimeters of the stream bed.

A stream is a dynamic but balanced environment (Huston, 1979; Minshall, 1988). The complexity of flow, substrates and communities is such that accurate prediction is difficult if not impossible, especially at the scale of the habitat. The distribution of oligochaetes in a stream surely is not coincidental. An oligochaetologist standing at the bank of a stream is able to name several species present without having seen them.

Already more than 10 years ago Statzner et al. (1988) pointed out that discharge and substratum are not enough to characterize the physical habitat of a stream. The question arises, which parameters are necessary to assist the oligochaetologist in determining which species will be present?

Discharge regimes shape the substrates (Verdonschot et al., 1998) and both compose the oligochaetes habitat. In this study, a number of relatively simple parameters are related to oligochaete distribution. The focus of this study is to determine how discharge affects the substrate of the streams and thus the distribution of oligochaetes within those substrates.

Study sites

All 10 streams in this study are soft-bottomed, lowland streams with a slope of about 0.5–5 m km^{-1}. They all are distributed over the eastern and southern part of The Netherlands. The streams are representative for the upper and middle courses of natural lowland streams. All streams are relatively unaltered, and represent different hydrological regimes.

The 10 streams are characterized by morphological features, below:

Stream code	Stream name	Stream type
SN	Springendal stream North	natural, spring-fed, small, sandy upper course
SZ	Springendal stream South	semi-natural, spring-fed, small, sandy upper course
BB	Forest stream	natural, organic small upper course
RB	Red stream	natural, sandy middle course
RO	Rosep	semi-natural, sandy middle course
RE	Reusel	semi-natural, sandy middle course
KB	Cold stream	natural, spring-fed, sandy upper course
OB	Old stream	natural, spring-fed, sandy upper course
TB	Tongerense stream	semi-natural, spring-fed, sandy upper course
FB	Frederik-Bernhard stream	natural, spring-fed, sandy upper course

The Netherlands has a temperate climate with a precipitation surplus of about 300 mm a year. The temperature of the streams range between 0 and 18 °C. The streams are either mainly rain-water fed or mainly fed by helocrene springs; direct runoff is also important. The width of the streams ranges from 0.5 to 4.0 m. Current velocities in these streams range from 0 to more than 60 cm s^{-1} ($\bar{x}$ = 20–30 cm s^{-1}). Stream depths range from 0 to 50 cm ($\bar{x}$ = ~10 cm). Diverse substrates are present in each stream. Values for several physical and chemical parameters measured during this study are presented in Table 1.

Materials and methods

Hydrology

The streams were studied over a 15-month period, from July 1997 through September 1998. Discharge was monitored continuously throughout the study period, with water level recorded every 15 min. Little reliable data was available for one stream. Based on one hydrological year, the discharge of the studied streams was summarized in proportion of discharge measurements exceeding percentiles of five percent each. The discharge values of these classes, expressed in m^3 per second, represent discharge percentiles. The 50% percentile (Q_{50}) is defined as base flow.

Discharge data were further summarized into two groups of parameters: the cumulative discharge over the period of half a year, and the bank full discharge (about the Q_{70}). The second group includes discharge dynamics. These parameters account for the number of low and high discharge events over a fixed period of half a year before oligochaete sampling. For low discharge events, classes are defined as the number of measurements below 0.3, 0.7 and 0.9 times base flow; for high discharge events, classes are the number of measurements above 2, 4 and 8 times base flow. Discharge can be continuous without peaks or drops, but can fluctuate somewhat more or even show high fluctuations. The classes of low and high discharge events were combined into four discharge dynamics classes: D1, without fluctuations (sum of measurements between 0.9 and 2 times base flow over a period of 35 days); D2, with slight fluctuations (number of measurements between 0.7 and 0.9 plus 2 and 4 times base flow); D3, with moderate fluctuations (number of measurements between 0.3 and 0.7 plus 4 and 8 times base flow); and D4, with high peaks and low drops (number of measurements below 0.3 plus number above 8 times base flow). All parameters are calculated as absolute divergence of base flow.

Table 1. General physico-chemical characteristics of all ten streams studied

Stream	SN		SZ		RO		RE		RB		BB		TB		OB		KB		FB	
	$\bar{x}$	n	$\bar{x}$	n	$\bar{x}$	n	$\bar{x}$	n	$\bar{x}$	n	$\bar{x}$	n	$\bar{x}$	n	$\bar{x}$	n	$\bar{x}$	n	$\bar{x}$	n
Ca mg l^{-1}	16.2	32	18.2	29	42.7	4	49.7	6	32.4	9	13.3	8	17.0	9	18.8	9				
Cl mg l^{-1}	25.3	57	20.4	54	23.3	73	34.8	88	27.5	10	11.1	8	10.1	9	15.9	9	20.9	12	22.3	59
CO_3 mg l^{-1}	1.1	32	1.0	29	2.0	1	3.0	2					2.0	9	2.0	9				
Fe^{2+} mg l^{-1}	0.02	19	0.04	17	2.20	5	2.08	5					0.01	9	0.01	9				
Total hardness	0.61	17	0.64	14	0.86	1		0					0.51	9	0.59	9				
HCO_3 mg l^{-1}	12.53	32	16.04	29	98.50	4	38.00	5	56.49	9	9.00	8	43.22	9	25.56	9				
EGV	221	56	216	54	315	3	510	6	323	9	126	8	131	9	174	9	244	12	286	57
K mg l^{-1}	4.6	56	6.6	53	8.0	16	16.9	19	5.4	9	1.5	8	1.7	9	2.6	9			9.9	4
Mg mg l^{-1}	8.3	32	6.3	29	4.7	1	12.5	7	5.2	9	1.6	8	2.0	9	3.0	9			4.6	4
Kj-N mg l^{-1}	0.53	57	0.84	54	1.30	38	1.77	41	2.16	9	1.00	8	0.91	9	0.85	9	0.99	12	1.56	59
Na mg l^{-1}	14.90	32	11.44	29	16.28	4	22.57	7	17.96	9	5.85	8	6.80	9	9.94	9			11.00	4
NH_4-N mg l^{-1}	0.10	57	0.11	54	0.24	107	0.50	123	2.10	10	0.34	8	0.04	9	0.04	9	0.08	12	0.25	46
NO_2-N mg l^{-1}	0.01	57	0.01	54	0.04	95	0.06	95	0.14	10	0.15	8	0.00	9	0.01	9	0.01	12	0.05	50
NO_3-N mg l^{-1}	10.31	57	9.11	54	3.27	95	8.78	95	1.34	10	0.43	8	0.62	9	5.85	9	5.68	12	5.20	59
o-P mg l^{-1}	0.03	56	0.04	54	0.20	1	0.09	4	0.07	10	0.05	8	0.02	9	0.02	9	0.02	12	0.09	45
P-totaal mg l^{-1}	0.07	57	0.12	54	0.83	1	0.09	117	0.24	10	0.10	8	0.08	9	0.07	9	0.07	12	0.19	55
SO_4 mg l^{-1}	27.0	57	33.8	54	32.7	107	101.9	123	49.1	10	33.1	8	14.8	9	17.0	9	29.3	12	29.7	59
pH	6.4	55	6.6	53	7.3	107	6.6	119	7.6	10	7.0	8	7.3	9	6.9	9	7.3	12		

Substratum

Every fifth week during the study period, the cover percentages of major substrates were estimated over a reach of 30 m in each stream. This field estimation was done per reach of two meters. So, for each site, 15 reaches were estimated. The major substrates were typified according to their dominant mineral (gravel, sand, silt) or organic component (vegetation type, leaves and fine, coarse, very coarse detritus) Samples for measurement of grain size and organic matter content also were taken from the major substrate types.

Oligochaeta

Oligochaetes were sampled 3 times in all 10 streams in autumn (1997), spring (1998) and autumn (1998). The samples were taken by means of a micro-macrofauna shovel of 10 by 15 cm (mesh size 0.5 mm, sampled surface area 150 cm^2). The sampling method is described by Tolkamp (1980). The five major substrate types were sampled at each site; samples were then sorted, with oligochaetes fixed in formalin (4%) prior to identification.

Data processing

All data were analyzed by means of statistical techniques. The oligochaete data were analyzed by multivariate analysis on habitat and stream level. The analyses of the relationship between oligochaetes and environmental parameters was done by ordination analysis using the program CANOCO (option: Detrended Canonical Correspondence Analysis; Ter Braak, 1989), an ordination (reciprocal averaging) technique which results in an ordination diagram.

Environmental variables were selected on the basis of the inter-set correlation (correlation >0.2) with axes, thus the correlation between a variable and an ordination axis. All variables selected by this procedure indicates the important environmental gradients.

For each ordination, the oligochaete taxa and the selected environmental variables were represented in the DCCA ordination diagrams. An environmental variable (indicated as an arrow in the diagram) points approximately in the direction of steepest increase of that variable across the diagram and the rate of change in that direction is equal to the length of the arrow. This means that the value of an important environmental variable to an oligochaete is visualized by its perpendi- cular projection on the environmental arrow or its imaginary extension (in both directions).

Some other parameters of an ordination are of interest. The eigenvalue ranges between zero and one and can be considered to be a measure of between-site variability or beta-diversity. The eigenvalues of the individual axes are regarded as a measure of their relative importance within one analysis. The species-environment correlation measures the strength of the relation between species and the environment for a particular axis. The percentage of variance of the species-abundance data accounted for by the species-site bi-plot indicates the goodness of fit of the diagram with respect to the distribution of species abundances. The percentage of variance of the species-environment data in the species-environment bi-plot indicates the goodness of fit of the environmental variables. These parameters never reach 100% because of noise in the data and are always relatively low in large data sets. The total inertia is the total variance in the species data as measured by the chi-square of the sample-by-species table divided by its total.

The distribution of the oligochaete taxa over the different substrate types was calculated by using the index of representation (*IR*: Hildrew & Townsend, 1978) and the statistical significance of this index was tested by using a chi-squared test (e.g., Lindgren & McElrath, 1970). The index of representation supposed under the null hypothesis that a taxon occurs in all substrate types in equal densities. The null hypothesis is accepted when the difference between observed and expected densities is too small to reach chi-squared values higher then the 5% significance level. A taxon shows a preference for a substrate type only in cases were the index shows an over-representation (positive *IR* with a value >5).

Results

Stream type

Important for a better understanding of stream ecology is knowledge about the type of the stream (Verdonschot, 1994). The type summarizes the key ecosystem characteristics into groups. Each group has a certain within homogeneity of ecological key features and the system within a group functions in a comparable way. A simply technique to define stream types is to plot the width *versus* depth. There is an increase in wet profile area (Fig. 1) whereby two small upper courses (SN, SZ), five upper courses (KB, OB, BB, FB, TB) and three middle courses (RB, RE, RO) of lowland streams are distinguished.

Hydrology

Discharge percentiles of all nine streams were plotted (Fig. 2) against percentage of each class. Within such a graph, a steep line indicates a high number of low and high flows thus a very dynamic or flashy stream. In a more constant discharging stream, all classes are about equal: a more horizontal line. A line with a steep slope in a range of low percentiles indicates a stream with low or no discharge in summer (an intermittent stream) and a line with a steep slope at the high percentiles indicates conditions of extreme peak discharges. As shown in Figure 2, based on discharge percentile distribution three constant (OB, TB, SN), three intermediate (RB, KB, BB) and three flashy (RE, RO, SZ) streams are included.

Discharge dynamics, expressed in classes D1–D4, are plotted in a graph (Fig. 3). Each of the streams is characterized on the basis of discharge dynamics as being constant (OB, SN, TB), intermediate (RB, KB, BB) or flashy (RO, RE, SZ).

Substratum

The average percentages of substrate cover for the five main substrate types were given for each stream. The streams are ordered according to stream type (Fig. 4). The proportion of each substrate type differs between all streams and no relation to stream type is seen.

What is important to oligochaetes are the dynamics in their habitat, thus the movement of the substrate. Therefore, the standard deviations of each substrate type per stream was cumulatively plotted (Fig. 5). The higher the bar the more change occurred in substrates. Two streams are quite constant (RB, SN), seven are intermediate (TB, OB, BB, SZ, KB, RE, FB) and one stream is very dynamic (RO) in its substrate pattern in time.

Oligochaeta

In total, 38 oligochaete taxa were collected in the 168 habitat samples. All taxa were identified, most to species level (Table 2). All data collected during 1997 and 1998 were ordinated to describe the variations in taxon distribution and abundances. The major ordination parameters are listed in Table 3. The DCCA-ordination (Fig. 6) shows the relationships between oligochaetes, streams and habitats. The variables significantly ($P = 0.05$; unrestricted permutation test) explain the oligochaete distribution. The

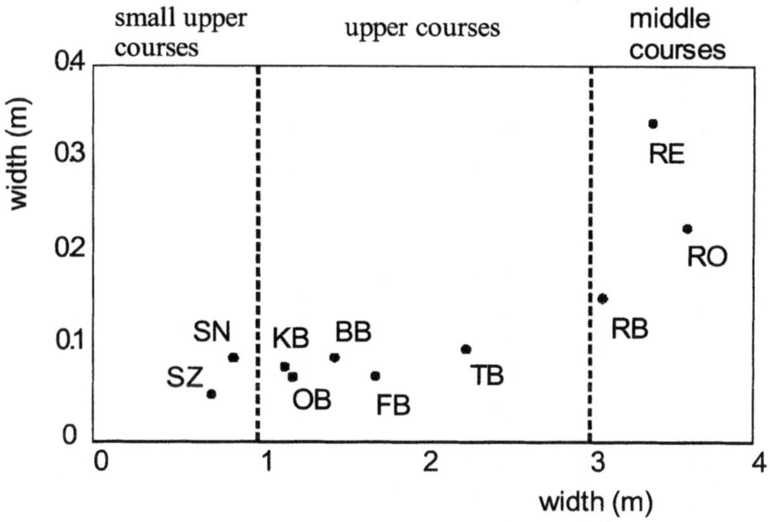

Figure 1. Width–depth relationship for the ten studied streams (for the explanation of abbreviations see text).

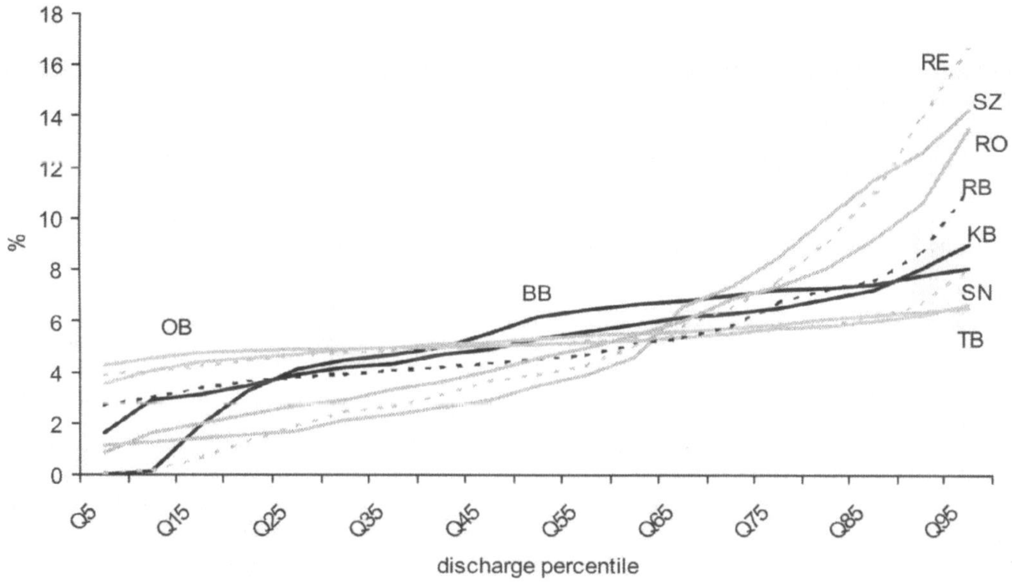

Figure 2. Discharge percentiles based on one hydrological year for nine studied streams (for the explanation of abbreviations see text).

individual streams were used as an explaining environmental variable. Six out of 10 streams occurred to be important explanatory variables in the analysis. Furthermore, several habitat variables are important. Both habitat variables and streams are not completely independent. Some habitats are more or less dominant in one or a few streams. Interesting species distribution patterns are e.g., *Psammoryctides barbatus* related to vegetation and depth, *Specaria josinae* related to total organic material, *Slavina appendiculata* related to leaves, *Nais communis* related to two certain streams (FB, SZ), and *Rhynchelmis tetratheca* related to *Berula*-vegetation and gravel.

A second DCCA-ordination (Fig. 7) was done by leaving streams out as explaining variables. This to test whether more specific variables explain the oligochaete distribution as well as the overall variable stream. The slight drop of the eigenvalue (Table 4)

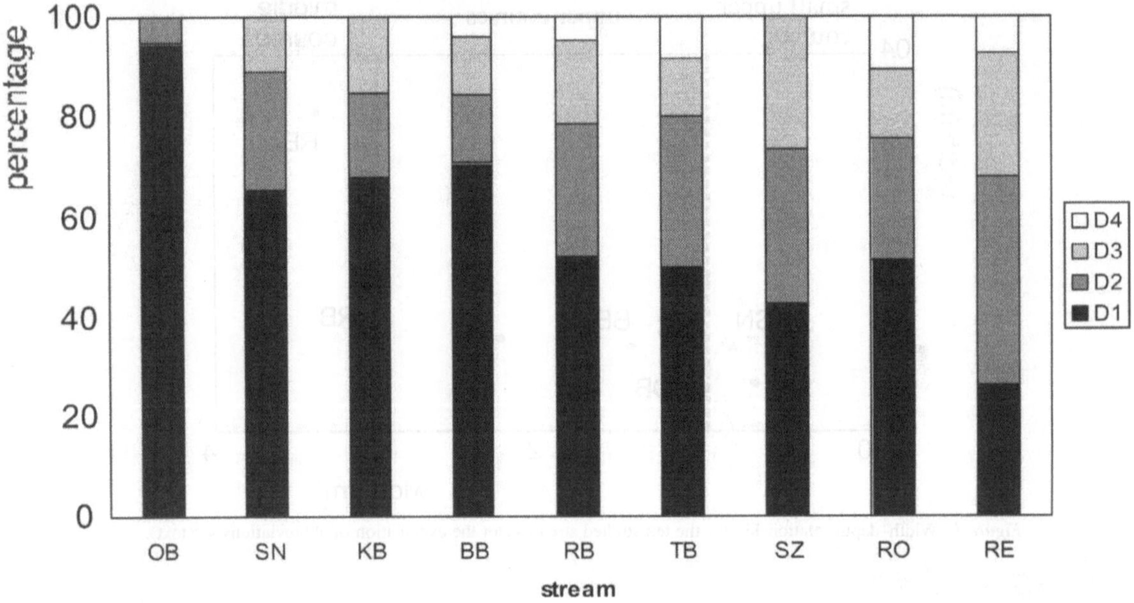

Figure 3. Percentage of occurrence over one hydrological year of each of the four discharge dynamics classses in the ten studied streams (for the explanation of abbreviations see text).

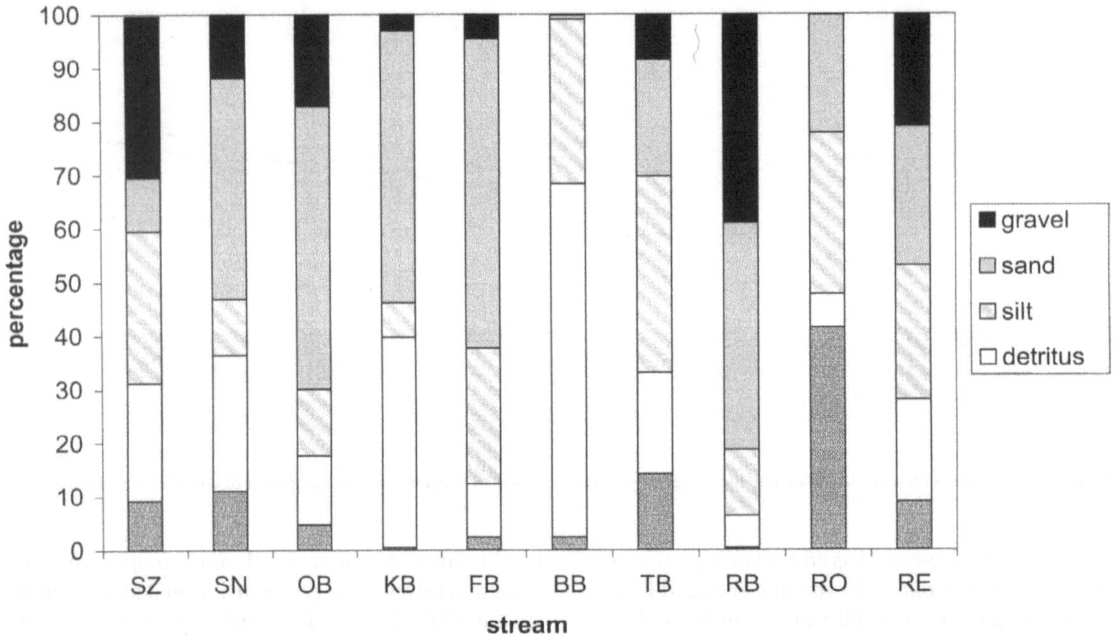

Figure 4. Percentage of occurrence over the study period of the major substrate types in the ten studied streams (for the explanation of abbreviations see text).

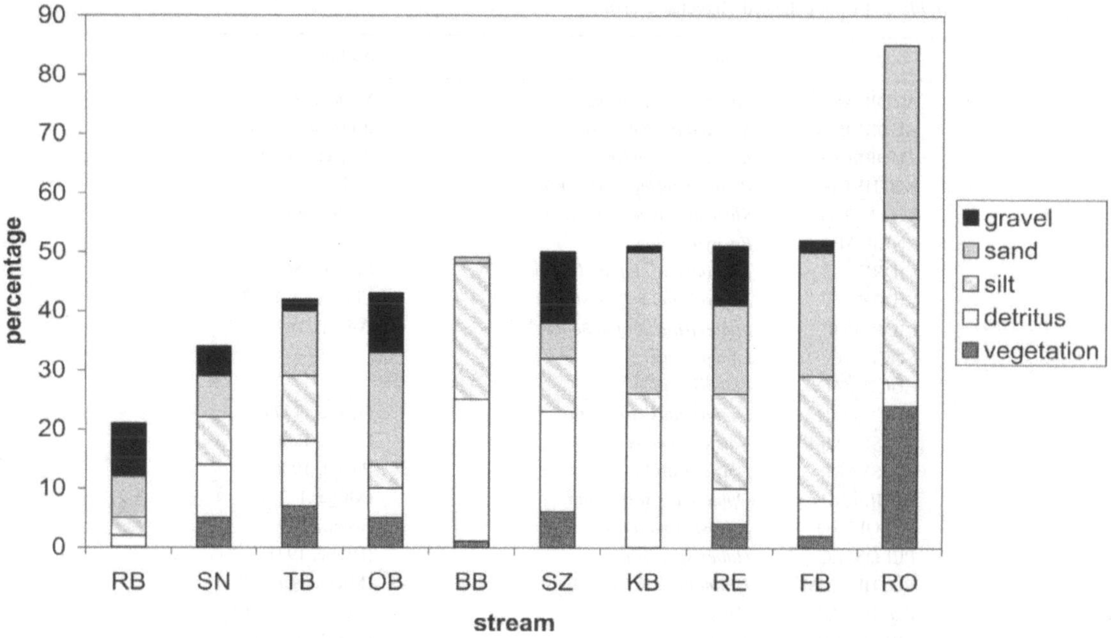

Figure 5. Cumulative percentage of the standard deviation of the occurrence of the major substrate types over the study period in the ten studied streams (for the explanation of abbreviations see text).

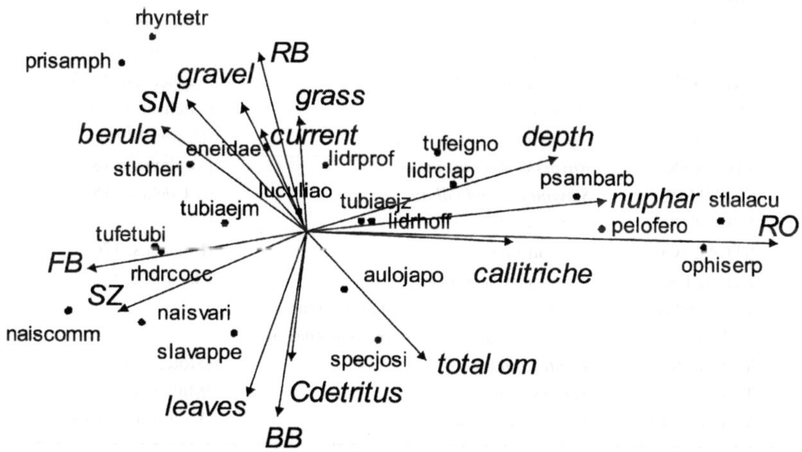

Figure 6. Ordination diagram (DCCA) for the axis 1 (horizontally) and 2 (vertically) of the oligochaete-habitat relation (with each stream as a nominal variable) over the 10 studied streams. The arrows indicate the important habitat variables, the dots indicate the oligochaete taxa (codes are eplained in Table 2 (oligochaetes) and Table 3 (variables).

indicates the importance of streams as explanatory variable. The species-environment correlation drops from 92 to 75%. Streams are thus important in the distribution of the oligochaetes. But the resulting diagram shows that the oligochaete distribution pattern remains the same over the habitat variables, confirming the relationship between certain streams and certain habitats. Therefore, a third ordination was performed.

The third DCCA-ordination (Fig. 8) was done at the level of the whole stream by combining all habitat samples per stream. Hydrology was included in this ordination, but stream was not. Hydrology as well as habitat are important in oligochaete distribu-

Table 2. List of collected oligochaete taxa

Code	Name	Author
AUDRJAPO	*Aulodrilus japonicus*	Yamaguchi, 1953
AUDRLIMN	*Aulodrilus limnobius*	Bretscher, 1899
AUDRPLUR	*Aulodrilus pluriseta*	(Piguet, 1906)
BOTHVEJD	*Bothrioneurum vejdovskyanum*	Stolč, 1888
RHYNTETR	*Rhynchelmis tetratheca*	Michaelsen, 1920
ENEIDAE	*Enchytraeidae*	
LIDRCLAP	*Limnodrilus claparedeianus*	Ratzel, 1868
LIDRHOFF	*Limnodrilus hoffmeisteri*	Claparède, 1862
LIDRPROF	*Limnodrilus profundicola*	(Verrill, 1871)
LUCIDAE	*Lumbricidae*	
LUCULIAO	*Lumbriculidae*	
LUCUVARI	*Lumbriculus variegatus*	(Müller, 1774)
NAISCOMM	*Nais communis*	Piguet, 1906
NAISVARI	*Nais variabilis*	Piguet, 1906
OPHISERP	*Ophidonais serpentina*	(Müller,1773)
PELOFERO	*Spirosperma ferox*	Eisen, 1879
PELOSPEC	*Haber speciosus*	(Hrabě, 1931)
POTHHAMM	*Potamothrix hammoniensis*	(Michaelsen, 1901)
POTHHEUS	*Potamothrix heuscheri*	(Bretscher, 1900)
PRISAMPH	*Pristina amphibiotica*	Lastočkin, 1927
PRISFORE	*Pristina foreli*	(Piguet, 1906)
PRISIDRE	*Pristinella jenkinae*	Lastočkin, 1927
PRISLONG	*Pristina longiseta*	Ehrenberg, 1828
PRISMENO	*Pristina menoni*	(Aiyer, 1906)
PRISOSBO	*Pristinella osborni*	(Walton, 1906)
PRISSYND	cf. *Pristina synclites*	Stephenson, 1925
PSAMALBI	*Psammoryctides albicola*	(Michaelsen, 1901)
PSAMBARB	*Psammoryctides barbatus*	(Grube, 1891)
RHDRCOCC	*Rhyacodrilus coccineus*	(Vejdovský, 1875)
SLAVAPPE	*Slavina appendiculata*	(d'Udekem, 1855)
SPECJOSI	*Specaria josinae*	(Vejdovský, 1883)
STLALACU	*Stylaria lacustris*	(Linnaeus, 1767)
STLOHERI	*Stylodrilus heringianus*	Claparède, 1862
TUBIAEJM	*Tubificidae* juvenile with hair chaetae	
TUBIAEJZ	*Tubificidae* juvenile without hair chaetae	
TUFEIGNO	*Tubifex ignotus*	(Stolč, 1886)
TUFETUBI	*Tubifex tubifex*	(Müller, 1774)
VEJDCOMA	*Vejdovskiella comata*	(Vejdovský, 1883)

tion. Oligochaetes respond to variables which together compose individual streams. The combination of discharge dynamics D3 and D4, and bank full discharge as total discharge – together with the habitat variables already shown in the habitat ordination – explain the oligochaete distribution. The relation between *Nais communis* and detritus thus becomes clear.

Substrates differed between streams and though we cannot yet explain fully their occurrence, variability and dynamics strongly strongly influence the oligochaete distribution. Therefore, the relation between oligochaetes and substratum expressed in the mineral material parameter 'grain size fraction' and the organic material parameter 'organic matter content', is calculated.

Table 3. Environmental variables used in the analyses

Code	Description	Unit
BB, FB, KB, OB, RB	stream	nominal
Md grain size	medium grain size	mm
total om	total organic matter content	%
spring	spring	nominal
autumn	autumn	nominal
depth	depth	cm
current	current velocity	cm s^{-1}
O_2	oxygen content	mg l^{-1}
sand	sand	average %
gravel	gravel	average%
Fdetritus	fine detritus	average%
Cdetritus	coarse detritus	average%
Vdetritus	very coarse detritus	average%
leaves	leaves	average%
berula	Berula-vegetation	average%
grass	overhanging grass-vegetation	average%
ceratophyllum	Ceratophyllum-vegetation	average%
nuphar	Nuphar-vegetation	average%
callitriche	Callitriche-vegetation	average%
slope	slope	m km^{-1}
D1	discharge dynamics class 1	number
D2	discharge dynamics class 2	number
D3	discharge dynamics class 3	number
D4	discharge dynamics class 4	number
Q-bank full	Bank full discharge (Q50)	m^3 s^{-1}
Q-cumulative	cumulative discharge over half a year	m^3

All combinations (30) of the following mineral and organic material classes are used:

Mineral material	Organic material
Coarse sand (>0.50)	Very high organic matter content (>10%)
Medium sand (0.25–0.50)	High organic matter content (4–10%)
Fine sand (0.125–0.25)	Medium organic matter content (1–4%)
Very fine sand (0.063–0.125)	Low organic matter content (<1%)
Silt (<0.063)	Leaves vegetation

Twelve taxa show a strong preference for one combination of mineral and organic material (Table 5). Note the preference of *Nais communis* for high organic matter content on fine sand, *Rhynchelmis tetratheca* in coarse substrates and *Slavina appendiculata* between leaves as was all already indicated in the ordination diagrams.

Another nine taxa show a somewhat wider preference range of combined substrate categories (Table 6). Note e.g., that *Specaria josinae* is preferring high organic matter content, and that *Psammoryctides barbatus* prefers vegetation. Both examples confirm the ordination results.

Discussion

Hydro-morphology

A review of the results of discharge percentiles and discharge dynamics suggests that all nine streams are classified by both parameters into the same classes. These three discharge classes are given in the columns of Table 7. Furthermore, substrate patterns differ

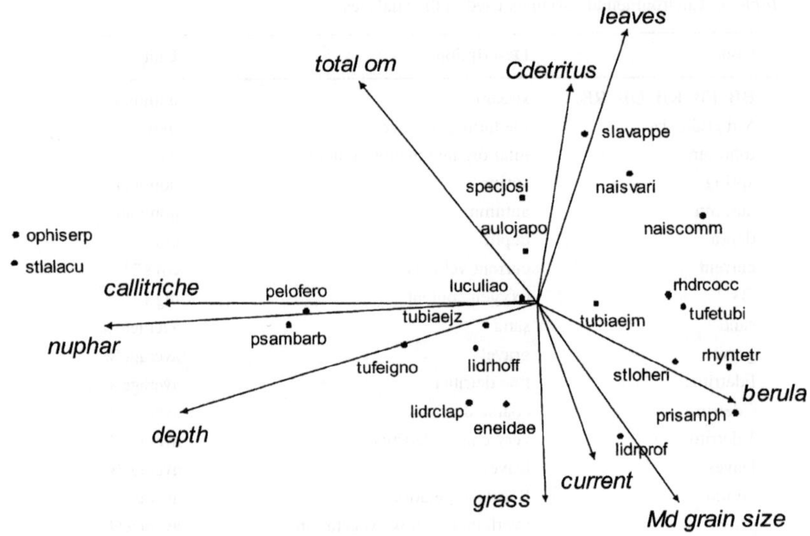

Figure 7. Ordination diagram (DCCA) for the axis 1 (horizontally) and 2 (vertically) of the oligochaete-habitat relation (without stream as a nominal variable) over the 10 studied streams. The arrows indicate the important habitat variables, the dots indicate the oligochaete taxa (codes are eplained in Table 2 (oligochaetes) and Table 3 (variables).

Table 4. Some multivariate characteristics of the ordination analyses

Ordination Analysis characteristics	Habitats with stream parameters		Habitats without stream parameters		Streams without stream parameters	
	axis 1	axis 2	axis 1	axis 2	axis 1	axis 2
Eigenvalue	0.29	0.13	0.20	0.10	0.29	0.18
Species-environment correlation	0.92	0.72	0.75	0.63	0.99	0.99
Cumulative percentage variance of species – abundance data	9	13	6	9	22	35
Cumulative percentage variance of species – environment data	29	43	31	46	23	38
Total inertia	3.17		3.17		1.32	
Sum all canonical eigenvalues	1.00		0.65		1.24	
P-value	0.005		0.005		0.05	

between all streams and show more or less dynamic patterns. The streams are also classified according to substrate dynamics classes (rows in Table 7). By combining both, Table 7 summarizes the discharge-substratum relationships, those I will term hydromorphological character. Discharge-substratum combinations do not appear to show simple linear relationship. As expected, neither flashy streams with a constant substrate pattern nor constantly discharging streams with a dynamic substrate pattern, occur. Streams can have a constant discharge over time and still show intermediate substrate dynamics. This can be explained by the fact that some substrates show seasonal patterns, like leaves and detritus, as well as from the influence of extensive maintenance in the 'Old stream' and the 'Tongerense stream' (one to two times a year). The 'Red stream' (RB) shows an intermediate discharge with a constant substrate pattern – the result of numerous stable gravel banks within the streambed. The 'Springendal stream South' (SZ) also

Table 5. Preference of oligochaete taxa for a specific organic matter – mineral substrate class

Organic matter	Mineral substrate class: Sand – Coarse	Sand – Medium	Sand – Fine	Sand – Very fine	Silt	Independent
Very high organic matter content			*Nais communis*			
High organic matter content			*Lumbriculus variegatus*		*Lumbricidae*	
Medium organic matter content	*Rhynchelmis tetratheca*		*Enchytraeidae* *Limnodrilus claparedeianus*	*Nais variabilis*		
Low organic matter content	*Limnodrilus udekemianus* *Pristina foreli*	*Aulodrilus pluriseta*				
Leaves						*Slavina appendiculata*
Vegetation						*Ophidonais sepentina*

Table 6. Preference of oligochaete taxa for a combined organic matter – mineral substrate class

Organic matter	Mineral substrate class: Gravel	Sand – Medium + fine	Sand – Fine + very fine	Sand – Very fine + silt
(Very) high organic matter content			*Specaria josinae*	*Spriosperma ferox* *Rhyacodrilus coccinea* *aulodrilus japonicus*
High + medium organic matter content			*Stylodrilus heringianus*	
Leaves	*Nais simplex*	*Nais bretscheri*		
vegetation			*Psammoryctides barbatus* *Stylaria lacustris*	

is dominated by gravel and thus is the streambed more stable despite the flashy discharge. The intermediate substrate dynamics of the 'Reusel' (RE) under a flashy discharge regime are a consequence of its wide and perennially sandy streambed. The latter observations do imply that this sandy streambed in fact is probably often moved and thus quite unstable.

Oligochaete distribution

Although oligochaete-habitat ordination excluding stream parameters when compared to ordination including stream parameters, showed a slight decrease in eigenvalues and species-environment correlations, the explanatory variables remained almost the same. The stream variables do not really add to the explanation

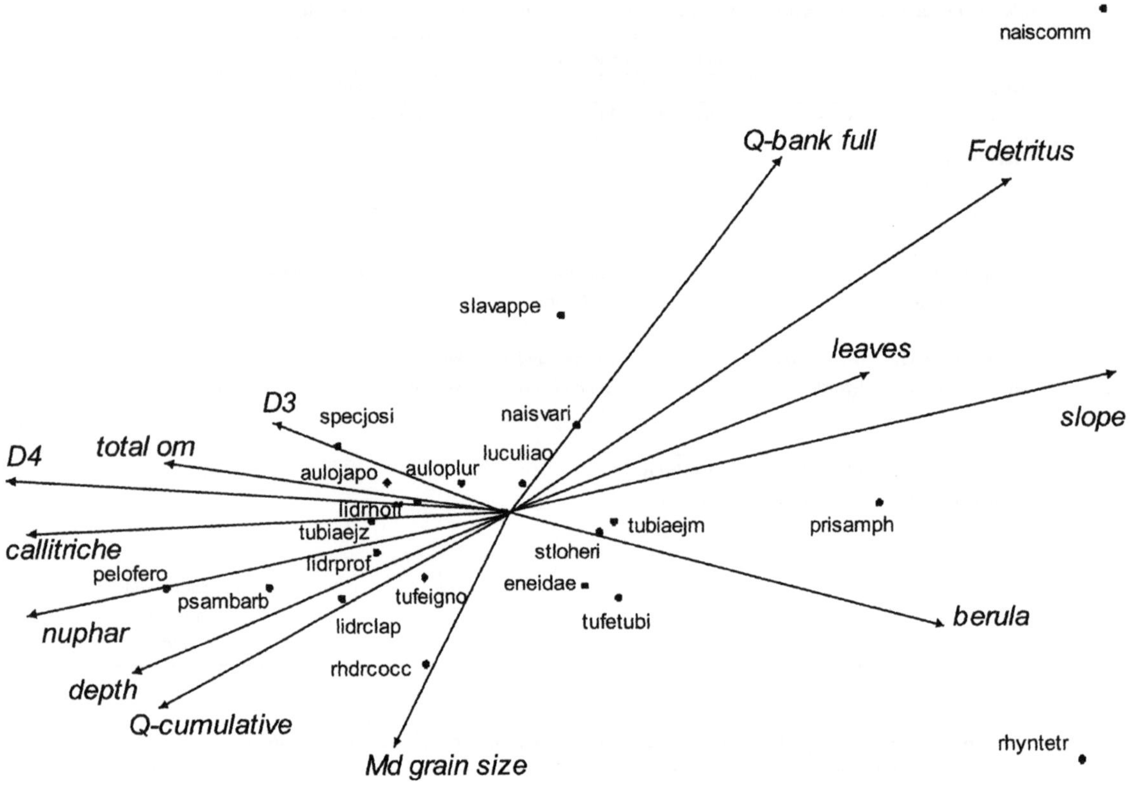

Figure 8. Ordination diagram (DCCA) for the axis 1 (horizontally) and 2 (vertically) of the oligochaete-stream relation over the ten studied streams. The arrows indicate the important stream variables, the dots indicate the oligochaete taxa (codes are eplained in Table 2 (oligochaetes) and Table 3 (variables).

of the first ordination, and they thus can be left out the ordination without causing a major change of the diagram. Only the parameter gravel was replaced by medium grain size, both parameters though describe a similar habitat feature. The oligochaete distribution is related to habitat – those which are strongly represented in but one or only a few streams. This major oligochaete distribution pattern is summarized in Table 8. Five groups of oligochaetes related to habitats and streams are distinguished from both habitat ordinations.

The ordination of streams included also hydrological variables. The major ordination pattern is the same and shows the same explanatory habitat variables combined with some hydrological ones. The stream level ordination shows that the group of *Nais communis*, *Nais variabilis*, *Tubifex tubifex* and *Rhyacodrilus coccineus* is a valid grouping, but the species can occur in different streams and habitats. The other four groupings show clear relationships between oligochaete species and the hydro-morphological features of streams and their specific habitats (Table 8).

Finally, the ordination results were combined with individual species preferences. From group 1, the species *Stylaria lacustris*, *Ophidonais serpentina* and *Psammoryctides barbatus* are confirmed. Their individual preferences confirm the group's abiotic characterization. The more common *Tubifex ignotus* and *Limnodrilus claparedeianus* do not show individual

Table 7. Number of streams in the different discharge – substrate dynamics classes (stream codes are explained in the text)

Substrate classes	Discharge classes: Constant	Intermediate	Flashy
Constant	SN	RB	
Intermediate	OB, TB	BB, KB	SZ, RE
Flashy			RO

Table 8. Grouping of oligochaetes in relation to habitat and stream based on the ordination results

Group	Species	Hydro-morphological variables	Best developed in stream (code)
1	*Stylaria lacustris*	depth	RO
	Ophidonais serpentina	nuphar	
	Spirosperma ferox	callitriche	
	Psammoryctides barbatus	D4	
	Tubifex ignotus	Q-cumulative	
	Limnodrilus claparedeianus		
2	*Rhynchelmis tetratheca*	berula	RB
	Pristina amphibiotica	gravel / Md grain size	SN
	Stylodrilus heringianus	current	
	Limnodrilus profundicola	grass	
		slope	
3	*Slavina appendiculata*	leaves	BB
		Cdetritus	
		Fdetritus	
		Q-bank full	
4	*Specaria josinae*	total om	
	Aulodrilus japonicus	D3	
5	*Nais communis*		FB
	Nais variabilis		SZ
	Tubifex tubifex		
	Rhyacodrilus coccineus		

preferences. *Spirosperma ferox* prefers a very fine sandy to silty substrate with a high organic matter content. This substratum could be present in the habitat associated with group 1, near the roots of the *Callitriche* vegetation. Because current velocity is usually low to negligible in vegetation, silt is continuously being deposited, providing a rather stable habitat for these species.

The two naidids, *Stylaria lacustris* and *Ophidonais serpentina*, are known to dwell between vegetation in standing or slowly flowing waters (e.g., Timm, 1970; Verdonschot, 1987). The three more common tubificids, *Psammoryctides barbatus*, *Tubifex ignotus* and *Limnodrilus claparedeianus*, inhabit sandy to silty, well-oxygenated substrates (e.g., Brinkhurst, 1964; Laakso, 1969).

The species in group 2 show a less defined relationship. Only *Rhynchelmis tetratheca* confirms the group's environment. *Pristina amphibiotica* and *Limnodrilus profundicola* are not significantly related to a certain structural habitat, and *Stylodrilus heringianus* shows a preference for fine to very fine sand with medium to high organic matter content. This substrate is not dominating group 2.

Rhynchelmis tetratheca, a species reported here as a new record for the Netherlands, is known to inhabit small natural streams. This habitat preference is also reported for *Stylodrilus heringianus*, especially in non-productive habitats (Brinkhurst, 1971). *Pristina amphibiotica* is more often reported from stagnant waters like bogs and marshes (Chekanovskaya, 1962).

The preference of *Slavina appendiculata* for leaves confirms the groups environmental description (group 3). This species is reported to occur among littoral plant thickets (Chekanovskaya, 1962), but also in mud in running waters (Schuster, 1915).

The same accounts for the preference of *Specaria josinae* and *Aulodrilus japonicus* for high organic matter load. The preference of *S. josinae* for organic mud was reported by Särkkä (1987). The collection of *A. japonicus, Pristinella jenkinae* and *P. osborni* during this research represent also new records for The Netherlands.

In general, most oligochaetes considered to be indicative prefer specific habitats. These habitats occur under specific conditions that – within these 10 streams – occur in one or only a few streams and depend on a complex interaction between different hydrological and substratum variables. Furthermore, the hydrological regime of each of these streams differed but add to the explanation of the oligochaete distribution. This clearly indicates that oligochaetes can fulfill a role as indicator for the hydro-morphological character of a lowland stream. Although their distribution is affected by a number of mutually interacting variables. A substrate preference of several oligochaete species is found, but to establish a relation to substrate dynamics as well as to discharge dynamics needs more attention. The substrate preference indicates a role of oligochaetes in habitat and stream assessment. Additional data from streams are necessary before hydro-morphological preferences of oligochaetes in streams can be defined. The relation to dynamics (stability) needs experimental research.

References

Boon, P. J., 1998. River restoration in five dimensions. Aquatic Conservation: Marine and Freshwat. Ecosystems 8: 257–264.

Brinkhurst, R. O., 1964. Observations on the biology of lake dwelling Tubificidae (Oligochaeta). Arch. Hydrobiol. 60: 385–418.

Brinkhurst, R. O., 1971. A guide for the identification of British aquatic Oligochaeta. Freshwat. Biol. Ass., Sc. Publ. 22.

Chekanovskaya, O. V., 1962. The aquatic oligochaete fauna of the U.S.S.R. Opred. Faune SSSR 78: 1–411.

Hildrew, A. G. & C. R. Townsend, 1977. The influence of substrate on the functional response of Plectrocnemia conspersa (Curtis) larvae (Trichoptera: Polycentropidae). Oecologia 31: 21–26.

Huston, M., 1979. A general hypothesis of species diversity. Am. Nat. 113: 81–101.

Laakso, M., 1969. Oligochaeta from brackish water near Tvärmine, south-west Finnland. Ann. zool. fenn. 6: 98–111.

Lindgren, B. W. & G. W. McElrath, 1970. Introduction to Probability and Statistics. MacMillan Comp., London (3rd edn.): 305 pp.

Minshall, G. W., 1988. Stream ecosystem theory: observed patterns and potential applications. J. n. am. Bentol. Soc. 7: 263–289.

Petersen, R. C., L. B. M. Petersen & J. Lacoursière, 1992. A building-block model for stream restoration. In Boon, P. J., Calow P. & G. E. Petts (eds), River Conservation and Management. John Wiley & Sons, Chichester: 293–310

Särkkä, J., 1987. The occurrence of oligochaetes in lake chains receiving pulp mill waste and their relation to eutrophication on the trophic scale. Hydrobiologia 155: 259–266.

Schuster, R. W., 1915. Morphologische und Biologische studien an Naididen in Sachsen und Böhmen. Julius Klinkhardt, Leipzig.

Statzner, B., J. A. Gore & V. H. Resh, 1988. Hydraulic stream ecology: observed patterns and potential applications. J. n. am. Bentol. Soc. 7(4): 307–360.

Ter Braak, C. J. F., 1987. CANOCO -A FORTRAN program for canonical community ordination by [partial] [detrended] [canonical] correspondence analysis, principal component analysis and redundancy analysis (version 2.1). TNO Institute of Applied Computer Science, Wageningen, The Netherlands: 1–95.

Timm, T., 1970. On the fauna of the Estonian Oligochaeta. Pedobiol. 10: 52–78.

Tolkamp, H. H., 1980. Organism-substrate relationships in lowland streams. Agricultural Research Reports 907, Doctoral thesis, Wageningen: 211 pp.

Verdonschot, P. F. M., 1987. Aquatic oligochaetes in ditches. Hydrobiologia 155: 215–222.

Verdonschot, P. F. M., 1994. Water typology: a tool for water management and nature conservation. Verh. int. Ver. Limnol. 25: 1911–1913.

Verdonschot, P. F. M., J. M. C. Driessen, H. G. Mosterdijk & J. A. Schot, 1998. The 5-S-Model, an integrated approach for stream rehabilitation. In Hansen, H. O. & B. L. Madsen (eds), River Restoration '96, Session Lectures Proceedings. National Environmental Research Institute, Denmark: 36–44.

Subject index

MIX
Papier aus verantwortungsvollen Quellen
Paper from responsible sources
FSC® C105338

If you have any concerns about our products,
you can contact us on
ProductSafety@springernature.com

In case Publisher is established outside the EU,
the EU authorized representative is:
Springer Nature Customer Service Center GmbH
Europaplatz 3, 69115 Heidelberg, Germany

Printed by Libri Plureos GmbH
in Hamburg, Germany